A TREATISE

ON THE

DIFFERENTIAL AND INTEGRAL

CALCULUS,

AND ON THE

CALCULUS OF VARIATIONS.

BY EDWARD H. COURTENAY, LL. D.

LATE PROFESSOR OF MATHEMATICS IN THE
UNIVERSITY OF VIRGINIA.

NEW YORK:
A. S. BARNES & BURR, 51 & 53 JOHN STREET.
SOLD BY BOOKSELLERS, GENERALLY, THROUGHOUT THE UNITED STATES.
1860.

JONES & DENYSE, Stereotypers,
183 William-street, New York.

EDWARD H. COURTENAY.

In the publication of the following Treatise on the Differential and Integral Calculus by Edward H. Courtenay, two Institutions have an equal interest — the Military Academy where he was graduated in the year 1821, and the University of Virginia, where he died in the Fall of 1853.

Mr. Courtenay was born in the City of Baltimore, on the 19th of November, 1803. He entered the Military Academy as a cadet in September, 1818, and was the youngest member of the Class of that year.

The Course of Study embraced a term of four years. In three years Mr. Courtenay made himself highly proficient in all the branches, and was graduated at the head of his class, in July, 1821.

In his initiatory examination he made a strong impression on the mind of the examiner, who remarked, when the examination was concluded, that "a boy from Baltimore, of spare frame, light complexion and light hair, would certainly take the first place in his class."

We transcribe the following record from the Register of the United States Military Academy.

"Edward H. Courtenay—Promoted Bvt. Second Lieut., Corps of Engineers, July 1, 1821.—Second Lieut. July 1, 1821.—Acting Asst. Professor of Natural and Experimental Philosophy, Military Academy, from July 23, 1821, to Sept.

1, 1822; and Asst. Professor of Engineering, from Sept 1, 1822, to Aug. 31, 1824.—Acting Professor of Natural and Experimental Philosophy, Military Academy, from Sept. 1, 1828, to Feb. 16, 1829; and Professor, from Feb. 16, 1829, to Dec. 31, 1834.—Resigned Lieutenancy of Engineers, Feb. 16, 1829; and Professorship of Natural and Experimental Philosophy, Dec. 31, 1834.—Professor of Mathematics, University of Pennsylvania, from 1834 to 1836.—Division Engineer, New York and Erie Railroad, 1836–37.—Civil Engineer, in the service of United States, employed in the construction of Fort Independence, Boston Harbor, from 1837 to 1841.*—Chief Engineer of Dry Dock, Navy Yard, Brooklyn, N. Y., 1841–42.—Professor of Mathematics, University of Virginia, since 1842.—Author of Elementary Treatise on Mechanics, translated from the French of M. Boucharlat, with additions and emendations, designed to adapt it to the use of the Cadets of the U. S. Military Academy," 1833.—Degree of A. M., conferred by University of Pennsylvania, 1834; and of LL.D., by Hampden Sidney College, Va., 1846."

---

* Mr. Courtenay, while employed as Engineer in the construction of the works in Boston Harbor, was associated with that distinguished officer, Colonel Sylvanus Thayer, of the Corps of Engineers.

The year before Mr. Courtenay entered the Military Academy, as a Cadet, Colonel Thayer had been appointed Superintendent. He was then engaged in laying the foundation of the system of instruction and discipline which has imparted so much reputation to that institution.

It was among the most agreeable and cherished remembrances of Mr. Courtenay's life that he enjoyed the entire confidence and friendship of so interesting and distinguished a man.

The relation of principal and pupil, in a public institution became the basis of a sincere and generous friendship; and when the news reached the north that Courtenay was dead, no eye was moistened by a tear of warmer sympathy than that of the Superintendent who had guided his youth and admired his life.

The author of this notice examined Mr. Courtenay when he entered the Military Academy, was associated with him in the Academic Board, and knew him intimately in all the situations which he subsequently filled; and yet feels quite incompetent to do justice to the memory of so perfect a man and so dear a friend.

The painter who has a faultless form to delineate or a perfect landscape to transfer to the canvas, is embarrassed by the very perfection of his subject. He has nothing to put in opposition to the beautiful—no shading that can give full effect to the living light. Characters which afford strong contrasts are easily drawn—it is the perfect character which it is difficult to sketch.

The intellectual faculties of Professor Courtenay were blended in such just proportions, that each seemed to aid and strengthen all the others. He examined the elements of knowledge with a microscopic power, and no distinction was so minute as to elude the vigilance of his search. He compared the elements of knowledge with a logic so scrutinizing that error found no place in his conclusions;—and he possessed, in an eminent degree, that marked characteristic of a great mind, the power of a just and profound generalization.

His mind was quick, clear, accurate and discriminating in its apprehensions—rapid, and certain, in its reasoning processes, and far-reaching and profound in its general views. It was admirably adapted both to acquire and use knowledge.

The intellectual faculties, however, are but the pedestal

and shaft of the column—the moral and social faculties are its entablature or crowning glory. It is these faculties which shed over the whole character a soft and attractive radiance, exhibiting in a favorable light the majesty of intellect and the divine attributes of truth, justice and beneficence.

It was the ardent desire and steady aim of Professor Courtenay, during his whole life, to be governed by these principles, and there are few cases in which the ideal and the actual have been brought more closely together. Modest and unassuming in his manners even to diffidence, he was bold, resolute and firm in asserting and maintaining the right. Liberal in his judgments of others, he was exacting in regard to himself. He could discriminate, reason, and decide justly even when his own interests were involved in the issue. His love of truth and justice was stronger than his love of self or of friends.

His intercourse with others was marked by the gentlest courtesies. He was an attentive and eloquent listener. Differences of opinion, appeared to excite regret rather than provoke argument, and his habitual respect for the opinions, wishes and feelings of others, imparted an indescribable charm to his manners.

As a professor he was a model. He was clear, concise, and luminous in his style and methods. Laborious in the preparation of his lectures, even to the minutest facts, he was at all times prepared to impart information. His manner, as a teacher, was highly attractive. He never by look, act, word, or emphasis disparaged the efforts or undervalued the acquirements of his pupils. His pleasant smile and kind voice, when he would say, "Is that answer *perfectly* correct?" gave hope to many minds struggling with the difficulties of science and have left the impression of affectionate recollection on many hearts.

At the Military Academy, on the banks of the Hudson, where Mr. Courtenay was educated, and where he first labored to advance the interest of instruction and science, his name is recorded on the list of distinguished graduates, and honorably enrolled among the most eminent Professors of that Institution. There his labors and memory will live long together.

At the University of Virginia he has left a name equally dear to that distinguished Faculty of which he was an ornament and to the many pupils whom he there taught. When these, in later years, shall revisit their Alma Mater, to revive early and cherished recollections—to strengthen the bonds of early friendships and renew their resolves to be good and great, they will find that a wide space has been made vacant. They will realize in sorrow that a favorite professor has been transferred from the halls of instruction to the grove of pines which borders the town, and which contains the remains of the revered dead. Thither they will go, in the twilight of the evening, to visit the grave of a man of science—their able teacher and faithful friend. In reviewing his life and contemplating his character, they will exclaim—

"Mark the perfect man and behold the upright; for the end of that man is peace."

FISHKILL LANDING,
March 10*th*, 1855.

NOTICE.

The following work was left by Professor Courtenay, in manuscript, in a highly finished condition; and yet, it must be regretted that it could not receive the final corrections of the author. A premature death, at the meridian of life, placed the work in other hands, and any slight inaccuracies of language which may now appear, would doubtless have been corrected, if the sheets could have passed under the eye of the author.

It is a cause of thankfulness, however, that the work was entirely completed by Professor Courtenay; and in its publication the plan, language, and even the punctuation, have been followed with a fidelity due to the memory of a friend.

The work will be found more full and extensive than any which has yet appeared in this country on the same subject; and the part which relates to the Calculus of Variations will be especially acceptable to the American public.

It is perhaps not improper to add, that the Publishers have generously offered to publish the work on very favorable terms, and that the profits, whatever they may be, will go to the family of the author.

CONTENTS.

THE DIFFERENTIAL CALCULUS. PART I.

PAGE.

CHAPTER VII.

CHAPTER VIII.

CHAPTER IX.

CHAPTER X.

CHAPTER XI.

CHAPTER XII.

THE DIFFERENTIAL CALCULUS. PART II.

APPLICATION OF THE DIFFERENTIAL CALCULUS TO THE THEORY OF PLANE CURVES.

PAGE.

CHAPTER I.

CHAPTER II.

THE DIFFERENTIAL CALCULUS. PART III.

THEORY OF CURVED SURFACES.

PAGE.

CHAPTER I.

CHAPTER II.

CHAPTER III.

CHAPTER IV.

THE INTEGRAL CALCULUS. PART I.

CHAPTER I.

CHAPTER II.

CHAPTER III.

CHAPTER IV.

PAGE.

CHAPTER V.

CHAPTER VI.

CHAPTER VII.

CHAPTER VIII.

CHAPTER IX.

CHAPTER X.

THE INTEGRAL CALCULUS. PART II.

PAGE.

RECTIFICATION OF CURVES. QUADRATURE OF AREAS. CUBATURE OF VOLUMES.

CHAPTER I.

CHAPTER II.

CHAPTER III.

CHAPTER IV.

THE INTEGRAL CALCULUS. PART III.

INTEGRATION OF FUNCTIONS OF TWO OR MORE VARIABLES.

CHAPTER I.

CHAPTER II.

CHAPTER III.

CHAPTER IV.

CALCULUS OF VARIATIONS.

DIFFERENTIAL CALCULUS.

CHAPTER I.

FIRST PRINCIPLES.

1. In all mathematical calculations, the quantities which are presented for our consideration belong to one of two remarkable classes: namely, *constant* quantities, which are such as preserve the same values throughout the limits of one investigation; or *variable* quantities, which may assume successively different values, the number of such values being unlimited.

The first letters of the alphabet, as a, b, c, &c., are usually employed to denote *constant* quantities, and the last letters z, y, x, &c. are used to represent such quantities as are *variable.*

2. When two quantities x and y are mutually dependent upon each other, so that a knowledge of the value of one will lead to that of the other, they are said to be *functions* of each other. Thus, in the equations

$$y = ax, \qquad y = bx^2 + cx + e, \qquad y = ax^3 + bx^2 - cx + e,$$

the value of y is determined as soon as that of x is known; and accordingly y is said to be a function of x.

In like manner, an assumed value of y will fix the corresponding values of x, and therefore x is a function of y. There is this difference, however, between the two cases: when the value of x is assumed, that of y is obtained by a simple substitution; whereas the determination of the value of x from that of y requires the solution of an equation. Hence, y is called an *explicit* function of x, but x is said to be an *implicit* function of y.

The general fact that y is an explicit function of x is written thus:

$$y = Fx, \qquad \text{or} \qquad y = \varphi x,$$

when the character F or φ stands as the representative of certain operations to be performed on the quantity x, the result of which operations will be a quantity equal in value to y. And when we wish to imply that the values of x and y are connected by an unresolved equation, or that y is an implicit function of x, we write

$$F(x, y) = 0, \qquad \text{or} \qquad \varphi(x, y) = 0.$$

For the purpose of illustration, let there be taken the three equations

$$y = ax + b \qquad (1),$$

$$y = ax^2 + bx + c \qquad (2),$$

$$y = ax^3 + bx^2 + cx + e \qquad (3),$$

and suppose x to receive an increment h in each equation, converting it into $x + h$, and causing y to assume a new value y_1. Then if the form of each function, or value of y, be supposed to remain unchanged, the three equations (1), (2), and (3), will become respectively

$$y_1 = a(x + h) + b \qquad (4),$$

$$y_1 = a(x + h)^2 + b(x + h) + c \qquad (5),$$

and $$y_1 = a(x + h)^3 + b(x + h)^2 + c(x + h) + e \qquad (6).$$

Subtracting (1) from (4) we obtain

$$y_1 - y = ah \quad (7).$$

From (2) and (5) we get

$$y_1 - y = a(2xh + h^2) + bh \quad (8).$$

And from (3) and (6)

$$y_1 - y = a(3x^2h + 3xh^2 + h^3) + b(2xh + h^2) + ch \quad (9).$$

From (7) we deduce, by division,

$$\frac{y_1 - y}{h} = a \qquad (10);$$

from (8)

$$\frac{y_1 - y}{h} = a(2x + h) + b \quad (11);$$

$$= 2ax + ah + b;$$

and from (9)

$$\frac{y_1 - y}{h} = a(3x^2 + 3xh + h^2) + b(2x + h) + c \qquad (12).$$

The results, (10), (11), and (12), express tne ratio between the increment h assigned to x, and the corresponding increment $y_1 - y$ imparted to y. The values of this ratio, in the three examples selected, present remarkable differences.

In the first example, this ratio retains the same value a, whatever may be the value assigned to the increment h. In the second example it consists of two parts,

one $= 2ax + b$,

entirely independent of h, and the other $= ah$,

which varies with h. If the value of h be supposed to diminish, the ratio

$$2ax + b + ah \qquad (11),$$

will become more and more nearly equal to $2ax + b$; and, finally, when h becomes indefinitely small, the ratio is reduced to this latter value.

The corresponding increments h and $y_1 - y$, when indefinitely small, are called the *differentials* of the quantities x and y, and the limiting value of the ratio

$$\frac{y_1 - y}{h}$$

is called the *differential coefficient*, because it is the multiplier of the differential of x necessary to produce the differential of y.

The differentials of x and y are written dx and dy, the character d being the symbol of an operation to be performed on x or y, not a factor; and the differential coefficient is written $\frac{dy}{dx}$.

Moreover, one of the variables (usually x) is called the *independent* variable, its increment dx (although small) being arbitrary; while the other y, whose increment dy depends on that of x, is called the *dependent* variable or simply the function.

In the third example, the ratio

$$\frac{y_1 - y}{h}$$

reduces, at the limit when $h = 0$, to

$$\frac{dy}{dx} = 3ax^2 + 2bx + c.$$

These examples illustrate the fact that two indefinitely small quantities may yet have a finite ratio; and they suffice to show that the form of the differential coefficient, which is usually a function of x, will depend very materially on the form of the original function y.

(3.) The considerations just presented analytically admit of geometrical illustration. For, whatever may be the relation between x and y, the former may be regarded as the abscissa, and the latter as the ordinate of a plane curve; and the determination of the relation between the corresponding increments of x and y, is reduced to finding the change in the length of the ordinate produced by an arbitrary change in the length of the abscissa.

It is the chief object of the Differential Calculus to investigate the laws of increase of functions having various forms, when such changes are produced by an arbitrary change in the value of the independent variable upon which the values of the functions depend.

Geometrical considerations will also point out very clearly how it happens that a given augmentation of the variable x will, in different stages of its magnitude, produce widely different increments of the function y.

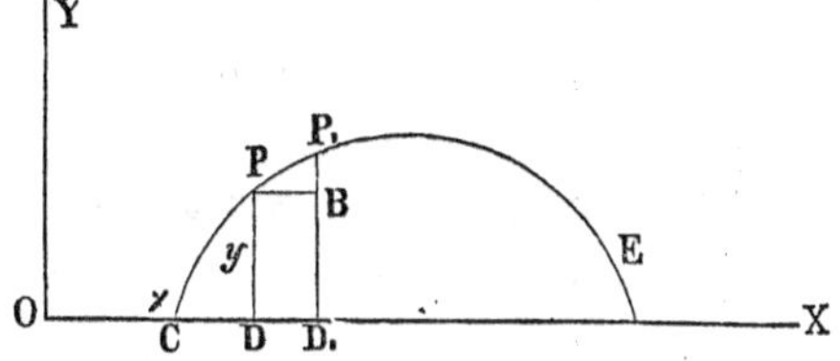

Referring to the annexed diagram, it will be apparent that near the vertex C of the curve CPE, a slight increase in the value of the abscissa x will produce a comparatively large increase in the value of the ordinate y; but when the tangent to the curve forms a smaller angle with the axis OX, as at P, the same increment in x will produce a much smaller increase of y; and if the tangent be nearly parallel to OX, the increment received by y will be very small in comparison with that given to x. Finally, by continuing to increase x, the ordinate y may first cease to increase, and may afterwards actually decrease, or the increment of y may become negative; and these different results will occur without any change in the form of the function y.

4. One of the first inquiries presented for consideration is the determination of the general form of the function $F(x+h)$; for, since we desire to compare

$$y = Fx \qquad \text{with} \qquad y_1 = F(x+h),$$

it is important to know what form $F(x+h)$ will assume when expanded into a series of terms involving x and h. Hence the following

Proposition. To determine the general form of the development of any function of the algebraic sum of two quantities, such as $F(x+h)$, arranged according to the powers of the second h.

1st. There must be one term in the development of the form Fx, and the other terms must contain h. For, since the development is supposed to be general, and therefore true for all values h, it ought to be applicable when $h=0$, in which case the undeveloped function $F(x+h)$ reduces to Fx. This condition will be satisfied by supposing the first term in the development to be Fx, and all the succeeding terms to contain powers of h, since the supposition $h=0$ will then give rise to an equation, $Fx = Fx$, which is identically true. And no other conceivable form of development would lead to this result.

We may therefore write

$$F(x+h) = Fx + Ah^a + Bh^b + Ch^c + \text{\&c.} \qquad (1),$$

in which the coefficients A, B, C, &c., will usually be functions of x, and the exponents a, b, c, &c., undetermined constants.

2d. None of the exponents, a, b, c, &c., can be negative. For if there could be a term of the form

$$Bh^{-b} \qquad \text{or} \qquad \frac{B}{h^b},$$

it would become infinite when $h=0$, thus rendering the developed expression infinite, while the undeveloped expression would become simply Fx, and this latter would probably be finite.

3d. None of the exponents can be fractional. For if there could be a term of the form

$$Eh^{\frac{r}{s}} \qquad \text{or} \qquad E\sqrt[s]{h^r},$$

such term would have as many different values as there are units in s; that is, it would have s values; and each of these values could be combined in succession with the aggregate of the other terms of the series.

Now if each of these other terms, except the first term Fx, be supposed to have but one value, the sum of all the terms containing h will have s different values. And if Fx be susceptible of n different values, the entire development will admit of $n \times s$ values, since each value of Fx may be combined, in succession, with each value of the remaining terms.

But $F(x+h)$ being of the *same form* with Fx, must have the same number n of values. Thus, for example, if

$$F(x+h) = (x+h)^{\frac{1}{3}}, \qquad \text{then} \qquad Fx = x^{\frac{1}{3}},$$

and both will have three values.

If $$F(x+h) = a(x+h)^2 + b(x+h)^{\frac{1}{5}},$$

then $$Fx = ax^2 + bx^{\frac{1}{5}},$$

and both will have five values, &c.

Thus, in the case supposed above, where there was one fractional exponent, $F(x+h)$ would have n values when undeveloped, but $n \times s$ values when developed—a manifest absurdity.

We conclude therefore that the exponents a, b, c, &c., in the general development, must be positive integers; and in order to make the development include every possible case, we write

$$F(x+h) = Fx + Ah + Bh^2 + Ch^3 + Dh^4, \text{ \&c.},$$

including every power of h. If in any particular case some of these terms should be unnecessary, it will suffice to suppose the corresponding coefficients A, B, C, &c., to reduce to zero.

We have a familiar example of the expansion of $F(x+h)$ in the well known binomial theorem. Thus, if

$$F(x+h) = (x+h)^n = x^n + nx^{n-1}h$$
$$+ \frac{n(n-1)}{1.2} \cdot x^{n-2}h^2 + \frac{n(n-1)(n-2)}{1.2.3} x^{n-3}h^3 + \text{\&c.},$$

we shall have

$$Fx=x^n,\ A=nx^{n-1},\ B=\frac{n(n-1)}{1\,.\,2}x^{n-2},$$

$$C=\frac{n(n-1)(n-2)}{1\,.\,2\,.\,3}x^{n-3},\ \text{\&c.},$$

where A, B, C, &c., are functions of x.

The following are likewise examples of the development as applied to particular cases.

2. Let $\quad Fx=(a+x)^{\frac{1}{2}}+bx^n:\quad$ then

$$F(x+h)=(a+x+h)^{\frac{1}{2}}+b(x+h)^n,$$

which expressions, when expanded by the binomial theorem, give

$$F(x+h)=(a+x)^{\frac{1}{2}}+\frac{1}{2}(a+x)^{-\frac{1}{2}}h-\frac{1}{8}(a+x)^{-\frac{3}{2}}h^2+\text{\&c.},$$

$$+bx^n+nbx^{n-1}h+\frac{n(n-1)}{1\,.\,2}bx^{n-2}h^2+\text{\&c.}$$

$$=Fx+\left[nbx^{n-1}+\frac{1}{2}(a+x)^{-\frac{1}{2}}\right]h$$

$$+\left[\frac{1}{2}bn(n-1)x^{n-2}-\frac{1}{8}(a+x)^{-\frac{3}{2}}\right]h^2+\text{\&c.}$$

which corresponds with the general form.

3. Let $\quad Fx=\log x:\quad$ then

$$F(x+h)=\log(x+h)=\log\left[x\left(1+\frac{h}{x}\right)\right]=\log x+\log\left(1+\frac{h}{x}\right)$$

$$=\log x+M\left(\frac{h}{x}-\frac{h^2}{2x^2}+\frac{h^3}{3x^3}-\frac{h^4}{4x^4}+\text{\&c.}\right)$$

where M denotes the modulus of the system of logarithms.

$$\therefore F(x+h)=Fx+\frac{M}{x}\cdot h-\frac{M}{2x^2}h^2+\frac{M}{3x^3}h^3-\frac{M}{4x^4}h^4+\text{\&c.}$$

which also corresponds to the general form.

It may be well to observe, that although the form of the development of $F(x+h)$ is always such as has been indicated while x retains its general value, yet it is possible (in some cases) to assign certain particular values to x which shall cause the development in this form to become impossible.

Thus, if in the second of the above examples, we put $x = -a$, the true development of $F(x+h)$ will become simply

$$F(x+h) = h^{\frac{1}{2}} + b\,(-a)^n + bn(-a)^{n-1}\,h + \&c.,$$

in which one fractional exponent appears.

The same supposition causes all the coefficients involving negative powers of $a+x$ to become infinite in the general expansion. It will be shown hereafter that the particular cases in which the general development is inapplicable, are always indicated by some of the terms of the development becoming infinite. At present it is sufficient to remark that the number of such cases is comparatively small, and that they will receive a special examination.

5. From the development of $F(x+h)$, we derive a direct and general method of finding the differential of any proposed function

$$y = Fx.$$

For, if we give to x an increment h, we shall have

$$y_1 = F(x+h) = Fx + Ah + Bh^2 + Ch^3 + \&c.$$

$$\therefore y_1 - y = F(x+h) - Fx = Ah + Bh^2 + Ch^3 + \&c.$$

$$\therefore \frac{y_1 - y}{h} = A + Bh + Ch^2 + \&c.$$

And by passing to the limit, when $h = 0$, we get

$$\frac{dy}{dx} = A, \qquad \text{whence} \qquad dy = Adx.$$

Thus it appears that the coefficient A of the 1st power of h in the development of $F(x + h)$ is the differential coefficient of the proposed function, and this multiplied by dx gives the required differential of y.

It will be found convenient, however, to form rules for differentiating functions of the various forms likely to arise, and to this investigation we proceed next.

CHAPTER II.

DIFFERENTIATION OF ALGEBRAIC FUNCTIONS.

6. *Prop.* To differentiate the product of two functions of a single variable.

Let
$$u = yz,$$

where y and z are given functions of the same independent variable x, and let x take an increment h, converting u, y, and z, into u_1, y_1, and z_1. Then, since y_1 and z_1 will each be a function of $x + h$, we shall have

$$y_1 = y + Ah + Bh^2 + Ch^3 + \&c.,$$

and
$$z_1 = z + A_1h + B_1h^2 + C_1h^3 + \&c.$$

$$\therefore\ u_1 = y_1z_1 = yz + (Az + A_1y)h + (Bz + B_1y + AA_1)h^2$$
$$+ (Cz + C_1y + AB_1 + A_1B)h^3 + \&c.$$

$$\therefore\ \frac{u_1 - u}{h} = \frac{y_1z_1 - yz}{h} = Az + A_1y + (Bz + B_1y + AA_1)h$$
$$+ (Cz + C_1y + AB_1 + A_1B)h^2 + \&c.$$

and when $h = 0$, this becomes

$$\frac{du}{dx} = Az + A_1y. = \frac{dy}{dx}\cdot z + \frac{dz}{dx}\cdot y,$$

since
$$A = \frac{dy}{dx} \quad \text{and} \quad A_1 = \frac{dz}{dx}.$$

And by multiplying by dx, we get

$$du = zdy + ydz.$$

Thus the differential of the product yz *of two functions is found by multiplying each function by the differential of the other function, and adding the results.*

7. *Prop.* To differentiate the product of several functions of a single variable.

1st. Let $u = vyz$, where v, y, and z, are functions of the independent variable x.

Put $yz = s$; then $u = vs$,

and by the last proposition,

$$du = vds + sdv, \quad \text{and also} \quad ds = ydz + zdy.$$

Substituting the values of s and ds in that of du, there results

$$du = v(ydz + zdy) + yzdv = vydz + vzdy + yzdv.$$

2d. Let $u = svyz$.

Put $yz = w$; then $u = svw$,

$$\therefore du = svdw + swdv + vwds = sv(ydz + zdy) + syzdv + vyzds,$$

or, $$du = svydz + svzdy + syzdv + vyzds;$$

and the same method could be applied to the product of a greater number of functions.

Hence we have the following rule for the differential of the product of several functions:

Multiply the differential of each factor by the continued product of all the other factors, and add the results.

8. *Prop.* To differentiate a fraction whose numerator and denominator are functions of a single variable.

Let $u = \frac{y}{z}$, where y and z are functions of x.

Then $y = uz$, and this differentiated by the rule for products, gives

$$dy = udz + zdu = \frac{y}{z} \cdot dz + zdu$$

$$\therefore\ zdy = ydz + z^2du,$$

and by reduction $$du = \frac{zdy - ydz}{z^2}.$$

Thus the rule is as follows:

Multiply the differential of the numerator by the denominator, and the differential of the denominator by the numerator; subtract the second product from the first, and divide the remainder by the square of the denominator.

9. *Prop.* To differentiate a power of a single variable.

1st. Let $u = x^n$, where n is a positive integer.

Regarding x^n as the product $x.\ x.\ x.\ x$, &c., of n equal factors each $= x$, and applying the rule for differentiating a product, we get

$$du = x^{n-1}dx + x^{n-1}dx + x^{n-1}dx + \text{\&c., to } n \text{ terms.}$$

$$\therefore\ du = nx^{n-1}dx,$$

and the rule in this case is the following:

Multiply the given power (x^n) *by the exponent* (giving nx^n); *then diminish the exponent by unity* (giving nx^{n-1}); *and finally, multiply by the differential of the root* (producing nx^{n-1}dx).

2d. Now suppose the exponent n to be a positive fraction $\frac{a}{c}$

Then $$u = x^{\frac{a}{c}}$$

$\therefore\ u^c = x^a$, where the exponents a and c are both positive integers.

Hence, by the application of the rule just established for such cases, we have

$$cu^{c-1}du = ax^{a-1}dx$$

$$\therefore\ du = \frac{ax^{a-1}}{cu^{c-1}}dx = \frac{a}{c} \cdot \frac{x^{a-1}}{\left(x^{\frac{a}{c}}\right)^{c-1}}dx = \frac{a}{c}x^{a-1-a+\frac{a}{c}}dx = \frac{a}{c}x^{\frac{a}{c}-1}dx.$$

and the rule for differentiating the power is the same as when the exponent is a positive integer.

3d. Let the exponent be a negative integer, or $u = x^{-n}$

Then
$$u = \frac{1}{x^n} = \frac{x}{x^{n+1}}$$

and this differentiated by the rule for fractions, gives

$$du = \frac{x^{n+1} - (n+1)x^{n+1}}{x^{2n+2}}dx = -\frac{ndx}{x^{n+1}} = -nx^{-n-1}dx.$$

And the rule is still the same.

4th. Let the exponent be a negative fraction, or let $u = x^{-\frac{a}{c}}$. Then $u^c = x^{-a}$, and by the first and third cases,

$$cu^{c-1}du = -ax^{-a-1}dx, \quad \text{or,} \quad du = -\frac{a}{c}\cdot\frac{x^{-a-1}}{u^{c-1}}dx.$$

$$\therefore\ du = -\frac{a}{c}\cdot\frac{x^{-a-1}dx}{\left(x^{-\frac{a}{c}}\right)^{c-1}} = -\frac{a}{c}x^{-\frac{a}{c}-1}dx,$$

and the formula is still the same.

We might have deduced the rule for differentiating a power, as alike applicable to all cases, by employing the binomial theorem; for, since the second term in the development of $(x+h)^n$, is $nx^{n-1}h$, for all values of n,

we must have
$$\frac{d\,(x^n)}{dx} = nx^{n-1}, \quad \text{or,} \quad d\,(x^n) = nx^{n-1}dx.$$

It is intended, however, to demonstrate the truth of the binomial theorem by the aid of the differential calculus, and hence the necessity of establishing the rules for differentiation, without reference to that theorem.

Remark. If the function which it is proposed to differentiate contain a constant factor, such factor will appear in the differential.

Thus $d\,(ax) = adx$, for when x takes the increment h, the function ax becomes

$$u_1 = a\,(x + h) \quad \text{and} \quad \therefore \frac{u_1 - u}{h} = a \quad \text{and} \quad \frac{du}{dx} = a.$$

Similarly if $u = a \,.\, Fx$, where F denotes any function,

then $$u_1 = aF\,(x + h) \quad \text{and} \quad du = ad\,(Fx).$$

10. *Prop.* To differentiate the algebraic sum of several functions of a single variable.

Let $$u = As + Bv - Cy + Dz,$$
where s, v, y, and z, are functions of x.

Then when x takes the increment h,

$$As \text{ becomes } As_1 = A\,(s + A_1h + B_1h^2 + C_1h^3 \text{ \&c.}).$$

$$Bv \text{ becomes } Bv_1 = B\,(v + A_2h + B_2h^2 + C_2h^3 \text{ \&c.}).$$

$$Cy \text{ becomes } Cy_1 = C\,(y + A_3h + B_3h^2 + C_3h^3 \text{ \&c.}).$$

$$Dz \text{ becomes } Dz_1 = D\,(z + A_4h + B_4h^2 + C_4h^4 \text{ \&c.}).$$

$$\therefore u \text{ becomes } u_1 = As + Bv - Cy + Dz$$
$$+ (AA_1 + BA_2 - CA_3 + DA_4)\,h + \text{\&c.}$$

$$\therefore du = (AA_1 + BA_2 - CA_3 + DA_4)\,dx.$$

But $$A_1dx = ds, \quad A_2dx = dv, \quad A_3dx = dy, \quad A_4dx = dz.$$

$$\therefore du = Ads + Bdv - Cdy + Ddz.$$

And the rule is as follows:

Differentiate the terms successively, and take the algebraic sum of the result.

Remark. If a constant be connected with a variable quantity by the sign + or −, such constant will disappear by differentiation.

Thus, when we have $u = a + Fx$, then

$$u_1 = a + F\,(x + h) = a + Fx + Ah + Bh^2, \text{ \&c.},$$
$$= u + Ah + Bh^2, \text{ \&c.}$$

$\therefore du = Adx$, the constant a having disappeared.

EXAMPLES.

11. 1. To differentiate

$$y = 4x^3 + 7x^2 - 8x + 5.$$

Applying the rule for powers to each term we obtain

$$dy = 4 \times 3\, x^2dx + 7 \times 2xdx - 8dx = (12x^2 + 14x - 8)dx.$$

$$\therefore \frac{dy}{dx} = 12x^2 + 14x - 8.$$

2. $$y = ax^2(bx + c) = abx^3 + acx^2.$$

Differentiating this as a product, we get

$$dy = 2ax(bx + c)dx + ax^2bdx = (3abx^2 + 2acx)dx.$$

Or by first performing the multiplication indicated, and then differentiating as a sum, the same result is obtained.

$$\therefore \frac{dy}{dx} = 3abx^2 + 2acx.$$

3. $$y = \frac{4x^3}{(b + x^2)^3}.$$

Differentiating by the rules for fractions and powers, we obtain

$$dy = \frac{12x^2(b + x^2)^3dx - 3(b + x^2)^2 \times 4x^3 \times 2xdx}{(b + x^2)^6}$$

$$= \frac{12x^2(b + x^2) - 24x^4}{(b + x^2)^4}dx = \frac{12x^2(b - x^2)}{(b + x^2)^4}dx.$$

$$\therefore \frac{dy}{dx} = \frac{12x^2(b - x^2)}{(b + x^2)^4}.$$

4. $$y = \sqrt{a + bx^2} = (a + bx^2)^{\frac{1}{2}}.$$

$$dy = \frac{1}{2}(a + bx^2)^{-\frac{1}{2}} \times 2bxdx \;\therefore\; \frac{dy}{dx} = \frac{bx}{\sqrt{a + bx^2}}.$$

5. $$u = x(1 + x^2)(1 + x^3).$$

$$\frac{du}{dx} = (1 + x^2)(1 + x^3) + x(1 + x^3) \times 2x + x(1 + x^2) \times 3x^2$$

$$= 1 + x^2 + x^3 + x^5 + 2x^2 + 2x^5 + 3x^3 + 3x^5$$

$$= 1 + 3x^2 + 4x^3 + 6x^5.$$

6. $$u = \sqrt{x + \sqrt{1 + x^2}} = \left[x + (1 + x^2)^{\frac{1}{2}}\right]^{\frac{1}{2}}$$

$$\frac{du}{dx} = \frac{1}{2}\left[x + (1 + x^2)^{\frac{1}{2}}\right]^{-\frac{1}{2}} \times \left[1 + (1 + x^2)^{-\frac{1}{2}} x\right]$$

$$= \frac{\frac{1}{2}(x + \sqrt{1 + x^2})}{\sqrt{x + \sqrt{1 + x^2}} \times \sqrt{1 + x^2}} = \frac{\sqrt{x + \sqrt{1 + x^2}}}{2\sqrt{1 + x^2}}.$$

7. $$u = bx^{\frac{8}{3}} \qquad \frac{du}{dx} = \frac{8}{3} bx^{\frac{5}{3}}.$$

8. $$u = \frac{c}{x^6} - b = cx^{-6} - b. \qquad \frac{du}{dx} = -6cx^{-7} = -\frac{6c}{x^7}.$$

9. $$u = \sqrt[3]{x} . \sqrt{\sqrt{x} + 1} = x^{\frac{1}{3}}(x^{\frac{1}{2}} + 1)^{\frac{1}{2}}.$$

$$du = \frac{1}{3} x^{-\frac{2}{3}} (x^{\frac{1}{2}} + 1)^{\frac{1}{2}} dx + \frac{1}{2} x^{\frac{1}{3}} (x^{\frac{1}{2}} + 1)^{-\frac{1}{2}} \times \frac{1}{2} x^{-\frac{1}{2}} dx.$$

$$\therefore \frac{du}{dx} = \frac{(x^{\frac{1}{2}} + 1)^{\frac{1}{2}}}{3x^{\frac{2}{3}}} + \frac{1}{4x^{\frac{1}{6}}(x^{\frac{1}{2}} + 1)^{\frac{1}{2}}} = \frac{7x^{\frac{1}{2}} + 4}{12\sqrt[3]{x^2}\sqrt{x^{\frac{1}{2}} + 1}}.$$

10. $$u = \frac{\sqrt{1 + x} + \sqrt{1 - x}}{\sqrt{1 + x} - \sqrt{1 - x}} = \frac{(\sqrt{1 + x} + \sqrt{1 - x})^2}{2x}$$

$$= \frac{1 + \sqrt{1 - x^2}}{x}.$$

$$\frac{du}{dx} = \frac{-x^2(1 - x^2)^{-\frac{1}{2}} - (1 + \sqrt{1 - x^2})}{x^2} = -\frac{1 + \sqrt{1 - x^2}}{x^2\sqrt{1 - x^2}}.$$

11. $$u = \frac{x}{x + \sqrt{1 + x^2}} = \frac{x(x - \sqrt{1 + x^2})}{x^2 - (1 + x^2)} = x\sqrt{1 + x^2} - x^2.$$

$$\frac{du}{dx} = \sqrt{1+x^2} + x^2(1+x^2)^{-\frac{1}{2}} - 2x = \frac{1+2x^2-2x\sqrt{1+x^2}}{\sqrt{1+x^2}}.$$

12. $$u = \sqrt[4]{\left[a - \frac{b}{\sqrt{x}} + \sqrt[3]{(c^2-x^2)^2}\right]^3}$$

$$= \left[a - bx^{-\frac{1}{2}} + (c^2-x^2)^{\frac{2}{3}}\right]^{\frac{3}{4}}$$

$$\frac{du}{dx} = \frac{3}{4}\left[a - bx^{-\frac{1}{2}} + (c^2-x^2)^{\frac{2}{3}}\right]^{-\frac{1}{4}} \times \left[\frac{1}{2}bx^{-\frac{3}{2}} - \frac{2}{3}(c^2-x^2)^{-\frac{1}{3}}.2x\right]$$

$$= \frac{\dfrac{3b}{2x\sqrt{x}} - \dfrac{4x}{\sqrt[3]{c^2-x^2}}}{4\sqrt[4]{a - \dfrac{b}{\sqrt{x}} + \sqrt[3]{(c^2-x^2)^2}}}$$

13. $u = \sqrt{a + x + \sqrt{a + x + \sqrt{a + x}}}$ &c., continued indefinitely.

Here $u = \sqrt{a+x+u}$, and $\therefore u^2 = a + x + u$,

or, $u^2 - u = a + x$, $\therefore u = \frac{1}{2} + \sqrt{a + x + \frac{1}{4}}$,

$$\therefore \frac{du}{dx} = \frac{1}{\sqrt{4a+4x+1}}.$$

The functions considered hitherto are called *algebraic* functions, because they require only the performance of the common algebraic operations of addition, subtraction, multiplication, division, raising of powers, and extraction of roots. There is a second and very extensive class of functions, in which the variable enters as an exponent, or in connection with logarithms, sines, cosines, tangents, circular arcs, &c., of which the following are examples: a^x, x^x, $\log x$, $\sin x$, $(\cos x)^{\sin x}$, $\sin^{-1} x$, $(\log x)^{\tan x}$, &c. These are called *transcendental* functions, and they will be considered in the next chapter.

CHAPTER III.

TRANSCENDENTAL FUNCTIONS.

12. *Prop.* To differentiate $u = \log x$.

Let x take the increment h_1, converting u into

$$u_1 = \log (x + h).$$

Then $u_1 = \log (x + h) = \log \left[x \left(1 + \frac{h}{x}\right)\right] = \log x + \log \left(1 + \frac{h}{x}\right)$.

or $$u_1 = u + M \left(\frac{h}{x} - \frac{h^2}{2x^2} + \frac{h^3}{3x^3} - \frac{h^4}{4x^4} \text{ \&c.}\right)$$

where M is the modulus of the system.

$$\therefore \frac{du}{dx} = \frac{d(\log x)}{dx} = \frac{M}{x} \quad \text{and} \quad du = \frac{M}{x}\, dx.$$

Hence the rule is as follows:

Multiply the differential of the variable by the modulus of the system in which the logarithm is taken, and divide the product by the variable.

If the logarithms belong to the Naperian system whose modulus is equal to unity, we shall have

$$d(\log x) = \frac{dx}{x}.$$

As the essential properties of logarithms are the same in all systems, while the form of the differential is simplest in the Naperian system, the logarithms employed throughout the Calculus will always be the Naperian, unless the contrary is distinctly specified, and the rule for differentiating a logarithm will be simply this:

Divide the differential of the quantity by the quantity itself.

13. *Prop.* To differentiate an exponential function as $u = a^x$, the base a being constant.

Passing to logarithms we have

$$\log u = x \log a.$$

$$\therefore\ d(\log u) = d(x \log a) \qquad \text{or} \qquad \frac{du}{u} = \log a \,.\, dx\,;$$

$$\therefore\ du = \log a \,.\, u \,.\, dx = \log a \,.\, a^x . \,dx \qquad \text{and} \qquad \frac{du}{dx} = \log a \,.\, a^x.$$

And the rule for differentiating an exponential is this:

Multiply the exponential (a^x) *by the differential of the exponent* (dx), *and that product by the Naperian logarithm of the base* (log a).

Cor. If $a = e$, the Naperian base, we shall have $\log e = 1$;

$$\therefore\ d(e^x) = e^x dx, \qquad \text{and} \qquad \frac{d(e^x)}{dx} = e^x.$$

Remark. The rule for differentiating logarithmic functions will often be found useful, even when the original function is algebraic, since by passing to logarithms we may give the function a simpler form.

Examples af Logarithmic and Exponential Functions.

14. 1. Let $\qquad u = \log\,(x + \sqrt{1 + x^2}).$

$$du = \frac{d(x + \sqrt{1 + x^2})}{x + \sqrt{1 + x^2}} = \frac{1 + (1 + x^2)^{-\frac{1}{2}} x}{x + \sqrt{1 + x^2}}\, dx$$

$$= \frac{x + \sqrt{1 + x^2}}{(x + \sqrt{1 + x^2})\sqrt{1 + x^2}}\, dx = \frac{dx}{\sqrt{1 + x^2}}. \quad \therefore\ \frac{du}{dx} = \frac{1}{\sqrt{1 + x^2}}.$$

2. $$u = x(a^2 + x^2)\sqrt{a^2 - x^2}.$$

Passing to logarithms we have

$$\log u = \log x + \log(a^2 + x^2) + \frac{1}{2}\log(a^2 - x^2).$$

$$\therefore \frac{du}{u} = \frac{dx}{x} + \frac{d(a^2 + x^2)}{a^2 + x^2} + \frac{1}{2}\frac{d(a^2 - x^2)}{a^2 - x^2}$$

$$= \frac{dx}{x} + \frac{2xdx}{a^2 + x^2} - \frac{xdx}{a^2 - x^2}$$

$$\therefore \frac{du}{dx} = (a^2 + x^2)\sqrt{a^2 - x^2} + 2x^2\sqrt{a^2 - x^2} - \frac{x^2(a^2 + x^2)}{\sqrt{a^2 - x^2}}$$

$$= \frac{a^4 + a^2x^2 - 4x^4}{\sqrt{a^2 - x^2}}.$$

3. $$u = \log\frac{\sqrt{x^2 + 1} - x}{\sqrt{x^2 + 1} + x}.$$

Multiplying numerator and denominator by the numerator we have

$$u = \log\frac{2x^2 + 1 - 2x\sqrt{x^2 + 1}}{x^2 + 1 - x^2} = \log(2x^2 + 1 - 2x\sqrt{x^2 + 1})$$

$$\frac{du}{dx} = \frac{4x - 2\sqrt{x^2 + 1} - 2x^2(x^2 + 1)^{-\frac{1}{2}}}{2x^2 + 1 - 2x\sqrt{x^2 + 1}} = -\frac{2}{\sqrt{x^2 + 1}}.$$

4. $u = x^{a\sqrt{-1}}$. Then $\log u = a\sqrt{-1}\log x$.

$$\therefore \frac{du}{u} = a\sqrt{-1}\,\frac{dx}{x},$$

and $$du = a\sqrt{-1}\,.\,x^{a\sqrt{-1}}\,.\,\frac{dx}{x} = a\sqrt{-1}\,x^{a\sqrt{-1}-1}\,dx.$$

Thus the rule for differentiating a power is still the same, when the exponent is imaginary

5. $$u = x^x. \quad \text{Then } \log u = x \log x.$$

$$\therefore \frac{du}{u} = \log x . dx + x . \frac{dx}{x} = (\log x + 1)dx$$

$$\therefore \frac{du}{dx} = x^x(\log x + 1).$$

6. $$u = x^{x^x}$$

This signifies that x is raised to a power whose exponent is x^x, and it must not be confounded with $(x^x)^x$, which latter implies that x^x is raised to the xth power.

Then $\log u = x^x \log x \therefore \frac{du}{u} = \log x(\log x + 1)x^x dx + x^x \frac{dx}{x}$

$$\therefore \frac{du}{dx} = x^{x^x} . x^x \left[\log x(\log x + 1) + \frac{1}{x}\right]$$

7. $u = e^{x^x}$ where e is the Naperian base.

$$\log u = x^x \log e = x^x \quad \therefore \frac{du}{dx} = e^{x^x} . x^x (\log x + 1).$$

8. $$u = x^{e^x}. \quad \text{Then} \quad \log u = e^x \log x$$

$$\therefore \frac{du}{dx} = x^{e^x}\left(\log x + \frac{1}{x}\right)e^x.$$

9. $$u = \log (nx). \quad \text{Then} \quad du = \frac{d(nx)}{nx} = \frac{dx}{x}.$$

This result is the same as when $u = \log x$, as might have been anticipated, since $\log (nx) = \log n + \log x$, and $\log n$ is constant.

10. $$u = \log (\log x). \quad \text{Then} \quad du = \frac{d (\log x)}{\log x} = \frac{dx}{x . \log x}.$$

$$\therefore \frac{du}{dx} = \frac{1}{x \log x}.$$

11. $u = (\log x)^n = \log^n x$. Then $du = n \log^{n-1} x \,.\, d(\log x)$

$$\therefore \frac{du}{dx} = \frac{n \,.\, \log^{n-1} x}{x}.$$

12. $u = e^{\log \sqrt{a^2+x^2}}$. Then $\log u = \log \sqrt{a^2 + x^2}$

$$\therefore u = \sqrt{a^2 + x^2} = (a^2 + x^2)^{\frac{1}{2}} \quad \text{and} \quad \frac{du}{dx} = \frac{x}{\sqrt{a^2 + x^2}}.$$

13. $u = e^{\log^n x}$. $\quad du = e^{\log^n x} \,.\, d\,(\log^n x)$

$$\therefore \frac{du}{dx} = \frac{n}{x} \cdot e^{\log^n x} \,.\, \log^{n-1} x.$$

14. $u = \frac{1}{4} x^4 \log^2 x - \frac{1}{8} x^4 \log x + \frac{1}{32} x^4.$

$$\frac{du}{dx} = x^3 \log^2 x + \frac{1}{2} x^3 \log x - \frac{1}{2} x^3 \log x - \frac{1}{8} x^3 + \frac{1}{8} x^3 = x^3 \log^2 x.$$

15. $u = e^x (x^4 - 4x^3 + 12x^2 - 24x + 24)$

$$\frac{du}{dx} = e^x (x^4 - 4x^3 + 12x^2 - 24x + 24)$$

$$+ e^x (4x^3 - 12x^2 + 24x - 24) = e^x \,.\, x^4.$$

Trigonometrical Functions.

15. The trigonometrical functions $\sin x$, $\cos x$, $\tan x$, &c. will next be considered, but the determination of the forms of their differentials will be facilitated by the following

Prop. The limit to the ratios $\frac{\text{arc}}{\sin}$, $\frac{\text{arc}}{\text{chord}}$, and $\frac{\text{arc}}{\text{tang}}$, when the arc is diminished indefinitely, is unity.

Proof. Since $\frac{\sin}{\tan} = \frac{\cos}{\text{radius}} = \frac{\text{rad} - \text{versin}}{\text{rad}} = 1 - \frac{\text{versin}}{\text{rad}}$, and since the last term in this equality can be rendered smaller than any assignable quantity by taking the arc sufficiently small, it follows that the limit to the ratio $\frac{\sin}{\tan}$ is unity.

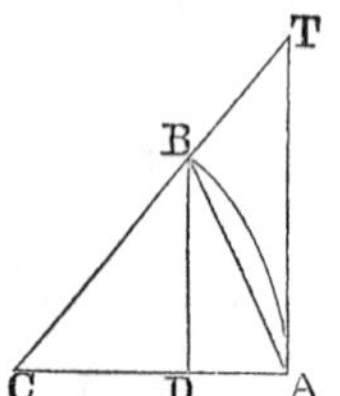

But both the chord AB and the arc AB are intermediate in value between the sine BD and the tangent AT. Hence at the limit, when the arc is indefinitely small,

$$\frac{\text{arc}}{\sin} = \frac{\text{arc}}{\text{chord}} = \frac{\text{arc}}{\tan} = \frac{\sin}{\tan} = 1.$$

16. *Prop.* To differentiate $y = \sin x$.

In the well known trigonometrical formula,

$$\sin a - \sin b = 2 \sin \frac{1}{2}(a - b) \cos \frac{1}{2}(a + b),$$

make $a = x + h$ and $b = x$.

Then $\frac{1}{2}(a - b) = \frac{1}{2}h$, and $\frac{1}{2}(a + b) = x + \frac{1}{2}h$.

$$\therefore \sin(x + h) - \sin x = 2 \sin \frac{1}{2}h \,.\, \cos(x + \frac{1}{2}h).$$

$$\therefore \frac{\sin(x + h) - \sin x}{h} = \frac{2 \sin \frac{1}{2}h \,.\, \cos(x + \frac{1}{2}h)}{h}$$

$$= \frac{\sin \frac{1}{2}h}{\frac{1}{2}h} \cdot \cos(x + \frac{1}{2}h).$$

But at the limit when $h = 0$,

$$\frac{\sin \frac{1}{2}h}{\frac{1}{2}h} = 1, \quad \text{and} \quad \cos(x + \frac{1}{2}h) = \cos x.$$

$$\therefore \frac{dy}{dx} = \frac{d(\sin x)}{dx} = \cos x, \quad \text{and} \quad d(\sin x) = \cos x \,.\, dx.$$

17. *Prop.* To differentiate $y = \cos x$.

Here
$$y = \cos x = \sin\left(\frac{1}{2}\pi - x\right)$$

where π = semi-circumference of the circle whose radius = 1.

$$\therefore dy = d\sin\left(\frac{1}{2}\pi - x\right) = \cos\left(\frac{1}{2}\pi - x\right) \cdot d\left(\frac{1}{2}\pi - x\right) = -\sin x dx$$

$$\therefore \frac{dy}{dx} = \frac{d\cos x}{dx} = -\sin x;$$

the negative sign prefixed to the value of this ratio signifies that the cosine decreases as the arc increases.

18. *Prop.* To differentiate $u = \tan x$.

$$du = d(\tan x) = d\frac{\sin x}{\cos x} = \frac{\cos x \,.\, d\sin x - \sin x \,.\, d\cos x}{\cos^2 x}$$

$$= \frac{\cos^2 x + \sin^2 x}{\cos^2 x} dx = \frac{dx}{\cos^2 x} = \sec^2 x \,.\, dx.$$

$$\therefore \frac{du}{dx} = \frac{d\tan x}{dx} = \sec^2 x.$$

19. *Prop.* To differentiate $u = \cot x$.

$$du = d(\cot x) = d\tan\left(\frac{1}{2}\pi - x\right) = \sec^2\left(\frac{1}{2}\pi - x\right) . \, d\left(\frac{1}{2}\pi - x\right)$$

$$= -\operatorname{cosec}^2 x \,.\, dx$$

$$\therefore \frac{du}{dx} = \frac{d\cot x}{dx} = -\operatorname{cosec}^2 x.$$

20. *Prop.* To differentiate $u = \sec x$.

Here $u = \sec x = \dfrac{1}{\cos x}$. $\therefore du = d\dfrac{1}{\cos x} = \dfrac{-d\cos x}{\cos^2 x} = \dfrac{\sin x \,.\, dx}{\cos^2 x}$

or, $du = \tan x \,.\, \sec x \,.\, dx$ and $\therefore \dfrac{du}{dx} = \dfrac{d\sec x}{dx} = \tan x \,.\, \sec x.$

21. *Prop.* To differentiate $u = \text{cosec}\, x$.

$$du = d(\text{cosec}\, x) = d\sec\left(\frac{1}{2}\pi - x\right)$$

$$= \tan\left(\frac{1}{2}\pi - x\right)\sec\left(\frac{1}{2}\pi - x\right)d\left(\frac{1}{2}\pi - x\right)$$

$$= -\cot x \,.\, \text{cosec}\, x dx.$$

$$\therefore \frac{du}{dx} = \frac{d\, \text{cosec}\, x}{dx} = -\cot x \,.\, \text{cosec}\, x.$$

22. *Prop.* To differentiate $u = \text{versin}\, x$.

$$du = d(\text{versin}\, x) = d(1 - \cos x) = \sin x dx.$$

$$\therefore \frac{du}{dx} = \frac{d\, \text{versin}\, x}{dx} = \sin x.$$

23. *Prop.* To differentiate $u = \text{coversin}\, x$.

$$du = d(\text{coversin}\, x) = d.\, \text{versin}\left(\frac{1}{2}\pi - x\right) = \sin\left(\frac{1}{2}\pi - x\right)d\left(\frac{1}{2}\pi - x\right)$$

$$= -\cos x \,.\, dx. \qquad \therefore \frac{du}{dx} = \frac{d\, \text{coversin}\, x}{dx} = -\cos x.$$

24. In each of these expressions, x represents the length of an arc described with a radius equal to unity, and the radius does not appear in the formulæ: but it is necessary to remember that, in each case, $R = 1$ must be understood to enter into the formula as often as may be required to make the two members of the equation homogeneous.

Geometrical Illustration.

25. The results just obtained may be illustrated geometrically in such a manner as to convey a more precise view of the comparative small changes imparted to the several trigonometrical functions, by an arbitrary small change in the arc upon which they depend.

Thus let ab represent an arc x described with rad $= 1$, and $bb_1 = dx$ a small increment given to x. Then

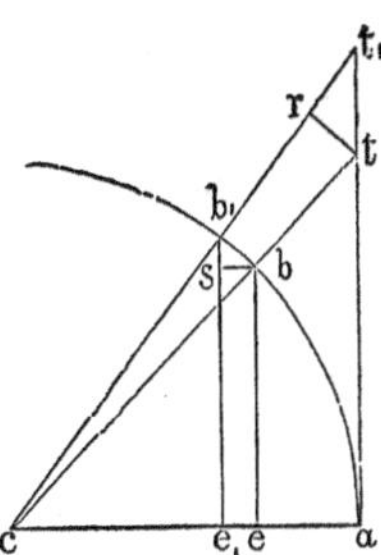

$eb = \sin x$, $ce = \cos x$, $at = \tan x$, $ct = \sec x$,

$sb_1 = d.\sin x$, $sb = d.\cos x$, $tt_1 = d.\tan x$,

$rt_1 = d.\sec x$.

Also when bb_1 is diminished continually, the small figures bsb_1 and trt_1 will continually approach to the forms of right angled triangles, becoming indefinitely near to such forms at the limit. Moreover, the two small triangles will then be similar to cbe. Hence we shall have the proportions

$$cb : ce :: bb_1 : b_1s \quad \text{or} \quad 1 : \cos x :: dx : d\sin x = \cos x dx.$$

$$cb : eb :: bb_1 : bs \quad \text{or} \quad 1 : \sin x :: dx : d\cos x = \sin x dx.$$

The latter result should be written $d\cos x = -\sin x . dx$, because the cosine diminishes as the arc increases.

Again we have the proportions

$$\left.\begin{array}{l} ca : ct :: rt : tt_1 \\ \text{and} \quad cb : ct :: bb_1 : rt \end{array}\right\} \begin{array}{l} \therefore ca \times cb : (ct)^2 :: bb_1 : tt_1 \\ \text{or } 1^2 : \sec^2 x :: dx : d\tan x. \end{array}$$

$$\therefore d\tan x = \sec^2 x dx.$$

$$\text{Also } \left.\begin{array}{l} ca : at :: rt : rt_1 \\ cb : ct :: bb_1 : rt \end{array}\right\} \begin{array}{l} \therefore ca \times cb : at \times ct :: bb_1 : rt_1. \\ \text{or } 1^2 : \tan x . \sec x :: dx : d \sec x. \end{array}$$

$$\therefore d\sec x = \tan x . \sec x . dx.$$

In the same manner, expressions for $d\cot x$, $d\,\text{cosec}\,x$, &c., could be obtained.

Circular Functions.

26. We will now consider the circular functions, $\sin^{-1}x$, $\tan^{-1}x$, &c., which expressions are read, the arc whose sine is x, the arc whose tangent is x, &c.

In these cases, it is the arc which is the function, or dependent variable, the independent variable being the sine, or the tangent, &c.

27. *Prop.* To differentiate $y = \sin^{-1}x$.

Since this notation is intended to imply that y is the arc whose sine is equal to x, we must have as an equivalent relation

$$x = \sin y$$

$$\therefore\ dx = \cos y \,.\, dy \quad \text{and} \quad \frac{dy}{dx} = \frac{1}{\cos y} = \frac{1}{\sqrt{1 - \sin^2 y}} = \frac{1}{\sqrt{1 - x^2}}$$

$$\therefore\ \frac{d \sin^{-1}x}{dx} = \frac{1}{\sqrt{1 - x^2}}.$$

28. *Prop.* To differentiate $y = \cos^{-1}x$.

Here $\quad x = \cos y, \qquad \therefore\ dx = -\sin y \,.\, dy$

$$\therefore\ \frac{dy}{dx} = -\frac{1}{\sin y} = -\frac{1}{\sqrt{1 - \cos^2 y}} = -\frac{1}{\sqrt{1 - x^2}}$$

$$\therefore\ \frac{d \cos^{-1}x}{dx} = -\frac{1}{\sqrt{1 - x^2}}.$$

29. *Prop.* To differentiate $u = \tan^{-1}x$.

$$x = \tan u, \qquad \therefore\ dx = \sec^2 u \,.\, du$$

$$\therefore\ \frac{du}{dx} = \frac{1}{\sec^2 u} = \frac{1}{1 + \tan^2 u} = \frac{1}{1 + x^2}$$

$$\therefore\ \frac{d \tan^{-1}x}{dx} = \frac{1}{1 + x^2}.$$

30. *Prop.* To differentiate $u = \cot^{-1}x$.

$$x = \cot u, \qquad \therefore\ dx = -\operatorname{cosec}^2 u \,.\, du$$

$$\therefore\ \frac{du}{dx} = -\frac{1}{\operatorname{cosec}^2 u} = -\frac{1}{1 + \cot^2 u} = -\frac{1}{1 + x^2}.$$

$$\therefore\ \frac{d \cot^{-1}x}{dx} = -\frac{1}{1 + x^2}.$$

31. *Prop.* To differentiate $u = \sec^{-1}x$.

$$x = \sec u, \qquad \therefore\ dx = \tan u \,.\, \sec u \,.\, du$$

$$\therefore\ \frac{du}{dx} = \frac{1}{\tan u \,.\, \sec u} = \frac{1}{\sec u\sqrt{\sec^2 u - 1}} = \frac{1}{x\sqrt{x^2 - 1}}$$

$$\therefore\ \frac{d \sec^{-1}x}{dx} = \frac{1}{x\sqrt{x^2 - 1}}.$$

32. *Prop.* To differentiate $u = \operatorname{cosec}^{-1}x$.

$$x = \operatorname{cosec} u, \qquad \therefore\ dx = -\cot u \,.\, \operatorname{cosec} u \,.\, du$$

$$\therefore\ \frac{du}{dx} = -\frac{1}{\cot u \,.\, \operatorname{cosec} u} = -\frac{1}{\operatorname{cosec} u\sqrt{\operatorname{cosec}^2 u - 1}}$$

$$= -\frac{1}{x\sqrt{x^2 - 1}}$$

$$\therefore\ \frac{d \operatorname{cosec}^{-1}x}{dx} = -\frac{1}{x\sqrt{x^2 - 1}}.$$

33. *Prop.* To differentiate $u = \operatorname{versin}^{-1}x$.

$$x = \operatorname{versin} u \quad \therefore\ dx = \sin u \,.\, du = \sqrt{2 \operatorname{versin} u - \operatorname{versin}^2 u}\ \ du$$

$$\therefore\ \frac{du}{dx} = \frac{1}{\sqrt{2 \operatorname{versin} u - \operatorname{versin}^2 u}} = \frac{1}{\sqrt{2x - x^2}}$$

or,
$$\frac{d \operatorname{versin}^{-1}x}{dx} = \frac{1}{\sqrt{2x - x^2}}.$$

34. *Prop.* To differentiate $u = \operatorname{coversin}^{-1}x$.

$$x = \operatorname{coversin} u$$

$$\therefore\ dx = -\cos u \,.\, du = -\sqrt{2 \operatorname{coversin} u - \operatorname{coversin}^2 u}\ \ du.$$

$$\therefore\ \frac{du}{dx} = -\frac{1}{\sqrt{2 \operatorname{coversin} u - \operatorname{coversin}^2 u}} = -\frac{1}{\sqrt{2x - x^2}}$$

or,
$$\frac{d \operatorname{coversin}^{-1}x}{dx} = -\frac{1}{\sqrt{2x - x^2}}.$$

35. The differentiation of trigonometrical and circular functions will now be illustrated by examples.

EXAMPLES.

1. $$u = 3\sin^4 x.$$

$$du = 3 \times 4\sin^3 x \,.\, d\sin x = 12\sin^3 x \,.\, \cos x \,.\, dx$$

$$\therefore \frac{du}{dx} = 12\sin^3 x \,.\, \cos x.$$

2. $$u = \cos nx.$$

$$du = -\sin nx \,.\, d(nx) = -n\sin nx \,.\, dx$$

$$\therefore \frac{du}{dx} = -n\sin nx.$$

3. $$u = \tan^n nx.$$

$$du = n\tan^{n-1} nx \,.\, d\tan nx = n^2\tan^{n-1} nx \,.\, \sec^2 nx \,.\, dx$$

$$\therefore \frac{du}{dx} = n^2\tan^{n-1} nx \,.\, \sec^2 nx.$$

4. $$u = \sin 3x \,.\, \cos 2x.$$

$$du = (3\cos 3x \,.\, \cos 2x - 2\sin 3x \,.\, \sin 2x)dx$$

$$\therefore \frac{du}{dx} = 3\cos 3x \,.\, \cos 2x - 2\sin 3x \,.\, \sin 2x = \cos 3x\cos 2x + 2\cos 5x.$$

5. $$u = (\sin x)^x. \quad \text{Then} \quad \log u = x \,.\, \log(\sin x)$$

$$\therefore \frac{du}{u} = [\log(\sin x) + x\cot x]dx \,.\, \therefore \frac{du}{dx} = (\sin x)^x . [\log(\sin x) + x\cot x].$$

6. $$u = (\cos x)^{\sin x}. \quad \text{Then} \quad \log u = \sin x\log(\cos x)$$

$$\therefore \frac{du}{dx} = (\cos x)^{\sin x}[\cos x\log(\cos x) - \sin x\tan x].$$

7. $$u = \sin(\cos x), \quad du = \cos(\cos x)d\cos x.$$

$$\therefore \frac{du}{dx} = -\sin x \,.\, \cos(\cos x).$$

8. $$u = \sin^{-1} \frac{x}{\sqrt{1+x^2}}.$$

$$du = \frac{d\,\dfrac{x}{\sqrt{1+x^2}}}{\sqrt{1-\dfrac{x^2}{1+x^2}}} = \frac{(1+x^2)^{\frac{1}{2}} - x^2(1+x^2)^{-\frac{1}{2}}}{(1+x^2)\left(1-\dfrac{x^2}{1+x^2}\right)^{\frac{1}{2}}}\,dx$$

$$\therefore \frac{du}{dx} = \frac{1}{1+x^2}.$$

9. $$u = \log \tan x.$$

$$\frac{du}{dx} = \frac{\sec^2 x}{\tan x} = \frac{1}{\sin x \,.\, \cos x} = \frac{2}{\sin 2x}.$$

10. $$u = \log\sqrt{\frac{1+\sin x}{1-\sin x}} = \frac{1}{2}\log(1+\sin x) - \frac{1}{2}\log(1-\sin x).$$

$$\frac{du}{dx} = \frac{1}{2}\left[\frac{\cos x}{1+\sin x} + \frac{\cos x}{1-\sin x}\right] = \frac{\cos x}{1-\sin^2 x} = \frac{1}{\cos x}.$$

11. $$u = \sin^{-1}(3x - 4x^3).$$

$$\frac{du}{dx} = \frac{3-12x^2}{\sqrt{1-(3x-4x^3)^2}} = \frac{3}{\sqrt{1-x^2}}.$$

12. $$u = \log(\cos x + \sqrt{-1}\,.\,\sin x).$$

$$\frac{du}{dx} = \frac{\sqrt{-1}\,.\,\cos x - \sin x}{\cos x + \sqrt{-1}\,.\,\sin x} = \sqrt{-1}.$$

13. $$u = \frac{1}{\sqrt{a^2-b^2}} \cdot \cos^{-1}\left(\frac{b + a\,.\cos x}{a + b\,.\cos x}\right).$$

$$du = -\frac{1}{\sqrt{a^2-b^2}} \cdot \frac{d\left(\dfrac{b + a\,.\cos x}{a + b\,.\cos x}\right)}{\sqrt{1-\left(\dfrac{b + a\,.\cos x}{a + b\,.\cos x}\right)^2}}$$

$$= \frac{a \sin x\,(a + b\cos x) - b\sin x\,(b + a\cos x)}{(a^2-b^2)^{\frac{1}{2}}(a + b\cos x)\left[(a+b\cos x)^2 - (b + a\cos x)^2\right]^{\frac{1}{2}}}\,dx$$

$$\therefore \frac{du}{dx} = \frac{(a^2 - b^2)\sin x}{(a^2 - b^2)^{\frac{1}{2}}(a + b\cos x)\left[(a^2 - b^2)(1 - \cos^2 x)\right]^{\frac{1}{2}}}$$

$$= \frac{1}{a + b\cos x}.$$

14. $$u = e^x \cos x.$$

$$\frac{du}{dx} = e^x \cos x - e^x \sin x = e^x (\cos x - \sin x).$$

15. $$u = \tan^{-1}(\sqrt{1 + x^2} - x).$$

$$\frac{du}{dx} = \frac{(1 + x^2)^{-\frac{1}{2}} x - 1}{1 + (\sqrt{1 + x^2} - x)^2} = -\frac{1}{2(1 + x^2)}.$$

16. $$u = \log\sqrt{\sin x} + \log\sqrt{\cos x}.$$

$$\frac{du}{dx} = \frac{1}{2}\left(\frac{\cos x}{\sin x} - \frac{\sin x}{\cos x}\right) = \frac{1}{\tan 2x}.$$

17. $$u = \log\sqrt[4]{\frac{1 + x}{1 - x}} + \frac{1}{2}\tan^{-1} x.$$

$$= \frac{1}{4}\log(1 + x) - \frac{1}{4}\log(1 - x) + \frac{1}{2}\tan^{-1} x.$$

$$\frac{du}{dx} = \frac{1}{4(1 + x)} + \frac{1}{4(1 - x)} + \frac{1}{2(1 + x^2)} = \frac{1}{1 - x^4}.$$

18. $$u = \frac{e^{ax}(a\sin x - \cos x)}{a^2 + 1}.$$

$$\frac{du}{dx} = \frac{1}{a^2 + 1}\left[ae^{ax}(a\sin x - \cos x) + ae^{au}\cos x + e^{ax}\sin x\right]$$

$$= e^{ax}\sin x.$$

CHAPTER IV.

SUCCESSIVE DIFFERENTIATION.

36. When we differentiate a function $u = Fx$, the differential coefficient $\frac{du}{dx}$ will usually be itself a function of x, and will therefore admit of being differentiated. This will simply be equivalent to examining the comparative rates of increase of the independent variable x and the variable ratio $\frac{du}{dx}$. This differentiation will give rise to a second differential coefficient, which may also be a function of x, and this, in its turn, being differentiated will give a third differential coefficient, &c.

37. To illustrate this subject, let $u = x^3$ be the proposed function. The first differential coefficient,

$$\frac{du}{dx} = 3x^2,$$

second differential coefficient,

$$= \frac{d\left(\frac{du}{dx}\right)}{dx} = 6x,$$

third differential coefficient,

$$= \frac{d\,\frac{d\,\frac{du}{dx}}{dx}}{dx} = 6.$$

As the third differential coefficient in this example proves constant, the fourth and all succeeding differential coefficients will be equal to zero.

38. The preceding notation of successive differential coefficients being inconvenient, it is replaced by the following:

For $$\frac{d\frac{du}{dx}}{dx}, \quad \text{we write} \quad \frac{d^2u}{dx^2};$$

for $$\frac{d\frac{d\frac{du}{dx}}{dx}}{dx} \quad \text{we write} \quad \frac{d^3u}{dx^3}, \text{ \&c.},$$

the symbols d^2, d^3, &c., indicating the repetition of the process of differentiation twice, thrice, &c., and not the formation of a power.

On the contrary, the expressions dx^2, dx^3, &c., represent powers of dx. The second differential coefficient $\frac{d^2u}{dx^2}$ may be obtained immediately from the first differential coefficient $\frac{du}{dx}$, by differentiating this latter as though dx was constant, $\left(\text{thus producing } \frac{d^2u}{dx}\right)$ and then dividing the result by dx.

Now since the law according to which the independent variable x changes, in different stages of its magnitude, is entirely arbitrary, we adopt, as most simple, that law by which the successive increments of x are supposed equal; that is, we make dx constant.

The same supposition will enable us to derive each successive differential coefficient from the preceding coefficient by a similar process of differentiation and division.

EXAMPLES.

39. 1. $u = x^n$. $\frac{du}{dx} = nx^{n-1}$, $\frac{d^2u}{dx^2} = n(n-1)x^{n-2}$

$$\frac{d^3u}{dx^3} = n(n-1)(n-2)x^{n-3}, \quad \frac{d^4u}{dx^4} = n(n-1)(n-2)(n-3)x^{n-4} \text{ \&c.}$$

This operation will terminate when n is a positive integer; but if n be a negative integer or a fraction, the number of variable differential coefficients will be unlimited.

2. $y = \log x.$ $\frac{dy}{dx} = \frac{1}{x}$, $\frac{d^2y}{dx^2} = -\frac{1}{x^2}$, $\frac{d^3y}{dx^3} = \frac{1.2}{x^3}$,

$$\frac{d^4y}{dx^4} = -\frac{1.2.3}{x^4} \text{ and by analogy } \frac{d^ny}{dx^n} = \pm\frac{1.2.3....(n-1)}{x^n}.$$

the upper sign will apply when n is odd, and the lower when n is even.

3. $u = \sin x.$

$$\frac{du}{dx} = \cos x, \quad \frac{d^2u}{dx^2} = -\sin x, \quad \frac{d^3u}{dx^3} = -\cos x, \quad \frac{d^4u}{dx^4} = \sin x,$$

and the succeeding differential coefficients will recur in the same order.

4. $y = \cos x.$

$$\frac{dy}{dx} = -\sin x, \quad \frac{d^2y}{dx^2} = -\cos x, \quad \frac{d^3y}{dx^3} = \sin x, \quad \frac{d^4y}{dx^4} = \cos x,$$

and the coefficients will now recur in the same order.

5. $u = \tan x.$

$$\frac{du}{dx} = \sec^2 x, \quad \frac{d^2u}{dx^2} = 2\sec^2 x . \tan x, \quad \frac{d^3u}{dx^3} = 4\sec^2 x \tan^2 x + 2\sec^4 x, \text{ \&c.}$$

Here the law of formation of the successive coefficients is not obvious.

6. $u = a^x.$

$$\frac{du}{dx} = a^x . \log a, \quad \frac{d^2u}{dx^2} = a^x . \log^2 a, \quad \frac{d^3u}{dx^3} = a^x \cdot \log^3 a, \text{ \&c.},$$

the law of the coefficients being very evident.

7.
$$u = e^x.$$

$$\frac{du}{dx} = e^x, \quad \frac{d^2u}{dx^2} = e^x, \quad \frac{d^3u}{dx^3} = e^x,$$

the coefficients being all equal.

8.
$$u = \sin(nx).$$

$$\frac{du}{dx} = n\cos(nx), \quad \frac{d^2u}{dx^2} = -n^2\sin(nx), \text{ \&c.}$$

The formation of successive differential coefficients will be found extremely useful in the expansion of functions by the methods which will be explained in the chapters immediately succeeding.

CHAPTER V.

MACLAURIN'S THEOREM.

40. The theory of Maclaurin is a very general and useful formula for the development or expansion of a function of a single variable, in a series involving the positive ascending powers of that variable, when such development is possible.

41. *Prop.* If $y = Fx$, where Fx denotes such a function of x as can be expanded in a series containing the positive ascending powers of x, then will the form of the development be the following:

$$y = (y) + \left(\frac{dy}{dx}\right)\frac{x}{1} + \left(\frac{d^2y}{dx^2}\right)\frac{x^2}{1.2} + \left(\frac{d^3y}{dx^3}\right)\frac{x^3}{1.2.3} + \&c.,$$

in which the parentheses are used to denote the particular values of the quantities y, $\frac{dy}{dx}$, $\frac{d^2y}{dx^2}$, &c., enclosed therein, when x is taken equal to zero.

Proof. By hypothesis, y can be expressed in the form

$$y = A + Bx + Cx^2 + Dx^3 + Ex^4 + \&c., \qquad (1).$$

in which A, B, C, &c., are unknown constants.

$$\therefore \frac{dy}{dx} = B + 2Cx + 3Dx^2 + 4Ex^3 + \&c.$$

$$\frac{d^2y}{dx^2} = 2C + 2.3Dx + 3.4Ex^2 + \&c.$$

$$\frac{d^3y}{dx^3} = 2.3D + 2.3.4Ex + \&c.$$

$$\frac{d^4y}{dx^4} = 2.3.4E + \&c.$$

&c. &c.

Now making $x = 0$ in each of these expressions, we obtain

$$(y) = A, \quad \left(\frac{dy}{dx}\right) = B, \quad \left(\frac{d^2y}{dx^2}\right) = 2C, \quad \left(\frac{d^3y}{dx^3}\right) = 2 \,.\, 3D,$$

$$\left(\frac{d^4y}{dx^4}\right) = 2 \,.\, 3 \,.\, 4E, \text{ \&c., \&c.}$$

$$\therefore\ A = (y), \quad B = \left(\frac{dy}{dx}\right), \quad C = \frac{1}{2}\left(\frac{d^2y}{dx^2}\right) = \frac{1}{1 \,.\, 2}\left(\frac{d^2y}{dx^2}\right),$$

$$D = \frac{1}{1 \,.\, 2 \,.\, 3}\left(\frac{d^3y}{dx^3}\right), \quad E = \frac{1}{1 \,.\, 2 \,.\, 3 \,.\, 4}\left(\frac{d^4y}{dx^4}\right), \text{ \&c., \&c.}$$

These values, being substituted in (1), reduce it to the form

$$y = (y) + \left(\frac{dy}{dx}\right)\frac{x}{1} + \left(\frac{d^2y}{dx^2}\right)\frac{x^2}{1 \,.\, 2} + \left(\frac{d^3y}{dx^3}\right)\frac{x^3}{1 \,.\, 2 \,.\, 3},$$

$$+ \left(\frac{d^4y}{dx^4}\right)\frac{x^4}{1 \,.\, 2 \,.\, 3 \,.\, 4} + \text{\&c.,} \qquad (2).$$

which agrees with the enunciation.

This formula, called Maclaurin's Theorem, may be written thus

$$Fx = (Fx) + \left(\frac{dFx}{dx}\right)\frac{x}{1} + \left(\frac{d^2Fx}{dx^2}\right)\frac{x^2}{1 \,.\, 2} + \left(\frac{d^3Fx}{dx^3}\right)\frac{x^3}{1 \,.\, 2 \,.\, 3}$$

$$+ \left(\frac{d^4Fx}{dx^4}\right)\frac{x^4}{1 \,.\, 2 \,.\, 3 \,.\, 4} + \text{\&c.,} \qquad (3);$$

or again, if we represent the 1st, 2d, 3d, &c., differential coefficients, which are functions of x, by F_1x, F_2x, F_3x, &c., the formula may be written

$$Fx = F0 + F_10\frac{x}{1} + F_20\frac{x^2}{1 \,.\, 2} + F_30\frac{x^3}{1 \,.\, 2 \,.\, 3}$$

$$+ F_40\frac{x^4}{1 \,.\, 2 \,.\, 3 \,.\, 4} + \text{\&c.} \qquad (4).$$

EXAMPLES.

42. 1. To expand $\quad y = (a + x)^n$.

Here $\frac{dy}{dx} = n(a + x)^{n-1}, \quad \frac{d^2y}{dx^2} = n(n-1)(a + x)^{n-2},$

$$\frac{d^3y}{dx^3} = n(n-1)(n-2)(a+x)^{n-3},$$

$$\frac{d^4y}{dx^4} = n(n-1)(n-2)(n-3)(a+x)^{n-4}, \text{ \&c., \&c.}$$

Hence, when $\qquad x = 0.$

$$(y) = a^n, \quad \left(\frac{dy}{dx}\right) = na^{n-1}, \quad \left(\frac{d^2y}{dx^2}\right) = n(n-1)a^{n-2},$$

$$\left(\frac{d^3y}{dx^3}\right) = n(n-1)(n-2)a^{n-3},$$

$$\left(\frac{d^4y}{dx^4}\right) = n(n-1)(n-2)(n-3)a^{n-4}, \text{ \&c., \&c.}$$

And, therefore, by substitution in Maclaurin's formula,

$$y = (a+x)^n = a^n + na^{n-1}x + \frac{n(n-1)}{1.2}a^{n-2}x^2$$
$$+ \frac{n(n-1)(n-2)}{1.2.3}a^{n-3}x^3$$
$$+ \frac{n(n-1)(n-2)(n-3)}{1.2.3.4}a^{n-4}x^4 + \text{\&c.}$$

Thus we have a simple proof of the *binomial theorem*, applicable to all values of the exponent, whether positive or negative, integral or fractional, real or imaginary.

2. To develop $\qquad y = \log(1+x),$

the modulus of the system being M,

$$\frac{dy}{dx} = \frac{M}{1+x}, \quad \frac{d^2y}{dx^2} = -\frac{M}{(1+x)^2}, \quad \frac{d^3y}{dx^3} = \frac{1.2M}{(1+x)^3},$$
$$\frac{d^4y}{dx^4} = -\frac{1.2.3M}{(1+x)^4}, \quad \text{\&c.}$$

$\therefore$ when $\qquad x = 0, \quad (y) = \log 1 = 0,$

$$\left(\frac{dy}{dx}\right) = \frac{M}{1}, \left(\frac{d^2y}{dx^2}\right) = -\frac{M}{1}, \left(\frac{d^3y}{dx^3}\right) = \frac{1.2M}{1}, \left(\frac{d^4y}{dx^4}\right) = -\frac{1.2.3M}{1}, \text{ \&c.}$$

And by substituting these values in Maclaurin's formula, we have

$$y = \log(1 + x) = M(x - \frac{1}{2}x^2 + \frac{1}{3}x^3 - \frac{1}{4}x^4 + \text{\&c.})$$

which is the fundamental theorem used in the computation of logarithms, and is, indeed, that which was employed in deducing the rule for differentiating logarithms.

3. To expand $\qquad y = \sin x.$

Here $\qquad Fx = \sin x$

$\therefore F_1x = \cos x, \quad F_2x = -\sin x, \quad F_3x = -\cos x, \quad F_4x = \sin x,$

and the succeeding coefficients recur in the same order.

$$\therefore F0 = \sin 0 = 0, \quad F_10 = \cos 0 = 1, \quad F_20 = 0, \quad F_30 = -1,$$
$$F_40 = 0, \quad F_50 = 1, \quad \text{\&c.}$$

$\therefore$ by substitution in (4) the third form of Maclaurin's theorem, we have

$$\sin x = x - \frac{x^3}{1.2.3} + \frac{x^5}{1.2.3.4.5} - \frac{x^7}{1.2.3.4.5.6.7} + \text{\&c.}$$

This series converges very rapidly when x is small.

4. To expand $\qquad y = \cos x.$

$Fx = \cos x, \ F_1x = -\sin x, \ F_2x = -\cos x, \ F_3x = \sin x, \ F_4x = \cos x,$

and the succeeding coefficients recur in the same order,

$\therefore F0 = 1, \ F_10 = 0, \ F_20 = -1, \ F_30 = 0, \ F_40 = 1, \ F_50 = 0,$ &c.

$$\therefore \cos x = 1 - \frac{x^2}{1.2} + \frac{x^4}{1.2.3.4} - \frac{x^6}{1.2.3.4.5.6} + \text{\&c.}$$

5. To develop $y = a^x$.

Employing Naperian logarithms, we have

$$Fx = a^x, \quad F_1x = a^x . \log a, \quad F_2x = a^x . \log^2 a, \quad F_3x = a^x . \log^3 a, \ \&c.$$

$$\therefore F0 = 1, \ F_1 0 = \log a, \ F_2 0 = \log^2 a, \ F_3 0 = \log^3 a, \ F_4 0 = \log^4 a, \ \&c.$$

$$\therefore \ a^x = 1 + \log a \frac{x}{1} + \log^2 a \frac{x^2}{1 . 2} + \log^3 a \frac{x^3}{1 . 2 . 3} + \log^4 a \frac{x^4}{1 . 2 . 3 . 4} + \&c.$$

This is called the exponential theorem.

Cor. If $a = e$ the Naperian base, then $\log a = \log e = 1$,

$$\therefore \ e^x = 1 + \frac{x}{1} + \frac{x^2}{1 . 2} + \frac{x^3}{1 . 2 . 3} + \frac{x^4}{1 . 2 . 3 . 4} + \&c;$$

and if $x = 1$ also,

$$e^x = e = 1 + \frac{1}{1} + \frac{1}{1 . 2} + \frac{1}{1 . 2 . 3} + \frac{1}{1 . 2 . 3 . 4} + \&c.,$$

a formula for the Naperian base.

Cor. If $x = 1$, but a not equal e, then

$$a^x = 1 + \log a + \frac{1}{1 . 2} \log^2 a + \frac{1}{1 . 2 . 3} \log^3 a + \frac{1}{1.2.3.4} \log^4 a + \&c.$$

a formula for a number in terms of its Naperian logarithm.

Prop. To express the sine and cosine of an arc in terms of imaginary exponentials.

In the series giving the value of e^x, put successively

$$z\sqrt{-1}, \quad \text{and} \quad -z\sqrt{-1} \ \text{for} \ x.$$

$$\therefore \ e^{z\sqrt{-1}} = 1 + \frac{z\sqrt{-1}}{1} - \frac{z^2}{1 . 2} - \frac{z^3\sqrt{-1}}{1 . 2 . 3} + \frac{z^4}{1 . 2 . 3 . 4} + \frac{z^5\sqrt{-1}}{1 . 2 . 3 . 4 . 5} - \&c.$$

and $$e^{-z\sqrt{-1}} = 1 - \frac{z\sqrt{-1}}{1} - \frac{z^2}{1.2} + \frac{z^3\sqrt{-1}}{1.2.3} + \frac{z^4}{1.2.3.4} - \frac{z^5\sqrt{-1}}{1.2.3.4.5} - \text{\&c.}$$

$$\therefore e^{z\sqrt{-1}} + e^{-z\sqrt{-1}} = 2\left[1 - \frac{z^2}{1.2} + \frac{z^4}{1.2.3.4} - \text{\&c.}\right]$$

$$e^{z\sqrt{-1}} - e^{-z\sqrt{-1}} = 2\sqrt{-1}\left[z - \frac{z^3}{1.2.3} + \frac{z^5}{1.2.3.4.5} - \text{\&c.}\right].$$

But the first series within the [ ] is the development of $\cos z$, and the second that of $\sin z$,

$$\therefore \cos z = \frac{e^{z\sqrt{-1}} + e^{-z\sqrt{-1}}}{2}, \quad \dots \quad (A),$$

$$\sin z = \frac{e^{z\sqrt{-1}} - e^{-z\sqrt{-1}}}{2\sqrt{-1}}, \quad \dots \quad (B),$$

These singular formulæ, discovered by Euler, are very useful in the higher branches of analysis, especially in the development of functions.

Cor. If we divide (B) by (A), there will result

$$\tan z = \frac{e^{z\sqrt{-1}} - e^{-z\sqrt{-1}}}{\sqrt{-1}[e^{z\sqrt{-1}} + e^{-z\sqrt{-1}}]} = \frac{e^{2z\sqrt{-1}} - 1}{\sqrt{-1}[e^{2z\sqrt{-1}} + 1]}. \quad \dots \quad (C).$$

Cor. If we make $z = x\sqrt{-1}$ in (A), (B), and (C), we can express the sine, cosine, and tangent of an imaginary arc in terms of real exponentials; thus:

$$\sin(x\sqrt{-1}) = \frac{e^{-x} - e^{x}}{2\sqrt{-1}} \dots (D), \quad \cos(x\sqrt{-1}) = \frac{e^{-x} + e^{x}}{2} \dots (E)$$

$$\tan(x\sqrt{-1}) = \frac{e^{-2x} - 1}{\sqrt{-1}(e^{-2x} + 1)} = \frac{1 - e^{2x}}{\sqrt{-1}(1 + e^{2x})}.$$

Cor. If we square (A) and (B) and add, there will result

$$\cos^2 z + \sin^2 z = \frac{e^{2z\sqrt{-1}} + 2 + e^{-2z\sqrt{-1}} - e^{2z\sqrt{-1}} + 2 - e^{-2z\sqrt{-1}}}{4} = 1.$$

And similarly $\sin^2(x\sqrt{-1}) + \cos^2(x\sqrt{-1}) = 1;$

two results obviously correct.

43. The applications of Maclaurin's theorem are often much restricted by the great labor necessary in forming the successive differential coefficients. This may sometimes be avoided by expanding the first differential coefficient by some of the algebraic processes. For example,

To expand $$u = \tan^{-1} x.$$

Here $$\frac{du}{dx} = \frac{1}{1 + x^2},$$

which gives by actual division, the quotient

$$1 - x^2 + x^4 - x^6 + x^8 - \&c.$$

$$\therefore\ Fx = \tan^{-1} x,$$
$$F_1 x = 1 - x^2 + x^4 - x^6 + x^8 - \&c.$$
$$F_2 x = -2x + 4x^3 - 6x^5 + 8x^7 - \&c.$$
$$F_3 x = -2 + 3.4x^2 - 5.6x^4 + 7.8x^6 - \&c.$$
$$F_4 x = 2.3.4x - 4.5.6x^3 + 6.7.8x^5 - \&c.$$
$$F_5 x = 2.3.4 - 3.4.5.6x^2 + 5.6.7.8x^4 - \&c.$$
$$F_6 x = -2.3.4.5.6x + 4.5.6.7.8x^3 - \&c.$$
$$F_7 x = -2.3.4.5.6 + 3.4.5.6.7.8x^2 - \&c.$$
$$F_8 x = 2.3.4.5.6.7.8x - \&c.$$
$$\&c., \qquad \&c.$$

$$\therefore\ F0 = \tan^{-1} 0 = 0,\quad F_1 0 = 1,\quad F_2 0 = 0,\quad F_3 0 = -1.2,$$
$$F_4 0 = 0,\quad F_5 0 = 1.2.3.4,\quad F_6 0 = 0,$$
$$F_7 0 = -1.2.3.4.5.6,\quad F_8 0 = 0,\ \&c.$$

Therefore, by substitution in Maclaurin's formula,

$$Fx = \tan^{-1}x = x - \frac{1}{3}x^3 + \frac{1}{5}x^5 - \frac{1}{7}x^7 + \frac{1}{9}x^9 - \&c.$$

If, in this formula, we make $u = \frac{1}{4}\pi =$ arc of 45°,

then $$x = \tan 45° = 1.$$

$$\therefore\ \frac{1}{4}\pi = (1 - \frac{1}{3} + \frac{1}{5} - \frac{1}{7} + \&c.),$$

and $$\pi = 4(1 - \frac{1}{3} + \frac{1}{5} - \frac{1}{7} + \&c.);$$

a formula for determining the ratio of the diameter to the circumference of a circle.

This series converges so very slowly, that even a tolerably accurate approximation to the value of π cannot be deduced from it, without employing a great number of terms.

44. *Prop.* To deduce Euler's more convergent series for the ratio of the diameter to the circumference.

If in the trigonometrical formula

$$\tan(a + b) = \frac{\tan a + \tan b}{1 - \tan a \,.\, \tan b},$$

we put $a + b = \frac{1}{4}\pi$, then $\tan(a + b) = 1$,

$$\therefore\ 1 - \tan a \,.\, \tan b = \tan a + \tan b;$$

whence we deduce $$\tan b = \frac{1 - \tan a}{1 + \tan a}.$$

And, therefore, if any value be assigned to $\tan a$, that of $\tan b$ can be determined.

Let $\tan a = \frac{1}{2}$, then $\tan b = \dfrac{1 - \frac{1}{2}}{1 + \frac{1}{2}} = \frac{1}{3}$.

$$\therefore\ \frac{1}{4}\pi = \tan^{-1}\frac{1}{2} + \tan^{-1}\frac{1}{3}.$$

But $$\tan^{-1}\frac{1}{2}=\frac{1}{2}-\frac{1}{3}\left(\frac{1}{2}\right)^3+\frac{1}{5}\left(\frac{1}{2}\right)^5-\frac{1}{7}\left(\frac{1}{2}\right)^7+\&c.$$

and $$\tan^{-1}\frac{1}{3}=\frac{1}{3}-\frac{1}{3}\left(\frac{1}{3}\right)^3+\frac{1}{5}\left(\frac{1}{3}\right)^5-\frac{1}{7}\left(\frac{1}{3}\right)^7+\&c.$$

$$\therefore \frac{1}{4}\pi=\frac{1}{2}-\frac{1}{3\,.\,2^3}+\frac{1}{5\,.\,2^5}-\frac{1}{7\,.\,2^7}+\&c.$$
$$+\frac{1}{3}-\frac{1}{3\,.\,3^3}+\frac{1}{5\,.\,3^5}-\frac{1}{7\,.\,3^7}+\&c.$$

By taking six terms in the first set, and four in the second, and multiplying by 4, we get the common approximation,

$$\pi=3.1416.$$

Cor. We might extend this method, obtaining series still more convergent. For if we take four arcs c_1, c_2, c_3, and c_4, such that $c_1+c_2=\tan^{-1}\frac{1}{2}$ and $c_3+c_4=\tan^{-1}\frac{1}{3}$. Then $c_1+c_2+c_3+c_4=\frac{1}{4}\pi$, and if we assume the values of $\tan c_1$ and $\tan c_3$, those of $\tan c_2$ and $\tan c_4$ can be determined. Moreover, the values of $\tan c_1$, $\tan c_2$, $\tan c_3$, and $\tan c_4$, can all be rendered less than $\frac{1}{3}$, and therefore the series for determining $\frac{1}{4}\pi$ will be more convergent.

45. *Prop.* To obtain more convergent series for the value of π.

If in the formula $$\tan 2a=\frac{2\tan a}{1-\tan^2 a},$$

we put $$\tan a=\frac{1}{5},\quad\text{then}$$

$$\tan 2a=\frac{2\times\frac{1}{5}}{1-\frac{1}{25}}=\frac{5}{12},$$

and $$\therefore \tan 4a=\frac{2\tan 2a}{1-\tan^2 2a}=\frac{2\times\frac{5}{12}}{1-\frac{25}{144}}=\frac{120}{119}.$$

Now this result is very little greater than unity, and therefore $4a$ must be slightly greater than 45°.

Put
$$4a - \frac{1}{4}\pi = z$$
where z is a very small arc.

Then
$$\tan z = \tan\left(4a - \frac{1}{4}\pi\right) = \frac{\tan 4a - \tan\frac{1}{4}\pi}{1 + \tan 4a \,.\, \tan\frac{1}{4}\pi}$$
$$= \frac{\frac{120}{119} - 1}{1 + \frac{120}{119}} = \frac{1}{239}.$$
$$\therefore \frac{1}{4}\pi = 4\tan^{-1}\frac{1}{5} - \tan^{-1}\frac{1}{239}$$
$$= 4\left(\frac{1}{5} - \frac{1}{3 \,.\, 5^3} + \frac{1}{5 \,.\, 5^5} - \frac{1}{7 \,.\, 5^7} + \frac{1}{9 \,.\, 5^9} - \&c.\right)$$
$$- \left(\frac{1}{239} - \frac{1}{3 \,.\, 239^3} + \frac{1}{5 \,.\, 239^5} - \&c.\right)$$

By taking three terms in the first line and one in the second, we get the common approximation $\pi = 3.1416$; and by taking eight terms of the first line and three of the second, we get
$$\pi = 3.14159265358979 3.$$

46. 1. To expand $u = \sin^{-1} x$.

$$Fx = \sin^{-1} x.$$
$$F_1x = \frac{1}{\sqrt{1 - x^2}} = (1 - x^2)^{-\frac{1}{2}}$$
$$= 1 + \frac{1}{1.2}x^2 + \frac{1.3}{1.2.2^2}x^4 + \frac{1.3.5}{1.2.3.2^3}x^6 + \&c.$$
$$F_2x = \frac{1.2}{1.2}x + \frac{1.3.4}{1.2.2^2}x^3 + \frac{1.3.5.6}{1.2.3.2^3}x^5 + \&c.$$
$$F_3x = \frac{1^2.2}{1.2} + \frac{1.3^2.4}{1.2.2^2}x^2 + \frac{1.3.5^2.6}{1.2.3.2^3}x^4 + \&c.$$

$$F_4x = \frac{1.2.3^2.4}{1.2.2^2}x + \frac{1.3.4.5^2.6}{1.2.3.2^3}x^3 + \text{\&c.}$$

$$F_5x = \frac{1^2.2.3^2.4}{1.2.2^2} + \frac{1.3^2.4.5^2.6}{1.2.3.2^3}x^2 + \text{\&c.}$$

$$F_6x = \frac{1.2.3^2\ 4.5^2.6}{1.2.3.2^3}x + \text{\&c.}$$

$$F_7x = \frac{1^2.2.3^2.4.5^2.6}{1.2.3.2^3} + \text{\&c.}$$

$$F_0 = 0,\ F_10 = 1,\ F_20 = 0,\ F_30 = 1^2,\ F_40 = 0,\ F_50 = 1^2.3^2,$$
$$F_60 = 0,\ F_70 = 1^2.3^2.5^2,\ \text{\&c.}$$

$$\therefore u = \sin^{-1}x = x + \frac{1^2.x^3}{1.2.3} + \frac{1^2.3^2.x^5}{1.2.3.4.5}$$
$$+ \frac{1^2.3^2.5^2.x^7}{1.2.3.4.5.6.7} + \text{\&c.}$$

CHAPTER VI.

TAYLOR'S THEOREM.

47. Taylor's Theorem is a general formula for the development of a function of the algebraic sum of two variables.

Prop. If $y = Fx$, and if x be supposed to receive an increment h, converting y into $y_1 = F(x+h)$; then will

$$y_1 = y + \frac{dy}{dx}\cdot\frac{h}{1} + \frac{d^2y}{dx^2}\cdot\frac{h^2}{1.2} + \frac{d^3y}{dx^3}\cdot\frac{h^3}{1.2.3} + \frac{d^4y}{dx^4}\cdot\frac{h^4}{1.2.3.4} + \&c.;$$

or
$$F(x+h) = Fx + \frac{dFx}{dx}\cdot\frac{h}{1} + \frac{d^2Fx}{dx^2}\cdot\frac{h^2}{1.2}$$
$$+ \frac{d^3Fx}{dx^3}\cdot\frac{h^3}{1.2.3} + \frac{d^4Fx}{dx^4}\cdot\frac{h^4}{1.2.3.4} + \&c.$$

To prove the truth of this formula, we first establish the following principle:

If in the expression $y_1 = F(x+h)$ we suppose first that x is variable and h constant, and then suppose h variable and x constant, the first differential coefficient will be the same in both cases;

that is,
$$\frac{dy_1}{dx} = \frac{dy_1}{dh}.$$

This is almost self-evident, for when a *given* increment is assigned to x, or to h the same increment must be imparted to $x+h$, and therefore $F(x+h) = y_1$ will undergo the same change in the one case as in the other. Hence the ratio of the corresponding changes of x and y_1 is equal to the ratio of the changes in h and y_1. This is true whatever may be the magnitudes of the increments im parted to x or h, provided that magnitude be the same in both cases. But when we suppose these increments indefinitely small, it is no longer *necessary* to consider them equal. For since the ratio $\frac{dy_1}{dx}$ does not contain dx, it will have the same value whether dx and dh be supposed equal or unequal.

$$\therefore \frac{dy_1}{dx} = \frac{dy_1}{dh}.$$

Similarly, $$\frac{d\left(\frac{dy_1}{dx}\right)}{dx} = \frac{d\left(\frac{dy_1}{dh}\right)}{dx} = \frac{d\left(\frac{dy_1}{dh}\right)}{dh} \quad \text{or} \quad \frac{d^2y_1}{dx^2} = \frac{d^2y_1}{dh^2}.$$

And generally, $$\frac{d^ny_1}{dx^n} = \frac{d^ny_1}{dh^n}.$$

Now assume

$$y_1 = F(x+h) = Fx + Ah + Bh^2 + Ch^3 + Dh^4 + \&c. \quad (1),$$

that being the general form in which $F(x+h)$ can be developed, as shown in Art. 4. The coefficients A, B, C, D, &c., are functions of x, but are independent of h.

If we differentiate (1) first with respect to h and then with respect to x, and place the resulting differential coefficients equal, we shall obtain

$$A + 2Bh + 3Ch^2 + 4Dh^3 + \&c.$$
$$= \frac{dFx}{dx} + \frac{dA}{dx}h + \frac{dB}{dx}h^2 + \frac{dC}{dx}h^3 + \&c.$$

which equation being true for all values of h, it follows, by the principle of *indeterminate coefficients*, that the coefficients of the like powers of h, in the two members of the equation, must be separately equal.

$$\therefore A = \frac{dFx}{dx}, \quad 2B = \frac{dA}{dx}, \quad 3C = \frac{dB}{dx}, \quad 4D = \frac{dC}{dx}, \ \&c.$$

$$\therefore \quad A = \frac{dFx}{dx}, \quad B = \frac{1}{2}\frac{dA}{dx} = \frac{1}{1.2}\cdot\frac{d^2Fx}{dx^2},$$

$$C = \frac{1}{3}\cdot\frac{dB}{dx} = \frac{1}{1.2.3}\cdot\frac{d^3Fx}{dx^3}, \quad D = \frac{1}{4}\cdot\frac{dC}{dx} = \frac{1}{1.2.3.4}\cdot\frac{d^4Fx}{dx^4}, \text{ \&c.}$$

Hence, by substitution in (1),

$$F(x+h) = Fx + \frac{dFx}{dx}\cdot\frac{h}{1} + \frac{d^2Fx}{dx^2}\cdot\frac{h^2}{1.2} + \frac{d^3Fx}{dx^3}\cdot\frac{h^3}{1.2.3}$$
$$+ \frac{d^4Fx}{dx^4}\cdot\frac{h^4}{1.2.3.4} + \text{\&c.},$$

or,
$$y_1 = y + \frac{dy}{dx}\cdot\frac{h}{1} + \frac{d^2y}{dx^2}\cdot\frac{h^2}{1.2} + \frac{d^3y}{dx^3}\cdot\frac{h^3}{1.2.3}$$
$$+ \frac{d^4y}{dx^4}\cdot\frac{h^4}{1.2.3.4} + \text{\&c.}$$

If we denote the successive differential coefficients by F_1x, F_2x, F_3x, F_4x, &c., the series may be written

$$F(x+h) = Fx + F_1x\frac{h}{1} + F_2x\frac{h^2}{1.2} + F_3x\frac{h^3}{1.2.3}$$
$$+ F_4x\frac{h^4}{1.2.3.4} + \text{\&c.} \qquad (2).$$

Cor. The formula of Maclaurin may be readily deduced from that of Taylor; for if we make $x = 0$ in (2), there will result

$$Fh = F0 + F_10\frac{h}{1} + F_20\frac{h^2}{1.2} + F_30\frac{h^3}{1.2.3}$$
$$+ F_40\frac{h^4}{1.2.3.4} + \text{\&c.},$$

which is Maclaurin's theorem.

EXAMPLES.

48. 1. To expand $\sin(x+h)$, in terms of the powers of the arc h.

$$F(x+h) = \sin(x+h),$$

$$\therefore\ Fx = \sin x, \quad F_1x = \cos x, \quad F_2x = -\sin x,$$
$$F_3x = -\cos x, \quad F_4x = \sin x, \text{ \&c.}$$

$\therefore$ By substitution in Taylor's formula

$$\sin(x+h) = \sin x + \cos x \frac{h}{1} - \sin x \frac{h^2}{1.2} - \cos x \frac{h^3}{1.2.3} + \text{\&c.}$$
$$= \sin x \left(1 - \frac{h^2}{1.2} + \frac{h^4}{1.2.3.4} - \text{\&c.}\right)$$
$$+ \cos x \left(h - \frac{h^3}{1.2.3} + \frac{h^5}{1.2.3.4.5} - \text{\&c.}\right)$$
$$= \sin x . \cos h + \cos x . \sin h, \text{ a well known formula.}$$

2. To expand $\cos(x+h)$, in terms of the powers of the arc h.

$$F(x+h) = \cos(x+h),$$
$$\therefore\ Fx = \cos x, \quad F_1x = -\sin x, \quad F_2x = -\cos x,$$
$$F_3x = \sin x, \quad F_4x = \cos x, \text{ \&c.}$$

$\therefore$ By substitution in Taylor's Theorem we have

$$\cos(x+h) = \cos x - \sin x \frac{h}{1} - \cos x \frac{h^2}{1.2} + \sin x \frac{h^3}{1.2.3}$$
$$+ \cos x \frac{h^4}{1.2.3.4} - \text{\&c.}$$
$$= \cos x \left(1 - \frac{h^2}{1.2} + \frac{h^4}{1.2.3.4} - \text{\&c.,}\right)$$
$$- \sin x \left(h - \frac{h^3}{1.2.3} + \frac{h^5}{1.2.3.4.5} - \text{\&c.,}\right)$$
$$= \cos x . \cos h - \sin x . \sin h, \ldots \text{ a well known formula.}$$

3. To expand $\log(x+h)$, where M is the modulus of the system.

$$Fx = \log x, \quad F_1x = \frac{M}{x}, \quad F_2x = -\frac{1M}{x^2},$$
$$F_3x = \frac{1.2M}{x^3}, \quad F_4x = -\frac{1.2.3M}{x^4}, \text{ \&c.,}$$

$$\therefore \log(x+h) = \log x + M\left(\frac{h}{x} - \frac{h^2}{2x^2} + \frac{h^3}{3x^3} - \frac{h^4}{4x^4} + \&c.\right)$$

4. To expand $u_1 = \tan^{-1}(x+h)$,

$$u = \tan^{-1}x = Fx, \quad F_1x = \frac{1}{1+x^2} = \frac{1}{\sec^2 u} = \cos^2 u.$$

$$F_2x = -2\sin u \,.\, \cos u \frac{du}{dx} = -\sin 2u \,.\, \cos^2 u.$$

$$F_3x = (-2\cos 2u \,.\, \cos^2 u + 2\sin 2u \,.\, \cos u \,.\, \sin u)\frac{du}{dx}$$

$$= -2\cos u \,.\, \cos 3u \frac{du}{dx} = -2\cos 3u \,.\, \cos^3 u.$$

$$F_4x = 2\,.\,3\,(\sin 3u \,.\, \cos^3 u + \cos 3u \,.\, \cos^2 u \,.\, \sin u)\frac{du}{dx}$$

$$= 2\,.\,3\cos^2 u \,.\, \sin 4u \frac{du}{dx} = 2\,.\,3\,.\,\sin 4u \,.\, \cos^4 u,$$

&c., &c.

$$\therefore \tan^{-1}(x+h) = u_1 = u + \cos^2 u \frac{h}{1} - \sin 2u \,.\, \cos^2 u \frac{h^2}{2}$$

$$- \cos 3u \,.\, \cos^3 u \frac{h^3}{3} + \sin 4u \,.\, \cos^4 u \frac{h^4}{4} + \cos 5u \,.\, \cos^5 u \frac{h^5}{5} - \&c.$$

5. To expand $u = \tan(x+h)$.

$$Fx = \tan x, \quad F_1x = \sec^2 x, \quad F_2x = 2\sec^2 x \,.\, \tan x,$$

$$F_3x = 2\sec^2 x\,(1 + 3\tan^2 x). \quad \&c., \quad \&c.$$

$$\therefore \tan(x+h) = \tan x + \sec^2 x \frac{h}{1} + 2\sec^2 x \,.\, \tan x \frac{h^2}{1\,.\,2}$$

$$+ 2\sec^2 x\,(1 + 3\tan^2 x)\frac{h^3}{1\,.\,2\,.\,3} + \&c.$$

Prop. Having given $u = Fy$, and $y = \varphi x$, to form the differential coefficient $\frac{du}{dx}$ of u with respect to x, without eliminating y between the equations, in which the characters F and φ, denote any functions whatever.

Let x take an increment h converting y into $y_1 = \varphi(x+h)$. Then if k denote the increment received by y, we shall have, by Taylor's theorem,

$$k = y_1 - y = \frac{dy}{dx}\cdot\frac{h}{1} + \frac{d^2y}{dx^2}\cdot\frac{h^2}{1.2} + \frac{d^3y}{dx^3}\cdot\frac{h^3}{1.2.3} + \&c. \quad (1).$$

Also when y takes the increment k, it imparts to $u = Fy$, an increment

$$u_1 - u = F(y+k) - Fy = \frac{du}{dy}\cdot\frac{k}{1} + \frac{d^2u}{dy^2}\cdot\frac{k^2}{1.2} + \frac{d^3u}{dy^3}\cdot\frac{k^3}{1.2.3} + \&c.,$$

or by substituting for k its value, (1).

$$u_1 - u = \frac{du}{dy}\left[\frac{dy}{dx}\cdot\frac{h}{1} + \frac{d^2y}{dx^2}\cdot\frac{h^2}{1.2} + \frac{d^3y}{dx^3}\cdot\frac{h^3}{1.2.3} + \&c.\right]$$
$$+ \frac{1}{1.2}\cdot\frac{d^2u}{dy^2}\left[\frac{dy}{dx}\cdot\frac{h}{1} + \frac{d^2y}{dx^2}\cdot\frac{h^2}{1.2} + \&c.\right]^2 + \&c.$$

Dividing both members by h, and then passing to the limit by making $h = 0$, in which case $\dfrac{u_1 - u}{h} = \dfrac{du}{dx}$ we get

$$\frac{du}{dx} = \frac{du}{dy}\cdot\frac{dy}{dx}. \quad (2).$$

Thus it appears that the differential coefficient of u with respect to x, is found by differentiating u as though y were the independent variable, then differentiating y as though x were the independent variable, and finally, multiplying the first of the coefficients so found by the second.

49. It might perhaps seem at first view that the equation (2) is necessarily and identically true, and therefore that the preceding investigation is unnecessary. But it must be borne in mind that the dy which appears in the coefficient $\dfrac{dy}{dx}$ and which represents the increment given to y by assigning an arbitrary small increment dx to the variable x, is not *necessarily* the same as dy which appears in

$\frac{du}{dy}$, since this latter increment of y is arbitrary (though likewise small).

1. $$u = a^y,\ y = b^x,\ \text{to find } \frac{du}{dx}.$$

Here $$\frac{du}{dy} = a^y . \log a,\quad \frac{dy}{dx} = b^x \log b.$$

$$\therefore \frac{du}{dx} = \frac{du}{dy} \cdot \frac{dy}{dx} = a^y . b^x . \log a . \log b . = a^{b^x} . b^x . \log a . \log b.$$

2. $$u = \log y.\quad y = \log x.$$

$$\frac{du}{dy} = \frac{1}{y},\quad \frac{dy}{dx} = \frac{1}{x},\quad \therefore \frac{du}{dx} = \frac{1}{y} \cdot \frac{1}{x} = \frac{1}{x \log x}.$$

50. Taylor's Theorem may be employed in approximating to the roots of numerical equations.

Let $Fx = 0$ be the given equation, and a an approximate value of one of its roots found by trial; then we may put $x = a + h$, in which h is a small fraction whose higher powers will be small in comparison with h, and may therefore be neglected without great error. But

$$Fx = F(a + h) = Fa + F_1 a . \frac{h}{1} + F_2 a \frac{h^2}{1.2} + F_3 a \frac{h^3}{1.2.3} + \&c. = 0.$$

$\therefore$ By neglecting the terms involving h_2, h^3, &c., we get

$$Fa + F_1 a \frac{h}{1} = 0 \quad \text{and} \quad \therefore h = -\frac{Fa}{F_1 a}.$$

Adding this approximate value of h to a, we have

$$x = a - \frac{Fa}{F_1 a} \text{ nearly.}$$

Call this value a_1 and put $x = a_1 + h_1$.

Then by similar reasoning we shall find

$$h_1 = -\frac{Fa_1}{F_1 a_1},\ \text{and}\ x = a_1 - \frac{Fa_1}{F_1 a_1} = a_2,$$ a nearer approximation,

and the same process may be repeated if necessary.

51. Find the positive root of the equation

$$x^4 - 12x^2 + 12x - 3 = 0,$$

to three places of decimals inclusive.

Here we find by trial that

$$x > 2.6 \quad \text{and} \quad x < 3.$$

Put $$a = 2.8.$$

$$\therefore Fa = a^4 - 12a^2 + 12a - 3$$
$$= (2.8)^4 - 12(2.8)^2 + 12(2.8) - 3 = -2.0144$$

$$F_1a = \frac{dFa}{da} = 4a^3 - 24a + 12 = 4(2.8)^3 - 24(2.8) + 12 = 32.608.$$

$$\therefore h = -\frac{-2.0144}{32.608} = 0.062 \text{ nearly.} \quad \therefore x = a + h = 2.862 \text{ nearly.}$$

To test the accuracy of this approximation, put

$$a_1 = 2.862 \quad \text{and} \quad x = a_1 + h_1.$$

$$Fa_1 = (2.862)^4 - 12(2.862)^2 + 12(2.862) - 3 = 0.144674 \text{ nearly.}$$

$$F_1a_1 = 4(2.862)^3 - 24(2.862) + 12 = 37.083072 \text{ nearly.}$$

$$\therefore h_1 = -\frac{0.144674}{37.083072} = -0.003901 \quad \text{nearly.}$$

$$\therefore x = a_1 + h_1 = 2.862 - 0.003901 = 2.858099 = 2.858$$

to three places of decimals.

If the process were repeated it would be found that

$$x = 2.85808;$$

so that the second approximation is true to four places of decimals, and the fifth place is slightly erroneous.

2. Given $$x^x = 100$$

to find the value of x to the place of hundredths.

Passing to the common logarithms, we have

$$x \log x = \log 100 = 2. \quad \therefore x \log x - 2 = 0.$$

Also $$x > 3, \quad \text{and} \quad x < 4.$$

Put $\quad a = 3.5, \quad$ and $\quad x = a + h.$

$$\therefore Fa = a \log a - 2, \quad F_1 a = \frac{dFa}{da} = \log a + M,$$

where $M =$ modulus of the common system $= .43429448$

$Fa \doteq 3.5 \log (3.5) - 2 = .544068 \times 3.5 - 2 = -0.095762$

$F_1 a = .544068 + .434294 = 0.978362.$

$$\therefore h = \frac{.095762}{.978362} = .098.$$

$$\therefore x = 3.5 + .098 = 3.598 \quad \text{or} \quad x = 3.60 \text{ nearly.}$$

We shall now apply Taylor's Theorem in deducing rules for the expansion and differentiation of functions of more complicated forms.

52. *Prop.* To establish a general rule for differentiating any function of two quantities p and q, which quantities are themselves functions of the single independent variable x.

Let $u = F(p, q)$, where $p = fx$, and $q = f_1 x$, the characters F, f, and f_1, denoting any function whatever, and let x take the increment h, converting p in $p + k = p_1$, q into $q + l = q_1$, and u into u_1.

Then $\quad u_1 = F(p + k, q + l) = F(p + k, q_1),$

which may be developed by Taylor's Theorem as a function of $p + k$, observing that q_1, which does not contain k, will appear in the development as would a constant:

$$\therefore u_1 = F(p + k, q_1) = F(p, q_1) + \frac{dF(p, q_1)}{dp} \cdot \frac{k}{1}$$
$$+ \frac{d^2F(p, q_1)}{dp^2} \cdot \frac{k^2}{1.2} + \&c. \quad (1).$$

But $F(p, q_1) = F(p, q + l)$, which developed as a function of $q + l$, gives

$$F(p, q+l) = F(p, q) + \frac{dF(p, q)}{dq} \cdot \frac{l}{1} + \frac{d^2F(p, q)}{dq^2} \cdot \frac{l^2}{1.2} + \&c.$$

And similarly the coefficient of k in the second term of (1).

$$\frac{dF(p, q_1)}{dp} = \frac{dF(p, q+l)}{dp},$$

may also be developed as a function of $q + l$, and will give

$$\frac{dF(p, q+l)}{dp} = \frac{dF(p, q)}{dp} + \frac{d}{dq}\left[\frac{dF(p, q)}{dp}\right]\frac{l}{1} + \&c.$$

And in like manner

$$\frac{d^2F(p, q_1)}{dp^2} = \frac{d^2F(p, q+l)}{dp^2} = \frac{d^2F(p, q)}{dp^2} + \frac{d}{dq}\left[\frac{d^2F(p, q)}{dp^2}\right] + \&c.$$

$\therefore$ By substitution in (1).

$$F(p+k, q+l) = F(p, q) + \frac{dF(p, q)}{dq} \cdot \frac{l}{1} + \frac{dF(p, q)}{dp} \cdot \frac{k}{1}$$

$$+ \text{ terms involving } k^2, kl, l^2, k^3, \&c.$$

But $\quad k = p_1 - p = \frac{dp}{dx} \cdot \frac{h}{1} + \frac{d^2p}{dx^2} \cdot \frac{h^2}{1.2} + \&c.,$

And $\quad l = q_1 - q = \frac{dq}{dx} \cdot \frac{h}{1} + \frac{d^2q}{dx^2} \cdot \frac{h^2}{1.2} + \&c.,$

$$\therefore u_1 - u = \frac{du}{dq}\left[\frac{dq}{dx} \cdot \frac{h}{1} + \frac{d^2q}{dx^2} \cdot \frac{h^2}{1.2} + \&c.,\right]$$

$$\frac{du}{dp}\left[\frac{dp}{dx} \cdot \frac{h}{1} + \frac{d^2p}{dx^2} \cdot \frac{h^2}{1.2} + \&c.,\right] + \&c.$$

Now dividing by h, and then passing to the limit, by making $h = 0$, in which case $\frac{u_1 - u}{h} = \frac{du}{dx}$, we obtain

$$\frac{du}{dx} = \frac{du}{dq} \cdot \frac{dq}{dx} + \frac{du}{dp} \cdot \frac{dp}{dx} \qquad (2).$$

$$\therefore \frac{du}{dx}dx = du = \frac{du}{dq} \cdot \frac{dq}{dx}dx + \frac{du}{dp} \cdot \frac{dp}{dx}dx. \qquad (3).$$

Thus it appears that we must differentiate u with respect to each function, as though the other functions were constant, and add the results.

53. It is very important that the precise signification of the notation here employed should be distinctly understood. By an attentive consideration of the manner in which the several expressions employed in the formulæ (2) and (3) arise, it will appear that the expression $\frac{du}{dx}$ in (2), represents the ratio of the change in x to the *entire* change in u, which latter is produced partly by the change imparted to p, and partly by that imparted to q: that the expression $\frac{du}{dq}\cdot\frac{dq}{dx}$ represents the ratio of the change in x to that part of the change in u which is communicated through q: and that $\frac{du}{dp}\cdot\frac{dp}{dx}$ represents the ratio of the change in x to that part of the change in u, which is communicated through p.

We must be careful, therefore, not to confound $\frac{du}{dq}\cdot\frac{dq}{dx}$, with $\frac{du}{dx}$, or to suppose that the first of these expressions can be brought to the form of the second by the ordinary process of algebraic reduction. This will appear evident, when it is recollected that the du which appears in $\frac{du}{dx}$ refers to the *total* change in u, while the du which occurs in $\frac{du}{dq}\cdot\frac{dq}{dx}$, refers only to so much of the change in u, as is communicated through q. Similarly, $\frac{du}{dp}\cdot\frac{dp}{dx}$, must not be confounded with $\frac{du}{dx}$, for a like reason.

54. To differentiate $u = F(p, q, r, s, \&c.)$ when p, q, r, s, &c. are functions of the same variable x.

By attributing to x an increment h, and reasoning as in the last proposition, we readily prove that

$$u_1 = u + \left[\frac{du}{dp}\cdot\frac{dp}{dx} + \frac{du}{dq}\cdot\frac{dq}{dx} + \frac{du}{dr}\cdot\frac{dr}{dx} + \frac{du}{ds}\cdot\frac{ds}{dx} + \&c.\right]\frac{h}{1}$$

$$+ \text{ terms in } h^2,\ h^3,\ \&c.$$

Transposing u, dividing by h, and then passing to the limit, we have

$$\frac{du}{dx} = \frac{du}{dp}\cdot\frac{dp}{dx} + \frac{du}{dq}\cdot\frac{dq}{dx} + \frac{du}{dr}\cdot\frac{dr}{dx} + \frac{du}{ds}\cdot\frac{ds}{dx} + \&c.$$

$$\therefore \frac{du}{dx}dx = du = \frac{du}{dp}\cdot\frac{dp}{dx}\cdot dx + \frac{du}{dq}\cdot\frac{dq}{dx}\cdot dx + \frac{du}{dr}\cdot\frac{dr}{dx}\cdot dx$$

$$+ \frac{du}{ds}\cdot\frac{ds}{dx}\cdot dx + \&c.$$

that is, we must differentiate u with respect to each of the functions, as if the other functions were constant, and add the results.

55. *Prop.* To differentiate $u = F(p, x)$, where $p = fx$.

Here u is *directly* a function of x, and also *indirectly* a function of x through p.

Now if in the equation $u = F(p, q)$, which gives

$$\frac{du}{dx} = \frac{du}{dp}\cdot\frac{dp}{dx} + \frac{du}{dq}\cdot\frac{dq}{dx},$$

we put $q = x$, there will result

$$u = F(p, x) \quad \text{and} \quad \frac{du}{dx} = \frac{du}{dp}\cdot\frac{dp}{dx} + \frac{du}{dx}\cdot\frac{dx}{dx}$$

$$\text{or} \quad \frac{du}{dx} = \frac{du}{dp}\cdot\frac{dp}{dx} + \frac{du}{dx} \quad (1), \quad \text{since} \quad \frac{dx}{dx} = 1.$$

The formula (1) is that required, but we must distinguish carefully between the differential coefficient $\frac{du}{dx}$ in the first member, and the similar expression in the second. The latter, called the *partial* differential coefficient of u with respect to x, refers only to that part of the change in u which results *directly* from a change in x, while p is supposed to remain constant; and the former, called the *total* differential coefficient of u with respect to x, refers to the entire change in u, which is partly the direct result of a change in x, and partly an indirect effect produced through p.

To distinguish the total from the partial differential coefficient, it has been agreed to enclose the former in a parenthesis; thus we write

$$\left[\frac{du}{dx}\right]=\frac{du}{dp}\cdot\frac{dp}{dx}+\frac{du}{dx} \quad \text{and} \quad \therefore\ du=\left[\frac{du}{dx}\right]dx=\frac{du}{dp}\cdot\frac{dp}{dx}\cdot dx+\frac{du}{dx}\,dx.$$

Here again there is a necessity for caution, so as not to confound $\frac{du}{dx}\cdot dx$ with du; the former being only a part of the change imparted to u by a change in x, while the latter is the symbol of the entire change.

Cor. If there were given $u=F(p,q,x)$

where p and q are functions of x, then

$$\left[\frac{du}{dx}\right]=\frac{du}{dp}\cdot\frac{dp}{dx}+\frac{du}{dq}\cdot\frac{dq}{dx}+\frac{du}{dx}.$$

and similar expressions would apply if there were a greater number of functions.

EXAMPLES.

56. 1. $u=\sin^{-1}(p-q)$, where $p=3x$ and $q=4x^3$.

$$\frac{du}{dp}=\frac{1}{\sqrt{1-(p-q)^2}},\quad \frac{du}{dq}=\frac{1}{\sqrt{1-(p-q)^2}},\quad \frac{dp}{dx}=3,\quad \frac{dq}{dx}=12x^2.$$

$$\therefore\ \frac{du}{dx}=\frac{du}{dp}\cdot\frac{dp}{dx}+\frac{du}{dq}\cdot\frac{dq}{dx}=\frac{3-12x^2}{\sqrt{1-(p-q)^2}}$$

$$=\frac{3-12x^2}{\sqrt{1-9x^2+24x^4-16x^6}}=\frac{3}{\sqrt{1-x^2}}.$$

2. $u=pq$, where $p=e^x$, and $q=x^4-4x^3+12x^2-24x+24$.

$$\frac{du}{dp}=q,\quad \frac{du}{dq}=p,\quad \frac{dp}{dx}=e^x,\quad \frac{dq}{dx}=4x^3-12x^2+24x-24;$$

$$\therefore \frac{du}{dx} = \frac{du}{dp} \cdot \frac{dp}{dx} + \frac{du}{dq} \cdot \frac{dq}{dx} = e^x \cdot x^4.$$

3. $$u = \frac{x^4p^2}{4} - \frac{x^4p}{8} + \frac{x^4}{32}, \quad \text{when} \quad p = \log x.$$

$$\frac{du}{dp} = \frac{x^4p}{2} - \frac{x^4}{8}, \quad \frac{dp}{dx} = \frac{1}{x}, \quad \frac{du}{dx} = x^3p^2 - \frac{x^3p}{2} + \frac{x^3}{8}.$$

$$\therefore \left[\frac{du}{dx}\right] = \frac{du}{dp} \cdot \frac{dp}{dx} + \frac{du}{dx} = x^4\left(\frac{p}{2} - \frac{1}{8}\right) \times \frac{1}{x} + x^3\left(p^2 - \frac{p}{2} + \frac{1}{8}\right)$$

$$= x^3(\log x)^2.$$

4. $$u = \frac{e^{ax}(p - q)}{a^2 + 1} \quad \text{where} \quad p = a \sin x, \quad \text{and} \quad q = \cos x.$$

$$\frac{du}{dp} = \frac{e^{ax}}{a^2 + 1}, \quad \frac{du}{dq} = -\frac{e^{ax}}{a^2 + 1}, \quad \frac{dp}{dx} = a \cos x, \quad \frac{dq}{dx} = -\sin x,$$

$$\frac{du}{dx} = \frac{ae^{ax}(p - q)}{a^2 + 1}.$$

$$\therefore \left[\frac{du}{dx}\right] = \frac{du}{dp} \cdot \frac{dp}{dx} + \frac{du}{dq} \cdot \frac{dq}{dx} + \frac{du}{dx}$$

$$= \frac{e^{ax}}{a^2 + 1}(a \cos x + \sin x + a^2 \sin x - a \cos x) = e^{ax} \sin x.$$

Differentiation of Implicit Functions.

57. In the various cases hitherto considered, we have supposed the function to be given *explicitly* in terms of the variable. It is now proposed to establish rules for differentiating *implicit* functions.

Prop. Having given $F(x, y) = 0$, to form the differential coefficient $\frac{dy}{dx}$ without solving the equation.

Put $u = F(x, y)$: then u will be a function of x directly, and also indirectly through y.

$$\therefore \left[\frac{du}{dx}\right] = \frac{du}{dy} \cdot \frac{dy}{dx} + \frac{du}{dx}.$$

But since u remains constantly equal to zero, the total differential coefficient of u with respect to x must be equal to zero also.

$$\therefore \frac{du}{dy}\cdot\frac{dy}{dx}+\frac{du}{dx}=0, \qquad \text{whence} \qquad \frac{dy}{dx}=-\frac{\frac{du}{dx}}{\frac{du}{dy}}.$$

Thus it appears that we must form the partial differential coefficients $\frac{du}{dx}$ and $\frac{du}{dy}$, then divide the former by the latter, and prefix the negative sign to the quotient.

Ex. 1. $y^2 - 2axy + x^2 - b^2 = 0$, to form the differential coefficient of y with respect to x.

$$u = y^2 - 2axy + x^2 - b^2, \ \frac{du}{dx} = -2ay + 2x, \ \frac{du}{dy} = 2y - 2ax.$$

$$\therefore \frac{dy}{dx} = -\frac{-2ay+2x}{2y-2ax} = \frac{ay-x}{y-ax}.$$

2. Given $x^3 + 3axy + y^3 = 0$, to form the 1st and 2d differential coefficients of y with respect to x.

$$\therefore u = x^3 + 3axy + y^3, \ \frac{du}{dx} = 3x^2 + 3ay, \ \frac{du}{dy} = 3ax + 3y^2.$$

$$\therefore \frac{dy}{dx} = -\frac{x^2+ay}{ax+y^2}.$$

Put $\frac{dy}{dx} = p_1$; then p_1 will be a function of x directly, and also indirectly through y.

$$\therefore \frac{d^2y}{dx^2} = \left[\frac{dp_1}{dx}\right] = \frac{dp_1}{dy}\cdot\frac{dy}{dx} + \frac{dp_1}{dx}.$$

But

$$\frac{dp_1}{dy} = -\frac{-a(ax+y^2)+2y(x^2+ay)}{(ax+y^2)^2}, \ \frac{dp_1}{dx} = \frac{-2x(ax+y^2)+a(x^2+ay)}{(ax+y^2)^2}.$$

Hence by substitution and reduction

$$\frac{d^2y}{dx^2} = \frac{2yx^2 + ay^2 - a^2x}{(ax+y^2)^2} \times \left(-\frac{x^2+ay}{ax+y^2}\right) + \frac{a^2y - ax^2 - 2xy^2}{(ax+y^2)^2}$$

$$= \frac{2a^3xy - 2xy(x^3 + 3axy + y^3)}{(ax+y^2)^3} = \frac{2a^3xy}{(ax+y^2)^3}.$$

58. Since it is possible to form the successive differential coefficients of y with respect to x, without solving the given equation, it will be possible to expand y in terms of x by Maclaurin's Theorem.

1. Given $$y^3 - 3y + x = 0,$$

to expand y in terms of the ascending powers of x.

$$u = y^3 - 3y + x = 0, \quad \frac{du}{dx} = 1, \quad \frac{du}{dy} = 3(y^2-1). \quad \therefore \frac{dy}{dx} = \frac{1}{3(1-y^2)}.$$

Expanding the last expression by actual division, we have

$$\frac{dy}{dx} = \frac{1}{3}(1 + y^2 + y^4 + \&c.)$$

$$\therefore \frac{d^2y}{dx^2} = \frac{1}{3}(2y + 4y^3 + 6y^5 + \&c.)\frac{dy}{dx} = \frac{1}{3^2}(2y + 6y^3 + 12y^5 + \&c.)$$

$$\frac{d^3y}{dx^3} = \frac{1}{3^2}(2 + 18y^2 + 60y^4 + \&c.)\frac{dy}{dx} = \frac{1}{3^3}(2 + 20y^2 + 80y^4 + \&c.)$$

$$\frac{d^4y}{dx^4} = \frac{1}{3^3}(40y + 320y^3 + \&c.)\frac{dy}{dx} = \frac{1}{3^4}(40y + 360y^3 + \&c.)$$

$$\frac{d^5y}{dx^5} = \frac{1}{3^4}(40 + 1080y^2 + \&c.)\frac{dy}{dx} = \frac{1}{3^5}(40 + 1120y + \&c.)\ \&c.$$

But when $$x = 0, \ [y] = 0,$$

$$\therefore \left[\frac{dy}{dx}\right] = \frac{1}{3}, \ \left[\frac{d^2y}{dx^2}\right] = 0, \ \left[\frac{d^3y}{dx^3}\right] = \frac{2}{3^3}, \ \left[\frac{d^4y}{dx^4}\right] = 0, \ \left[\frac{d^5y}{dx^5}\right] = \frac{40}{3^5}, \ \&c.$$

$\therefore$ By substitution in Maclaurin's formula,

$$y = \frac{x}{3} + \frac{x^3}{3^4} + \frac{x^5}{3^6} + \&c.$$

2. To expand y in terms of the *descending* powers of x, from the relation

$$ay^3 - x^3y - ax^3 = 0.$$

Put $\quad x^3 = \frac{1}{v}; \quad$ then $\quad ay^3v - y - a = 0.$

$$\therefore u = ay^3v - y - a, \quad \frac{du}{dv} = ay^3, \quad \frac{du}{dy} = 3ay^2v - 1, \quad \frac{dy}{dv} = \frac{ay^3}{1 - 3ay^2v}.$$

$$\frac{d^2y}{dv^2} = \frac{3ay^2(1 - 3ay^2v)\frac{dy}{dv} + (6ayv\frac{dy}{dv} + 3ay^2)ay^3}{(1 - 3ay^2v)^2}, \text{ \&c., \&c.}$$

But when $\quad v = 0, \ [y] = -a, \quad \left[\frac{dy}{dv}\right] = -a^4, \quad \left[\frac{d^2y}{dv^2}\right] = -6a^7$, &c.

$$\therefore y = -a - \frac{a^4v}{1} - \frac{6a^7v^2}{1.2} \text{ \&c.}$$

or by replacing v by $\frac{1}{x^3}$,

$$y = -a - \frac{a^4}{x^3} - \frac{3a^7}{x^6} \text{ \&c.}$$

The use of this method is much restricted by the great labor usually required in forming the successive differential coefficients.

CHAPTER VII.

ESTIMATION OF THE VALUES OF FUNCTIONS HAVING THE INDETERMINATE FORM.

59. It frequently occurs that the substitution of a particular value for a variable x in a fractional expression will cause that expression to assume the indeterminate form $\frac{0}{0}$. Such expressions are often called Vanishing Fractions, and they may be regarded as limits to the values of the ratios expressed by these fractions, when the variable value of x is caused to approach indefinitely near to some particular value.

Thus in the example $u = \frac{x^4 - 1}{x^3 - 1}$, the value of which can usually be determined when that of x is given, by a simple substitution, we find that it assumes the form $\frac{0}{0}$ when $x = 1$. But the value of u is even then determinate; for if we divide the numerator and denominator of the fraction by $x - 1$, before making $x = 1$, we get

$$u = \frac{x^3 + x^2 + x + 1}{x^2 + x + 1}$$

as a general value of u, and this becomes

$$\frac{1 + 1 + 1 + 1}{1 + 1 + 1} = \frac{4}{3} \quad \text{when} \quad x = 1.$$

Here we see plainly that it is the presence of the common factor $x - 1$ in the numerator and denominator which causes the fraction to assume the indeterminate form. In this, and in all similar cases, the removal of the common factor serves to determine the value of u. But it usually occurs that the discovery of this factor is attended with considerable difficulty, and hence the necessity of some more general method by which to estimate the values of fractions which assume the indeterminate form $\frac{0}{0}$, when the variable x takes a particular value. Such a method is readily supplied by the Differential Calculus.

It should be observed, however, that there are other indeterminate forms besides $\frac{0}{0}$, such as the following:

$$\frac{\infty}{\infty}, \quad \infty \times 0, \quad \infty - \infty, \quad 0^0, \quad \infty^0. \quad 1^{\pm\infty},$$

each of which will be considered in succession.

60. *Prop.* To determine the value of a function which takes the form $\frac{0}{0}$ for a particular value of the variable.

Let $u = \frac{P}{Q} = \frac{Fx}{\varphi x}$ be a function which takes the form $\frac{0}{0}$ when $x = a$; that is, let $Fa = 0$, and $\varphi a = 0$: let it be proposed to find the particular value $[u]$ assumed by u when $x = a$.

Suppose x to take an increment h, converting u, P, and Q into u_1, P_1, and Q_1, respectively, and let $P_1 = F(x + h)$ and $Q_1 = \varphi(x + h)$ be expanded by Taylor's Theorem: then denoting the successive differential coefficients Fx by F_1x, F_2x, &c., and those of φx by $\varphi_1 x$, $\varphi_2 x$, &c., we have

$$u_1 = \frac{P_1}{Q_1} = \frac{F(x+h)}{\varphi(x+h)} = \frac{Fx + F_1x\frac{h}{1} + F^2x\frac{h^2}{1\,.\,2} + F^3x\frac{h^3}{1.2.3} + \&c.}{\varphi x + \varphi_1 x\frac{h}{1} + \varphi_2 x\frac{h^2}{1\,.\,2} + \varphi_3 x\frac{h^3}{1.2.3} + \&c.}$$

or when $x = a$

$$u_1 = \frac{Fa + F_1a\frac{h}{1} + F_2a\frac{h^2}{1\,.\,2} + F_3a\frac{h^3}{1\,.\,2\,.\,3} + \&c.}{\varphi a + \varphi_1 a\frac{h}{1} + \varphi_2 a\frac{h^2}{1\,.\,2} + \varphi_3 a\frac{h^3}{1\,.\,2\,.\,3} + \&c.}$$

But by hypothesis, $Fa = 0$, and $\varphi a = 0$. $\therefore$ Omitting the first term in the numerator and denominator, and then dividing each by h. we get

$$u_1 = \frac{F_1 a + F_2 a \dfrac{h}{1.2} + F_3 a \dfrac{h^2}{1.2.3} + \&c.}{\varphi_1 a + \varphi_2 a \dfrac{h}{1.2} + \varphi_3 a \dfrac{h^2}{1.2.3} + \&c.} \quad . \; . \; . \quad (1)$$

Now making $h = 0$, we convert u_1 into $[u]$, and thus obtain

$$[u] = \frac{F_1 a}{\varphi_1 a}.$$

Hence it appears that, in order to determine the value of a function $\frac{Fx}{\varphi x}$ which takes the form $\frac{0}{0}$ when $x = a$, we must replace Fx and φx by the values of their first differential coefficients, and then make $x = a$ in each.

It will sometimes occur that this substitution will reduce to zero both $F_1 a$ and $\varphi_1 a$, in case

$$[u] = \frac{F_1 a}{\varphi_1 a} = \frac{0}{0} \text{ remains still undetermined.}$$

we then omit $F_1 a$ and $\varphi_1 a$ in equation, (1) and divide the numerator and denominator by $\frac{h}{1.2}$, thus obtaining

$$u_1 = \frac{F_2 a + F_3 a \dfrac{h}{3} + \&c.}{\varphi_2 a + \varphi_3 a \dfrac{h}{3} + \&c.} \quad . \; . \; . \quad (2)$$

which becomes $$[u] = \frac{F_2 a}{\varphi_2 a}$$

when $$h = 0.$$

$\therefore$ when the first differential coefficients both reduce to zero, they must be replaced by the second differential coefficients. If $F_2 a$ and

$\varphi_2 a$ both become zero also, we omit them in (2), then divide by $\frac{h}{3}$, and finally make $h = 0$, obtaining

$$[u] = \frac{F_3 a}{\varphi_3 a}.$$

And since the same reasoning may be extended, we have the following rule for finding the value of $[u] = \frac{Fa}{\varphi a} = \frac{0}{0}$, viz.:

Substitute for Fx *and* φx *their first, second, third, &c., differential coefficients, and make* x = a *in each result, until a pair of coefficients is obtained, both of which do not reduce to zero; the fraction thus found will be the true value of* [u].

EXAMPLES.

61. 1. $$u = \frac{x^5 - 1}{x - 1} = \frac{0}{0} \quad \text{when} \quad x = 1.$$

$Fx = x^5 - 1$, and $\varphi x = x - 1$. $\therefore F_1 x = 5x^4$, and $\varphi_1 x = 1$.

$$\therefore F_1 a = 5, \quad \varphi_1 a = 1, \quad \text{and} \quad [u] = \frac{F_1 a}{\varphi_1 a} = \frac{5}{1} = 5.$$

This result is easily verified by division, before making $x = 1$; thus by actual division

$$\frac{x^5 - 1}{x - 1} = x^4 + x^3 + x^2 + x + 1 = 5 \quad \text{when} \quad x = 1.$$

2. $$u = \frac{a^x - b^x}{x} = \frac{0}{0} \quad \text{when} \quad x = 0.$$

$$\frac{F_1 x}{\varphi_1 x} = \frac{\log a \,.\, a^x - \log b \,.\, b^x}{1}. \quad \therefore \frac{F_1 a}{\varphi_1 a} = \log a - \log b = [u].$$

This result is easily verified by expanding a^x and b^x.

Thus $$\frac{a^x - b^x}{x}$$

$$= \frac{1 + \log a \cdot \frac{x}{1} + \log^2 a \frac{x^2}{1.2} + \text{\&c.} - 1 - \log b \cdot \frac{x}{1} - \log^2 b \cdot \frac{x^2}{1.2} + \text{\&c.}}{x}$$

$$\therefore u = \log a - \log b + \frac{x}{1.2}(\log^2 a - \log^2 b) + \&c.$$

$$\therefore [u] = \log a - \log b \quad \text{by making} \quad x = 0.$$

3. $$u = \frac{a - \sqrt{a^2 - x^2}}{x^2} = \frac{0}{0} \quad \text{when} \quad x = 0.$$

$$\frac{F_1 x}{\varphi_1 x} = \frac{x(a^2 - x^2)^{-\frac{1}{2}}}{2x}; \qquad \therefore \frac{F_1 a}{\varphi_1 a} = \frac{0}{0}.$$

Here the first differential coefficients prove equal to zero, and therefore they must be replaced by the second differential coefficients. But

$$\frac{F_2 x}{\varphi_2 x} = \frac{(a^2 - x^2)^{-\frac{1}{2}} + x^2(a^2 - x^2)^{-\frac{3}{2}}}{2}.$$

$$\therefore \frac{F_2 a}{\varphi_2 a} = \frac{(a^2)^{-\frac{1}{2}}}{2} = \frac{1}{2a} = [u].$$

4. $$u = \frac{ax^2 + ac^2 - 2acx}{bx^2 - 2bcx + bc^2} = \frac{0}{0} \quad \text{when} \quad x = c.$$

$$\frac{F_1 x}{\varphi_1 x} = \frac{2ax - 2ac}{2bx - 2bc}. \qquad \therefore \frac{F_1 a}{\varphi_1 a} = \frac{2ac - 2ac}{2bc - 2bc} = \frac{0}{0}.$$

Then $$\frac{F_2 x}{\varphi_2 x} = \frac{2a}{2b}. \qquad \therefore \frac{F_2 a}{\varphi_2 a} = \frac{2a}{2b} = \frac{a}{b} = [u].$$

5. $$u = \frac{x^3 - ax^2 - a^2 x + a^3}{x^2 - a^2} = \frac{0}{0} \quad \text{when} \quad x = a.$$

$$\frac{F_1 x}{\varphi_1 x} = \frac{3x^2 - 2ax - a^2}{2x}. \qquad \therefore \frac{F_1 a}{\varphi_1 a} = \frac{0}{2a} = 0 = [u].$$

6. $$u = \frac{ax - x^2}{a^4 - 2a^3 x + 2ax^3 - x^4} = \frac{0}{0} \quad \text{when} \quad x = a.$$

$$\frac{F_1 x}{\varphi_1 x} = \frac{a - 2x}{-2a^3 + 6ax^2 - 4x^3}. \qquad \therefore \frac{F_1 a}{\varphi_1 a} = \frac{a - 2a}{-2a^3 + 6a^3 - 4a^3}.$$

or $$\frac{F_1 a}{\varphi_1 a} = -\frac{a}{0} = -\infty = [u].$$

7.
$$u = \frac{\log x}{(1-x)^{\frac{1}{2}}} = \frac{0}{0} \quad \text{when} \quad x = 1.$$

$$\frac{F_1 x}{\varphi_1 x} = \frac{\frac{1}{x}}{-\frac{1}{2}(1-x)^{-\frac{1}{2}}} = -\frac{2(1-x)^{\frac{1}{2}}}{x},$$

$$\therefore \frac{F_1 a}{\varphi_1 a} = -\frac{2(1-1)^{\frac{1}{2}}}{1} = 0 = [u].$$

8. $u = \dfrac{e^{nx} - e^{na}}{(x-a)^s} = \dfrac{0}{0}$ when $x = a$, (s being an integer.)

Differentiating s times, we get

$$\frac{F_s x}{\varphi_s x} = \frac{n^s e^{nx}}{s(s-1)(s-2)\ldots\ldots 3.2.1},$$

$$\therefore \frac{F_s a}{\varphi_s a} = \frac{n^s e^{na}}{s(s-1)(s-2)\ldots\ldots 3.2.1} = [u].$$

9.
$$u = \frac{\tan x - \sin x}{\sin^3 x} = \frac{0}{0} \quad \text{when} \quad x = 0.$$

$$\frac{F_1 x}{\varphi_1 x} = \frac{\sec^2 x - \cos x}{3\sin^2 x \,.\, \cos x}, \quad \therefore \frac{F_1 a}{\varphi_1 a} = \frac{\sec^2 0 - \cos 0}{3 \sin^2 0 \,.\, \cos 0} = \frac{0}{0}.$$

$$\frac{F_2 x}{\varphi_2 x} = \frac{2\sec^2 x \,.\, \tan x + \sin x}{6 \sin x \,.\, \cos^2 x - 3\sin^3 x},$$

$$\therefore \frac{F_2 a}{\varphi_2 a} = \frac{2\sec^2 0 \,.\, \tan 0 + \sin 0}{6\sin 0 \,.\, \cos^2 0 - 3\sin^3 0} = \frac{0}{0}.$$

$$\frac{F_3 x}{\varphi_3 x} = \frac{4\sec^2 x \,.\, \tan^2 x + 2\sec^4 x + \cos x}{6\cos^3 x - 12 \sin^2 x \,.\, \cos x - 9 \sin^2 x \,.\, \cos x},$$

$$\therefore \frac{F_3 a}{\varphi^3 a} = \frac{4\sec^2 0 \,.\, \tan^2 0 + 2\sec^4 0 + \cos 0}{6\cos^3 0 - 12\sin^2 0 \,.\, \cos 0 - 9\sin^2 0 \,.\, \cos 0} = \frac{3}{6} = \frac{1}{2} = [u].$$

62. The method just explained and illustrated, ceases to be applicable when we obtain a differential coefficient whose value becomes infinite by making $x = a$; for such a result shows the impossibility of developing the corresponding function $F(x+h)$ by Taylor's Theorem, for that particular value of x, and therefore the process founded on such development fails.

The expedient adopted in such cases, is that of substituting $a+h$ for x, then expanding numerator and denominator by the common algebraic methods, then dividing numerator and denominator by the lowest power of h found in either, and finally making $h=0$. A few examples will illustrate this method.

63. 1. $$u=\frac{(a^2-x^2)^{\frac{3}{2}}}{(a-x)^{\frac{3}{2}}}=\frac{0}{0} \text{ when } x=a.$$

Here the first differential coefficients reduce to zero, and all succeeding coefficients become infinite when $x=a$. We therefore put $a+h$ for x and expand.

$$\therefore u_1=\frac{(a^2-a^2-2ah-h^2)^{\frac{3}{2}}}{(a-a-h)^{\frac{3}{2}}}=\cdot\frac{(2a+h)^{\frac{3}{2}}.(-h)^{\frac{3}{2}}}{(-h)^{\frac{3}{2}}}$$

$$=(2a+h)^{\frac{3}{2}}=(2a)^{\frac{3}{2}}+\frac{3}{2}(2a)^{\frac{1}{2}}h+ \&c.$$

$$\therefore [u]=(2a)^{\frac{3}{2}}.$$

2. $$u=\frac{\sqrt{x}-\sqrt{a}+\sqrt{x-a}}{\sqrt{x^2-a^2}}=\frac{0}{0} \text{ when } x=a.$$

Put $a+h$ for x

$$\therefore u_1=\frac{(a+h)^{\frac{1}{2}}-a^{\frac{1}{2}}+(a+h-a)^{\frac{1}{2}}}{(a^2+2ah+h^2-a^2)^{\frac{1}{2}}}=\frac{h^{\frac{1}{2}}+\frac{1}{2}a^{-\frac{1}{2}}h-\&c.}{h^{\frac{1}{2}}(2a+h)^{\frac{1}{2}}}$$

$$=\frac{1+\frac{1}{2}a^{-\frac{1}{2}}h^{\frac{1}{2}}\ \&c.}{(2a)^{\frac{1}{2}}+\frac{1}{2}(2a)^{-\frac{1}{2}}h\ \&c.}$$

$$\therefore [u]=\frac{1}{(2a)^{\frac{1}{2}}}.$$

Remark. This method may be used even in those cases to which the method of differentiation is applicable. We will now consider the other indeterminate forms.

64. *Prop.* To find the value of the function $u = \frac{P}{Q} = \frac{Fx}{\varphi x}$, which assumes the form $\frac{\infty}{\infty}$ when $x = a$.

Put $P = \frac{1}{p}$ and $Q = \frac{1}{q}$. Then we have

$$u = \frac{\frac{1}{p}}{\frac{1}{q}} = \frac{q}{p} = \frac{0}{0} \text{ when } x = a.$$

Thus the function being reduced to the ordinary form $\frac{0}{0}$, its value may be found by the methods already explained.

Now since $p = \frac{1}{P}$, $\frac{dp}{dx} = -\frac{1}{P^2} \cdot \frac{dP}{dx} = -\frac{F_1 x}{(Fx)^2}$

And similarly $\frac{dq}{dx} = -\frac{\varphi_1 x}{(\varphi x)^2}$

$$\therefore [u] = \frac{Fa}{\varphi a} = \frac{\frac{\varphi_1 a}{(\varphi a)^2}}{\frac{F_1 a}{(Fa)^2}} = \frac{(Fa)^2}{(\varphi a)^2} \times \frac{\varphi_1 a}{F_1 a}.$$

$$\therefore 1 = \frac{Fa}{\varphi a} \times \frac{\varphi_1 a}{F_1 a} \text{ whence } \frac{Fa}{\varphi a} = \frac{F_1 a}{\varphi_1 a} = [u].$$

Hence it appears that the ordinary direct process of substituting for numerator and denominator, their first differential coefficients will apply when the function takes the form $\frac{\infty}{\infty}$. But since when $P = \infty$ and $Q = \infty$, their differential coefficients will also be infinite, the reduced fraction will still take the form $\frac{\infty}{\infty}$, and therefore will not serve to determine the true value of u, unless we can discover a factor common to the numerator and denominator, or can trace some relation between the numerator and denominator of the new fraction, which will facilitate the determination of its value.

65. *Prop.* To find the value of the function $u = P \times Q = Fx \times \varphi x$ which takes the form $\infty \times 0$ when $x = a$.

Put $P = \frac{1}{p}$. Then $u = \frac{Q}{p} = \frac{0}{0}$ when $x = a$, the common form.

Now since $p = \frac{1}{P}$, we have $\frac{dp}{dx} = -\frac{1}{P^2} \cdot \frac{dP}{dx} = -\frac{F_1 x}{(Fx)^2}$

$$\therefore [u] = -(Fa)^2 . \frac{\varphi_1 a}{F_1 a}$$

But since when $P = Fx = \infty$, its differential coefficients will also be infinite, the value of u will take the form $\frac{\infty}{\infty}$, unless the infinite factor should disappear by division.

66. *Prop.* To find the value of the function $u = P - Q = Fx - \varphi x$, which takes the form $\infty - \infty$ when $x = a$.

Put $$P = \frac{1}{p} \text{ and } Q = \frac{1}{q}. \text{ Then}$$

$$u = \frac{1}{p} - \frac{1}{q} = \frac{q-p}{pq} = \frac{0}{0} \quad \text{when} \quad x = a,$$

and the value is to be found by the ordinary method.

67. *Prop.* To find the value of the function $u = P^Q = (Fx)^{\varphi x}$ which takes either of the forms 0^0, ∞^0, or $1^{\pm\infty}$, when $x = a$.

1st. Let the form be $u = 0^0$. Passing to logarithms we have

$$\log u = Q . \log P = \varphi x . \log (Fx)$$

$$\therefore [\log u] = \varphi a . \log (Fa) = -0 \times \infty .$$

which is one of the forms already provided for.

Thus, having found $\log u$, we have $u = e^{\log u}$.

2d. Let the form be $u = \infty^0$. Then $\log u = Q . \log P = \varphi x . \log (Fx)$.

$$\therefore [\log u] = \varphi a . \log (Fa) = 0 \times \infty .$$

a form already considered.

3d. Let the form be $1^{\pm\infty}$.

Then $$\log u = Q \log P = \varphi x . \log(Fx).$$

$$\therefore\ [\log u] = \varphi a . \log(Fa) = \pm \infty \times 0,$$

and the form is still the same.

EXAMPLES.

68. 1. $u = (1 - x) . \tan\left(x \cdot \frac{\pi}{2}\right) = 0 \times \infty$ when $x = 1$.

Here $Fx = \tan\left(x \cdot \frac{\pi}{2}\right)$ and $\varphi x = 1 - x$,

$$\therefore\ F_1 x = \frac{\pi}{2} \sec^2\left(x \cdot \frac{\pi}{2}\right), \quad \varphi_1 x = -1,$$

$$\therefore\ [u] = -(Fa)^2 \frac{\varphi_1 a}{F_1 a} = \tan^2\left(1 \times \frac{\pi}{2}\right) \frac{1}{\frac{\pi}{2} \cdot \sec^2\left(1 \times \frac{\pi}{2}\right)}$$

$$= \frac{\sec^2 \frac{1}{2}\pi - 1}{\frac{\pi}{2} \sec^2 \frac{1}{2}\pi} = \frac{2}{\pi}\left(1 - \cos^2 \frac{1}{2}\pi\right) = \frac{2}{\pi}.$$

2. $u = e^{\frac{1}{x}} . \sin x = \infty \times 0$ when $x = 0$.

$$Fx = e^{\frac{1}{x}}, \quad \varphi x = \sin x, \quad \therefore\ F_1 x = -\frac{1}{x^2} e^{\frac{1}{x}}, \quad \varphi_1 x = \cos x.$$

$$\therefore\ [u] = -(Fa)^2 \frac{\varphi_1 a}{F_1 a} = e^{\frac{2}{0}} . \frac{0^2}{e^{\frac{1}{0}}} = e^{\frac{1}{0}} \times 0^2.$$

Here the function still takes the form $\infty \times 0$; but the true value is easily found by expanding $e^{\frac{1}{x}}$.

For $$e^{\frac{1}{x}} . x^2 = \left(1 + \frac{1}{x} + \frac{1}{1 . 2 . x^2} + \frac{1}{1 . 2 . 3 . x^3} + \&c.\right) x^2$$

$$= x^2 + x + \frac{1}{1 . 2} + \frac{1}{1 . 2 . 3\, x} + \&c.$$

$$\therefore\ e^{\frac{1}{0}} \times 0^2 = \infty = [u].$$

3. $u = \frac{x^n}{e^x} = \frac{\infty}{\infty}$ when $x = \infty$.

Differentiating n times, we get

$$[u] = \frac{n(n-1)(n-2)\ldots.3.2.1}{e^\infty} = 0.$$

4. $u = \frac{\log x}{x^n} = \frac{\infty}{\infty}$ when $x = \infty$.

$$F_1 x = \frac{1}{x}, \qquad \varphi_1 x = nx^{n-1}.$$

$$\therefore \quad [u] = \frac{F_1 a}{\varphi_1 a} = \frac{1}{n \cdot \infty^n} = 0.$$

5. $u = \frac{1}{1-x} - \frac{2}{1-x^2} = \infty - \infty$ when $x = 1$.

$$p = \frac{1}{Fx} = 1 - x, \quad q = \frac{1}{\varphi x} = \frac{1-x^2}{2}.$$

$$\therefore u = \frac{q-p}{pq} = \frac{\frac{1}{2}(1-x^2) - (1-x)}{\frac{1}{2}(1-x^2)(1-x)}$$

$$= \frac{\frac{1}{2}(1+x) - 1}{\frac{1}{2}(1-x^2)} = \frac{0}{0} \text{ when } x = 1.$$

$$\therefore \quad [u] = \frac{\frac{1}{2}}{-1} = -\frac{1}{2}.$$

3. $u = \frac{x}{x-1} - \frac{1}{\log x} = \infty - \infty$ when $x = 1$.

$$p = \frac{1}{Fx} = \frac{x-1}{x}, \quad q = \log x.$$

$$\therefore u = \frac{q-p}{pq} = \frac{x \log x - x + 1}{(x-1)\log x} = \frac{0}{0} \text{ when } x = 1.$$

$$\therefore \quad [u] = \frac{\log 1 + 1 - 1}{\log 1 + 1 - 1} = \frac{0}{0}.$$

Differentiating numerator and denominator of the valu(of u a second time, and making $x = 1$, we get

$$[u] = \frac{1}{1+1} = \frac{1}{2}.$$

7. $u = x^x = 0^0$ when $x = 0$.

$$\log u = x \,.\, \log x = -0 \times \infty, \quad \text{when} \quad x = 0.$$

or, $$\log u = \frac{\log x}{\frac{1}{x}} = -\frac{\infty}{\infty} \text{ when } x = 0.$$

Then $$Fx = \log x, \qquad \text{and} \qquad \varphi x = \frac{1}{x}.$$

$$\therefore \frac{F_1 x}{\varphi_1 x} = \frac{\frac{1}{x}}{-\frac{1}{x^2}} = -x. \quad \therefore [\log u] = \frac{F_1 a}{\varphi_1 a} = 0, \text{ and } [u] = 1.$$

8. $u = x^{\sin x} = 0^0$, when $x = 0$.

Since $\frac{\sin x}{x} = 1$ when $x = 0$, $\therefore x^{\sin x} = x^x = 1$ when $x = 0$.

And similarly $\sin x^{\sin x} = x^x = 1$ when $x = 0$.

Again, since $\sin x \,.\, \log x = \sin x \,.\, \log x \,.\, \log e$.

$$\therefore x^{\sin x} = e^{\sin x \cdot \log x} = 1 \text{ when } x = 0.$$

$$\therefore \sin x \,.\, \log x = 0, \text{ when } x = 0.$$

And similarly $\sin x \,.\, \log \sin x = 0$ when $x = 0$.

9. $u = \cot x^{\sin x} = \infty^0$ when $x = 0$.

$$\log u = \sin x \,.\, \log \frac{\cos x}{\sin x} = \sin x\,(\log \cos x - \log \sin x).$$

$$= 0 - 0 = 0 \quad \text{when} \quad x = 0, \qquad \therefore [u] = 1.$$

10. $u = (1 + nx)^{\frac{1}{x}} = 1^\infty$ when $x = 0$.

$$\log u = \frac{\log (1 + nx)}{x} = \frac{0}{0} \text{ when } x = 0.$$

$\therefore$ By differentiation, $[\log u] = \frac{n}{1} = n$, and $[u] = e^n$.

This result is easily verified; for by expanding $(1 + nx)^{\frac{1}{x}}$ by the binomial theorem, we obtain

$$u = 1 + \frac{1}{x}(nx) + \frac{1}{x}\left(\frac{1}{x} - 1\right)\frac{(nx)^2}{1.2} + \frac{1}{x}\left(\frac{1}{x} - 1\right)\left(\frac{1}{x} - 2\right)\frac{(nx)^3}{1.2.3} + \&c.$$

$$= 1 + n + \frac{n^2}{1.2}(1 - x) + \frac{n^3}{1.2.3}(1 - 3x - 2x^2) \&c.$$

$$\therefore \quad [u] = 1 + n + \frac{n^2}{1.2} + \frac{n^3}{1.2.3} + \&c.,$$

which series is the expansion of e^n.

11. $u = (\cos ax)^{(\operatorname{cosec} cx)^2} = 1^\infty$, when $x = 0$.

$$\log u = \operatorname{cosec}^2 cx . \log \cos ax = \frac{\log \cos ax}{\sin^2 cx} = \frac{0}{0} \text{ when } x = 0.$$

Put $\log \cos ax = Fx$ and $\sin^2 cx = \phi x$.

$\therefore F_1 x = - a . \tan ax$, $\phi_1 x = 2c . \sin cx . \cos cx = c . \sin 2 cx$.

$$\therefore \frac{F_1 a}{\phi_1 a} = \frac{0}{0}.$$

Differentiating again we get

$$F_2 x = - a^2 . \sec^2 ax, \quad \phi_2 x = 2c^2 . \cos 2cx.$$

$$\therefore \frac{F_2 a}{\phi_2 a} = -\frac{a^2}{2c^2}. \quad \therefore [u] = e^{-\frac{a^2}{2c^2}}.$$

CHAPTER VIII.

MAXIMA AND MINIMA FUNCTIONS OF A SINGLE VARIABLE.

69. If u be a function whose value depends on that of a variable x, so that $u = Fx$, and if, when x takes a certain value a, the corresponding value u_1 of u be greater than the values which immediately precede and follow it, then the value u_1 is called a *maximum;* but if the intermediate value be less than those which precede and follow it immediately, the value u_1 is said to be a *minimum.*

Suppose for example that when $x = a$, the general value $u = Fx$ becomes $u_1 = Fa$, that when $x = a \pm h$ u becomes $u_2 = F(a + h)$, or $u_3 = F(a - h)$, and suppose that for some small but finite value of h, and for all values between that and zero, the corresponding values of both u_2 and u_3 shall be less than u_1, then will u_1 be a maximum; but if u_2 and u_3 be both greater than u_1, then the latter will be a minimum.

70. In order to discover the conditions necessary to render a function ($u = Fx$) either a maximum or minimum, the following principle will be established.

Prop. In any series $Ah^a + Bh^b + Ch^c +$ &c., arranged according to the positive ascending powers of h, a value may be assigned to h so small as to render the first term Ah^a, (which contains the lowest power of h), greater than the sum of all the succeeding terms.

Proof. Assume $A > Bh^{b-a} + Ch^{c-a} +$ &c., a condition always possible, since by diminishing h the second member may be rendered less than any assignable quantity. Multiply each member by h^a; and there will result $Ah^a > Bh^b + Ch^c +$ &c., as stated in the enunciation of the proposition.

Cor. The value of h may be taken so small that the sign of the first term shall control that of the entire series.

71. *Prop.* To determine the conditions necessary to render a function u of a single variable x, either a maximum or minimum.

Let $u = Fx$, and suppose x to receive successively an increment and a decrement h. Then developing by Taylor's Theorem, we get

$$u_2 = F(x+h) = Fx + \frac{dFx}{dx}\cdot\frac{h}{1} + \frac{d^2Fx}{dx^2}\cdot\frac{h^2}{1.2} + \frac{d^3Fx}{dx^3}\cdot\frac{h^3}{1.2.3} + \&\text{c.} \quad (1)$$

$$u_3 = F(x-h) = Fx - \frac{dFx}{dx}\cdot\frac{h}{1} + \frac{d^2Fx}{dx^2}\cdot\frac{h^2}{1.2} - \frac{d^3Fx}{dx^3}\cdot\frac{h^3}{1.2.3} + \&\text{c.} \quad (2).$$

Now in order that Fx may exceed both $F(x+h)$ and $F(x-h)$, it is obviously necessary that the algebraic sum of the terms succeeding the first term in each of the series (1) and (2) shall be negative; that is, we must have by employing the usual notation,

$$F_1x\cdot\frac{h}{1} + F_2x\frac{h^2}{1.2} + F_3x\frac{h^3}{1.2.3} + \&\text{c.} < 0 \ldots . (3),$$

and $$-F_1x\cdot\frac{h}{1} + F_2x\frac{h^2}{1.2} - F_3x\frac{h^3}{1.2.3} + \&\text{c.} < 0 \ldots . (4).$$

Now the sign of the first term in each of the series (3) and (4) will control that of the entire series when h is taken sufficiently small, and since the first terms of (3) and (4) have contrary signs, it is impossible that both of these series shall be negative, so long as the term $F_1x\,.\,\frac{h}{1}$ has a finite value. Hence the first condition necessary to render Fx a maximum is that $F_1x\,.\,\frac{h}{1} = 0$, or since h is finite

$$F_1x = \frac{dFx}{dx} = 0 \ldots . (5).$$

Now omitting the first terms of (3) and (4), we have

$$F_2x\frac{h^2}{1.2}+F_3x\frac{h^3}{1.2.3}+\&c. < 0 \quad \ldots \quad (6),$$

and

$$F_2x\frac{h^2}{1.2}-F_3x\cdot\frac{h^3}{1.2.3}+\&c. < 0 \quad \ldots \quad (7).$$

The signs of the series (6) and (7) will be controlled by those of their first terms, which *terms* have the positive sign in both series; and therefore each series will be negative when $F_2x\frac{h^2}{1.2}$ is an essentially negative quantity, or when F_2x is essentially negative, (since $\frac{h^2}{1.2}$ is always positive).

Thus the two conditions which usually characterize a maximum value of Fx are

$$\frac{dFx}{dx}=0, \quad \text{and} \quad \frac{d^2Fx}{dx^2} < 0.$$

On the contrary, when Fx is a minimum, we must have Fx less than $F(x+h)$ and $F(x-h)$, and therefore by a similar course of reasoning, the necessary conditions are

$$\frac{dFx}{dx}=0, \quad \text{and} \quad \frac{d^2Fx}{dx^2} > 0.$$

The conditions here obtained are those usually applicable: the exceptions will now be considered.

72. The results obtained in the last proposition indicate the following as the ordinary rule by which to discover those values of the independent variable x, which will render any proposed function u a maximum or minimum.

1st. Form the first differential coefficient $\frac{du}{dx}$, place its value equal to zero, and then solve the equation thus formed, obtaining the several values of x.

2d. Form the second differential coefficient $\frac{d^2u}{dx^2}$, and substitute for x, in the value of that coefficient, each of the values found above. Then all those values of x which render $\frac{d^2u}{dx^2}$ negative, will correspond to maximum values of u; but those values of x which render $\frac{d^2u}{dx^2}$ positive, will correspond to minimum values of u. And when the proper values of x have been ascertained, the maximum or minimum values of u are found by simple substitution in the equation $u = Fx$.

73. 1. In the application of the preceding method, it may occur that a value of x, obtained by making $\frac{du}{dx} = 0$, will, when substitued in $\frac{d^2u}{dx^2}$, cause that coefficient to reduce to zero also. In that case, the signs of the series, (6) and (7), in the last proposition, will depend on the terms which contain the third differential coefficients; and since these terms have contrary signs in the two series, the value of x which renders $\frac{du}{dx} = 0$, and $\frac{d^2u}{dx^2} = 0$, cannot render u either a maximum or minimum, unless it should happen to render $\frac{d^3u}{dx^3} = 0$ also. When this occurs, we must examine the sign of the fourth differential coefficient, which now controls the sign of each series, and if this be negative, the value of u will be a maximum; but if positive, a minimum.

And since the same reasoning could be extended when other differential coefficients reduce to zero, we have the following more general rule for the discovery of maximum and minimum values of a function of a single variable.

1st. Form the first differential coefficient, place its value equal to zero, and deduce the corresponding values of x.

2d. Substitute each of these values in the succeeding differential coefficients, stopping at the first coefficient which does not reduce to zero. If this coefficient be of an odd degree, the corresponding value of u will be neitner a maximum nor a minimum; but if it be of an even degree, the value of u will be a maximum or mininum, according as the sign of that coefficient is negative or positive.

The annexed diagram will illustrate the fact that the same function may have several maximum and several minimum values; and that one minimum may exceed another maximum. Thus, if the curve $CDEFGH$ be the locus of the equation $y = Fx$, then will DQ and FS represent maximum values of the ordinates y, while CP, ER, and GX will be minimum values of the same. Also the minimum GX exceeds the maximum DQ.

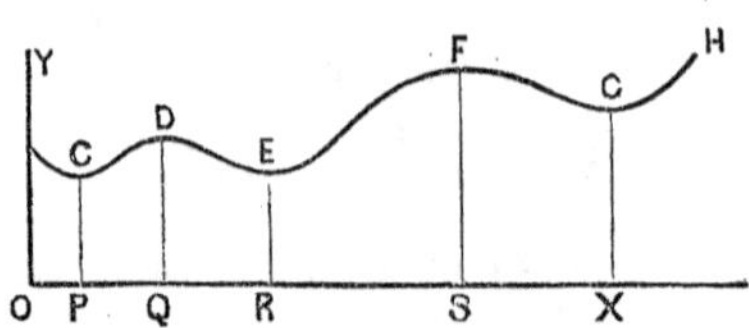

74. The substitution of a value $x = a$, derived from the equation $\frac{du}{dx} = 0$, in the succeeding differential coefficients, will sometimes cause the first of these coefficients which does not reduce to zero, to become infinite.

This happens only when the development of $F(x + h)$ in the ordinary form (by Taylor's Theorem) is not possible for that particular value of x. We must then find by other methods (such as algebraic development) the true value of the term which cannot be obtained by Taylor's Theorem. If it be found to contain a power of h, which will change sign with h, such as $h^{\frac{11}{3}}$ or $h^{\frac{1}{5}}$, the value of u will be neither a maximum nor a minimum; but if the power o h be such as will not change with h, as $h^{\frac{8}{3}}$ or $h^{\frac{12}{5}}$, the value of will be a maximum when that term is essentially negative, and a minimum when the term is essentially positive.

75. Finally, it may occur that when x has a particular value a, the first differential coefficient $\frac{du}{dx}$ will become infinite, and, therefore,

in order to complete the search for maximum and minimum values of u, we ought to solve the equation $\frac{du}{dx} = \infty$, and if a be a root of that equation, we must substitute $a + h$, and $a - h$ for x in $u = Fx$. Then if the term containing the lowest power of h be found to change sign with h, there will be neither maximum nor minimum; but if not, there will be a maximum when that term is negative, and a minimum when it is positive.

76. *Prop.* To determine the maximum and minimum values of an implicit function of a single variable x.

Let $F(x, y) = 0$ be the relation connecting x and y,

Put $$u = F(x,y) = 0; \quad \text{then} \quad \frac{dy}{dx} = -\frac{\frac{du}{dx}}{\frac{du}{dy}}.$$

But when y is a maximum or minimum, $\frac{dy}{dx} = 0$; $\therefore \frac{du}{dx} = 0$ also, and we have the two following conditions by which to determine the values of x and y, viz.:

$$\frac{du}{dx} = \frac{dF(x \cdot y)}{dx} = 0 \ldots (1), \quad \text{and} \quad u = F(x, y) = 0 \ldots. (2).$$

Having found the values of x and y which correspond to either a maximum or minimum, we distinguish one from the other by substituting the same values in the successive differential coefficients, and stopping at the first which does not reduce to zero. If this be negative, y will be a maximum; if positive, a minimum.

The successive differential coefficients are formed without difficulty from the value of $\frac{dy}{dx}$ already found, and their particular values, when $\frac{dy}{dx} = 0$, become much simplified.

Thus, put $\frac{du}{dx} = p_1$ $\frac{du}{dy} = q_1$ $\frac{d^2u}{dx_2} = p_2$, $\frac{d^2u}{dy_2} = q_2$, &c., and employ the [ ] to represent the particular values of the quantities enclosed, when $\frac{du}{dx} = p_1 = 0$. Then observing that $p_1\, q_1$ &c., are usually functions of both x and y, we have

$$\frac{dy}{dx} = -\frac{p_1}{q_1} = 0. \quad \therefore p_1 = 0.$$

$$\frac{d^2y}{dx^2} = -\frac{q_1\left(\frac{dp_1}{dx} + \frac{dp_1}{dy}\cdot\frac{dy}{dx}\right) - p_1\left(\frac{dq_1}{dx} + \frac{dq_1}{dy}\cdot\frac{dy}{dx}\right)}{q_1^{\,2}}$$

$$= -\frac{q_1\left(p_2 - \frac{dp_1}{dy}\cdot\frac{p_1}{q_1}\right) - p_1\left(q_2 - \frac{dq_1}{dy}\cdot\frac{p_1}{q_1}\right)}{q_1^{\,2}}$$

$$\therefore \left[\frac{d^2y}{dx^2}\right] = -\frac{[p_2]}{[q_1]} = -\frac{\left[\frac{d^2u}{dx^2}\right]}{[q_1]}.$$

And in a similar manner the higher differential coefficients can be formed, although the operation is more laborious.

77. The following considerations will facilitate the application of the preceding principles to particular examples:

1st. If a quantity which is a maximum or minimum contain a constant factor, that factor may be omitted and the result will still be a maximum or minimum.

2d. If u be a maximum or minimum, then $u \pm a$ is also a maximum or minimum, but $a - u$ will be a minimum when u is a maximum, and a maximum when u is a minimum.

3d. If u be a maximum, $\frac{1}{u}$ will be a minimum; and if u be a minimum, $\frac{1}{u}$ will be a maximum.

4th. If u be a maximum or minimum and positive, then u^2, u^3, and in general u^n, will be a maximum or minimum where u is any positive integer: but if u be negative, u^2, u^4, and in general u^{2n}, will be a maximum when u is a minimum; and a minimum when u is a maximum.

5th. If u be a maximum or minimum and positive, $\log u$ will also be a maximum or minimum.

6th. If the power u^{2n} be a maximum or minimum, the root u is not necessarily either a maximum or minimum; for it may be imaginary; and even when $u^{2n} = 0$ and a maximum, the corresponding root $u = 0$, although real, is not admissible as a maximum, because the adjacent values of u are imaginary.

7th. The value $x = \infty$ cannot correspond to a maximum or minimum value of u, because x cannot have a preceding and a succeeding value; but $u = \infty$ may be a maximum provided the preceding and succeeding values of u have like signs.

8th. In determining whether u is a maximum or minimum by the sign of $\frac{d^2u}{dx^2}$, when $\frac{du}{dx}$ has the form of a product $v_1 . v_2 . v_3 \ldots . v_n$, and $x = a$ causes one factor v_n to become equal to zero, the only term in $\frac{d^2u}{dx^2}$ necessary to be examined is that involving $\frac{dv_n}{dx}$, since the other terms disappear with v_n.

78. 1. To determine the values of the variable x which render the function $u = 6x + 3x^2 - 4x^3$ a maximum or minimum, and the corresponding values of the function u.

Here $\quad u = 6x + 3x^2 - 4x^3. \quad \therefore \frac{du}{dx} = 6 + 6x - 12x^2 = 0,$

or $\quad x^2 - \frac{1}{2}x = \frac{1}{2}. \quad \therefore x = \frac{1}{4} \pm \frac{3}{4} = +1 \text{ or } -\frac{1}{2}.$

Hence if u have maximum or minimum values, they must occur when $x = 1$ or when $x = -\frac{1}{2}$.

To discover whether these values are maxima or minima, we form the second differential coefficient: thus

$$\frac{d^2u}{dx^2} = 6 - 24x = 6 - 24 = -18 \quad \text{when} \quad x = 1$$

$$= 6 + 12 = +18 \quad \text{when} \quad x = -\frac{1}{2}.$$

$\therefore$ when $\qquad x = 1, \quad u = 6 + 3 - 4 = 5$ a maximum,

when $\qquad x = -\frac{1}{2}, \quad u = -3 + \frac{3}{4} + \frac{1}{2} = -\frac{7}{4}$ a minimum.

2. $u = x^4 - 8x^3 + 22x^2 - 24x + 12$, a maximum or minimum.

$$\frac{du}{dx} = 4x^3 - 24x^2 + 44x - 24 = 0 \quad \text{or} \quad x^3 - 6x^2 + 11x - 6 = 0.$$

The value $x = 1$ is obviously a root of this equation, and by dividing the first member by $x - 1$ we have for the depressed equation

$$x^2 - 5x + 6 = 0. \quad \therefore x = 2, \quad \text{or} \quad x = 3.$$

Hence the values requiring examination are

$$x = 1, \quad x = 2, \quad \text{and} \quad x = 3.$$

But $\qquad \frac{d^2u}{dx^2} = 12x^2 - 48x + 44 = +8$ when $x = 1$,

$\qquad\qquad = -4$ when $x = 2$,

$\qquad\qquad = +8$ when $x = 3$.

. when $\qquad x = 1, \quad u = 3$ a minimum,

when $\qquad x = 2, \quad u = 4$ a maximum,

when $\qquad x = 3, \quad u = 3$ a minimum.

3. $\qquad u = x^5 - 5x^4 + 5x^3 + 1$ a maximum or minimum

$$\frac{du}{dx} = 5x^4 - 20x^3 + 15x^2 = 0 \quad \text{or} \quad x^4 - 4x^3 + 3x^2 = 0 \; . \; . \; . \; (1).$$

$$\therefore x^2 = 0, \quad \text{or} \quad x^2 - 4x + 3 = 0,$$

and the four roots of (1) are 0, 0, 1, and 3.

$\frac{d^2u}{dx^2} = 20x^3 - 60x^2 + 30x = 0$ when $x = 0$; $\left(\therefore \text{let us examine } \frac{d^3u}{dx^3}\right)$.

$\qquad = -10$ " $x = 1$; then, $u = 2$, a max.

$\qquad = +90$ " $x = 3$; then, $u = -26$, a min.

$$\frac{d^3u}{dx^3} = 60x^2 - 120x + 30 = 30 \quad \text{when} \quad x = 0.$$

$\therefore u = 1$ is neither a maximum nor a minimum.

79. In each of the preceding examples, the condition $\frac{du}{dx} = \infty$, renders $x = \infty$, and therefore not applicable to a maximum or minimum.

Remark. In forming the second differential coefficient, it will save labor to omit any positive numerical factor common to every term of the first differential coefficient, and the sign of the second differential coefficient will not be affected by such omission.

80. *Ex.* 1. $u = \frac{(x+3)^3}{(x+2)^2}$ a maximum or minimum.

$$\frac{du}{dx} = \frac{3(x+3)^2(x+2) - 2(x+3)^3}{(x+2)^3} = 0, \quad \text{or,} \quad \frac{du}{dx} = \infty.$$

But, when $\frac{du}{dx} = 0$ we have $3(x+3)^2(x+2) - 2(x+3)^3 = 0$.

$\therefore x + 3 = 0$, or, $3(x+2) = 2(x+3)$, $\therefore x = -3$, or, $x = 0$.

$$\frac{d^2u}{dx^2} = \frac{6(x+3)(x+2)^2 - 12(x+3)^2(x+2) + 6(x+3)^3}{(x+2)^4}$$

$$= \frac{9}{8} \text{ when } x = 0, \text{ and } \therefore u = \frac{27}{4} \text{ a minimum.}$$

$$= 0 \quad \text{"} \quad x = -3.$$

Now, without actually forming the 3d differential coefficient, it is easily seen that it will contain one term (and only one) which will not reduce to zero when $x = -3$; and, therefore, the corresponding value of u is neither a maximum nor minimum.

The value $x = 0$, gives $u = \frac{27}{4}$ a minimum.

Now taking the equation $\frac{du}{dx} = \frac{3(x+3)^3(x+2) - 2(x+3)^3}{(x+2)^3} = \infty$,

we get $\quad x + 2 = 0$, or, $x = -2$,

and by putting successively $x = -2 + h$ and $x = -2 - h$, in the value of the original function u, there results

$$u_2 = F(a+h) = \frac{(-2+h+3)^3}{(-2+h+2)^2} = \frac{(1+h)^3}{h^2}$$

$$u_3 = F(a-h) = \frac{(-2-h+3)^3}{(-2-h+2)^2} = \frac{(1-h)^3}{h^2}$$

and since both of these are positive values, and less than that corresponding to $x = -2$, we have $u = \infty$ a maximum.

2. $u = \dfrac{(x-1)^2}{(x+1)^3}$, a maximum or minimum.

$$\frac{du}{dx} = \frac{2(x-1)(x+1)^3 - 3(x+1)^2(x-1)^2}{(x+1)^6} = 0, \text{ or, } \frac{du}{dx} = \infty,$$

$$\therefore\ x-1=0, \text{ or, } 2(x+1) - 3(x-1) = 0, \text{ or, } x+1=0.$$

$$\therefore\ x = 1, \text{ or, } x = 5, \text{ or, } x = -1.$$

$$\frac{d^2u}{dx^2} = \frac{2(x+1)^2 - 12(x+1)(x-1) + 12(x-1)^2}{(x+1)^5}$$

$$\therefore\ \frac{d^2u}{dx^2} = \frac{1}{4} \text{ when } x = 1, \text{ and } u = 0, \text{ a minimum.}$$

$$= -\frac{1}{324} \text{ when } x = 5, \text{ and } u = \frac{2}{27} \text{ a maximum.}$$

When $x = -1$, $u = \infty$, which is neither a maximum nor a minimum, for

$$u_2 = F(a+h) = \frac{(-2+h)^2}{(+h)^3} > 0,$$

but

$$u_3 = F(a-h) = \frac{(-2-h)}{(-h)^3} < 0,$$

3. $u = b + (x-a)^{\frac{3}{2}}$, a maximum or minimum.

$$\frac{du}{dx} = \frac{3}{2}(x-a)^{\frac{1}{2}} = 0, \quad \therefore\ x = a, \text{ and } u = b.$$

But $\dfrac{d^2u}{dx^2} = \dfrac{3}{4}(x-a)^{-\frac{1}{2}} = \infty$, when $x = a$.

Hence we cannot develop by Taylor's Theorem. Put $a \pm h$, for x in the value of u.

$$\therefore\ u_2 = b + (a + h - a)^{\frac{3}{2}} = b + (+h)^{\frac{3}{2}} > b,$$

and $$u_3 = b + (a - h - a)^{\frac{3}{2}} = b + (-h)^{\frac{3}{2}}.$$

This last value is imaginary, and therefore, $u = b$ is neither a maximum nor a minimum.

4. $u = b + (x - a)^{\frac{4}{3}}$, a maximum or minimum.

$$\frac{du}{dx} = \frac{4}{3}(x - a)^{\frac{1}{3}} = 0, \quad \therefore\ x = a, \quad \text{and} \quad u = b.$$

$$\frac{d^2u}{dx^2} = \frac{4}{9}(x - a)^{-\frac{2}{3}} = \infty, \text{ when } x = a. \quad \text{Then put } x = a \pm h.$$

$$\therefore\ u_2 = b + (a + h - a)^{\frac{4}{3}} = b + (+h)^{\frac{4}{3}} > b,$$

and $$u_3 = b + (a - h - a)^{\frac{4}{3}} = b + (-h)^{\frac{4}{3}} > b, \text{ also.}$$

$$\therefore\ x = a \text{ gives } u = b, \text{ a minimum.}$$

5. $u = b - (a - x)^{\frac{8}{5}}$, a maximum or minimum.

$$\frac{du}{dx} = \frac{8}{5}(a - x)^{\frac{3}{5}} = 0, \quad \therefore\ x = a, \quad \text{and} \quad u = b.$$

$$\frac{d^2u}{dx^2} = -\frac{24}{25}(a - x)^{-\frac{2}{5}} = -\infty, \text{ when } x = a. \quad \text{Then put } x = a \pm h.$$

$$\therefore\ u_2 = b - (-h)^{\frac{8}{5}} < b, \quad \text{and} \quad u_3 = b - (+h)^{\frac{8}{5}} < b, \text{ also.}$$

$$\therefore\ x = a \quad \text{gives} \quad u = b, \text{ a maximum.}$$

6. $u = b + (x - a)^{\frac{5}{3}} + c(x - a)^2$, a maximum or minimum.

$$\frac{du}{dx} = \frac{5}{3}(x - a)^{\frac{2}{3}} + 2c(x - a) = 0.$$

$$\therefore\ x - a = 0, \quad \text{or,} \quad \frac{5}{3} + 2c(x - a)^{\frac{1}{3}} = 0.$$

$$\therefore x = a, \text{ and } u = b, \text{ or, } x = a - \frac{125}{216c^3}, \text{ and } u = b - \frac{3125}{46656c^5}.$$

$$\text{But } \frac{d^2u}{dx^2} = \frac{10}{9}(x-a)^{-\frac{1}{3}} + 2c = \infty, \quad \text{when } x = a$$

$$= \frac{2}{3}c > 0 \text{ when } x = a - \frac{125}{216c^3}.$$

Hence when $x = a - \dfrac{125}{216c^3}$, we have $u = b - \dfrac{3125}{46656c^5}$, a minimum.

In order to examine the value of $x = a$, put $a \pm h$ for x in the original value of u.

$$\therefore u_2 = b + (+h)^{\frac{5}{3}} + c(+h)^2 > b, \quad u_3 = b + (-h)^{\frac{5}{3}} + c(-h)^2 < b.$$

$\therefore u = b$ is neither a maximum nor a minimum.

7. To inscribe the greatest rectangle in a given circle.

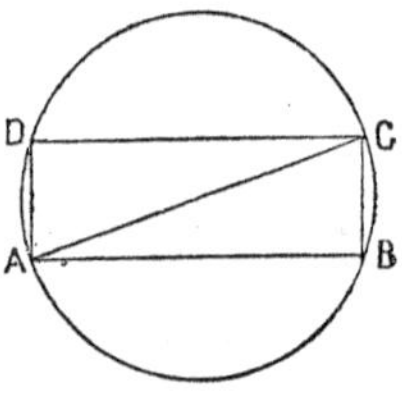

Put the diameter $AC = a$, and the side $AB = x$; then

$$AD = \sqrt{a^2 - x^2} \text{ and } AB \times AD = x\sqrt{a^2 - x^2}.$$

$$\therefore u = (AB \times AD)^2 = x^2(a^2 - x^2) = \text{a max.}$$

$$\therefore \frac{du}{dx} = 2a^2x - 4x^3 = 0. \quad \therefore x = 0, \quad \text{or} \quad x = a\sqrt{\frac{1}{2}}$$

$$\frac{d^2u}{dx^2} = 2a^2 - 12x^2 = 2a^2 \quad \text{when} \quad x = 0$$

$$= -4a^2 \quad \text{“} \quad x = a\sqrt{\frac{1}{2}}.$$

$$\therefore AD = \sqrt{a^2 - \frac{1}{2}a^2} = a\sqrt{\frac{1}{2}} = AB,$$

and the rectangle must be a square. Its area is $\frac{1}{2}a^2$.

8. To inscribe a maximum cylinder in a given right cone having a circular base.

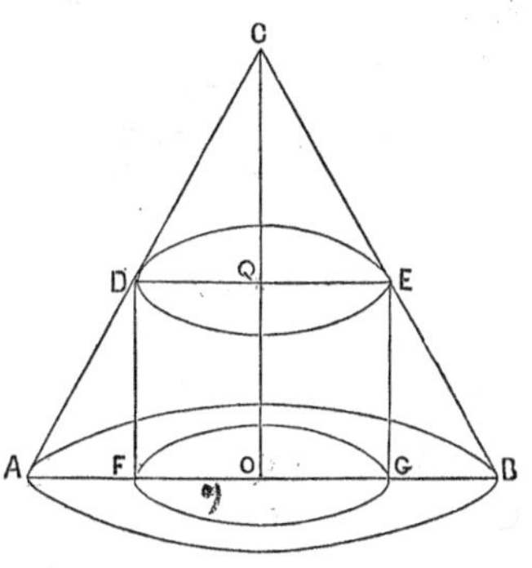

Put AO the radius of the cone's base $= b$, CO the altitude of the cone $= a$, DF the altitude of the cylinder $= x$.

Then from the similar triangles COA and CQD we have

$$CO : CQ :: OA : QD = \frac{CQ \times OA}{CO}$$

$$= \frac{(a-x)b}{a}.$$

$\therefore$ volume of cylinder $= \dfrac{\pi(a-x)^2 b^2 x}{a^2}$.

$$\therefore u = (a-x)^2 x = a^2 x - 2ax^2 + x^3 = \text{maximum}.$$

$$\therefore \frac{du}{dx} = a^2 - 4ax + 3x^2 = 0, \quad \text{or} \quad x^2 - \frac{4}{3}ax = -\frac{1}{3}a^2.$$

$$x = \frac{2}{3}a \pm \frac{1}{3}a = a \quad \text{or} \quad = \frac{1}{3}a.$$

$$\frac{d^2u}{dx^2} = -4a + 6x = 2a \quad \text{when} \quad x = a$$

$$= -2a \quad \text{"} \quad x = \frac{1}{3}a.$$

Hence the altitude of the cylinder is one-third of the cone, and consequently

$$\text{volume of cylinder} = \frac{4}{27}\pi ab^2 = \frac{4}{9} \text{ volume of cone}.$$

9. Find the greatest and least ordinates of the curve whose equation is $a^2y - ax^2 + x^3 = 0$.

Put $\quad u = a^2y - ax^2 + x^3 = 0 \; \ldots\ldots \; (1).$

Then $\quad \dfrac{du}{dx} = -2ax + 3x^2 = 0 \; \ldots\ldots \; (2).$

Combining (1) and (2), we get

$$x = 0 \quad \text{and} \quad y = 0, \quad \text{or} \quad x = \frac{2}{3}a \quad \text{and} \quad y = \frac{4}{27}a.$$

But $$\frac{du}{dy} = a^2, \quad \frac{d^2u}{dx^2} = -2a + 6x.$$

$$\therefore \left[\frac{d^2y}{dx^2}\right] = -\left[\frac{d^2u}{dx^2}\right] \div \left[\frac{du}{dy}\right] = \frac{2}{a} \quad \text{when} \quad x = 0$$

$$= -\frac{2}{a} \quad \text{"} \quad x = \frac{2}{3}a.$$

$\therefore$ When $x = 0$, $y = 0$, a min., and when $x = \frac{2}{3}a$, $y = \frac{4}{27}a$ = a max.

10. To find a number x such that its xth root shall be a maximum.

$$u = x^{\frac{1}{x}} = \text{a maximum.} \quad \frac{du}{dx} = x^{\frac{1}{x}-2}(1 - \log x) = 0.$$

$$\therefore x^{\frac{1}{x}-2} = 0 \quad \text{or} \quad 1 - \log x = 0.$$

The first of these equations give $x = 0$; the second $\log x = 1$; whence $x = e$ and $u = e^{\frac{1}{e}}$ = maximum.

In this and in many similar examples, we may draw the final inference without forming the second differential coefficient, it being obvious from the nature of the question that there is one, and only one maximum, and it being easy to decide which of the values of x is that applicable to the maximum.

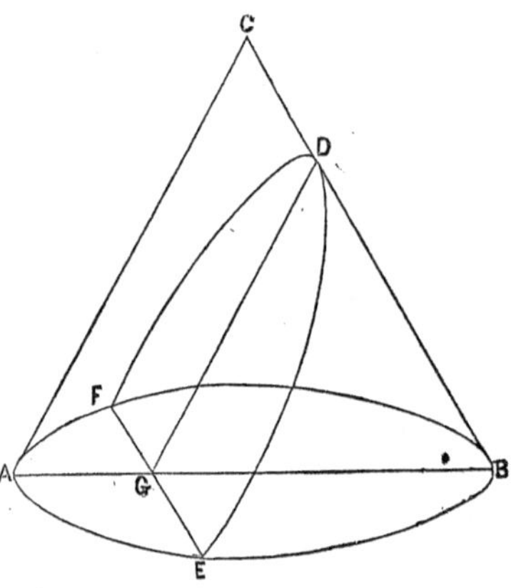

11. To cut the greatest parabola from a given right cone having a circular base.

Put AB the diameter of the base $= a$, AC the slant height $= b$, and $BG = x$.

Then $AG = a - x$, and by the property of the circle,

$$FE = 2FG = 2\sqrt{x(a - x)}.$$

Also by the similar triangles BAC and BGD, we have

$$BA : AC :: BG : GD = \frac{AC \times BG}{BA} = \frac{bx}{a}.$$

But the area of the parabola

$$FDE = \frac{2}{3} FE \times GD = \frac{4}{3} \cdot \frac{bx}{a} \sqrt{ax - x^2}.$$

$\therefore\ u = ax^3 - x^4 = \text{max.}$; $\frac{du}{dx} = 3ax^2 - 4x^3 = 0$, and $x = 0$ or $x = \frac{3}{4} a$,

the second value being obviously that required, since when $x = 0$ the area of the parabola $= 0$.

$\therefore$ area of maximum parabola $= \frac{1}{4} ab\sqrt{3}$.

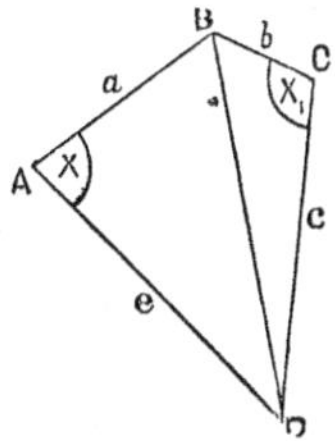

12. To form the greatest quadrilateral with four given lines taken in a given order.

Put $AB = a$, $BC = b$, $CD = c$, $DA = e$, angle $BAD = x$, and $BCD = x_1$, the latter angle x_1 being obviously a function of x, since the two are connected by the relation

$$[BD]^2 = a^2 + e^2 - 2ae \cos x = b^2 + c^2 - 2bc . \cos x_1 \ldots . (1).$$

But area $\quad ABCD = \Delta ABD + \Delta BCD = \frac{1}{2} ae \sin x + \frac{1}{2} bc \sin x_1$.

$\therefore u = ae \sin x + bc . \sin x_1 = \text{a max.}$, and $\frac{du}{dx} = ae \cos x + bc \cos x_1 \frac{dx_1}{dx} = 0$.

Now by differentiating (1), we have

$$ae . \sin x = bc . \sin x_1 \frac{dx_1}{dx}.$$

$$\therefore \frac{dx_1}{dx} = \frac{ae . \sin x}{bc . \sin x_1}. \quad \therefore \frac{du}{dx} = ae . \cos x + bc . \cos x_1 \frac{ae \sin x}{bc \sin x_1} = 0.$$

$$\therefore \sin x \cos x_1 + \sin x_1 \cos x = 0, \quad \text{or} \quad \sin (x + x_1) = 0.$$

$$\therefore x + x_1 = 0, \quad \text{or} \quad x + x_1 = 180°.$$

The latter is plainly the required solution, and consequently the quadrilateral must be such as can be inscribed in a circle.

13. To find the greatest quadrilateral that can be contained within a given perimeter.

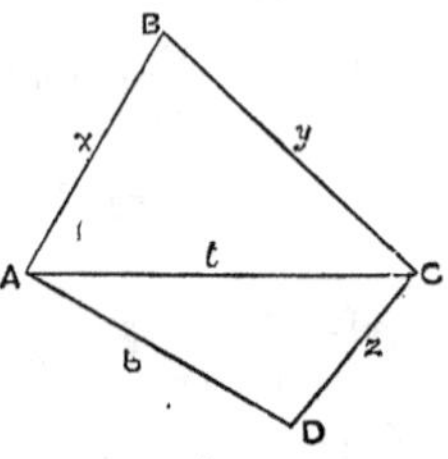

Suppose $ABCD$ to be the required figure, and suppose two of the sides x and y to vary, while the other two sides v and z and the diagonal t remain unchanged. Then, since $ABCD$ is supposed to be the greatest quadrilateral which can be formed with the given perimeter, the triangle ABC must be greater than any other triangle having the same base t, and the sum of the sides $= x + y = b$ a constant.

But if
$$x + y + t = s,$$

the area of the $\Delta ABC = \sqrt{\frac{1}{2}s\left(\frac{1}{2}s - x\right)\left(\frac{1}{2}s - y\right)\left(\frac{1}{2}s - t\right)}$

Therefore, by squaring and omitting the constant factors

$$\frac{1}{2}s \text{ and } \frac{1}{2}s - t, \text{ we have}$$

$$u = \left(\frac{1}{2}s - x\right)\left(\frac{1}{2}s - y\right) = \left(\frac{1}{2}s - x\right)\left(\frac{1}{2}s - b + x\right) = \text{a maximum.}$$

$$\therefore \frac{du}{dx} = \left(\frac{1}{2}s - x\right) - \left(\frac{1}{2}s - b + x\right) = 0, \text{ or } b - 2x = 0, \text{ and } \therefore x = \frac{1}{2}b.$$

$$\therefore \ y = b - x = x = \frac{1}{2}b.$$

that is, the sides AB and BC must be equal. Similarly it may be shown that $x = v$, $v = z$, $z = y$.

Hence the figure must be equilateral, and, consequently, either a rhombus or square. But, since the lengths of the sides are now given, the quadrilateral must admit of being inscribed in a circle.

$\therefore$ The figure must be a square.

14. To find the greatest figure of n sides contained within a given perimeter.

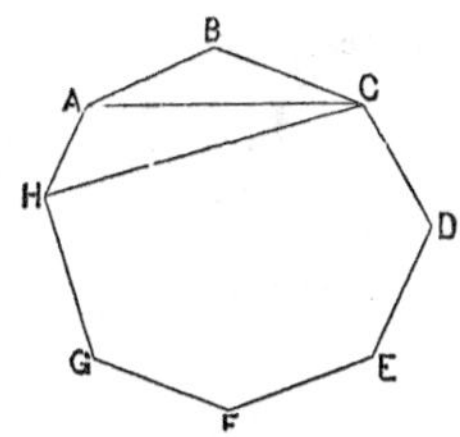

By supposing two sides AB and BC to vary, while the other sides remain fixed in magnitude and position, we prove, as in the last example, that $AB = BC$; and similarly that $BC = CD$, $CD = DE$, &c. Therefore the required figure must be equilateral.

Then, supposing the three equal sides HA, AB, and BC, to vary in position, while the other sides remain fixed, we show, as in a preceding example, that the circumference of a circle can be described through H, A, B, and C; and, similarly, that a circumference can be drawn through A, B, C, and D. But only one circumference can be drawn through the same three points A, B, and C. Therefore the same circumference passes through H, A, B, C, and D. And, similarly, it may be shown that this circumference passes through E, F, G, &c. $\therefore$ The polygon must be equiangular, and, consequently, regular.

15. To divide a line a into n parts, x, x_1, x_2, &c., and determine the relations between those parts when the continued product of their numerical values shall be a maximum.

Let two of these parts x and x_1 vary, while x_2, x_3, &c., remain constant.

Put $x_2 + x_3 + \text{\&c.} = b$, and $x_2 \times x_3 \times x_4 \text{ \&c.} = c$

Then $x + x_1 = a - b$, and $xx_1 \cdot x_2 \cdot x_3 \cdot \text{\&c.} = x(a - b - x)\,c$.

$\therefore\ u = x(a - b - x) = \text{a maximum}, \quad \dfrac{du}{dx} = a - b - 2x = 0$

$\therefore\ x = \dfrac{1}{2}(a - b), \text{ and } x_1 = a - b - x = \dfrac{1}{2}(a - b) = x.$

Similarly, $x_2 = x$, $x_3 = x$, &c., and, therefore, the parts are all equal.

16. To determine the number of equal parts into which a given number a must be divided, so that their continued product may be a maximum.

Let $x =$ required number of parts; then $\frac{a}{x} =$ value of one part.

$$\therefore \frac{a}{x} \times \frac{a}{x} \times \frac{a}{x} \text{ \&c. to } x \text{ factors} = \left(\frac{a}{x}\right)^x = \text{a maximum.}$$

$$\therefore u = \log\left(\frac{a}{x}\right)^x = x(\log a - \log x) = \text{a max. } \frac{du}{dx} = \log a - \log x - 1 = 0$$

$$\therefore \log\frac{a}{x} = 1. \quad \frac{a}{x} = e, \quad \text{and} \quad x = \frac{a}{e}.$$

This is a solution in the *arithmetical* sense only when a is a multiple of e, for otherwise n would not be an integer.

The general solution belongs to the following problem. To find a number x such that the xth power of $\frac{a}{x}$ shall be a maximum.

17. To determine the point P, in the line joining the centres C and C_1 of two unequal spheres, from which the greatest amount of spherical surface can be seen.

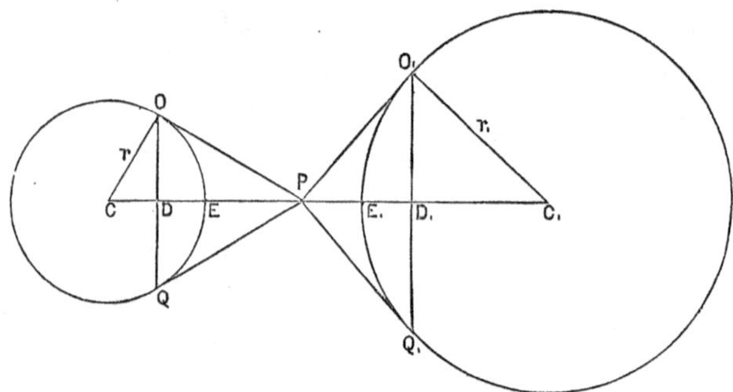

Put $CO = r$, $C_1O_1 = r_1$, $CC_1 = a$ $CP = x$, $C_1P = x_1 = a - x$.

$$\therefore CD = \frac{r^2}{x}, \; C_1D_1 = \frac{r_1^2}{x_1}, \; DE = r - \frac{r^2}{x} = \frac{r(x-r)}{x}$$

and similarly $$D_1E_1 = \frac{r_1(x_1 - r_1)}{x_1}.$$

$$\therefore \text{Surface of zone } OEQ = 2\pi r\frac{r(x-r)}{x}$$

$$O_1E_1Q_1 = 2\pi r_1\frac{r_1(x_1 - r_1)}{x_1}$$

$$\therefore u = \frac{r^2x - r^3}{x} + \frac{r_1{}^2x_1 - r_1{}^3}{x_1} = r^2 + r_1{}^2 - \frac{r^3}{x} - \frac{r_1{}^3}{a - x} = \max.$$

$$\therefore \frac{du}{dx} = \frac{r^3}{x^2} - \frac{r_1{}^3}{(a-x)^2} = 0 \quad \therefore \frac{r^{\frac{3}{2}}}{x} = \frac{r_1^{\frac{3}{2}}}{a - x}$$

$$\therefore x = \frac{ar^{\frac{3}{2}}}{r^{\frac{3}{2}} + r_1^{\frac{3}{2}}}, \text{ and surface seen} = 2\pi\left[r^2 + r_1{}^2 - \frac{(r^{\frac{3}{2}} + r_1^{\frac{3}{2}})^2}{a}\right],$$

which is always less than the entire surface of the two spheres.

18. A right prism, whose base is a regular hexagon, is truncated by three planes drawn through the alternate vertices of the upper base, and intersecting at a common point in the axis prolonged. Required the inclinations of the planes to the axis, when the truncate prism shall (with a given volume) be contained under the least surface.

Let $ABCDEF$ be the lower base of the prism, and $abcdf$, the upper base.

Join fb, bd, and df, and through these lines draw planes intersecting the axis Rr prolonged at some point v.

The plane fvb intersects the edge Aa at a_1, cutting off from the prism, the triangular pyramid $fbaa_1$. From r, the centre of the upper base, draw rf, ra, and rb. Then $fabr$ is a rhombus, whose diagonals bisect each other at o perpendicularly. Join va_1; it will be perpendicular to fb, and will pass through o. Then $ao = or$, and $\therefore aa_1 = rv$.

$\therefore$ Pyramid $fbaa_1$, is equal to the pyramid $fbrv$.

$\therefore$ The volume of the prism, when terminated by the three planes which intersect at v, is equal to the volume of the original prism, for all inclinations of the planes.

Put $\quad Bb = a, \quad AB = b, \quad$ the angle $rvo = x$.

Then $\quad ro = oa = \frac{1}{2}b, \quad aa_1 = \frac{1}{2}b\cot x, \quad oa_1 = ov = \frac{1}{2}b\operatorname{cosec} x,$

$$of = ob = \frac{1}{2}b\sqrt{3}, \quad Aa_1 = a - \frac{1}{2}b\cot x.$$

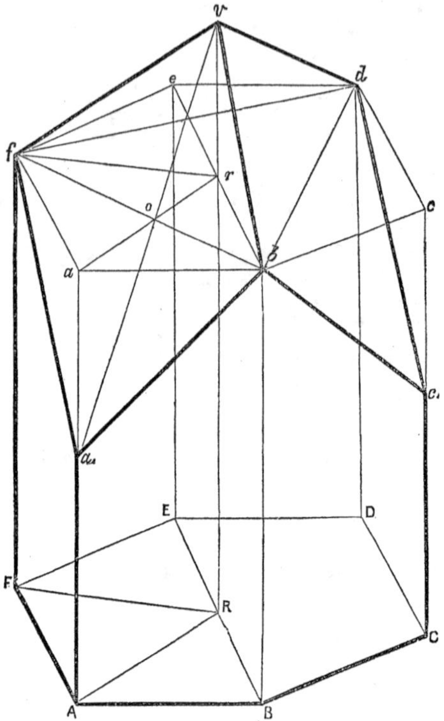

$$\therefore \text{ Surface } ABba_1 = \frac{1}{2}b\,(2a - \frac{1}{2}b\,.\cot x).$$

$$\text{Surface } a_1bvf = fb \times vo = \frac{1}{2}b^2\sqrt{3}\operatorname{cosec} x.$$

Hence by the nature of the question, we shall have

$$6ABba_1 + 3a_1bvf = 3b(2a - \frac{1}{2}b\cot x) + \frac{3}{2}b^2\sqrt{3}\operatorname{cosec} x = \text{a min.}$$

$$\therefore\ u = 2a - \frac{1}{2}b\cot x + \frac{1}{2}b\sqrt{3}\operatorname{cosec} x = \text{a minimum.}$$

$$\frac{du}{dx} = \frac{1}{2}b(\operatorname{cosec}^2 x - \sqrt{3}\operatorname{cosec} x\cot x) = 0. \quad \therefore \cos x = \frac{1}{\sqrt{3}}$$

$$\therefore\ x = 54^\circ\ 44'\ 08''.$$

This is the celebrated problem relating to the form of the cells of the bee.

CHAPTER IX.

FUNCTIONS OF TWO INDEPENDENT VARIABLES.

81. Hitherto it has been supposed that the function u depended, either directly or indirectly, on a single variable x. But the value of u may depend on the values of two or more variables, entirely independent of each other. Thus, if there were given

$$u = xy + y^2, \quad \ldots \ldots (1).$$

we might suppose x to vary and y to be constant; or y to be variable and x constant; or, lastly, x and y may vary simultaneously. These three suppositions lead to three essentially different changes in the function u.

Thus when x becomes $x + h$, and y is constant, u becomes

$$u_1 = xy + hy + y^2.$$

When y becomes $y + k$, and x is constant, u becomes

$$u_2 = xy + xk + y^2 + 2yk + k^2.$$

And finally, when x and y become respectively $x + h$ and $y + k$, u becomes $u_3 = xy + hy + kx + y^2 + 2ky + k^2 + hk.$

The general case is presented in the following proposition.

82. *Prop.* Having given $u = F(x, y) \ldots . (1).$, to develop $u_3 = F(x + h \;\; y + k)$, the variables x and y being independent of each other.

Since x and y are supposed to have no mutual dependence, they may be supposed to vary successively.

Let x take an increment h; then u becomes $u_1 = F(x + h, y)$ which, developed as a function of $x + h$ by Taylor's Theorem, gives

$$u_1 = u + \frac{du}{dx}\cdot\frac{h}{1} + \frac{d^2u}{dx^2}\cdot\frac{h^2}{1.2} + \frac{d^3u}{dx^3}\cdot\frac{h^3}{1.2.3} + \&c. \quad (2).$$

in which u, $\frac{du}{dx}$, $\frac{d^2u}{dx^2}$, &c., are functions of both x and y.

Now, if in every term of (2), we replace y by $y + k$, we shall convert $u_1 = F(x + h, y)$ into $u_3 = F(x + h, y + k)$, and, since each term in the second member of (2) will then be a function of $y + k$, we must replace

$$u \text{ by } u + \frac{du}{dy}\cdot\frac{k}{1} + \frac{d^2u}{dy^2}\cdot\frac{k^2}{1.2} + \&c.$$

$$\frac{du}{dx} \text{ by } \frac{du}{dx} + \frac{d\frac{du}{dx}}{dy}\cdot\frac{k}{1} + \frac{d^2\frac{du}{dx}}{dy^2}\cdot\frac{k^2}{1.2} + \&c.$$

$$\frac{d^2u}{dx^2} \text{ by } \frac{d^2u}{dx^2} + \frac{d\frac{d^2u}{dx^2}}{dy}\cdot\frac{k}{1} + \frac{d^2\frac{d^2u}{dx^2}}{dy^2}\cdot\frac{k^2}{1.2} + \&c.$$

$$\frac{d^3u}{dx^3} \text{ by } \frac{d^3u}{dx^3} + \frac{d\frac{d^3u}{dx^3}}{dy}\cdot\frac{k}{1} + \frac{d^2\frac{d^3u}{dx^3}}{dy^2}\cdot\frac{k^2}{1.2} + \&c.$$

But we put for convenience $\frac{d\frac{du}{dx}}{dy} = \frac{d^2u}{dxdy}$, indicating thereby that two differentiations of u have been performed, the first with respect to x, and the second with respect to y. Similarly we put $\frac{d^2\frac{du}{dx}}{dy^2} = \frac{d^3u}{dxdy^2}$, and $\frac{d\frac{d^2u}{dx^2}}{dy} = \frac{d^3u}{dx^2dy}$; the first expression indicating one differentiation with respect to x, followed by two with respect to y; and the second implying two differentiations with regard to x, followed by one with regard to y. And generally, we denote the result of n differentiations with respect to x, followed by m differentiations with respect to y, by the symbol.

$$\frac{d^{n+m}u}{dx^ndy^m}.$$

Now let the necessary substitutions be made in (2), and we shall get

$$u_3 = u + \frac{du}{dx}\cdot\frac{h}{1} + \frac{du}{dy}\cdot\frac{k}{1} + \frac{d^2u}{dx^2}\cdot\frac{h^2}{1.2} + \frac{d^2u}{dxdy}\cdot\frac{hk}{1} + \frac{d^2u}{dy^2}\cdot\frac{k^2}{1.2}$$

$$+ \frac{d^3u}{dx^3}\cdot\frac{h^3}{1.2.3} + \frac{d^3u}{dx^2dy}\cdot\frac{h^2k}{1.2} + \frac{d^3u}{dxdy^2}\cdot\frac{hk^2}{1.2} + \frac{d^3u}{dy^3}\cdot\frac{k^3}{1.2.3} + \&c.$$

which is the proposed expansion.

If we had supposed the variable y to receive its increment first, we should have obtained the following series for u_3.

$$u_3 = u + \frac{du}{dy}\cdot\frac{k}{1} + \frac{du}{dx}\cdot\frac{h}{1} + \frac{d^2u}{dy^2}\cdot\frac{k^2}{1.2} + \frac{d^2u}{dydx}\cdot\frac{kh}{1} + \frac{d^2u}{dx^2}\cdot\frac{h^2}{1.2}$$

$$+ \frac{d^3u}{dy^3}\cdot\frac{k^3}{1.2.3} + \frac{d^3u}{dy^2dx}\cdot\frac{k^2h}{1.2} + \frac{d^3u}{dydx^2}\cdot\frac{kh^2}{1.2} + \frac{d^3u}{dx^3}\cdot\frac{h^3}{1.2.3} + \&c.$$

The two series must obviously give equal results, and being true for all values of h and k, the coefficients of the like powers and products of h and k must be equal.

$$\therefore \frac{d^2u}{dxdy} = \frac{d^2u}{dydx}, \quad \frac{d^3u}{dx^2dy} = \frac{d^3u}{dydx^2}, \quad \frac{d^3u}{dxdy^2} = \frac{d^3u}{dy^2dx} \ \&c.$$

Hence the result of n differentiations with respect to x, followed by m differentiations with respect to y, will be the same as that produced by performing the differentiations in a contrary order.

EXAMPLES OF DIFFERENTIAL COEFFICIENTS.

83. 1. $$u = x^3y + ay^2.$$

$$\frac{du}{dx} = 3x^2y, \quad \frac{du}{dy} = x^3 + 2ay, \quad \frac{d^2u}{dx^2} = 6xy, \quad \frac{d^2u}{dy^2} = 2a$$

$$\frac{d^2u}{dxdy} = 3x^2 \text{ and } \frac{d^2u}{dydx} = 3x^2 \text{ also.}$$

$$\frac{d^3u}{dx^3} = 6y, \quad \frac{d^3u}{dy^3} = 0, \quad \frac{d^3u}{dx^2dy} = 6x = \frac{d^3u}{dydx^2}, \quad \frac{d^3u}{dxdy^2} = 0 = \frac{d^3u}{dy^2dx},$$

$$\frac{d^4u}{dx^4} = 0, \quad \frac{d^4u}{dx^3dy} = 6 = \frac{d^4u}{dydx^3}, \quad \frac{d^4u}{dx^2dy^2} = 0 = \frac{d^4u}{dy^2dx^2}, \ \&c., \&c.$$

8

2. $u = \tan^{-1}\frac{x}{y}.\quad \frac{du}{dx} = \frac{y}{x^2+y^2},\quad \frac{du}{dy} = -\frac{x}{x^2+y^2},$

$$\frac{d^2u}{dxdy} = \frac{x^2-y^2}{(x^2+y^2)^2} = \frac{d^2u}{dydx}, \text{ \&c., \&c.}$$

3. $u = \sin x . \cos y.\quad \frac{du}{dx} = \cos x . \cos y\ \frac{du}{dy} = -\sin x . \sin y.$

$$\frac{d^2u}{dxdy} = -\cos x . \sin y = \frac{d^2u}{dydx},\quad \frac{d^2u}{dx^2} = -\sin x \cos y$$

$$\frac{d^3u}{dx^2dy} = \sin x \sin y = \frac{d^3u}{dydx^2} = \frac{d^3u}{dxdydx} \text{ \&c.}$$

In general the order of the differentiations is immaterial, provided we always differentiate the same number of times with respect to the same variable.

The expressions $\frac{du}{dx}$ and $\frac{du}{dy}$ are called *partial differential coefficients:* $\frac{du}{dx}dx$ and $\frac{du}{dy}dy$ are called *partial* differentials, and $du = \frac{du}{dx}dx + \frac{du}{dy}dy$ is the *total* differential of u.

84. Similarly, if $u = F(x, y, z)$, where x, y, and z, are independent variables, then

$$du = \frac{du}{dx}dx + \frac{du}{dy}dy + \frac{du}{dz}dz.$$

And generally, to differentiate a function of several independent variables, we must differentiate successively with respect to each, and add the results.

85. If it were proposed to develop $u_1 = F(x + h, y + k, z + l)$, where $u = F(x, y, z)$, we should obtain, by supposing x, y, and z to vary, and reasoning as in the expansion of $F(x + h, y + k)$,

$$u_1 = u + \frac{du}{dx}\cdot\frac{h}{1} + \frac{du}{dy}\cdot\frac{k}{1} + \frac{du}{dz}\cdot\frac{l}{1} + \frac{d^2u}{dx^2}\cdot\frac{h^2}{1.2} + \frac{d^2u}{dxdy}\cdot\frac{hk}{1} + \frac{d^2u}{dy^2}\cdot\frac{k^2}{1.2}$$

$$+ \frac{d^2u}{dxdz}\cdot\frac{hl}{1} + \frac{d^2u}{dz^2}\cdot\frac{l^2}{1.2} + \frac{d^2u}{dydz}\cdot\frac{kl}{1} + \frac{d^3u}{dx^3}\cdot\frac{h^3}{1.2.3}$$

$$+ \frac{d^3u}{dx^2dy}\cdot\frac{h^2k}{1.2} + \frac{d^3u}{dxdy^2}\cdot\frac{hk^2}{1.2} + \frac{d^3u}{dy^3}\cdot\frac{k^3}{1.2.3} + \frac{d^3u}{dx^2dz}\cdot\frac{h^2l}{1.2} + \text{\&c.}$$

Remark. The formula $du = \frac{du}{dx}dx + \frac{du}{dy}dy + \frac{du}{dz}dz +$ &c., for differentiating a function of several variables, may be deduced immediately from the preceding development.

For put $k = rh$, $l = r_1h$, &c. where r, r_1 &c. are arbitrary, since x, y, z, &c., are independent of each other. Then by substitution and reduction,

$$\frac{u_1 - u}{h} = \frac{du}{dx} + r\frac{du}{dy} + r_1\frac{du}{dz} \text{ &c.} + \text{terms in } h, h^2, \text{ &c.}$$

and by passing to the limit, making $h = 0$, neglecting terms containing h, h^2, &c., and finally making

$u_1 - u = du$, $h = dx$, $rh = k = dy$, $r_1h = l = dz$, &c., we get

$$du = \frac{du}{dx}dx + \frac{du}{dy}dy + \frac{du}{dz}dz + \text{&c.}$$

86. *Prop.* To differentiate successively $u = F(x.y)$.

We have already found the first differential

$$du = \frac{du}{dx}dx + \frac{du}{dy}dy.$$

Differentiating this and observing that $\frac{du}{dx}$, and $\frac{du}{dy}$ are usually functions of both x and y, but that dx and dy are constant, we get

$$d^2u = \frac{d^2u}{dx^2}dx^2 + \frac{d^2u}{dxdy}\cdot dxdy + \frac{d^2u}{dydx}\cdot dydx + \frac{d^2u}{dy^2}\cdot dy^2,$$

or $$d^2u = \frac{d^2u}{dx^2}dx^2 + 2\frac{d^2u}{dxdy}\cdot dxdy + \frac{d^2u}{dy^2}\cdot dy^2:$$

and by differentiating again, we have

$$d^3u = \frac{d^3u}{dx^3}dx^3 + 3\frac{d^3u}{dx^2dy}\cdot dx^2dy + 3\frac{d^3u}{dxdy^2}dxdy^2 + \frac{d^3u}{dy^3}\cdot dy^3.$$

$$d^4u = \frac{d^4u}{dx^4}\cdot dx^4 + 4\frac{d^4u}{dx^3dy}\cdot dx^3dy + 6\frac{d^4u}{dx^2dy^2}dx^2dy^2 + 4\frac{d^4u}{dxdy^3}dxdy^3$$
$$+ \frac{d^4u}{dy^4}dy^4.:$$

and similarly may d^5u, d^6u, &c., be found, the numerical coefficients of the several terms proving the same as in the powers of the binomial.

Implicit Functions of two Independent Variables.

87. *Prop.* Let $F(x, y, z)=0$, so that z shall be an implicit function of the two independent variables x and y, and let it be proposed to form expressions dz, d^2z, &c., without solving the equation with respect to z.

Put $u = F(x, y, z) = 0$; then, observing that u is directly a function of the independent variables x and y, and also indirectly a function of x and y through z, we shall have for the total differential coefficient $\left[\frac{du}{dx}\right]$ and $\left[\frac{du}{dy}\right]$

$$\left[\frac{du}{dx}\right]=\frac{du}{dz}\cdot\frac{dz}{dx}+\frac{du}{dx}=0\ldots(1).\text{ and }\left[\frac{du}{dy}\right]=\frac{du}{dz}\cdot\frac{dz}{dy}+\frac{du}{dy}=0\ldots(2).$$

$$\therefore\ \frac{dz}{dx}=-\frac{\frac{du}{dx}}{\frac{du}{dz}},\quad \frac{dz}{dy}=-\frac{\frac{du}{dy}}{\frac{du}{dz}}.$$

$$\therefore\ dz=\frac{dz}{dx}dx+\frac{dz}{dy}\cdot dy=-\frac{\frac{du}{dx}}{\frac{du}{dz}}dx-\frac{\frac{du}{dy}}{\frac{du}{dz}}dy.$$

Next to form d^2z, we have

$$d^2z=\frac{d^2z}{dx^2}\cdot dx^2+2\frac{d^2z}{dxdy}\cdot dxdy+\frac{d^2z}{dy^2}dy^2.$$

But by differentiating (1) with respect to x, (2) with respect to y, and (1) or (2) with regard to y or x, respectively, and observing that $\frac{du}{dx}, \frac{du}{dy}, \frac{du}{dz}$, are functions of x, y, and z, we get

$$2\frac{d^2u}{dxdz}\cdot\frac{dz}{dx}+\frac{d^2u}{dz^2}\cdot\frac{dz^2}{dx^2}+\frac{du}{dz}\cdot\frac{d^2z}{dx^2}+\frac{d^2u}{dx^2}=0$$

$$2\frac{d^2u}{dydz}\cdot\frac{dz}{dy}+\frac{d^2u}{dz^2}\cdot\frac{dz^2}{dy^2}+\frac{du}{dz}\cdot\frac{d^2z}{dy^2}+\frac{d^2u}{dy^2}=0$$

$$\frac{d^2u}{dydz}\cdot\frac{dz}{dx}+\frac{d^2u}{dz^2}\cdot\frac{dz}{dy}\cdot\frac{dz}{dx}+\frac{du}{dz}\cdot\frac{d^2z}{dxdy}+\frac{d^2u}{dxdy}+\frac{d^2u}{dxdz}\cdot\frac{dz}{dy}=0.$$

whence $\frac{d^2z}{dx^2}$, $\frac{d^2z}{dy^2}$, and $\frac{d^2z}{dxdy}$ may be found in terms of the partial differential coefficients of the first and second orders of u, with respect to x, y, and z, all of which are easily formed.

88. *Prop.* Having given $u = \phi z$, and $z = F(x, y)$, to differentiate u without previously eliminating z.

If we suppose x alone to increase, it will impart a change to u through z; and a similar change will be transmitted to u, when y alone varies; thus we shall have

$$\frac{du}{dx}=\frac{du}{dz}\cdot\frac{dz}{dx} \quad\text{and}\quad \frac{du}{dy}=\frac{du}{dz}\cdot\frac{dz}{dy}$$

$$\therefore du=\frac{du}{dx}\cdot dx+\frac{du}{dy}dy=\frac{du}{dz}\cdot\frac{dz}{dx}dx+\frac{du}{dz}\cdot\frac{dz}{dy}\cdot dy.$$

Elimination by Differentiation.

89. When a constant is connected with a function by the sign + or −, it disappears by differentiation; but when it is a coefficient of the function, it will appear in the differential also.

Thus, if $u=F(x, y)=0$ . . . (1) be a relation connecting x and y, into which the constant a enters as a factor, then a will also be found in the equation.

$$\left[\frac{du}{dx}\right]=\frac{du}{dx}+\frac{du}{dy}\cdot\frac{dy}{dx}=F_1\left(x,\,y,\,\frac{dy}{dx}\right)=0\ldots.\ (2).$$

Now a may be eliminated between (1) and (2), and the resulting equation, called a *differential equation*, will contain x, y, and $\frac{dy}{dx}$.

If it were required to eliminate two constants, we might differentiate twice, thus obtaining *three* equations, including the primitive, (1), with which the elimination could be effected, and the resulting equation would contain x, y, $\frac{dy}{dx}$, and $\frac{d^2y}{dx^2}$. Surds and transcendental quantities may also be eliminated by a similar process.

90. **1.** Given $y^2 = 2ax$, or $u = y^2 - 2ax = 0$, to eliminate $2a$.

$$\left[\frac{du}{dx}\right] = 2y\frac{dy}{dx} - 2a = 0. \quad \therefore\ 2a = \frac{y^2}{x} = 2y\frac{dy}{dx},$$

or
$$y - 2x\frac{dy}{dx} = 0,$$

an equation in which $2a$ does not appear, but which implies the same relation between x and y.

2. Eliminate the surd from the equation $y = (a^2 + x^2)^{\frac{3}{2}} \ldots . (1)$.

$$\frac{dy}{dx} = \frac{3}{2}(a^2 + x^2)^{\frac{1}{2}} . 2x = \frac{3x\,(a^2 + x^2)^{\frac{3}{2}}}{a^2 + x^2} = \frac{3xy}{a^2 + x^2}$$

$$\therefore\ (a^2 + x^2)\frac{dy}{dx} = 3xy.$$

3. Eliminate a and b from the equation $y = ax^2 + bx \ldots . (1)$.

$$\frac{dy}{dx} = 2ax + b \ldots (2) \quad \frac{d^2y}{dx^2} = 2a \ldots .. (3).$$

$\therefore$ By combining (1), (2), and (3).

$$y = x^2 . \frac{1}{2}\frac{d^2y}{dx^2} + x\left(\frac{dy}{dx} - x\frac{d^2y}{dx^2}\right)$$

or
$$\frac{d^2y}{dx^2} - \frac{2}{x}.\frac{dy}{dx} + \frac{2y}{x^2} = 0.$$

4. Eliminate the exponential from the equation $y = 2ae^{cx}$.

$$\frac{dy}{dx} = 2ace^{cx} = cy.$$

5. Eliminate a and b from the equation $y = a\cos 2x + b\sin 2x$.

$$\frac{dy}{dx} = -2a\sin 2x + 2b\cos 2x,\quad \frac{d^2y}{dx^2} = -4a\cos 2x - 4b\sin 2x = -4y.$$

$$\therefore\ \frac{d^2y}{dx^2} + 4y = 0.$$

91. *Prop.* Let $u = Fz$, and $z = \varphi(x, y)$, where x and y are independent variables, and let it be proposed to eliminate the function F.

Differentiate u first with respect to x, and then with respect to y.

$$\therefore\ \frac{du}{dx} = \frac{du}{dz}\cdot\frac{dz}{dx} = \frac{dFz}{dz}\cdot\frac{d\varphi(x, y)}{dx} \quad \ldots\ldots (1).$$

and
$$\frac{du}{dy} = \frac{du}{dz}\cdot\frac{dz}{dy} = \frac{dFz}{dz}\cdot\frac{d\varphi(x, y)}{dy} \quad \ldots\ldots (2).$$

Now divide (1) by (2), observing that the common factor $\frac{dFz}{dz}$ will disappear;

$$\therefore\ \frac{\frac{du}{dx}}{\frac{du}{dy}} = \frac{\frac{d\varphi(x, y)}{dx}}{\frac{d\varphi(x, y)}{dy}} \quad \text{or,}\quad \frac{du}{dx}\cdot\frac{d\varphi(x, y)}{dy} = \frac{du}{dy}\cdot\frac{d\varphi(x, y)}{dx},$$

in which equation F does not appear.

1. Eliminate the function F from the equation $u = F(ax^3 + by^2)$.

Put $ax^3 + by^2 = z$. $\therefore\ \frac{dz}{dx} = 3ax^2$, and $\frac{dz}{dy} = 2by$.

$$\therefore\ \frac{du}{dx}\cdot 2by = \frac{du}{dy}\cdot 3ax^2.$$

2. Eliminate the function F from the equation $u = \frac{1}{x}F\left(\frac{x}{y}\right)$.

$$\frac{du}{dx} = -\frac{1}{x^2}F\left(\frac{x}{y}\right) + \frac{1}{x}F_1\left(\frac{x}{y}\right)\cdot\frac{1}{y}, \quad \text{and}\quad \frac{du}{dy} = -\frac{1}{x}F_1\left(\frac{x}{y}\right)\frac{x}{y^2}.$$

$$\therefore\ \left[\frac{du}{dx} + \frac{u}{x}\right]\times\left[\frac{1}{-y^2}\right] = -\frac{1}{xy^3}F_1\left(\frac{x}{y}\right) = \frac{du}{dy}\times\frac{1}{xy}.$$

$$\therefore\ u + x\frac{du}{dx} + y\frac{du}{dy} = 0.$$

3. Prove that if $y = a \sin x + b \sin 2x$, then

$$\frac{d^4y}{dx^4} + 5\frac{d^2y}{dx^2} + 4y = 0.$$

$y = a \sin x + b \sin 2x. \ldots (1).$ $\quad \frac{dy}{dx} = a \cos x + 2b \cos 2x.$

$\frac{d^2y}{dx^2} = -a \sin x - 4b \sin 2x. \ldots (2).$ and $\frac{d^4y}{dx^4} = a \sin x + 16b \sin 2x.. (3)$

Multiply (1) by 4, and (2) by 5, and add the results to (3); thus

$$\frac{d^4y}{dx^4} + 5\frac{d^2y}{dx^2} + 4y = 0.$$

92. *Prop.* To determine whether any proposed combination of x and y, as $F(x, y)$, is a function of some other combination, as $\varphi(x, y)$.

Put $u = F(x, y)$, and $z = \varphi(x, y)$; then if u be a function of z, we must have

$$\frac{du}{dx} = \frac{du}{dz} \cdot \frac{dz}{dx} \quad \text{and} \quad \frac{du}{dy} = \frac{du}{dz} \cdot \frac{dz}{dy}.$$

And $$\therefore \frac{du}{dx} \cdot \frac{dz}{dy} = \frac{du}{dy} \cdot \frac{dz}{dx}.$$

which is the required test.

1. Is $u = x^3 - 6x^2y + 12xy^2 - 8y^3$, a function of $z = 2y + a - x$?

$$\frac{du}{dx} = 3x^2 - 12xy + 12y^2. \quad \frac{du}{dy} = -6x^2 + 24xy - 24y^2.$$

$$\frac{dz}{dx} = -1, \quad \text{and} \quad \frac{dz}{dy} = 2.$$

$$\therefore \frac{du}{dx} \cdot \frac{dz}{dy} = 6x^2 - 24xy + 24y^2 = \frac{du}{dy} \cdot \frac{dz}{dx}.$$

Hence u *is* a function of z.

2. Is $u = \log(x^2) - 2 \log y$, a function of $z = \sin\left(a + \frac{y}{x}\right)$?

$$\frac{du}{dx} = \frac{2}{x}, \quad \frac{du}{dy} = -\frac{2}{y}, \quad \frac{dz}{dx} = -\cos\left(a + \frac{y}{x}\right) \cdot \frac{y}{x^2}, \quad \frac{dz}{dy} = \cos\left(a + \frac{y}{x}\right) \cdot \frac{1}{x}.$$

$$\therefore\ \frac{du}{dx}\cdot\frac{dz}{dy}=\frac{2}{x^2}\cdot\cos\left(a+\frac{y}{x}\right)=\frac{du}{dy}\cdot\frac{dz}{dx}.$$

Hence u is a function of z.

3. Is $u = x^2 + y^2$, a function of $z = \tan(x+y)$?

$$\frac{du}{dx}=2x,\quad \frac{du}{dy}=2y,\quad \frac{dz}{dx}=\sec^2(x+y),\quad \frac{dz}{dy}=\sec^2(x+y).$$

$$\therefore\ \frac{du}{dx}\cdot\frac{dz}{dy}=2x\sec^2(x+y),\quad\text{and}\quad \frac{du}{dy}\cdot\frac{dz}{dx}=2y\sec^2(x+y).$$

Hence u is not a function of z.

Development of Functions of Two Independent Variables.

93. *Prop.* To extend Maclaurin's Theorem to functions of two independent variables.

If, in the general development of $F(x+h,\ y+k)$, we make $x=0$, and $y=0$, and denote the particular resulting values by the use of the [ ], changing h and k into x and y respectively, we shall obtain

$$u=F(x,y)=[u]+\left[\frac{du}{dx}\right]\cdot\frac{x}{1}+\left[\frac{du}{dy}\right]\cdot\frac{y}{1}+\left[\frac{d^2u}{dx^2}\right]\cdot\frac{x^2}{1.2}$$
$$+\left[\frac{d^2u}{dxdy}\right]\cdot\frac{xy}{1}+\left[\frac{d^2u}{dy^2}\right]\cdot\frac{y^2}{1.2}+\left[\frac{d^3u}{dx^3}\right]\cdot\frac{x^3}{1.2.3}$$
$$+\left[\frac{d^3u}{dx^2dy}\right]\cdot\frac{x^2y}{1.2}+\left[\frac{d^3u}{dxdy^2}\right]\cdot\frac{xy^2}{1.2}+\left[\frac{d^3u}{dy^3}\right]\cdot\frac{y^3}{1.2.3}+\ \&c.$$

Example. Expand $u=e^x\sin y$.

$$\frac{du}{dx}=e^x\sin y,\quad \frac{du}{dy}=e^x\cos y,\quad \frac{d^2u}{dx^2}=e^x\sin y,\quad \frac{d^2u}{dxdy}=e^x\cos y,$$

$$\frac{d^2u}{dy^2}=-\,e^x\sin y,\quad \frac{d^3u}{dx^3}=e^x\sin y,\quad \frac{d^3u}{dx^2dy}=e^x\cos y,$$

$$\frac{d^3u}{dxdy^2}=-\,e^x\sin y,\quad \frac{d^3u}{dy^3}=-\,e^x\cos y,\quad \&c.,\quad \&c.$$

$$\therefore \ [u]=0,\ \left[\frac{du}{dx}\right]=0,\ \left[\frac{du}{dy}\right]=1,\ \left[\frac{d^2u}{dx^2}\right]=0,\ \left[\frac{d^2u}{dxdy}\right]=1,$$

$$\left[\frac{d^2u}{dy^2}\right]=0,\ \left[\frac{d^3u}{dx^3}\right]=0,\ \left[\frac{d^3u}{dx^2dy}\right]=1,\ \left[\frac{d^3u}{dxdy^2}\right]=0,$$

$\left[\dfrac{d^3u}{dy^3}\right]=-1$, &c., the law being quite apparent.

$$\therefore u = e^x \sin y = y + xy + \frac{x^2y}{1\cdot 2} - \frac{y^3}{1\cdot 2\cdot 3} + \frac{x^3y}{1\cdot 2\cdot 3} - \frac{xy^3}{1\cdot 2\cdot 3}$$

$$+ \frac{x^4y}{1\cdot 2\cdot 3\cdot 4} + \text{\&c.}$$

94. In a similar manner we might apply the general formula deduced in the last proposition to the expansion of any function of two variables, x and y, but among these functions there is one of peculiar interest, in consequence of its frequent occurrence in the application of the Calculus to Physical Astronomy. The formula for the expansion referred to, is known as Lagrange's Theorem. It will be deduced in the next proposition.

Prop. Having given $u = Fz$, and $z = y + x\varphi z$, where F and φ denote any function, and y is independent of x, to expand u in terms of the ascending powers of the variable x.

We observe first that u is a function of x, and therefore if we denote, as usual, by $[u]$, $\left[\dfrac{du}{dx}\right]$, $\left[\dfrac{d^2u}{dx^2}\right]$ &c., the particular values assumed by u, $\dfrac{du}{dx}$, $\dfrac{d^2u}{dx^2}$, &c., when $x = 0$ we shall have, by Maclaurin's Theorem,

$$u = [u] + \left[\frac{du}{dx}\right]\frac{x}{1} + \left[\frac{d^2u}{dx^2}\right]\frac{x^2}{1\cdot 2} + \left[\frac{d^3u}{dx^3}\right]\frac{x^3}{1\cdot 2\cdot 3} + \text{\&c.} \cdots\cdot (A).$$

Now since $\qquad z = y + x\varphi z, \cdots\cdot (1).$

$\therefore$ when $\quad x = 0, \quad [z] = y, \quad$ and $\quad \therefore [u] = Fy.$

Also $$\frac{du}{dx} = \frac{du}{dz} \cdot \frac{dz}{dx} \quad \text{and} \quad \frac{du}{dy} = \frac{du}{dz} \cdot \frac{dz}{dy}.$$

But by differentiating (1) with respect to x we get

$$\frac{dz}{dx} = \varphi z + x \frac{d\varphi z}{dz} \cdot \frac{dz}{dx} \quad \text{whence} \quad \frac{dz}{dx} = \frac{\varphi z}{1 - x \cdot \dfrac{d\varphi z}{dz}} \quad . \; . \; . \; (2).$$

And by differentiating (1) with respect to y we have

$$\frac{dz}{dy} = 1 + x \frac{d\varphi z}{dz} \cdot \frac{dz}{dy} \quad \text{whence} \quad \frac{dz}{dy} = \frac{1}{1 - x \dfrac{d\varphi z}{dz}} \quad . \; . \; (3).$$

Dividing (2) by (3) and reducing, we obtain

$$\frac{dz}{dx} = \varphi z \cdot \frac{dz}{dy}.$$

$$\therefore \frac{du}{dx} = \frac{du}{dz} \cdot \frac{dz}{dx} = \frac{du}{dz} \cdot \varphi z \cdot \frac{dz}{dy} = \varphi z \cdot \frac{du}{dy}.$$

Hence when $x = 0$, and $z = y$, $\left[\dfrac{du}{dx}\right] = \varphi y \dfrac{dFy}{dy}$.

Now assume u_1 such that $\varphi z \cdot \dfrac{du}{dy} = \dfrac{du_1}{dy}$.

$$\therefore \frac{du}{dx} = \frac{du_1}{dy} \quad \text{and} \quad \frac{d^2u}{dx^2} = \frac{d^2u_1}{dydx} = \frac{d^2u_1}{dxdy} = \frac{d\left(\dfrac{du_1}{dx}\right)}{dy}.$$

But $$\frac{du_1}{dx} = \frac{du_1}{dz} \cdot \frac{dz}{dx} = \frac{du_1}{dz} \cdot \varphi z \cdot \frac{dz}{dy} = \varphi z \cdot \frac{du_1}{dy} = (\varphi z)^2 \cdot \frac{du}{dy}.$$

$$\therefore \frac{d^2u}{dx^2} = \frac{d\left[(\varphi z)^2 \cdot \dfrac{du}{dy}\right]}{dy} \quad \text{and} \quad \left[\frac{d^2u}{dx^2}\right] = \frac{d\left[(\varphi y)^2 \cdot \dfrac{dFy}{dy}\right]}{dy}$$

And similarly it may be shown that

$$\left[\frac{d^3u}{dx^3}\right] = \frac{d^2\left[(\varphi y)^3 \cdot \dfrac{dFy}{dy}\right]}{dy^2}, \quad \left[\frac{d^4u}{dx^4}\right] = \frac{d^3\left[(\varphi y)^4 \cdot \dfrac{dFy}{dy}\right]}{dy^3}, \text{ \&c.}$$

But to show that this law of formation of the differential coefficients is general, suppose that it has been proved that

$$\frac{d^{n-1}u}{dx^{n-1}} = \frac{d^{n-2}\left[(\varphi z)^{n-1}.\frac{du}{dy}\right]}{dy^{n-2}} \quad \ldots\ldots (4).$$

Put $$(\varphi z)^{n-1}.\frac{du}{dy} = \frac{du_{n-1}}{dy}. \quad \therefore \frac{d^{n-1}u}{dx^{n-1}} = \frac{d^{n-1}u_{n-1}}{dy^{n-1}}.$$

$$\therefore \frac{d^n u}{dx^n} = \frac{d^n u_{n-1}}{dy^{n-1}dx} = \frac{d^n u_{n-1}}{dx dy^{n-1}} = \frac{d^{n-1}\left(\frac{du_{n-1}}{dx}\right)}{dy^{n-1}}.$$

But $$\frac{du_{n-1}}{dx} = \frac{du_{n-1}}{dz}.\frac{dz}{dx} = \frac{du_{n-1}}{dz}.\varphi z.\frac{dz}{dy} = \varphi z.\frac{du_{n-1}}{dy} = (\varphi z)^n.\frac{du}{dy}.$$

$$\therefore \frac{d^n u}{dx^n} = \frac{d^{n-1}\left[(\varphi z)^n.\frac{du}{dy}\right]}{dy^{n-1}} \quad \ldots\ldots (5)$$

Thus the form (4), if true for any value of n, is also true for the next higher value. Now it has been shown true for $n = 1$ and $n = 2$; and hence it is true when $n = 3$, $n = 4$, &c., or it is universally true.

Now making $x = 0$ and $z = y$ and the expression (5) becomes

$$\left[\frac{d^n u}{dx^n}\right] = \frac{d^{n-1}\left[(\varphi y)^n\frac{dFy}{dy}\right]}{dy^{n-1}}.$$

Making the substitutions for $[u]$, $\left[\frac{du}{dx}\right]$, $\left[\frac{d^2u}{dx^2}\right]$, &c., in the expansion (A), we get

$$u = Fy + \varphi y.\frac{dFy}{dy}.\frac{x}{1} + \frac{d\left[(\varphi y)^2.\frac{dFy}{dy}\right]}{dy}.\frac{x^2}{1.2}$$
$$+ \frac{d^2\left[(\varphi y)^3.\frac{dFy}{dy}\right]}{dy^2}.\frac{x^3}{1.2.3} + \&c. \ldots (L).$$

This formula is called Lagrange's Theorem.

Cor. Let $u = Fz = z$; then $Fy = y$, and $\frac{dFy}{dy} = 1$.

$$\therefore z = y + \varphi y\frac{x}{1} + \frac{d[(\varphi y)^2]}{dy}\cdot\frac{x^2}{1.2} + \frac{d^2[(\varphi y)^3]}{dy^2}\cdot\frac{x^3}{1.2.3} + \&c.. (M)$$

a formula for the expansion of z when we have given $z = y + x\varphi z$.

EXAMPLES.

95. 1. Given $z^3 - az + b = 0$ to express z in terms of a and b.

Here $z = \frac{b}{a} + \frac{1}{a}z^3$, which corresponds to the form $z = y + x\varphi z$, when we make

$$\frac{b}{a} = y,\quad \frac{1}{a} = x,\quad \text{and}\quad z^3 = \varphi z.$$

Hence by substitution $\varphi y = y^3 = \frac{b^3}{a^3}$,

$$\frac{d[(\varphi y)^2]}{dy} = \frac{d(y^6)}{dy} = 6y^5 = 6\frac{b^5}{a^5},\ \frac{d^2[(\varphi y^3]}{dy^2} = \frac{d^2(y^9)}{dy^2} = 8.9y^7 = 8.9\frac{b^7}{a^7},\ \&c.$$

Introducing these values into the formula (M), it becomes

$$z = \frac{b}{a} + \frac{b^3}{a^3}\cdot\frac{1}{a} + 6\frac{b^5}{a^5}\cdot\frac{1}{1.2a^2} + 8.9\frac{b^7}{a^7}\cdot\frac{1}{1.2.3.a^3} + 10.11.12\frac{b^9}{a^9}\cdot\frac{1}{1.2.3.4a^4} + \&c.$$

$$= \frac{b}{a}[1 + \frac{b^2}{a^3} + 3\frac{b^4}{a^6} + 12\frac{b^6}{a^9} + 55\frac{b^8}{a^{12}}\ \&c.]$$

If b be very small in comparison with a this series will converge very rapidly.

2. Given $y = z + z^2 + z^3 + z^4 + \&c.$,

to revert the series, that is to express z in terms of y.

By transposition, $z = y - (z^2 + z^3 + z^4 + \&c.)$

Put $x = -1$; $\varphi z = z^2 + z^3 + z^4 + \&c.$

Then $$\varphi y = y^2 + y^3 + y^4 + \&c.$$

$$\frac{d[(\varphi y)^2]}{dy} = \frac{d[(y^2+y^3+y^4+\&c.)^2]}{dy}$$

$$=2(y^2+y^3+y^4+\&c.)(2y+3y^2+4y^3+\&c.)$$

$$=2(2y^3+5y^4+9y^5+14y^6+\&c.)$$

$$(\varphi y)^3 = (y^2 + y^3 + y^4 + \&c.)^3 = y^6 + 3y^7 + 6y^8 + \&c.$$

$$\therefore \frac{d^2[(\varphi y)^3]}{dy^2} = 5 . 6y^4 + 3 . 6 . 7y^5 + 6 . 7 . 8y^6 + \&c.$$

$$(\varphi y)^4 = (y^2 + y^3 + y^4 + \&c.)^4 = y^8 + 4y^9 + \&c.$$

$$\therefore \frac{d^3[(\varphi y)^4]}{dy^3} = 6 . 7 . 8y^5 + 4 . 7 . 8 . 9y^6 + \&c.$$

$$(\varphi y)^5 = (y^2 + y^3 + y^4 + \&c.)^5 = y^{10} + \&c.$$

$$\therefore \frac{d^4[(\varphi y)^5]}{dy^4} = 7 . 8 . 9 . 10y^6 + \&c. \qquad \&c., \&c.$$

$\therefore$ By substitution in formula (M).

$$z = y - \frac{1}{1}[y^2 + y^3 + y^4 + y^5 + y^6 + \&c.]$$

$$+ \frac{1}{1 . 2}[2 . 2y^3 + 2 . 5y^4 + 2 . 9y^5 + 2 . 14y^6 + \&c.]$$

$$- \frac{1}{1 . 2 . 3}[5 . 6y^4 + 3 . 6 . 7y^5 + 6 . 7 . 8y^6 + \&c.]$$

$$+ \frac{1}{1 . 2 . 3 . 4}[6 . 7 . 8y^5 + 4 . 7 . 8 . 9y^6 + \&c.]$$

$$- \frac{1}{1 . 2 . 3 . 4 . 5}[7 . 8 . 9 . 10y^6 + \&c.]$$

$$+ \&c. = y - y^2 + y^3 - y^4 + y^5 - y^6 + \&c.$$

3. Given $1 - z + e^z = 0$ to expand z^n.

Here $z = 1 + e^z$. Put $x = 1$, $y = 1$, $\varphi z = e^z$, $Fz = z^n$,

$$\therefore \varphi y = e^y.\ Fy = y^n,\ \varphi y \cdot \frac{dFy}{dy} = e^y \cdot \frac{d\,(y^n)}{dy} = ny^{n-1}\,e^y = ne.$$

$$\frac{d}{dy}\left[(\varphi y)^2 \cdot \frac{dFy}{dy}\right] = \frac{d\,(e^{2y} n y^{n-1})}{dy}$$

$$= 2ne^{2y} \cdot y^{n-1} + n\,(n-1)\, e^{2y} \cdot y^{n-2} = n\,(n+1)\, e^2.$$

$$\frac{d^2}{dy^2}\left[(\varphi y)^3 \cdot \frac{dFy}{dy}\right] = \frac{d^2\,(e^{3y} \cdot n y^{n-1})}{dy^2}$$

$$= 9ne^{3y} \cdot y^{n-1} + 6n\,(n-1)\, e^{3y} \cdot y^{n-2} + n\,(n-1)\,(n-2)\, e^{3y} \cdot y^{n-3}$$

$$= n\,(n^2 + 3n + 5)\, e^3. \quad \&c.\ \&c.$$

Hence, by substitution in formula (L), we have

$$z^n = 1 + ne + \frac{n\,(n+1)}{1\,.\,2}\, e^2 + \frac{n\,(n^2 + 3n + 5)}{1\,.\,2\,.\,3}\, e^3 + \&c.$$

4. Given $z = y + e \,.\, \sin z$, to expand z and $\sin z$.

Put $\qquad x = e, \; \varphi z = \sin z, \; Fz = \sin z.$

$$\therefore \; \varphi y = \sin y\,, (\varphi y)^2 = \sin^2 y, \; (\varphi y)^3 = \sin^3 y \;\&c.$$

$$\therefore \; \frac{d\,[(\varphi y)^2]}{dy} = 2 \sin y \,.\, \cos y. = \sin 2y.$$

$$\frac{d^2\,[(\varphi y)^3]}{dy^2} = \frac{d\,(3 \sin^2 y \,.\, \cos y)}{dy} = 6 \sin y \,.\, \cos^2 y - 3 \sin^3 y.$$

$$= 3 \sin y\,(1 + \cos 2y - \frac{1}{2} + \frac{1}{2} \cos 2y)$$

$$= \frac{3}{2} \sin y + \frac{9}{2}\left(\frac{1}{2} \sin 3y - \frac{1}{2} \sin y\right)$$

$$= \frac{9}{4} \sin 3y - \frac{3}{4} \sin y. \quad \&c.$$

Hence by substitution in (M).

$$z = y + \sin y \frac{e}{1} + \sin 2y \frac{e^2}{1\,.\,2} + \left(\frac{9}{4} \sin 3y - \frac{3}{4} \sin y\right) \frac{e^3}{1\,.\,2\,.\,3} + \&c.$$

Again $\; Fy = \sin y. \quad \therefore \; \varphi y \,.\, \dfrac{dFy}{dy} = \sin y \,.\, \cos y = \dfrac{1}{2} \sin 2y.$

$$\frac{d}{dy}\left[(\varphi y)^2 \cdot \frac{dFy}{dy}\right] = \frac{d(\cos y \,.\, \sin^2 y)}{dy} = \frac{d\left(\frac{1}{2}\sin y \,.\, \sin 2y\right)}{dy}$$

$$= \frac{d\left(\frac{1}{4}\cos y - \frac{1}{4}\cos 3y\right)}{dy} = \frac{3}{4}\sin 3y - \frac{1}{4}\sin y.$$

$$\frac{d^2}{dy^2}\left[(\varphi y)^3 \cdot \frac{dFy}{dy}\right] = \frac{d^2(\cos y \,.\, \sin^3 y)}{dy^2} = \frac{d^2\left[\frac{1}{2}\sin 2y\left(\frac{1}{2} - \frac{1}{2}\cos 2y\right)\right]}{dy^2}$$

$$= \frac{d^2\left(\frac{1}{4}\sin 2y - \frac{1}{8}\sin 4y\right)}{dy^2} = 2\sin 4y - \sin 2y. \quad \&c. \ \&c.$$

Hence by substitution in formula (L).

$$\sin z = \sin y + \frac{1}{2}\sin 2y \,.\, \frac{e}{1} + \left(\frac{3}{4}\sin 3y - \frac{1}{4}\sin y\right)\frac{e^2}{1\,.\,2}$$

$$+ (2\sin 4y - \sin 2y)\frac{e^3}{1\,.\,2\,.\,3} + \&c.$$

5. Given $u + \frac{du}{dx} \cdot \frac{h}{1} + \frac{d^2u}{dx^2} \cdot \frac{h^2}{1\,.\,2} + \frac{d^3u}{dx^3} \cdot \frac{h^3}{1\,.\,2\,.\,3} + \&c. = 0$, to find h in terms of u and its differential coefficients.

Put $\qquad \frac{du}{dx} = p_1, \ \frac{d^2u}{dx^2} = p_2, \ \frac{d^3u}{dx^3} = p_3$

$$\therefore h = -\frac{u}{p_1} - \frac{1}{p_1}\left(\frac{p_2 h^2}{1\,.\,2} + \frac{p_3 h^3}{1\,.\,2\,.\,3} + \&c.\right)$$

$$\therefore y = -\frac{u}{p_1}, \ x = -\frac{1}{p_1}, \ z = h, \ \varphi z = \varphi h = \frac{p_2 h^2}{1\,.\,2} + \frac{p_3 h^3}{1\,.\,2\,.\,3} + \&c.$$

$$\therefore h = -\left(\frac{u}{p_1} + \frac{p_2}{p_1^3} \cdot \frac{u^2}{1\,.\,2} + \frac{3p_2^2 - p_1 p_3}{p_1^5} \cdot \frac{u^3}{1\,.\,2\,.\,3} + \&c.\right)$$

Now if a be a root of the equation $u = 0$, and x an approximate value of a, so that $x + h = a$, we may use this series in finding a more exact value of x. Thus, if $x = \frac{3}{2} = 1\,.\,5$ be an approximate root of the equation $u = x^4 - 2x^2 + 4x - 8 = 0$,

then $1.5 + h = a$ and

$$h = -\left(\frac{u}{2^2(x^3-x+1)} + \frac{3x^2-1}{2^4(x^3-x+1)^3} \cdot \frac{u^2}{1.2} + \frac{21x^4-12x^2-6x+3}{2^6(x^3-x+1)} \cdot \frac{u^3}{1.2.3} + \&c.\right)$$

Here $u = -\frac{23}{16}$ $\therefore$ $h = .11$ and $a = 1.5 + .11 = 1.61$ nearly. And if we repeat the operation by putting $x = 1.61$, a nearer approximation will be obtained.

CHAPTER X.

MAXIMA AND MINIMA FUNCTIONS OF TWO INDEPENDENT VARIABLES.

96. A function u of two independent variables x and y, is said to be a maximum when its value exceeds all those other values obtained by replacing x by $x \pm h$ and y by $y \pm k$, when h and k may take any values between zero and certain small but finite quantities; and u is said to be a minimum when its value is less than all other values determined by the conditions above described.

97. *Prop.* Having given $u = F(x, y)$, when x and y are independent variables, to determine the values of x and y which shall render u a maximum or minimum.

Suppose x to receive an increment $\pm h$, and y an increment $\pm k$, the value h and k being small but finite and entirely independent of each other; and denote by u_2 the value assumed by u, so that

$$u_2 = F(x \pm h, \quad y \pm k).$$

9

Then, by Taylor's Theorem, as applied to functions of two independent variables, we have

$$u_2 = u + \frac{du}{dx}\cdot\frac{(\pm h)}{1} + \frac{du}{dy}\cdot\frac{(\pm k)}{1} + \frac{d^2u}{dx^2}\cdot\frac{(\pm h)^2}{1.2}$$
$$+ \frac{d^2u}{dxdy}\cdot\frac{(\pm h)}{1}\cdot\frac{(\pm k)}{1} + \frac{d^2u}{dy^2}\cdot\frac{(\pm k)^2}{1.2} + \&c.$$

Now in order that u may exceed u_2 for all small values of h and k, whether positive or negative, it is obviously necessary to have

$$\frac{du}{dx}\cdot\frac{(\pm h)}{1} + \frac{du}{dy}\cdot\frac{(\pm k)}{1} + \frac{d^2u}{dx^2}\cdot\frac{(\pm h)^2}{1.2} + \frac{d^2u}{dxdy}\cdot\frac{(\pm h)}{1}\cdot\frac{(\pm k)}{1}$$
$$+ \frac{d^2u}{dy^2}\cdot\frac{(\pm k)^2}{1.2} + \&c. < 0 \;.\;.\;.\;.\; (1);$$

in which series we must be at liberty to make h and k both positive, or both negative, or one positive and the other negative: or, finally, either may be taken equal to zero, the other remaining finite.

Now when $k = 0$ the series (1) reduces to

$$\frac{du}{dx}\cdot\frac{(\pm h)}{1} + \frac{d^2u}{dx^2}\cdot\frac{(\pm h)^2}{1.2} + \frac{d^3u}{dx^3}\cdot\frac{(\pm h)^3}{1.2.3} + \&c. < 0 \ldots (2);$$

in which h may be taken so small that the sign of the first term $\frac{du}{dx}\cdot\frac{(\pm h)}{1}$, which contains the lowest power of h, shall control the sign of the series. But this term obviously changes sign with h, since $\frac{du}{dx}$ does not contain h; and as we are at liberty to make h alternately positive and negative, it is impossible that the series (2) should remain negative so long as $\frac{du}{dx}\cdot\frac{(\pm h)}{1}$ has any value other than zero.

We have then, as a first condition necessary to render u a maximum,

$$\frac{du}{dx}\cdot\frac{(\pm h)}{1} = 0 \quad \text{or simply} \quad \frac{du}{dx} = 0 \;.\;.\;.\;.\; (A).$$

Omitting the first term in (2) we have

$$\frac{d^2u}{dx^2}\cdot\frac{(\pm h)^2}{1.2}+\frac{d^3u}{dx^3}\cdot\frac{(\pm h)^3}{1.2.3}+\&c. < 0 \;\ldots\ldots (3).$$

Here again the sign of the series will depend on that of the first term when h is small, and since that term does not change sign when we substitute $-h$ for $+h$, the series (3) will remain negative for small values of h, when

$$\frac{d^2u}{dx^2}\cdot\frac{(\pm h)^2}{1.2} < 0, \quad \text{or simply when} \quad \frac{d^2u}{dx^2} < 0.$$

Hence

$$\frac{d^2u}{dx^2} < 0 \;\ldots\ldots (B)$$

is a second condition necessary for a maximum.

98. Returning to the series (1) and supposing $h = 0$ while k remains finite, we prove, by a course of reasoning entirely similar, that the following conditions are also necessary, viz.:

$$\frac{du}{dy} = 0 \;\ldots\ldots (C) \quad \text{and} \quad \frac{d^2u}{dy^2} < 0 \;\ldots\ldots (D).$$

Now omitting the terms in (1) which contain the first powers of h and k, and which it has been seen must reduce to zero, we obtain

$$\frac{d^2u}{dx^2}\cdot\frac{(\pm h)^2}{1.2}+\frac{d^2u}{dxdy}\cdot\frac{(\pm h)}{1}\cdot\frac{(\pm k)}{1}+\frac{d^2u}{dy^2}\cdot\frac{(\pm k)^2}{1.2}+\frac{d^3u}{dx^3}\cdot\frac{(\pm h)^3}{1.2.3}$$

$$+\frac{d^3u}{dx^2dy}\cdot\frac{(\pm h)^2.(\pm k)}{1.2}+\frac{d^3u}{dxdy^2}\cdot\frac{(\pm h)(\pm k)^2}{1.2}$$

$$+\frac{d^3u}{dy^3}\cdot\frac{(\pm k)^3}{1.2.3}+\&c. < 0,$$

or, by making $\frac{k}{h} = r$, where r is entirely arbitrary, since h and k have no necessary dependence upon each other.

$$\frac{h^2}{1.2}\left[\frac{d^2u}{dx^2}\pm 2r\frac{d^2u}{dxdy}+r^2\frac{d^2u}{dy^2}\right]$$

$$\pm\frac{h^3}{1.2.3}\cdot\left[\frac{d^3u}{dx^3}\pm 3r\frac{d^3u}{dx^2dy}+3r^2\frac{d^3u}{dxdy^2}\pm r^3\frac{d^3u}{dy^3}\right]+\&c.<0 \;\ldots\ldots (4),$$

in which, when h is small, the sign of the series will depend on that of

$$\frac{d^2u}{dx^2} \pm 2r\frac{d^2u}{dxdy} + r^2\frac{d^2u}{dy^2},$$

and this must be negative for all values of r, whether positive or negative, when u is a maximum.

We must now search for the condition necessary to render

$$\frac{d^2u}{dx^2} \pm 2r\frac{d^2u}{dxdy} + r^2\frac{d^2u}{dy^2} < 0 \quad . \; . \; . \; . \quad (5) \text{ for all values of } r.$$

Put for brevity $\frac{d^2u}{dx^2} = A$, $\frac{d^2u}{dydx} = B$, and $\frac{d^2u}{dy^2} = C$.

Then A, B, and C, must, if possible, be so related to each other that $A \pm 2Br + Cr^2$ shall be negative for every real value of r.

Now it is known, from the theory of equations, that if we solve the equation $A \pm 2Br + Cr^2 = 0$ with respect to r, and obtain the values

$$r_1 = \frac{\mp B + \sqrt{B^2 - AC}}{C}, \quad \text{and} \quad r_2 = \frac{\mp B - \sqrt{B^2 - AC}}{C},$$

and then substitute in the polynomial $A \pm 2Br + Cr^2$, for r values alternately a little greater and somewhat less than r_1 or r_2, the sign of the polynomial will undergo a change. If therefore the proposed substitution be possible, the condition $A \pm 2Br + Cr^2 < 0$ for all values of r will be impossible.

And so long as the values of r_1 and r_2 are real and unequal, this substitution can be made; but if those values of r prove imaginary, it will no longer be possible to substitute for r, real quantities alternately greater and less than such values, and therefore the polynomial cannot change its sign.

Now by examining the values of r_1 and r_2 it will be seen that the condition necessary to render r_1 and r_2 imaginary is $B^2 < AC$. Hence we have a fifth condition necessary for a maximum, viz.:

$$\frac{d^2u}{dx^2} \cdot \frac{d^2u}{dy^2} - \left(\frac{d^2u}{dxdy}\right)^2 > 0 \quad . \; . \; . \; . \quad (E),$$

when this condition is satisfied, the condition (5) will also be satisfied, since (5) is true when $r = 0$.

It ought to be remarked, however, that when $B^2 = AC$, the two roots r_1 and r_2 become real and equal, and therefore in passing over one of these roots, we necessarily pass over both. Thus the sign of polynomial will not change, so that the fifth condition would be more correctly stated as follows:

$$\frac{d^2u}{dx^2}\cdot\frac{d^2u}{dy^2} - \left(\frac{d^2u}{dxdy}\right)^2 \overset{=}{>} 0 \;.\;.\;.\;.\; (E).$$

By a course of reasoning entirely similar, we can prove that the five conditions necessary to render u a minimum are the following:

$$\frac{du}{dx} = 0,\; \frac{du}{dy} = 0,\; \frac{d^2u}{dx^2} > 0,\; \frac{d^2u}{dy^2} > 0,\; \frac{d^2u}{dx^2}\cdot\frac{d^2u}{dy^2} - \left(\frac{d^2u}{dxdy}\right)^2 \overset{=}{>} 0.$$

99. If $\frac{d^2u}{dx^2} = 0$, when $\frac{du}{dx} = 0$, there can be neither maximum nor minimum, unless $\frac{d^3u}{dx^3} = 0$ also; and similarly, if $\frac{d^2u}{dy^2} = 0$, when $\frac{du}{dy} = 0$, there can be neither maximum nor minimum unless $\frac{d^3u}{dy^3} = 0$.

There are other conditions likewise necessary to render u a maximum or minimum in such cases, but they are usually of so complicated a character as to be unfit for use.

EXAMPLES.

100. 1. To determine the values of x and y which render $u = x^3 + y^3 - 3axy$ a maximum or minimum.

$$\frac{du}{dx} = 3x^2 - 3ay = 0, \;.\;.\;.\;.\; (1). \qquad \frac{du}{dy} = 3y^2 - 3ax = 0, \;.\;.\;.\;.\; (2).$$

From (1), $y = \frac{x^2}{a}$, and this substituted in (2), gives

$$x^4 - a^3x = 0; \quad \therefore x = 0, \quad \text{or}, \quad x = a,$$

the two other roots being imaginary.

But when $x = 0$, $y = \frac{x^2}{a} = 0$,

and when $x = a$, $y = a$.

Now forming the second differential coefficients, we get

$$\frac{d^2u}{dx^2} = 6x = 0 \text{ when } x = 0, \qquad \frac{d^2u}{dy^2} = 6y = 0 \text{ when } y = 0,$$
$$= 6a \text{ when } x = a, \qquad = 6a \text{ when } y = a,$$

$$\frac{d^2u}{dxdy} = -3a, \quad \frac{d^2u}{dx^2} \cdot \frac{d^2u}{dy^2} - \left(\frac{d^2u}{dxdy}\right)^2 = -9a^2 \text{ when } x = 0 \text{ and } y = 0.$$
$$= 27a^2 \text{ when } x = a \text{ and } y = a.$$

$\therefore$ The five conditions necessary for a minimum are fulfilled when $x = a$ and $y = a$, viz.

$$\frac{du}{dx} = 0, \ \frac{du}{dy} = 0, \ \frac{d^2u}{dx^2} > 0, \ \frac{d^2u}{dy^2} > 0, \text{ and } \frac{d^2u}{dx^2} \cdot \frac{d^2u}{dy^2} - \left(\frac{d^2u}{dxdy}\right)^2 > 0,$$

$$\therefore u = a^3 + a^3 - 3a^3 = -a^3.$$

But when $x = 0$ and $y = 0$, $\frac{d^2u}{dx^2}$ and $\frac{d^2u}{dy^2}$ reduce to zero, while $\frac{d^3u}{dx^3}$ and $\frac{d^3u}{dy^3}$ do not reduce to zero. Hence the value $u = 0$, is neither a maximum nor a minimum.

2. To find the lengths of the three edges of a rectangular parallelopipedon which shall contain a given volume, a^3, under the least surface.

Let x, y, and z, be the required edges, $\therefore xyz = a^3$. . . . (1).

And $u = 2(xy + xz + yz) =$ surface $=$ a minimum.

But from (1), $xz = \frac{a^3}{y}$, and $yz = \frac{a^3}{x}$,

$$\therefore u = 2\left(xy + \frac{a^3}{y} + \frac{a^3}{x}\right) \ldots . (2).$$

$$\therefore \frac{du}{dx} = 2y - \frac{2a^3}{x^2} = 0, \ldots (3). \text{ and } \frac{du}{dy} = 2x - \frac{2a^3}{y^2} = 0, \ldots . (4)$$

$\therefore x^2y = a^3 = xy^2$, $\therefore x = y$, and consequently $x^3 = a^3$,

$$\therefore\ x = a, \quad y = a, \quad \text{and} \quad z = a.$$

$$\frac{d^2u}{dx^2} = \frac{4a^3}{x^3} = 4 > 0, \quad \frac{d^2u}{dy^2} = \frac{4a^3}{y^3} = 4 > 0, \quad \frac{d^2u}{dxdy} = 2.$$

$$\frac{d^2u}{dx^2} \cdot \frac{d^2u}{dy^2} - \left(\frac{d^2u}{dxdy}\right)^2 = 12 > 0,$$

$\therefore\ u = 2(a^2 + a^2 + a^2) = 6a^2 =$ a minimum, and the parallelopipedon must be a cube.

3. Given $x + y + z = \pi$, to find the values of x, y, and z, when $\cos x \cos y \cos z = u =$ a maximum.

Regarding x and y as independent variables, and z a function of x and y, we obtain by differentiating

$x + y + z = \pi$, with respect to x and y successively.

$$1 + \frac{dz}{dx} = 0, \quad \text{and} \quad 1 + \frac{dz}{dy} = 0 \ldots\ldots (1).$$

But since $u =$ maximum,

$$\log u = \log\cos x + \log\cos y + \log\cos z = \text{maximum},$$

$$\therefore\ \left(\frac{d \log u}{dx}\right) = -\tan x - \tan z \frac{dz}{dx} = 0.$$

and
$$\left(\frac{d \log u}{dy}\right) = -\tan y - \tan z \frac{dz}{dy} = 0.$$

or, by replacing $\frac{dz}{dx}$ and $\frac{dz}{dy}$ by their values derived from equations (1).

$$-\tan x + \tan z = 0, \quad -\tan y + \tan z = 0,$$

$$\therefore\ \tan x = \tan z = \tan y, \quad \text{and} \quad x = y = z = \frac{1}{3}\pi.$$

$$u = \cos^3 \frac{1}{3}\pi = \left(\frac{1}{2}\right)^3 = \frac{1}{8}.$$

4. To find the greatest rectangular parallelopipedon which can be inscribed in a given ellipsoid.

Let a, b, and c, be the semi-axes of the ellipsoid, x, y, and z, the co-ordinates of one of the vertices of the parallelopipedon.

Then $2x$, $2y$, and $2z$, are the three edges of the parallelopipedon, and, therefore, $2x \,.\, 2y \,.\, 2z =$ maximum, or

$$u = xyz = \text{maximum.} \quad \ldots \ldots (1).$$

But, since each vertex is in the surface of the ellipsoid, the co-ordinates x, y, z, must satisfy the equation of the surface.

$$\therefore \quad \frac{x^2}{a^2} + \frac{y^2}{b^2} + \frac{z^2}{c^2} = 1. \quad \ldots \ldots (2).$$

Differentiating (2) with respect to x and y successively, regarding z as a function of the independent variables x and y, we get

$$\frac{2x}{a^2} + \frac{2z}{c^2} \cdot \frac{dz}{dx} = 0, \quad \frac{2y}{b^2} + \frac{2z}{c^2} \cdot \frac{dz}{dy} = 0. \quad \ldots \ldots (3).$$

But, from (1) we have

$$\left(\frac{du}{dx}\right) = yz + xy\frac{dz}{dx} = 0, \quad \left(\frac{du}{dy}\right) = xz + xy\frac{dz}{dy} = 0,$$

or, by introducing the values of $\frac{dz}{dx}$ and $\frac{dz}{dy}$ from equations (3).

$$yz - xy\frac{c^2x}{a^2z} = 0, \quad \text{and} \quad xz - xy\frac{c^2y}{b^2z} = 0.$$

$$\therefore \; a^2z^2 = c^2x^2 \quad \text{and} \quad b^2z^2 = c^2y^2 \quad \therefore \; \frac{x^2}{a^2} = \frac{z^2}{c^2} = \frac{y^2}{b^2}$$

Hence from (2), $\frac{x^2}{a^2} + \frac{x^2}{a^2} + \frac{x^2}{a^2} = 1$ and $x = \frac{a}{\sqrt{3}}$: in like manner it may be shown that, $y = \frac{b}{\sqrt{3}}$, and $z = \frac{c}{\sqrt{3}}$.

Thus the edges of the parallelopipedon must be proportional to the axes to which they are parallel. In each of the last two examples, the formation of the second differential coefficient has been omitted as unnecessary, it being easily seen that the proposed question admitted of the maximum or minimum sought, and also that the values found were the only suitable values.

CHAPTER XI.

CHANGE OF THE INDEPENDENT VARIABLE.

101. Hitherto we have employed the differential coefficients $\frac{dy}{dx}, \frac{d^2y}{dx^2}$, &c. or $\frac{du}{dx}, \frac{d^2u}{dx^2}$, &c. exclusively upon the hypothesis that x was the independent variable. But there are many cases in which it is more convenient to adopt some other quantity t upon which both x and y, or x and u depend as the independent variable, and perhaps to pass from one supposition to the other within the limits of the same investigation.

It then becomes necessary to express $\frac{dy}{dx}, \frac{d^2y}{dx^2}$, &c. or $\frac{du}{dx}, \frac{d^2u}{dx^2}$, &c. in terms of the differential coefficients of x and y, or those of x and u taken with respect to the new variable t.

102. *Prop.* Given $y = \varphi x$, and $x = Ft$, to express $\frac{dy}{dx}$, and $\frac{d^2y}{dx^2}$ in terms of $\frac{dx}{dt}, \frac{d^2x}{dt^2}, \frac{dy}{dt}, \frac{d^2y}{dt^2}$, &c.

Since y is a function of x, and x a function of t, we have

$$\frac{dy}{dt} = \frac{dy}{dx} \cdot \frac{dx}{dt} \cdots\cdots (1) \quad \text{and} \quad \therefore \frac{dy}{dx} = \frac{\frac{dy}{dt}}{\frac{dx}{dt}} \cdots\cdots (A)$$

Now differentiating (1), and observing that $\frac{dy}{dx}$ is a function of t through x, we get

$$\frac{d^2y}{dt^2} = \frac{d^2y}{dx^2} \cdot \frac{dx^2}{dt^2} + \frac{dy}{dx} \cdot \frac{d^2x}{dt^2},$$

$$\therefore \frac{d^2y}{dx^2} = \frac{\frac{d^2y}{dt^2} - \frac{dy}{dx}\cdot\frac{d^2x}{dt^2}}{\frac{dx^2}{dt^2}} = \frac{\frac{d^2y}{dt^2}\cdot\frac{dx}{dt} - \frac{d^2x}{dt^2}\cdot\frac{dy}{dt}}{\frac{dx^3}{dt^3}} \dots\dots (B)$$

The two formulæ (A) and (B) resolve the problem.

Cor. In a similar manner we might form expressions for $\frac{d^3y}{dx^3}, \frac{d^4y}{dx^4}$ upon the same hypothesis, but they are seldom required.

Cor. If y be taken as the independent variable, then

$$t = y, \quad \frac{dy}{dt} = \frac{dy}{dy} = 1 \quad \text{and} \quad \therefore \frac{d^2y}{dt^2} = \frac{d\left(\frac{dy}{dy}\right)}{dt} = 0.$$

$$\therefore \frac{dy}{dx} = \frac{1}{\frac{dx}{dt}} = \frac{1}{\frac{dx}{dy}}. \quad \text{and} \quad \frac{d^2y}{dx^2} = -\frac{\frac{d^2x}{dy^2}}{\frac{dx^3}{dy^3}}$$

Cor. If x be the independent variable, then

$$t = x, \quad \frac{dx}{dt} = \frac{dx}{dx} = 1 \quad \text{and} \quad \frac{d^2x}{dt^2} = 0, \text{ and } (A) \text{ and } (B) \text{ reduce}$$

to $\frac{dy}{dx} = \frac{dy}{dx}$, and $\frac{d^2y}{dx^2} = \frac{d^2y}{dx^2}$, the ordinary forms.

EXAMPLES.

103. 1. Transform the differential equation

$$\frac{d^2y}{dx^2} - \frac{x}{1-x^2}\cdot\frac{dy}{dx} + \frac{y}{1-x^2} = 0,$$ so as to render θ the independent variable, having given $\theta = \cos^{-1}x$.

Here $x = \cos\theta$. $\therefore \frac{dx}{d\theta} = -\sin\theta, \frac{d^2x}{d\theta^2} = -\cos\theta = -x,$

$$\therefore \frac{dy}{dx} = \frac{\frac{dy}{d\theta}}{\frac{dx}{d\theta}} = -\frac{1}{\sin\theta}\cdot\frac{dy}{d\theta}$$

$$\frac{d^2y}{dx^2} = \frac{\frac{d^2y}{d\theta^2}\cdot\frac{dx}{d\theta} - \frac{d^2x}{d\theta^2}\cdot\frac{dy}{d\theta}}{\frac{dx^3}{d\theta^3}} = \frac{1}{\sin^2\theta}\cdot\frac{d^2y}{d\theta^2} - \frac{\cos\theta}{\sin^3\theta}\cdot\frac{dy}{d\theta}$$

Hence by substitution in the given equation,

$$\frac{1}{\sin^2\theta}\cdot\frac{d^2y}{d\theta^2} - \frac{\cos\theta}{\sin^3\theta}\cdot\frac{dy}{d\theta} + \frac{\cos\theta}{\sin^3\theta}\cdot\frac{dy}{d\theta} + \frac{y}{\sin^2\theta} = 0$$

or

$$\frac{d^2y}{d\theta^2} + y = 0.$$

This example illustrates the important fact, that a change of the independent variable will sometimes simplify the form of the differential equation.

2. Transform $\frac{d^2u}{dx^2} + \frac{d^2u}{dy^2} = 0$, so as to render r the independent variable, where $r^2 = x^2 + y^2$.

Here $x^2 = r^2 - y^2 \quad \therefore \frac{dx}{dr} = \frac{r}{x}, \quad \therefore \frac{d^2x}{dr^2} = \frac{d}{dr}\left(\frac{r}{x}\right)$

$$= \frac{1}{x} - \frac{r}{x^2}\cdot\frac{dx}{dr} = \frac{1}{x} - \frac{r^2}{x^3} = -\frac{y^2}{x^3}$$

And similarly $\frac{dy}{dr} = \frac{r}{y}, \quad \frac{d^2y}{dr^2} = -\frac{x^2}{y^3}.$

$$\therefore \frac{d^2u}{dx^2} = \frac{\frac{d^2u}{dr^2}\cdot\frac{dx}{dr} - \frac{d^2x}{dr^2}\cdot\frac{du}{dr}}{\frac{dx^3}{dr^3}} = \frac{x^2}{r^2}\cdot\frac{d^2u}{dr^2} + \frac{y^2}{r^3}\cdot\frac{du}{dr}.$$

And

$$\frac{d^2u}{dy^2} = \frac{y^2}{r^2}\cdot\frac{d^2u}{dr^2} + \frac{x^2}{r^3}\cdot\frac{du}{dr}.$$

$\therefore$ By substitution in the given relation $\frac{d^2u}{dx^2} + \frac{d^2u}{dy^2} = 0$, and reduction,

$$\frac{d^2u}{dr^2} + \frac{1}{r}\cdot\frac{du}{dr} = 0.$$

104. *Prop.* Having given $u = F(x, y)$ when $x = \varphi(r, \theta)$ and $y = f(r, \theta)$, to express $\frac{du}{dx}$ and $\frac{du}{dy}$ in terms of r and θ.

Since u is a function of x and y, each of which is a function of r, we have

$$\frac{du}{dr} = \frac{du}{dx}\cdot\frac{dx}{dr} + \frac{du}{dy}\cdot\frac{dy}{dr}\cdot \cdots\cdots (1)$$

And similarly, x and y being functions of θ,

$$\frac{du}{d\theta} = \frac{du}{dx}\cdot\frac{dx}{d\theta} + \frac{du}{dy}\cdot\frac{dy}{d\theta}\cdot \cdots\cdots (2)$$

Multiply (1) by $\frac{dx}{d\theta}$, and (2) by $\frac{dx}{dr}$ and subtract; then multiply (1) by $\frac{dy}{d\theta}$ and (2) by $\frac{dy}{dr}$ and subtract. We shall then obtain

$$\frac{du}{dr}\cdot\frac{dx}{d\theta} - \frac{du}{d\theta}\cdot\frac{dx}{dr} = -\frac{du}{dy}\left(\frac{dx}{dr}\cdot\frac{dy}{d\theta} - \frac{dy}{dr}\cdot\frac{dx}{d\theta}\right)$$

and
$$\frac{du}{dr}\cdot\frac{dy}{d\theta} - \frac{du}{d\theta}\cdot\frac{dy}{dr} = \frac{du}{dx}\left(\frac{dx}{dr}\cdot\frac{dy}{d\theta} - \frac{dy}{dr}\cdot\frac{dx}{d\theta}\right).$$

$$\therefore \frac{du}{dx} = \frac{\frac{du}{dr}\cdot\frac{dy}{d\theta} - \frac{du}{d\theta}\cdot\frac{dy}{dr}}{\frac{dx}{dr}\cdot\frac{dy}{d\theta} - \frac{dy}{dr}\cdot\frac{dx}{d\theta}} \quad \text{and} \quad \frac{du}{dy} = -\frac{\frac{du}{dr}\cdot\frac{dx}{d\theta} - \frac{du}{d\theta}\cdot\frac{dx}{dr}}{\frac{dx}{dr}\cdot\frac{dy}{d\theta} - \frac{dy}{dr}\cdot\frac{dx}{d\theta}}.$$

105. These formulæ become much simplified when we have $x = r\cos\theta$, $y = r\sin\theta$, the common formula for passing from rectangular to polar co-ordinates. For we then have

$$\frac{dx}{dr} = \cos\theta,\quad \frac{dy}{dr} = \sin\theta,\quad \frac{dx}{d\theta} = -r\,.\sin\theta,\quad \frac{ay}{d\theta} = r\cos\theta.$$

$$\therefore \frac{dx}{dr}\cdot\frac{dy}{d\theta} - \frac{dy}{dr}\cdot\frac{dx}{d\theta} = r(\cos^2\theta + \sin^2\theta) = r.$$

$$\therefore \frac{du}{dx} = \cos\theta\,\frac{du}{dr} - \frac{\sin\theta}{r}\cdot\frac{du}{d\theta} \quad \text{and} \quad \frac{du}{dy} = \sin\theta\,\frac{du}{dr} + \frac{\cos\theta}{r}\cdot\frac{du}{d\theta}.$$

Ex. Having given $x\frac{du}{dy} - y\frac{du}{dx} = a$, to transform the equation to the variables r and θ, where $x = r\cos\theta$, $y = r\sin\theta$.

$$x\frac{du}{dy} - y\frac{du}{dx} = r\cos\theta\left(\sin\theta\frac{du}{dr} + \frac{\cos\theta}{r}\cdot\frac{du}{d\theta}\right) - r\sin\theta\left(\cos\theta\frac{du}{dr} - \frac{\sin\theta}{r}\frac{du}{d\theta}\right)$$

$$= (\cos^2\theta + \sin^2\theta)\frac{du}{d\theta} = \frac{du}{d\theta} = a.$$

CHAPTER XII.

FAILURE OF TAYLOR'S THEOREM.

106. It has been shown that the *general* development of $F(x+h)$, so long as the value of h remains unassigned, is of the form

$$F(x+h) = Fx + Ah + Bh^2 + Ch^3 + \&c. \dots . (1),$$

containing none but the positive integral powers of h.

But although this be true for the general value of x, it is possible in some cases, to assign certain particular values to x, which shall cause fractional powers of h to appear in the development; and to such cases Taylor's Theorem does not apply, because its proof depends upon the assumption that the series (1) holds true. If, for example, we assign to x such a value as shall cause fractional powers of h to appear in the undeveloped function, we may expect to find similar powers in the development, and we therefore cannot expect Taylor's Theorem to give the correct expansion. Now when the particular value $x = a$ introduced into the undeveloped function the fractional power $h^{\frac{m}{n}}$, there must have been in the general expression for Fx (before a was substituted for x) a term of the form $(x-a)^{\frac{m}{n}}$ which becomes $(x-a+h)^{\frac{m}{n}}$ in $F(x+h)$, and reduces to $h^{\frac{m}{n}}$ when $x = a$.

When this occurs some of the differential coefficients will certainly become infinite, if we make $x = a$.

To illustrate this fact, take the example

$$u = Fx = b + (x - a)^3 + (x - a)^{\frac{m}{n}},$$

and suppose x to receive the increment h, converting u into

$$u_1 = F(x + h) = b + (x - a + h)^3 + (x - a + h)^{\frac{m}{n}}.$$

Now, for the particular value $x = a$, u_1 becomes $b + h^3 + h^{\frac{m}{n}}.$ But by forming the successive differential coefficients of u with respect to x, we get

$$\frac{du}{dx} = 3(x - a)^2 + \frac{m}{n}(x - a)^{\frac{m}{n}-1},$$

$$\frac{d^2u}{dx^2} = 2.3(x - a) + \frac{m}{n}\left(\frac{m}{n} - 1\right)(x - a)^{\frac{m}{n}-2},$$

$$\frac{d^3u}{dx^3} = 1.2.3 + \frac{m}{n}\left(\frac{m}{n} - 1\right)\left(\frac{m}{n} - 2\right)(x - a)^{\frac{m}{n}-3},$$

$$\frac{d^4u}{dx^4} = \frac{m}{n}\left(\frac{m}{n} - 1\right)\left(\frac{m}{n} - 2\right)\left(\frac{m}{n} - 3\right)(x - a)^{\frac{m}{n}-4}, \text{ \&c., \&c.,}$$

and since the exponent of $x - a$ is diminished by unity at each differentiation, it must eventually become negative, rendering the coefficient infinite when $x = a$. Moreover, all the succeeding differential coefficients will likewise become infinite.

It may be observed also that if the lowest (and therefore the first) fractional exponent which appears in the development, be intermediate in value between the integers r and $r + 1$; then the first differential coefficient which becomes infinite will be the $(r + 1)th$.

It appears then that this peculiarity will arise whenever the value assigned to x causes a surd (such as $(x - a)^{\frac{m}{n}}$) to disappear in Fx, while the corresponding surd $[(x - a + h)^{\frac{m}{n}}]$ continues to appear in $F(x + h)$ in the form of a fractional power of h. This inapplicability of Taylor's Theorem, improperly called a *failure* of the theorem, occurs precisely when the development is impossible in the general form, and therefore does not result from any defect in the theorem itself.

Again, it has been shown that the general development does not contain negative powers of h, because we would have, (if there were such a term Ch^{-c}) $F(x+h) = Fx = \infty$ when $h = 0$, an obvious absurdity. But when we assign to x such a value a as shall render $Fx = \infty$, the above argument ceases to be conclusive. In this case $Fx = \infty$, and the differential coefficients will be infinite also. Thus Taylor's Theorem will be inapplicable.

Here also we see that the presence of a negative power of h in the development must result from a term of the form $\dfrac{B}{(x-a)^n}$ in Fx, which becomes $\dfrac{B}{(x-a+h)}$ in $F(x+h)$ and reduces to $\dfrac{B}{h^n} = Bh^{-n}$ when $x = a$.

We conclude, therefore, that there are two cases in which Taylor's Theorem is not applicable, viz.:

1st. When $x = a$ causes a surd to disappear in Fx, thereby introducing a fractional power of h into $F(x+h)$.

2d. When $x = a$ renders $Fx = \infty$.

EXAMPLES.

107. 1*st. Case.* Given $u = b + (x+c)^2 + (x-a)^{\frac{3}{2}} = Fx$, to expand $u_1 = F(x+h) = b + (x+c+h)^2 + (x-a+h)^{\frac{3}{2}}$.

$$\frac{du}{dx} = 2(x+c) + \frac{3}{2}(x-a)^{\frac{1}{2}}, \qquad \frac{d^2u}{dx^2} = 1.2 + \frac{1}{2}\cdot\frac{3}{2}(x-a)^{-\frac{1}{2}}$$

$$\frac{d^3u}{dx^3} = -\frac{1}{2}\cdot\frac{1}{2}\cdot\frac{3}{2}(x-a)^{-\frac{3}{2}} \qquad \text{\&c., \&c.}$$

$\therefore$ By substitution in Taylor's Theorem,

$$u_1 = b + (x + c)^2 + (x - a)^{\frac{3}{2}} + \left[2(x + c) + \frac{3}{2}(x - a)^{\frac{1}{2}}\right]\frac{h}{1}$$

$$+ \left[1.2 + \frac{1}{2}\cdot\frac{3}{2}(x - a)^{-\frac{1}{2}}\right]\frac{h^2}{1.2}$$

$$-\frac{1}{2}\cdot\frac{1}{2}\cdot\frac{3}{2}(x - a)^{-\frac{3}{2}}\cdot\frac{h^3}{1.2.3} + \&c. \dots . (1).$$

Now this development is entirely true for all values of x, except $x = a$, which renders the term $\left[1.2 + \frac{1}{2}\cdot\frac{3}{2}(x - a)^{-\frac{1}{2}}\right]\frac{h^2}{1.2}$, and all succeeding terms, infinite; the true development in this case being $u_1 = b + (a + c + h)^2 + h^{\frac{3}{2}} = b + (a + c)^2 + 2(a + c)h + h^{\frac{3}{2}} + h^2$, which agrees with the series (1), only so far as to include the term

$$\left[2(x + c) + \frac{3}{2}(x - a)^{\frac{1}{2}}\right]\frac{h}{1}.$$

2d. *Case.* Given $u = b + \sin x + \dfrac{c}{(x - a)^2} = Fx$, to expand

$$u_1 = F(x + h) = b + \sin(x + h) + \frac{c}{(x - a + h)^2}.$$

$$\frac{du}{dx} = \cos x - \frac{1.2c}{(x - a)^3}, \quad \frac{d^2u}{dx^2} = -\sin x + \frac{1.2.3c}{(x - a)^4},$$

$$\frac{d^3u}{dx^3} = -\cos x - \frac{1.2.3.4c}{(x - a)^5}, \qquad \&c. \qquad \&c.$$

∴ By substitution in Taylor's Theorem,

$$u_1 = b + \sin x + \frac{c}{(x - a)^2} + \left[\cos x - \frac{1.2c}{(x - a)^3}\right]\frac{h}{1}$$

$$+ \left[-\sin x + \frac{1.2.3c}{(x - a)^4}\right]\frac{h^2}{1.2} + \left[-\cos x - \frac{1.2.3.4c}{(x - a)^5}\right]\frac{h^3}{1.2.3} \ \&c.$$

This development is correct except when $x = a$, the true development then being (Art. 48)

$$u_1 = b + \sin(a + h) + \frac{c}{h^2} = b + \sin a + ch^{-2} + \cos a.h - \sin a\frac{h^2}{1.2} \ \&c.$$

Here the very first term given by Taylor's formula, viz.:

$$Fx = b + \sin a + \frac{c}{(a-a)^2}, \quad \text{is incorrect.}$$

108. *Prop.* If the true development of $F(x+h)$ contain positive integral powers of h to the $(n-1)th$ power inclusive, followed by a term containing h^s where s is a fraction intermediate in value between $n-1$ and n, the first n terms of the expansion will be given correctly by Taylor's Theorem, but the $(n+1)th$ term will not be given correctly.

Proof. Let the true development of $F(x+h)$, when $x = a$, be

$$F(x+h) = A + Bh + Ch^2 + Dh^3 \ldots\ldots + Nh^{n-1} + Ph^s + \&c.,$$

where s denotes a fraction greater than $n-1$ and less than n.

Then, since the differential coefficients of $F(x+h)$, taken first with respect to x, and afterwards with respect to h, are equal, we have

$$\frac{dF(x+h)}{dx} = \frac{dF(x+h)}{dh} = 1B + 2Ch + 3Dh^2 \ldots.$$
$$+ (n-1)Nh^{n-2} + sPh^{s-1} + \&c.$$

$$\frac{d^2F(x+h)}{dx^2} = \frac{d^2F(x+h)}{dh^2} = 1 \,.\, 2C + 2 \,.\, 3Dh \ldots\ldots$$
$$+ (n-2)(n-1)Nh^{n-3} + (s-1)sPh^{s-2} + \&c.$$

$$\frac{d^3F(x+h)}{dx^3} = 1 \,.\, 2 \,.\, 3D \ldots\ldots + (n-3)(n-2)(n-1)Nh^{n-4}$$
$$+ (s-2)(s-1)sPh^{s-3} + \&c.$$

. . . .

$$\frac{d^{n-1}F(x+h)}{dx^{n-1}} = 1 \,.\, 2 \,.\, 3 \ldots (n-3)(n-2)(n-1)N$$
$$+ (s-n+2)(s-n+3) \ldots. (s-2)(s-1)sPh^{s-n+1} + \&c.$$

$$\frac{d^nF(x+h)}{dx^n} = (s-n+1)(s-n+2)(s-n+3) \ldots.$$
$$(s-2)(s-1)sPh^{s-n} + \&c.$$

Now when $h = 0$, the preceding expressions reduce to

$$Fx = A, \quad \frac{dFx}{dx} = 1B, \quad \frac{d^2Fx}{dx^2} = 1 \,.\, 2C, \quad \frac{d^3Fx}{dx^3} = 1 \,.\, 2 \,.\, 3D \ldots..$$

$$\frac{d^{n-1}Fx}{dx^{n-1}} = 1.2.3\ldots\ldots(n-3)(n-2)(n-1)N.$$

$$\frac{d^nFx}{dx^n} = (s-n+1)(s-n+2)(s-n+3)\ldots\ldots(s-2)(s-1)s\frac{P}{0} = \infty.$$

$$\therefore A = Fx, \quad B = \frac{dFx}{dx}, \quad C = \frac{1}{1.2}\frac{d^2Fx}{dx^2}, \quad \&c.$$

Thus each of the terms A, Bh, Ch^2, &c., of the true development will be given correctly by Taylor's Theorem as far as the term Nh^{n-1} inclusive (that is to n terms), but the $(n+1)th$ term of the true expansion is Ph^s, while by Taylor's series it would appear to be infinite.

The results established in this proposition are important, because it frequently occurs that the first or leading terms of an expansion, are those only which we have occasion to consider.

PART II.

APPLICATION OF THE DIFFERENTIAL CALCULUS TO THE THEORY OF PLANE CURVES.

CHAPTER I.

TANGENTS TO PLANE CURVES.—NORMALS.—ASYMPTOTES.

109. In the application of the Differential Calculus to the investigation of the properties of plane curves, we regard the two variable co-ordinates x and y or θ and r, which serve to fix the position of a point on the curve, as the independent variable and the dependent function respectively.

These two quantities are connected by a general relation called the equation of the curve,

Such as $y = Fx$ or $r = \phi\theta$, $F(x, y) = 0$, or $\phi(r, \theta) = 0$.

When the form of this equation is given, we can readily determine the values of the differential coefficients $\frac{dy}{dx}$, $\frac{d^2y}{dx^2}$, &c., or $\frac{dr}{d\theta}$, $\frac{d^2r}{d\theta^2}$, &c., in terms of the co-ordinates, and these values will be found extremely serviceable in the discussion of the properties of the curves.

110. The first application of the Calculus to Geometry which it is proposed to make, is the determination of the tangents to plane curves.

Prop. To find the general differential equation of a line which is tangent to a plane curve at a given point $x_1\ y_1$.

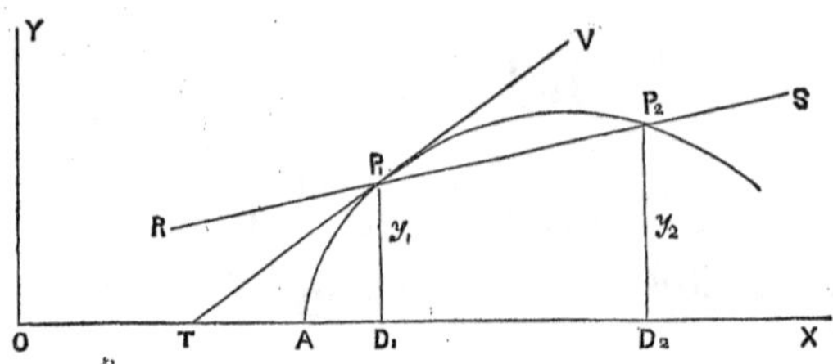

The equation of the secant line RS, passing through the points $x_1\ y_1$ and $x_2\ y_2$, is

$$y - y_1 = \frac{y_2 - y_1}{x_2 - x_1}(x - x_1) \quad \ldots . \quad (1).$$

But if the secant RS be caused to revolve about the point P_1, approaching to coincidence with the tangent TV, the point P_2 will approach P_1, and the differences $y_2 - y_1$ and $x_2 - x_1$ will also diminish, so that at the limit, when RS and TV coincide, $\frac{y_2 - y_1}{x_2 - x_1}$ will reduce to $\frac{dy_1}{dx_1}$, and the equation (1) will take the form

$$y - y_1 = \frac{dy_1}{dx_1}(x - x_1) \quad \ldots . \quad (2),$$

which is the required equation of the tangent line at the point $x_1\ y_1$.

111. To apply (2) to any particular curve we substitute for $\frac{dy_1}{dx_1}$ its value deduced from the equation of the curve and expressed in terms of the co-ordinates of the point of tangency.

Cor. The differential coefficient $\frac{dy_1}{dx_1}$ represents the trigonometrical tangent of the angle P_1TX formed by the tangent line with the axis of x.

Cor. To find the value of the subtangent D_1T, we make $y = 0$ in (2). The corresponding value of x will be the distance OT, and therefore $x - x_1$ will represent the subtangent D_1T, this latter being reckoned *from* D_1 the foot of the ordinate. Thus

$$\text{subtan } D_1T = x - x_1 = -\frac{y_1}{\frac{dy_1}{dx_1}} \quad \ldots . (3).$$

In the formula (3), x represents the independent variable, but if we take y as the independent variable, this formula may be simplified. For it has been shown that $\frac{dy}{dx} = \frac{1}{\frac{dx}{dy}}$ or $\frac{1}{\frac{dy}{dx}} = \frac{dx}{dy}$. Hence (3) may be written

$$\text{subtan } D_1T = -y_1\frac{dx_1}{dy_1} \quad \ldots . (4).$$

112. *Prop.* To determine the general differential equation of a line which is normal to a plane curve at a given point $x_1\,y_1$.

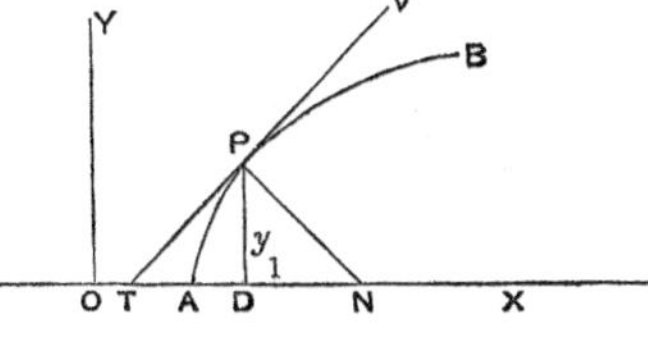

The equation of the normal PN, which passes through the point $x_1\,y_1$, will be of the form

$y - y_1 = t_1\,(x - x_1) \ldots (5)$,

where t_1 denotes the unknown tangent of the angle PNX formed by PN with the axis of x.

But since the normal PN is perpendicular to the tangent PT, we must have, by the condition of perpendicularity of lines in a plane,

$$1 + tt_1 = 0 \quad \text{or} \quad t_1 = -\frac{1}{t} \quad \text{where} \quad t = \frac{dy_1}{dx_1} = \text{tan. angle } PTD.$$

Replacing t_1 by its value in (5) there results

$$y - y_1 = -\frac{1}{\frac{dy_1}{dx_1}}(x - x_1) = -\frac{dx_1}{dy_1}(x - x_1) \quad \ldots . (6).$$

To apply (6) we substitute for $\frac{dx_1}{dy_1}$ its value derived from the equation of the given curve.

Cor. To find the value of the subnormal DN, we make $y = 0$ in (6) and thus obtain ON as the corresponding value of x.

$$\therefore DN = x - x_1 = y_1 \frac{dy_1}{dx_1} \quad \dots \quad (7),$$

when either the subtangent or subnormal has been determined, the tangent and normal can be readily constructed.

APPLICATIONS.

113. 1. Let the curve be the common parabola, whose equation is

$$y^2 = 2px.$$

$$\therefore \frac{dy}{dx} = \frac{p}{y}, \quad \frac{dy_1}{dx_1} = \frac{p}{y_1}, \quad \text{and} \quad \frac{dx_1}{dy_1} = \frac{y_1}{p}.$$

Hence the equation of the tangent is

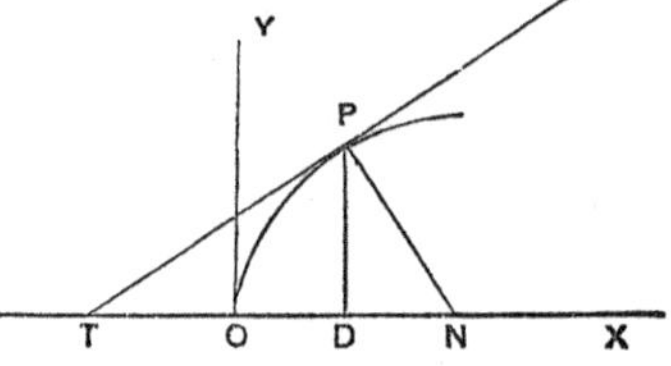

$$y - y_1 = \frac{p}{y_1}(x - x_1)$$

or $yy_1 - y_1^2 = p(x - x_1)$,

whence

$$yy_1 = p(x - x_1) + 2px_1 = p(x + x_1).$$

And that of the normal is

$$y - y_1 = -\frac{y_1}{p}(x - x_1).$$

Also, $\quad \text{subtan } DT = -\frac{y_1}{p} \cdot y_1 = -\frac{2px_1}{p} = -2x_1,$

and $\quad \text{subnorm } DN = y_1 \frac{p}{y_1} = p.$

Thus it appears that the subtangent of the parabola is negative and equal to twice the abscissa; and the subnormal is positive and constant, being equal to the semi-parameter.

2. The Ellipse, $a^2y^2 + b^2x^2 = a^2b^2$

$$\frac{dy}{dx} = -\frac{b^2x}{a^2y}, \quad \therefore \quad \frac{dy_1}{dx_1} = -\frac{b^2x_1}{a^2y_1}, \quad \text{and} \quad \frac{dx_1}{dy_1} = -\frac{a^2y_1}{b^2x_1}$$

$\therefore$ The equation of the tangent is

$$y - y_1 = -\frac{b^2x_1}{a^2y_1}(x - x_1), \qquad \text{or,} \qquad a^2yy_1 + b^2xx_1 = a^2b^2.$$

$$\text{Also subtangent} = -y_1\frac{dx_1}{dy_1} = \frac{a^2y_1{}^2}{b^2x_1} = \frac{a^2}{x_1} - x_1.$$

$$\text{And subnormal} = y_1\frac{dy_1}{dx_1} = -\frac{b^2}{a^2}x_1.$$

3. The logarithmic curve, whose equation is $y = a^x$.

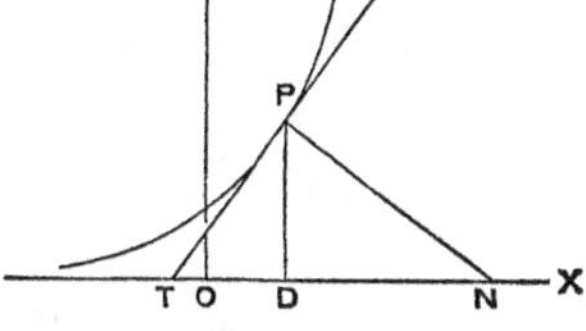

$$\frac{dy}{dx} = \log a \,.\, a^x.$$

$$\therefore \quad \text{subtan} = -\frac{y_1}{\log a \,.\, a^{x_1}} = -\frac{1}{\log a} = -m,$$

where m is the modulus of the system of logarithms whose base is a.

$$\text{Also subnorm} = \log a \,.\, a^{x_1}y_1 = \frac{y_1{}^2}{m} = \frac{a^{2x_1}}{m}.$$

In this curve, the values of the abscissas are the logarithms of the values of the corresponding ordinates in the system whose base is a.

114. *Prop.* To determine expressions for the tangent, the normal, and the perpendicular from the origin to the tangent of a plane curve.

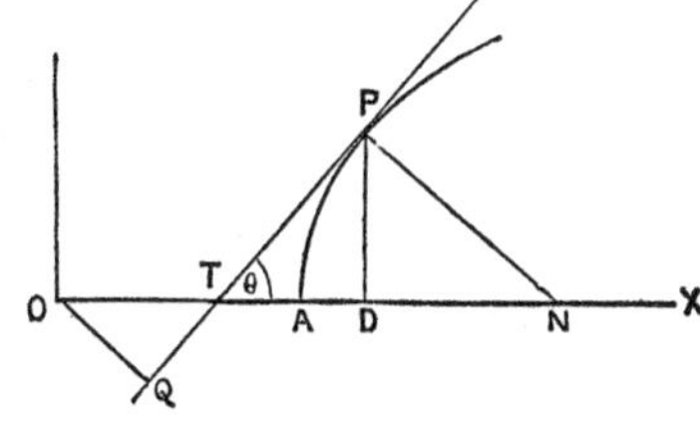

For the tangent PT, we have

$$PT = \sqrt{PD^2 + DT^2}$$

$$= y_1\sqrt{1 + \frac{dx_1{}^2}{dy_1{}^2}}.$$

For the normal PN, we have

$$PN = \sqrt{PD^2 + DN^2} = y_1\sqrt{1 + \frac{dy_1^2}{dx_1^2}}.$$

For the perpendicular OQ, we have

$$OQ = OT.\sin OTQ = OT\frac{1}{\operatorname{cosec}\theta} = OT\sqrt{\frac{1}{1 + \cot^2\theta}} = \frac{x_1 - y_1\frac{dx_1}{dy_1}}{\left[1 + \frac{dx_1^2}{dy_1^2}\right]^{\frac{1}{2}}}$$

or,
$$OQ = \frac{x_1 dy_1 - y_1 dx_1}{(dx_1^2 + dy_1^2)^{\frac{1}{2}}}.$$

Ex. The general equation of all parabolas.

The general name of *parabola* is applied to all curves included in the equation $y^m = a^{m-1}x$, in which m may represent any positive number either whole or fractional. When $m = 2$, the curve becomes the common parabola.

Here $y^m = a^{m-1}x$, $\therefore \frac{dy}{dx} = \frac{a^{m-1}}{my^{m-1}}$, and $\frac{dx}{dy} = \frac{my^{m-1}}{a^{m-1}} = \frac{mx_1}{y_1}$.

$$\therefore \text{subtan} = -y_1\frac{dx_1}{dy_1} = -\frac{my_1^m}{a^{m-1}} = -mx_1$$

$$\text{subnorm} = y_1\frac{dy_1}{dx_1} = \frac{a^{m-1}}{my_1^{m-2}} = \frac{y_1^2}{mx_1},$$

$$\tan = y_1\sqrt{1 + \frac{dx_1^2}{dy_1^2}} = \sqrt{y_1^2 + m^2x_1^2},$$

$$\text{norm} = y_1\sqrt{1 + \frac{dy_1^2}{dx_1^2}} = \sqrt{y_1^2 + \frac{y_1^4}{m^2x_1^2}}$$

and
$$\text{perp} = \frac{x_1 - mx_1}{\left[1 + \frac{m^2x_1^2}{y_1^2}\right]^{\frac{1}{2}}} = \frac{x_1y_1(1 - m)}{[y_1^2 + m^2x_1^2]^{\frac{1}{2}}}.$$

115. *Prop.* To obtain expressions for the polar subtangent, subnormal, tangent, normal, and perpendicular to the tangent of a plane curve, when it is referred to polar co-ordinates.

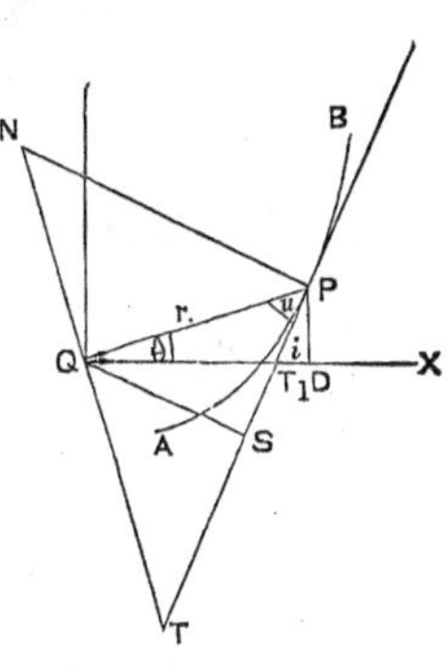

Let AB be the curve, Q the pole, P the point to be referred, QX the fixed axis from which the variable angle PQX is reckoned, QP the radius vector, TQN a line drawn through the pole Q, perpendicular to the radius vector PQ, and limited by the tangent PT, and the normal PN, QS a perpendicular on the tangent from the pole. Then QT is called the polar subtangent, and QN the polar subnormal.

Put $QP = r$, angle $PQX = \theta$, angle $QPT = u$, angle $PT_1X = i$, $QD = x$, $DP = y$.

Then $QT = QP \, . \tan QPT = r \tan u = r \, . \tan(i - \theta)$

$$= r \, . \frac{\tan i - \tan \theta}{1 + \tan i \tan \theta}.$$

But $\tan i = \dfrac{dy}{dx}$, $\tan \theta = \dfrac{y}{x}$, $\therefore \tan u = \dfrac{\dfrac{dy}{dx} - \dfrac{y}{x}}{1 + \dfrac{y}{x} \cdot \dfrac{dy}{dx}}.$

Now if we change the independent variable from x to θ, we must employ the formula $\dfrac{dy}{dx} = \dfrac{\dfrac{dy}{d\theta}}{\dfrac{dx}{d\theta}}$,

$$\therefore \tan u = \frac{x\dfrac{dy}{d\theta} - y\dfrac{dx}{d\theta}}{x\dfrac{dx}{d\theta} + y\dfrac{dy}{d\theta}}, \quad . \; . \; . \; . \; (1).$$

And from the formula for passing from rectangular to polar co-ordinates, we have $x = r \cos \theta$, $y = r \sin \theta$, which being differentiated with respect to θ, observing that r is a function of θ, we get

$$\therefore \frac{dx}{d\theta} = \frac{dr}{d\theta} \cdot \cos \theta - r \, . \sin \theta, \quad \frac{dy}{d\theta} = \frac{dr}{d\theta} \sin \theta + r \cos \theta.$$

and these substituted in (1) give

$$\tan u = \frac{r\cos\theta\left(\frac{dr}{d\theta}\cdot\sin\theta + r\cos\theta\right) - r\sin\theta\left(\frac{dr}{d\theta}\cdot\cos\theta - r\sin\theta\right)}{r\cos\theta\left(\frac{dr}{d\theta}\cdot\cos\theta - r\sin\theta\right) + r\sin\theta\left(\frac{dr}{d\theta}\cdot\sin\theta + r\cos\theta\right)}$$

$$= \frac{r}{\frac{dr}{d\theta}} = r\frac{d\theta}{dr}.$$

$$\therefore \text{ subtangent } QT = r\tan u = \frac{r^2}{\frac{dr}{d\theta}} = r^2\frac{d\theta}{dr}.$$

$$\text{Also subnormal } QN = \frac{QP^2}{QT} = \frac{dr}{d\theta}.$$

$$\text{Tangent } PT = \sqrt{QP^2 + QT^2} = r\sqrt{1 + r^2\frac{d\theta^2}{dr^2}}.$$

$$\text{Normal } PN = \sqrt{QP^2 + QN^2} = \sqrt{r^2 + \frac{dr^2}{d\theta^2}}.$$

$$\text{Perpendicular } QS = \frac{PQ \times QT}{PT} = \frac{r^2}{\sqrt{r^2 + \frac{dr^2}{d\theta^2}}}.$$

EXAMPLES.

116. 1. The spiral of Archimedes whose equation is $r = a\theta$.

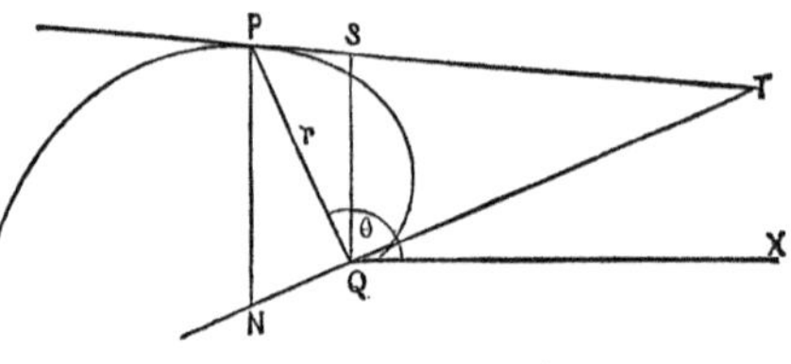

$$\frac{dr}{d\theta} = a, \quad \frac{d\theta}{dr} = \frac{1}{a},$$

$$\therefore \text{ subtan } QT = r^2\frac{d\theta}{dr} = \frac{r^2}{a}, \quad \text{subnorm } QN = \frac{dr}{d\theta} = a,$$

$$\tan PT = r\sqrt{1 + \frac{r^2}{a^2}}, \text{ norm } PN = \sqrt{r^2 + a^2}, \text{ perp } QS = \frac{r^2}{\sqrt{r^2 + a^2}}.$$

2. The logarithmic spiral $r = a^\theta$.

In this curve, the numerical value θ of the arc which measures the variable angle is the logarithm of the value of the radius vector r, in the system whose base is a.

$$\frac{dr}{d\theta} = \log a a^\theta = r \,.\, \log a. \qquad \therefore \text{ Subtan} = \frac{r}{\log a} = mr, \text{ where}$$

$$m = \text{modulus.} \quad \text{Subnormal} = r \,.\, \log a = \frac{r}{m}.$$

This curve cuts every radius vector under the same angle; that is, the tangent at any point is inclined to the radius vector at that point in a constant angle.

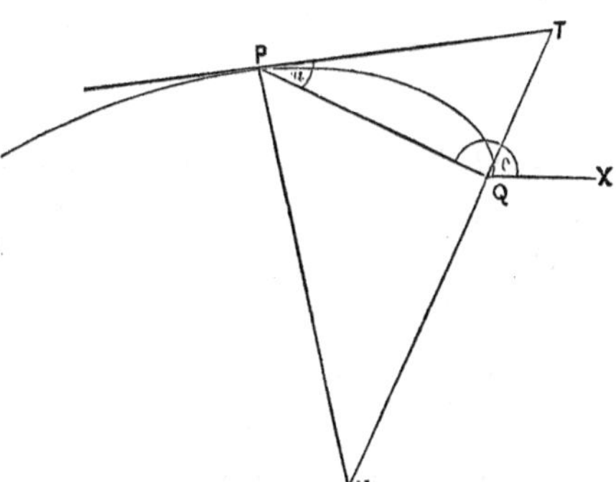

For $\quad \tan u = r\dfrac{d\theta}{dr}$

$$= \frac{r}{r \log a} = \frac{1}{\log a} = m.$$

If $a = e$ the Naperian base, then $\log a = 1$, $\tan u = 1$ and $u = 45°$, and $QT = QN = r$.

3. The lemniscata of Bernouilli, $r^2 = a^2 \cos 2\theta$.

$$r\frac{dr}{d\theta} = -a^2 \sin 2\theta. \quad \therefore \text{ subtan} = \frac{-r^3}{a^2 \sin 2\theta}, \text{ subnorm} = \frac{-a^2}{r} \cdot \sin 2\theta.$$

$$\text{perp} = \frac{r^3}{\sqrt{r^4 + a^4 \sin^2 2\theta}} = \frac{r^3}{\sqrt{a^4 \cos^2 2\theta + a^4 \sin^2 2\theta}} = \frac{r^3}{a^2}.$$

This curve has the form of the figure 8, is perpendicular to the axis AB at A and B, and forms angles of 45° with AB at the pole Q. For when $\theta = 0$,

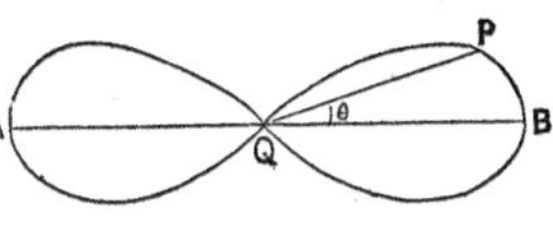

or $\theta = \pi$, $r = a$, and $\dfrac{dr}{d\theta} = 0$. And when $\theta = 45°$, or 135°, or 225°, or 315°, then $r = 0$.

Rectilinear Asymptotes.

117. A rectilinear asymptote to a curve is a line which touches the curve at a point infinitely distant from the origin, and yet passes within a finite distance of the origin.

118. If in the differential equation of a tangent line

$$y - y_1 = \frac{dy_1}{dx_1}(x - x_1), \text{ we make successively } x=0, \text{ and } y=0,$$

we shall obtain for the distances intercepted on the axes,

$$y' = y_1 - x_1\frac{dy_1}{dx_1}, \quad \text{and} \quad x' = x_1 - y_1\frac{dx_1}{dy_1}.$$

Now if when either x_1 or y_1 becomes infinite, one or both of these values should prove finite, the curve will have an asymptote whose position will be determined by the values of x' and y'.

If $x' = a$, and $y' = b$ when a and b are both finite, the asymptote will cut both axes: if $x' = a$ and $y' = \infty$, the asymptote will be parallel to the axis of y; and, finally, if $x' = \infty$ and $y' = b$, the asymptote will be parallel to the axis of x.

119. When the curve is referred to polar co-ordinates, there will be an asymptote whenever the subtangent (which is then equal to the perpendicular from the pole upon the tangent) becomes finite for an infinite value of the radius vector. Its position will be fixed also, since it will be parallel to the radius vector; that is, it will form with the radius vector an indefinitely small angle. The existence of an asymptote may be ascertained from the equation of the curve by finding what value of θ will render r infinite. If the same value of θ makes $r^2\frac{d\theta}{dr}$ either finite or zero, there will be an asymptote parallel to the radius vector, and passing through the extremity of the subtangent.

120. 1. The hyperbola $a^2y^2 - b^2x^2 = -a^2b^2$.

$$\frac{dy_1}{dx_1} = \frac{b^2x_1}{a^2y_1} = \frac{b^2}{a^2}\cdot\frac{a}{b}\cdot\sqrt{\frac{b^2+y_1^2}{y_1^2}} = \frac{b}{a}\sqrt{1+\frac{b^2}{y_1^2}} = \frac{b}{a} \text{ when } y_1 = \infty.$$

Also $y_1 = \frac{b}{a}\sqrt{x_1^2 - a^2} = \frac{bx_1}{a}\sqrt{1 - \frac{a^2}{x_1^2}} = \frac{bx_1}{a}$ when x_1 or $y_1 = \infty$.

$$\therefore\ y' = y_1 - x_1\frac{dy_1}{dx_1} = \frac{bx_1}{a} - \frac{bx_1}{a} = 0$$

and $$x' = x_1 - y_1\frac{dx_1}{dy_1} = x_1 - \frac{bx_1}{a}\cdot\frac{a}{b} = x_1 - x_1 = 0.$$

$\therefore$ The hyperbola has an asymptote passing through the origin, and forming with the axis of x an angle whose tangent $= \pm\frac{b}{a}$.

2. The logarithmic curve $y = a^x$.

$$\frac{dy}{dx} = \log a \,.\, a^x,\ x' = x_1 - y_1\frac{dx_1}{dy_1} = x_1 - \frac{a^{x_1}}{\log a\,.\,a^{x_1}} = x_1 - m.$$

$$y' = y_1 - x_1\frac{dy_1}{dx_1} = a^{x_1} - \frac{x_1 a^{x_1}}{m}.$$

Now when $x_1 = +\infty$, $y_1 = +\infty$, $\therefore\ x' = \infty$ and $y' = \infty$ and the corresponding tangent is not an asymptote.

But when $x_1 = -\infty$, $y_1 = 0$. $\therefore\ x' = -\infty$ and $y' = 0$, and therefore the axis of x is an asymptote.

3. The cissoid whose equation is $y^2 = \frac{x^3}{2r - x}$ or $2ry^2 - y^2x - x^3 = 0$

$$\frac{dy_1}{dx_1} = \frac{y_1^2 + 3x_1^2}{4ry_1 - 2x_1y_1},\quad \therefore\ x' = x_1 - y_1\frac{4ry_1 - 2x_1y_1}{y_1^2 + 3x_1^2} = \frac{2x_1 - 4r}{1 + \frac{3x_1^2}{y_1^2}} + x_1$$

$$\therefore\ x' = 2r,\quad \text{when}\quad x_1 = 2r\quad \text{and}\quad y_1 = \infty.$$

Also $$y' = y_1 - x_1\frac{y_1^2 + 3x_1^2}{4ry_1 - 2x_1y_1} = y_1 - \frac{y_1(6r - 2x_1)}{4r - 2x_1}$$

$$= -\frac{2ry_1}{4r - 2x_1} = \infty \text{ when } x_1 = 2r.$$

$\therefore$ The cissoid has an asymptote parallel to y, at a distance $2r$ from the origin.

4. The parabola $y^2 = 2px$.

$$\frac{dy_1}{dx_1} = \frac{p}{y_1} \quad \therefore x' = x_1 - y_1 \frac{y_1}{p} = x_1 - 2x_1 = -x_1 = \infty \text{ when } x_1 = \infty.$$

Also $y' = y_1 - x_1 \frac{p}{y_1} = y_1 - \frac{1}{2} y_1 = \frac{1}{2} y_1 = \infty$ when $y_1 = \infty$ or $x_1 = \infty$,

$\therefore$ The parabola has no asymptote.

5. To find the equation of the asymptote to the curve $y^3 = ax^2 + x^3$.

$$y_1 = \infty, \quad \text{when} \quad x_1 = \infty.$$

$$\frac{dy_1}{dx_1} = \frac{2ax_1 + 3x_1^2}{3(ax_1^2 + x_1^3)^{\frac{2}{3}}} = \frac{\frac{2a}{x_1} + 3}{3\left[\frac{a}{x_1} + 1\right]^{\frac{2}{3}}} = 1 \quad \text{when} \quad x_1 = \infty.$$

Also
$$y' = y_1 - x_1 \frac{2ax_1 + 3x_1^2}{3(ax_1^2 + x_1^3)^{\frac{2}{3}}} = (ax_1^2 + x_1^3)^{\frac{1}{3}} - \frac{2ax_1^2 + 3x_1^3}{3(ax_1^2 + x_1^3)^{\frac{2}{3}}}$$

$$= \frac{ax_1^2}{3(ax_1^2 + x_1^3)^{\frac{2}{3}}} = \frac{a}{3\left(\frac{a}{x_1} + 1\right)^{\frac{2}{3}}} = \frac{a}{3} \quad \text{when} \quad x_1 = \infty.$$

$$\therefore y = x + \frac{1}{3} a \text{ the equation of the asymptote.}$$

Polar Curves. 1. The hyperbolic spiral $r\theta = a$.

$$\frac{d\theta}{dr} = -\frac{a}{r^2}. \quad \therefore \text{subtan} = r^2 \frac{a}{r^2} = a, \quad \text{for all values of } r.$$

$\therefore$ There is an asymptote which passes at a distance a from the origin. Also, since $r = \infty$ when $\theta = 0$, the asymptote is parallel to the fixed axis from which θ is reckoned.

2. The spiral of Archimedes $r = a\theta$.

$$\frac{d\theta}{dr} = \frac{1}{a}, \quad \text{subtan} = \frac{r^2}{a} = \infty \quad \text{when} \quad r = \infty.$$

$\therefore$ The curve has no asymptote.

3. The logarithmic spiral $r = a^{\theta}$.

$$\frac{d\theta}{dr} = \frac{m}{r}, \quad \text{subtan} = \frac{mr^2}{r} = mr = \infty \quad \text{when} \quad r = \infty.$$

$\therefore$ There is no asymptote.

4. The Lituus $r\theta^{\frac{1}{2}} = a$.

$$\frac{d\theta}{dr} = -\frac{2a^2}{r^3}. \quad \therefore \text{subtan} = \frac{2a^2r^2}{r^3} = \frac{2a^2}{r} = 0 \quad \text{when } r = \infty.$$

Also $r = \infty$ when $\theta = 0$. $\therefore$ The fixed axis is an asymptote.

Circular Asymptotes.

121. When the equation of the curve has such a form as will render $r =$ a finite value when $\theta = \infty$, the curve will make an infinite number of revolutions about the pole before becoming tangent to a circle whose radius $= a$. This circle is therefore called a *circular* asymptote. If $r > a$ for every finite value of θ, the curve will lie wholly exterior to the circle; but if $r < a$ for all finite values of θ, the curve will lie entirely within the circle.

1. Let the equation be $(r^2 - ar)\theta^2 = 1$ or $\theta = \dfrac{1}{\sqrt{r^2 - ar}}$.

Then $\theta = \infty$ when $r = a$. And θ is real when $r > a$, but imaginary when $r < a$.

$\therefore$ The circle with radius $= a$ is an asymptote, and lies within the spiral.

2. The curve $(ar - r^2)\theta^2 = 1$.

$$\theta = \frac{1}{\sqrt{ar - r^2}} = \infty \quad \text{when} \quad r = a.$$

Also θ is real when $r < a$, and imaginary when $r > a$.

$\therefore$ The circle with radius $= a$ is an asymptote and encloses the curve within it.

CHAPTER II.

CURVATURE AND OSCULATION OF PLANE CURVES.

122. As introductory to the discussion of the subject of the curvature of plane curves, the following proposition will be found useful:

Prop. To show that the limit to the ratio of the chord and arc of any plane curve, when that arc is diminished indefinitely, is unity, and to deduce an expression for the differential of the arc of a plane curve, in terms of the differentials of the co-ordinates.

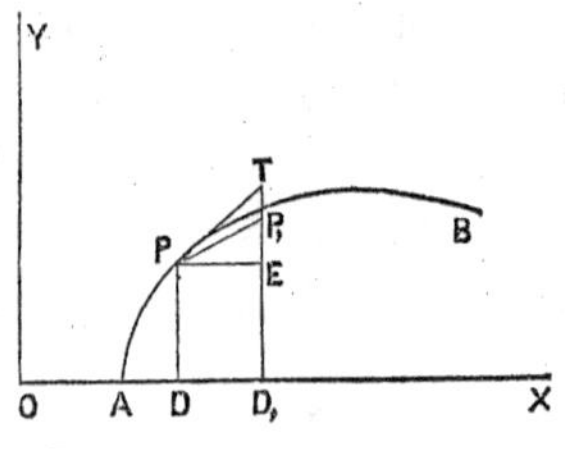

Let PP_1 be an arc of a plane curve APB, whose equation is $y = Fx$.

Put $OD=x$, $DP=y$, $DD_1=h$, $D_1P_1=y_1$, $AP=s$, $AP_1=s_1$.

Then $$y_1 = F(x+h).$$

The arc PP_1 is intermediate in length between the chord $PP_1=C$, and the broken line $PTP_1 = B$. If, therefore, we can prove that the limit to the ratio $\frac{B}{C}$ is unity, it will follow that the limit to the ratio of the chord and arc is unity, and therefore at that limit the expression for the chord PP_1 will be a suitable expression for the arc PP_1 which will then become the differential of s.

But $\frac{B}{C} = \frac{PT + P_1T}{PP_1} = \frac{\sqrt{PE^2 + ET^2} + P_1T}{\sqrt{PE^2 + EP_1^2}}$

$$= \frac{\sqrt{h^2 + h^2\frac{dy^2}{dx^2}} + h\frac{dy}{dx} - (y_1 - y)}{\sqrt{h^2 + (y_1 - y)^2}}$$

$$\overset{*}{=} \frac{h\sqrt{1 + \frac{dy^2}{dx^2}} + h\frac{dy}{dx} - \left(h\frac{dy}{dx} + \frac{h^2}{1.2}\cdot\frac{d^2y}{dx^2} + \frac{h^3}{1.2.3}\cdot\frac{d^3y}{dx^3} + \&c.\right)}{\sqrt{h^2 + \left(h\frac{dy}{dx} + \frac{h^2}{1.2}\cdot\frac{d^2y}{dx^2} + \&c.\right)^2}}$$

and by dividing numerator and denominator by h

$$\frac{B}{C} = \frac{\sqrt{1 + \frac{dy^2}{dx^2}} - h\left[\frac{1}{1.2}\frac{d^2y}{dx^2} + \frac{h}{1.2.3}\cdot\frac{d^3y}{dx^3} + \&c.\right]}{\sqrt{1 + \frac{dy^2}{dx^2} + \frac{h}{1}\cdot\frac{dy}{dx}\cdot\frac{d^2y}{dx^2} + \frac{h^2}{1.3}\cdot\frac{dy}{dx}\cdot\frac{d^3y}{dx^3} + \&c.}} = 1, \text{ when } h = 0.$$

$\therefore$ at the limit,

$$\frac{\text{arc}}{\text{chord}} = 1, \quad \text{or} \quad \frac{ds}{dx} = \frac{\text{chord}}{dx} = \frac{\tan PT}{dx} = \frac{h\sqrt{1 + \frac{dy^2}{dx^2}}}{h}$$

or $\frac{ds}{dx} = \sqrt{1 + \frac{dy^2}{dx^2}}$. $\therefore ds = dx\sqrt{1 + \frac{dy^2}{dx^2}} = \sqrt{dx^2 + dy^2}$.

Also $$ds = dy\sqrt{1 + \frac{dx^2}{dy^2}}.$$

123. In the first of these expressions x is the independent variable; in the second, y.

Cor. If we wish to employ some other quantity t upon which s, x and y depend, as the independent variable, we must use the formulæ for changing the independent variable, viz.:

$$\frac{ds}{dx} = \frac{\frac{ds}{dt}}{\frac{dx}{dt}} \quad \text{and} \quad \frac{dy}{dx} = \frac{\frac{dy}{dt}}{\frac{dx}{dt}}.$$

which, substituted in the value of $\frac{ds}{dx}$, give

$$\frac{ds}{dt} = \sqrt{\frac{dx^2}{dt^2} + \frac{dy^2}{dt^2}}.$$

124. We proceed now to consider the osculation of plane curves.

Let $Y = Fx$ (1), and $y = \varphi x$ (2) be the equations of two plane curves, the first of which is given in species, magnitude, and position, but the latter in species only.

Then the constants or *parameters* which enter into equation (1) are fixed and determinate, but those which appear in (2) entirely arbitrary, and may therefore be so assumed as to fulfil as many independent conditions as there are constants to be determined.

If, when the abscissa x is supposed the same in both curves, the condition $y = Y$ is satisfied, the curves will have a common point P, but will usually intersect at that point.

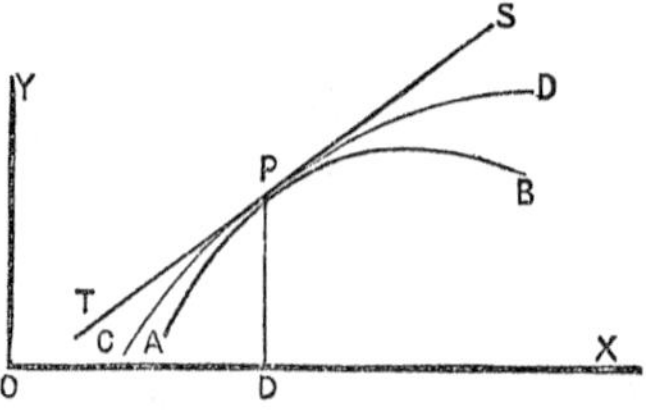

If the condition $\frac{dy}{dx} = \frac{dY}{dx}$ be true also, the curves will have a common tangent such as SPT; and the contact is then said to be of the first order: if the second differential coefficients be also equal, viz., $\frac{d^2y}{dx^2} = \frac{d^2Y}{dx^2}$, the contact is said to be of the second order; if $\frac{d^3y}{dx^3} = \frac{d^3Y}{dx^3}$, the contact is of the third order, &c. &c.

125. In order to show that the contact will be more intimate as the number of corresponding equal differential coefficients becomes greater, let x take the arbitrary increment h, converting y and Y into y_1 and Y_1 respectively.

Then $$Y_1 = Y + \frac{dY}{dx} \cdot \frac{h}{1} + \frac{d^2Y}{dx^2} \cdot \frac{h^2}{1.2} + \frac{d^3Y}{dx^3} \cdot \frac{h^3}{1.2.3} + \&c.$$

and $$y_1 = y + \frac{dy}{dx}\cdot\frac{h}{1} + \frac{d^2y}{dx^2}\cdot\frac{h^2}{1.2} + \frac{d^3y}{dx^3}\cdot\frac{h^3}{1.2.3} + \&c.$$

$$\therefore\ Y_1 - y_1 = \left[\frac{dY}{dx} - \frac{dy}{dx}\right]\frac{h}{1} + \left[\frac{d^2Y}{dx^2} - \frac{d^2y}{dx^2}\right]\frac{h^2}{1.2}$$
$$+ \left[\frac{d^3Y}{dx^3} - \frac{d^3y}{dx^3}\right]\frac{h^3}{1.2.3} + \&c.$$

Now the value of this difference, which expresses the distance by which the one curve departs from the other, measured on the line parallel to y, will depend, when h is small, chiefly upon the terms containing the lowest powers of h.

If, then, the first differential coefficients derived from the equations of three curves (A), (B) and (C) be equal, at a common point, and if the second differential coefficients derived from the equations of (A) and (B) be also equal, but those derived from (A) and (C) unequal, the curves (A) and (B) will separate more slowly than (A) and (C), because the expression for the difference of the ordinates of (A) and (C) corresponding to the abscissa $x + h$, will contain a term including the second power of h, but the difference of the ordinates of (A) and (B) will contain no power of h lower than the third.

126. The order of closest possible contact between one curve entirely given, and another given only in species, will depend on the number of arbitrary parameters contained in the equation of the second curve.

Thus a contact of the first order requires two conditions, viz.:

$$y = Y \text{ and } \frac{dy}{dx} = \frac{dY}{dx}:$$

the first of these conditions being employed in giving the curves a common point, and the second in giving their tangents at that point a common direction. Hence there must be at least two arbitrary parameters.

A contact of the second order requires three parameters; one of the third order, four parameters, &c. Hence the straight line, whose equation $y = ax + b$ has two parameters, a and b, can have contact of the first order only.

The circle $(x - a)^2 + (y - b)^2 = r^2$ having in its equation three parameters, can have contact of the second order.

The parabola can have contact of the third order; the ellipse or hyperbola a contact of the fourth order, &c.

The curve of a given species, which has the most intimate contact possible with a given curve at a given point, is called the osculatory curve of that species.

The osculatory circle is employed to measure the curvature of plane curves, and its radius is called the radius of curvature of the given curve.

127. *Prop.* To determine the radius of curvature of a given curve at a given point, and also the co-ordinates of the centre of the osculatory circle.

Let the equation of the given curve be $y = Fx$ (1), and that of the required circle $(x - a)^2 + (y - b)^2 = r^2$ (2), the quantities a, b and r being those which it is proposed to determine.

There being three disposable parameters, a, b, and r, in equation (2), we can impose the three conditions

$$y = Y, \ \frac{dy}{dx} = \frac{dY}{dx} \quad \text{and} \quad \frac{d^2y}{dx^2} = \frac{d^2Y}{dx^2}.$$

with which determine a, b, and r, and the contact will be of the second order.

Denote the first and second differential coefficients derived from the equation of the given curve by p' and p'', that is, put

$$\frac{dY}{dx} = p' \quad \text{and} \quad \frac{d^2Y}{dx^2} = p''.$$

Then, since the corresponding differential coefficients derived from the equation of the osculatory circle must have the same values, we shall have

$$\frac{dy}{dx}=p' \quad \text{and} \quad \frac{d^2y}{dx^2}=p''.$$

Now let (2) be differentiated twice successively, replacing

$$\frac{dy}{dx} \quad \text{and} \quad \frac{d^2y}{dx^2} \text{ by } p' \text{ and } p''.$$

$\therefore\ (x-a)+(y-b)p'=0 \ldots (3)$, and $1+p'^2+(y-b)p''=0 \ldots (4)$.

The equations (2), (3), and (4), will just suffice to determine a, b, and r.

Thus, from (4) $y-b=-\dfrac{1+p'^2}{p''} \cdots (5)$ or $b=y+\dfrac{1+p'^2}{p''} \cdots (6)$

and from (3) and (4) $x-a=-(y-b)p'=\dfrac{p'(1+p'^2)}{p''} \ldots (7)$

$$\therefore\ a=x-\frac{p'(1+p'^2)}{p''} \ldots (8).$$

Now combining (2), (5), and (7), we get

$$r^2=\frac{(1+p'^2)^2}{p''^2}+\frac{p'^2(1+p'^2)^2}{p''^2}=\frac{(1+p'^2)^3}{p''^2}.$$

$$\therefore\ r=\pm\frac{(1+p'^2)^{\frac{3}{2}}}{p''} \ldots (9).$$

The equations (6), (8), and (9), resolve the problem. To apply them to a particular case, we form the differential coefficients p' and p'' from the equation of the given curve, and substitute their values in (6), (8), and (9).

Cor. Since $1+p'^2$ or $1+\dfrac{dy^2}{dx^2}=\dfrac{ds^2}{dx^2}$, (Art. 122) the value of r may be written thus

$$r=\pm\frac{\dfrac{ds^3}{dx^3}}{\dfrac{d^2y}{dx^2}} \ldots (10).$$

Remark. We may omit the double sign $\pm$ in (9) and (10) and regard the radius of curvature as an essentially positive quantity in all cases. This double sign is sometimes employed to indicate the direction of the curvature, being positive when the curve presents its convexity to the axis of x, and negative in the contrary case. But it seems more simple to consider r essentially positive, and to fix the direction of the curvature by the sign of $\frac{d^2y}{dx^2}$. It will now be shown that the sign of this second differential will always be determined by the direction of the curvature.

If the curve be *convex* towards the axis of x, as in Fig. 1, and if an increment h be given to the abscissa $OD = x$, the ordinate y will take an increment

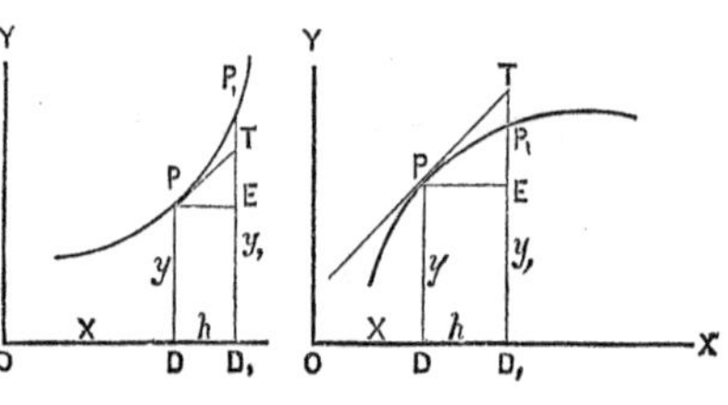

$$EP_1 = \frac{dy}{dx} \cdot \frac{h}{1} + \frac{d^2y}{dx^2} \cdot \frac{h^2}{1\,.\,2} + \frac{d^3y}{dx^3} \cdot \frac{h^3}{1\,.\,2\,.\,3} + \&c.,$$

and the ordinate of the tangent will take a corresponding increment $ET = \frac{dy}{dx} \cdot \frac{h}{1}$, and the former of these two increments will be the greater since the tangent lies between the curve and the axis of x.

$$\therefore EP_1 - ET = \frac{d^2y}{dx^2} \cdot \frac{h^2}{1\,.\,2} + \frac{d^3y}{dx^3} \cdot \frac{h^3}{1\,.\,2\,.\,3} + \&c. > 0.$$

or since the sign of this series depends, when h is small, on that of the first term, we must have $\frac{d^2y}{dx^2} > 0$.

But when the curve is *concave* towards the axis of x, as in Fig. 2,

$$EP_1 - ET < 0, \quad \text{and} \quad \therefore \frac{d^2y}{dx^2} < 0.$$

Again, since the arc s and the abscissa x may always be supposed to increase together, $\frac{ds^3}{dx^3}$ may be considered as essentially positive, and therefore the sign of r in (10) would be controlled by that of $\frac{d^2y}{dx^2}$. It is in this way that the sign of r may be regarded as indicating the direction of the curvature.

EXAMPLES.

128. 1. To find the radius of curvature of the common parabola $y^2 = 2px$, at a given point.

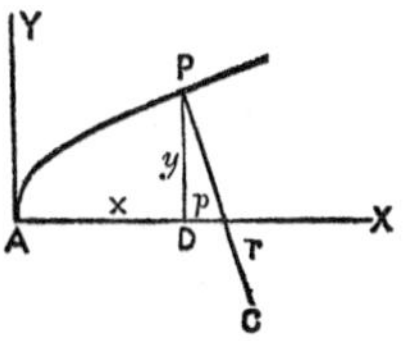

Here $$p' = \frac{dy}{dx} = \frac{p}{y},$$

and $$p'' = \frac{d^2y}{dx^2} = -\frac{p}{y^2}\cdot\frac{dy}{dx} = -\frac{p^2}{y^3}.$$

$$\therefore r = \frac{(1+p'^2)^{\frac{3}{2}}}{p''} = \frac{(y^2+p^2)^{\frac{3}{2}}}{p^2}$$

or $$r = \frac{(\text{normal})^3}{(\text{semi-parameter})^2}.$$

At the vertex, $y = 0$, and $\therefore$ $r = p$ the semi-parameter; and $y = \infty$, $r = \infty$ also.

2. The ellipse $A^2y^2 + B^2x^2 = A^2B^2$.

$$p' = -\frac{B^2x}{A^2y}, \quad p'' = -\frac{B^2A^2y - A^2B^2xp'}{A^4y^2}$$

$$= -\frac{B^2(A^2y^2 + B^2x^2)}{A^4y^3} = -\frac{B^4}{A^2y^3}.$$

$$\therefore r = \frac{\left[1 + \frac{B^4x^2}{A^4y^2}\right]^{\frac{3}{2}}}{\frac{B^4}{A^2y^3}} = \frac{[A^4y^2 + B^4x^2]^{\frac{3}{2}}}{A^4B^4}.$$

At the extremity of the transverse axis $x = A$ and $y = 0$. $\therefore r = \frac{B^2}{A}$,

and " " " conjugate " $x = 0$ and $y = B$. $\therefore r = \frac{A^2}{B}$.

3. The logarithmic curve $y = a^x$.

$$p' = \log a \,.\, a^x = \frac{y}{m}, \quad p'' = \frac{1}{m} \cdot \frac{dy}{dx} = \frac{y}{m^2}, \quad \text{where } m = \text{modulus.}$$

$$\therefore\ r = \frac{(1 + p'^2)^{\frac{3}{2}}}{p''} = \frac{\left[1 + \frac{y^2}{m^2}\right]^{\frac{3}{2}}}{\frac{y}{m^2}} = \frac{[m^2 + y^2]^{\frac{3}{2}}}{my}.$$

When $y = 0,\ r = \infty$; and when $y = \infty,\ r = \infty$ also.

4. The cubical parabola $y^3 = a^2x$.

$$p' = \frac{a^2}{3y^2}, \quad p'' = -\frac{3a^2 \,.\, 2y}{9y^4} \cdot \frac{a^2}{3y^2} = -\frac{2a^4}{9y^5}.$$

$$\therefore\ r = \frac{\left[1 + \frac{a^4}{9y^4}\right]^{\frac{3}{2}}}{\frac{2a^4}{9y^5}} = \frac{(9y^4 + a^4)^{\frac{3}{2}}}{6a^4y}.$$

When $y = 0,\ r = \infty$, and when $y = \pm\infty,\ r = \infty$.

5. The cycloid, or curve generated by the motion of a point on the circumference of a circle, while the circle rolls on a straight line.

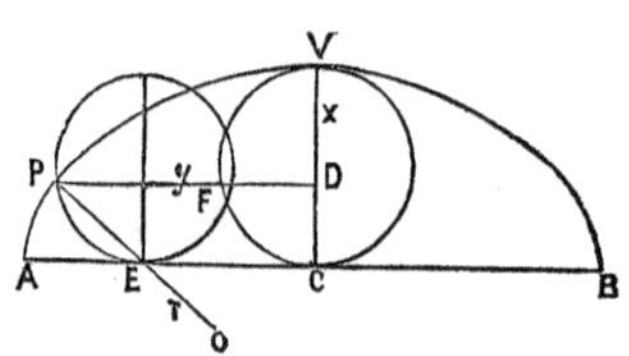

Let the radius of the generating circle $= a$. Place the origin at V, the vertex of the cycloid. Put $VD = x$, $DP = y$, the point P being that which describes the curve $APVB$, while the circle rolls on the line ACB.

Then $PD = DF + FP = DF + EC$ since EP and CF are parallel. Also, since each point of the semi-circumference CFV has been in contact with the semi-base CA we must have arc $CFV = CA$ and similarly arc $EP = EA =$ arc CF.

$\therefore$ By subtraction

$CA - EA = CFV - CF$ or $CE = FV$; and $\therefore PD = DF + FV$.

But $\quad DF = \sqrt{2ax - x^2}, \quad$ and $\quad FV = a \text{ versin}^{-1}\frac{x}{a}.$

Hence the equation of the cycloid is

$$y = \sqrt{2ax - x^2} + a \, . \, \text{versin}^{-1}\frac{x}{a}.$$

$$\therefore \; p' = \frac{a - x}{\sqrt{2ax - x^2}} + \frac{a\frac{1}{a}}{\sqrt{2\frac{x}{a} - \frac{x^2}{a^2}}} = \sqrt{\frac{2a - x}{x}}.$$

$$p'' = -\frac{a}{x\sqrt{2ax - x^2}}$$

$$\therefore \; r = \frac{\left[1 + \frac{2a - x}{x}\right]^{\frac{3}{2}}}{\frac{a}{x\sqrt{2ax - x^2}}} = \frac{(2a)^{\frac{3}{2}}\sqrt{2ax - x^2}}{a\sqrt{x}} = 2\sqrt{2a(2a - x)},$$

or, $$r = 2 \text{ chord } PE.$$

129. *Prop.* At the points of greatest and least curvature of any curve, the osculatory circle has contact of the third order.

The condition which characterises these points, is that the differential coefficient $\frac{dr}{dx}$ shall reduce to zero, since r is a minimum when the curvature is greatest, and a maximum when it is least.

But by the general formula for the radius of curvature, $r = \frac{(1 + p'^2)^{\frac{3}{2}}}{p''}$, we have, by putting $\frac{d^3y}{dx^3} = p'''$,

$$\frac{dr}{dx} = \frac{\frac{3}{2}(1 + p'^2)^{\frac{1}{2}} . 2p'p''^2 - p'''(1 + p'^2)^{\frac{3}{2}}}{p''^2} = 0.$$

$$\therefore \; p''' = \frac{3p'p''^2}{1 + p'^2} \; . \; . \; . \; . \; (1).$$

This is the value of the third differential $\frac{d^3y}{dx^3}$, at the points of greatest and least curvature, of any curve; and if it can be shown that the third differential coefficient in the osculatory circle has the same value, it will follow that the contact must be of the third order.

But in the circle we have already found $y - b = -\dfrac{1+p'^2}{p''}$,

$$\therefore \frac{dy}{dx} = p' = -\frac{2p'p''^2 - p'''(1+p'^2)}{p''^2}, \quad \therefore p''' = \frac{3p'p''^2}{1+p'^2} \cdots (2).$$

which being identical with (1), the contact must be of the third order.

130. *Prop.* If two curves have contact of an even order, they will intersect at the point of contact; but if the order of their contact be odd, they will not intersect at that point.

If $Y = Fx$, and $y = \varphi x$, be the equations of the two curves, the difference of their ordinates corresponding to the abscissa $x + h$, will be expressed by

$$Y_1 - y_1 = \left(\frac{dY}{dx} - \frac{dy}{dx}\right)\left(\frac{\pm h}{1}\right) + \left(\frac{d^2Y}{dx^2} - \frac{d^2y}{dx^2}\right)\frac{(\pm h)^2}{1 \, . \, 2}$$
$$+ \left(\frac{d^3Y}{dx^3} - \frac{d^3Y}{dx^3}\right)\frac{(\pm h)^3}{1 \, . \, 2 \, . \, 3} + \&c.$$

Now when the order of contact is even, the first term of this difference which does not reduce to zero, must contain an odd power of $\pm h$, and must therefore change sign with h, thus imparting a change of sign to $Y_1 - y_1$, in passing through the point x,y.

Hence the first curve will lie alternately above and below the second, intersecting it at the point x,y.

But if the order of contact be odd, the first term in the difference will contain an even power of $\pm h$, which will not change sign with h, and therefore there will be no intersection; the first curve lying entirely above or entirely below the second.

Cor. The osculatory circle intersects the curve, except at the points of greatest and least curvature.

For usually, the circle has contact of the second order—but at the points of greatest and least curvature, the contact is of the third order.

Cor. At those points of a curve where $p'' = 0$, a straight line may have contact of the second order, and it will intersect the curve. If $p''' = 0$, also there will be no intersection unless $p'''' = 0$, also.

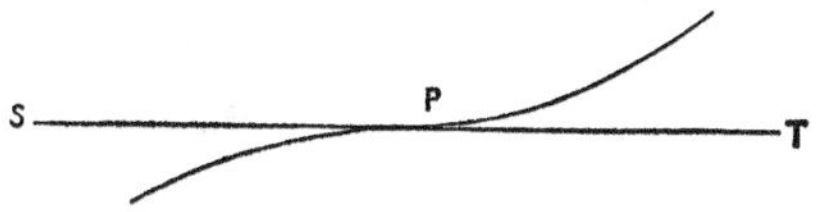

131. *Prop.* To find a formula for the radius of curvature, when any quantity t, other than the abscissa x, is taken as the independent variable.

To effect this object, we must substitute in the value of r, already found, the values of $p' = \frac{dy}{dx}$, and $p'' = \frac{d^2y}{dx^2}$, given by the formula for changing the independent variable, viz.:

$$\frac{dy}{dx} = \frac{\frac{dy}{dt}}{\frac{dx}{dt}}, \quad \text{and} \quad \frac{d^2y}{dx^2} = \frac{\frac{d^2y}{dt^2}\cdot\frac{dx}{dt} - \frac{d^2x}{dt^2}\cdot\frac{dy}{dt}}{\frac{dx^3}{dt^3}}.$$

We thus obtain

$$r = \frac{\left(1 + \frac{dy^2}{dx^2}\right)^{\frac{3}{2}}}{\frac{d^2y}{dx^2}} = \frac{\left(\frac{dx^2}{dt^2} + \frac{dy^2}{dt^2}\right)^{\frac{3}{2}}}{\left(\frac{dx^2}{dt^2}\right)^{\frac{3}{2}}} \div \frac{\frac{d^2y}{dt^2}\cdot\frac{dx}{dt} - \frac{d^2x}{dt^2}\cdot\frac{dy}{dt}}{\frac{dx^3}{dt^3}}$$

$$= \frac{\left(\frac{dx^2}{dt^2} + \frac{dy^2}{dt^2}\right)^{\frac{3}{2}}}{\frac{d^2y}{dt^2}\cdot\frac{dx}{dt} - \frac{d^2x}{dt^2}\cdot\frac{dy}{dt}} = \frac{\frac{ds^3}{dt^3}}{\frac{d^2y}{dt^2}\cdot\frac{dx}{dt} - \frac{d^2x}{dt^2}\cdot\frac{dy}{dt}}. \quad \ldots\ldots (1)$$

Cor. If x be the independent variable, $\frac{dx}{dt} = \frac{dx}{dx} = 1$,

and $\frac{d^2x}{dt^2} = 0$, $\quad \therefore\ r = \frac{\frac{ds^3}{dx^3}}{\frac{d^2y}{dx^2}}$, the common formula.

If y be the independendent variable, $\frac{dy}{dt} = \frac{dy}{dy} = 1$, and $\frac{d^2y}{dt^2} = 0$.

$$\therefore \; r = \frac{-\dfrac{ds^3}{dy^3}}{\dfrac{d^2x}{dy^2}}.$$

If s be the independent variable, $\frac{ds}{dt} = \frac{ds}{ds} = 1$,

$$\therefore \; r = \frac{1}{\dfrac{d^2y}{ds^2} \cdot \dfrac{dx}{ds} - \dfrac{d^2x}{ds^2} \cdot \dfrac{dy}{ds}}, \quad \ldots \ldots (2).$$

But $\frac{dx^2}{ds^2} + \frac{dy^2}{ds^2} = \frac{ds^2}{ds^2} = 1$, which, being differentiated with respect to s, gives

$$\frac{dx}{ds} \cdot \frac{d^2x}{ds^2} + \frac{dy}{ds} \cdot \frac{d^2y}{ds^2} = 0, \quad \ldots \ldots (3).$$

$$\therefore \; r = \frac{1}{-\dfrac{\dfrac{d^2x}{ds^2} \cdot \dfrac{dx^2}{ds^2}}{\dfrac{dy}{ds}} - \dfrac{d^2x}{ds^2} \cdot \dfrac{dy}{ds}} = -\frac{\dfrac{dy}{ds}}{\dfrac{d^2x}{ds^2}}, \quad \ldots (4).$$

and similarly
$$r = \frac{\dfrac{dx}{ds}}{\dfrac{d^2y}{ds^2}}. \quad \ldots \ldots (5).$$

And, finally, by squaring the equation (2), and adding to the denominator of the second member, the square of (3), which is equal to zero, there results, by reduction,

$$r^2 = \frac{1}{\left(\dfrac{d^2y}{ds^2}\right)^2 + \left(\dfrac{d^2x}{ds^2}\right)^2} \quad \text{and} \quad \therefore \; r = \frac{1}{\sqrt{\left(\dfrac{d^2y}{ds^2}\right)^2 + \left(\dfrac{d^2x}{ds^2}\right)^2}}$$

132. *Prop.* To obtain a formula for the radius of curvature of curves when referred to polar co-ordinates.

Adopting the variable angle θ as the independent variable, denoting the radius vector by r, and the radius of curvature by R, we have, from the formulæ for the transformation of co-ordinates,

$$x = r\cos\theta,\ y = r\sin\theta,\ \therefore\ \frac{dx}{d\theta} = -r\sin\theta + \cos\theta\frac{dr}{d\theta},$$

$$\frac{dy}{d\theta} = r\cos\theta + \sin\theta\frac{dr}{d\theta}.$$

$$\frac{d^2x}{d\theta^2} = -r\cos\theta - 2\sin\theta\frac{dr}{d\theta} + \cos\theta\frac{d^2r}{d\theta^2},$$

$$\frac{d^2y}{d\theta^2} = -r\sin\theta + 2\cos\theta\frac{dr}{d\theta} + \sin\theta\frac{d^2r}{d\theta^2}.$$

Put $\frac{dr}{d\theta} = p_1$ and $\frac{d^2r}{d\theta^2} = p_2$ and substitute in the general value of the radius of curvature.

$$\therefore\ R = \frac{\left[\frac{dx^2}{d\theta^2} + \frac{dy^2}{d\theta^2}\right]^{\frac{3}{2}}}{\frac{d^2y}{d\theta^2}\cdot\frac{dx}{d\theta} - \frac{d^2x}{d\theta^2}\cdot\frac{dy}{d\theta}} = \frac{[r^2 + p_1^2(\sin^2\theta + \cos^2)]^{\frac{3}{2}}}{(r^2 + 2p_1^2 - rp_2)(\sin^2\theta + \cos^2\theta)}$$

$$= \frac{(r^2 + p_1^2)^{\frac{3}{2}}}{r^2 + 2p_1^2 - rp_2} = \frac{N^3}{r^2 + 2p_1^2 - rp_2}$$

where N is the polar normal.

EXAMPLES.

133. 1. The logarithmic Spiral $r = a^{\theta}$.

$$p_1 = \log a\,.\,a^{\theta} = \frac{r}{m},\quad p_2 = \log^2 a\cdot a^{\theta} = \frac{r}{m^2}.$$

$$\therefore\ R = \frac{N^3}{r^2 + 2\frac{r^2}{m^2} - \frac{r^2}{m^2}} = \frac{N^3m^2}{r^2(m^2+1)} = \frac{N(r^2 + m^2r^2)}{r^2(1+m^2)} = N.$$

$\therefore$ The radius of curvature of the logarithmic spiral is always equal to the polar normal.

2. The spiral of Archimedes $r = a\theta$.

$$p_1 = a, \ p_2 = 0 \ \therefore \ R = \frac{(r^2 + a^2)^{\frac{3}{2}}}{r^2 + 2a^2}.$$

When $r = 0$, $R = \frac{1}{2}a$, and when $r = \infty$, $R = r = \infty$

3. The hyperbolic spiral $r\theta = a$.

$$p_1 = -\frac{a}{\theta^2} = -\frac{r^2}{a}, \ p_2 = \frac{2a}{\theta^3} = \frac{2r^3}{a^2}.$$

$$\therefore R = \frac{\left(r^2 + \frac{r^4}{a^2}\right)^{\frac{3}{2}}}{r^2 + 2\frac{r^4}{a^2} - \frac{2r^4}{a^2}} = \frac{r(a^2 + r^2)^{\frac{3}{2}}}{a^3},$$

when $r = 0$, $R = 0$, and when $r = \infty$, $R = \infty$.

4. The lituus $r^2\theta = a^2$.

$$p_1 = -\frac{1}{2}a\theta^{-\frac{3}{2}} = -\frac{r^3}{2a^2}, \ p_2 = \frac{3}{4}a\theta^{-\frac{5}{2}} = \frac{3r^5}{4a^4}$$

$$\therefore R = \frac{\left[r^2 + \frac{r^6}{4a^4}\right]^{\frac{3}{2}}}{r^2 + 2\frac{r^6}{4a^4} - \frac{3r^6}{4a^4}} = \frac{r(4a^4 + r^4)^{\frac{3}{2}}}{2a^2(4a^4 - r^4)}$$

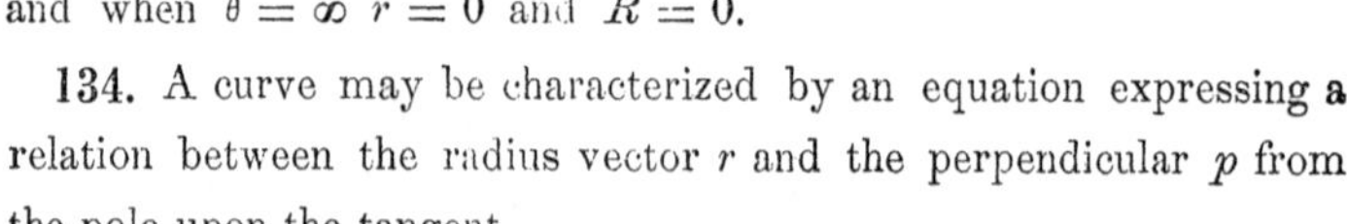

When $\theta = 0$, $r = \infty$ and $R = \infty$; when $\theta = 1$, $r = a$, and $R = a\frac{\sqrt{125}}{6}$; when $r = a\sqrt{2}$, or $r^4 = 4a^4$, $R = \infty$ and when $\theta = \infty$ $r = 0$ and $R = 0$.

134. A curve may be characterized by an equation expressing a relation between the radius vector r and the perpendicular p from the pole upon the tangent.

Thus the equation of the circle referred to the co-ordinates r and p is $r = p$, the pole being at the centre. That of the logarithmic spiral is $r = cp$, &c.

135. *Prop.* To obtain a formula for the radius of curvature of curves referred to the radius vector and the perpendicular upon the tangent.

From the general value of the perpendicular when the curve is referred to the ordinary polar co-ordinates r and θ, viz.: (Art. 115.)

$$p = \frac{r^2}{\sqrt{r^2 + \frac{dr^2}{d\theta^2}}} \quad \text{we obtain} \quad \frac{dr^2}{d\theta^2} = \frac{r^4}{p^2} - r^2 = p_1{}^2$$

which, differentiated with respect to θ, gives

$$2\frac{dr}{d\theta}\cdot\frac{d^2r}{d\theta^2} = \frac{4r^3}{p^2}\cdot\frac{dr}{d\theta} - \frac{2r^4}{p^3}\cdot\frac{dp}{d\theta} - 2r\frac{dr}{d\theta}.$$

Substituting for $\frac{dp}{d\theta}$ its value $\frac{dp}{dr}\cdot\frac{dr}{d\theta}$ and divide by $2\frac{dr}{d\theta}$.

$$\therefore \frac{d^2r}{d\theta^2} = \frac{2r^3}{p^2} - \frac{r^4}{p^3}\cdot\frac{dp}{dr} - r. = p_2.$$

Now substituting the values of $\frac{dr}{d\theta}$ and $\frac{d^2r}{d\theta^2}$ in that of R, we get

$$R = \frac{(r^2 + p_1{}^2)^{\frac{3}{2}}}{r^2 + 2p_1{}^2 - rp_2} = \frac{\left(r^2 + \frac{r^4}{p^2} - r^2\right)^{\frac{3}{2}}}{r^2 + 2\left(\frac{r^4}{p^2} - r^2\right) - r\left(\frac{2r^3}{p^2} - \frac{r^4}{p^3}\cdot\frac{dp}{dr} - r\right)}$$

$$= \frac{r}{\frac{dp}{dr}} = r\frac{dr}{dp}.$$

Ex. The involute of the circle whose equation referred to p and r is $p^2 = r^2 - a^2$.

$$\frac{dr}{dp} = \frac{p}{r}, \quad \therefore R = r\frac{dr}{dp} = r\frac{p}{r} = p = \sqrt{r^2 - a^2}.$$

CHAPTER III.

EVOLUTES AND INVOLUTES.

136. The curve which is the locus of the centres of all the osculatory circles applied to every point of a given curve, is called the *evolute* of that curve, the latter being termed the *involute* of the former.

137. *Prop.* To determine the evolute of a given curve $y = Fx$.

If in the formulæ for the co-ordinates of the centre of the osculatory circle, viz.: (Art. 127.)

$$a = x - p'\frac{1 + p'^2}{p''} \dots (1) \quad \text{and} \quad b = y + \frac{1 + p'^2}{p''} \dots (2),$$

we substitute the values of p' and p'', derived from the equation of the curve $y = Fx$ (3), we shall have the three equations (1), (2), and (3), involving the four variable quantities x, y, a, and b; and by eliminating x and y the result will be a general relation between a and b, the co-ordinates of the required evolute. This equation being independent of x and y will apply to every point in the desired curve.

138. In most cases the necessary elimination is quite difficult; the following are comparatively simple examples.

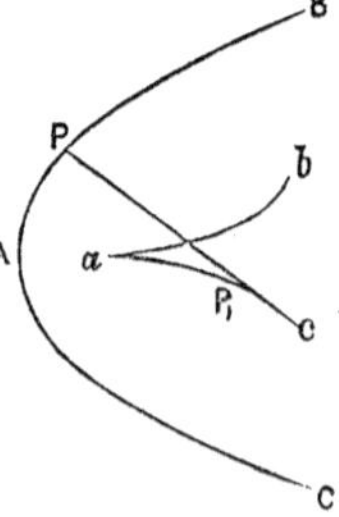

1. The evolute of the common parabola.

Here we have $y^2 = 2px \dots (1)$.

$$\therefore p' = \frac{p}{y}, \quad \text{and} \quad p'' = -\frac{p^2}{y^3}.$$

$$\therefore a = x + \frac{p}{y} \cdot \frac{y^2 + p^2}{y^2} \cdot \frac{y^3}{p^2} = x + \frac{2px + p^2}{p} = 3x + p \; . \; . \; . \; . \; (2).$$

$$b = y - \frac{y^2 + p^2}{y^2} \cdot \frac{y^3}{p^2} = y - \frac{y^3}{p^2} - y = -\frac{y^3}{p^2} \; . \; . \; . \; . \; (3).$$

From (2) and (3) we get $x = \dfrac{a-p}{3}$, and $y^2 = p^{\frac{4}{3}} b^{\frac{2}{3}}$;

and these values substituted in (1) give

$$p^{\frac{4}{3}} b^{\frac{2}{3}} = 2p \cdot \frac{a-p}{3}. \quad \therefore b^2 = \frac{8}{27p}(a-p)^3,$$

the equation of the semi-cubical parabola, whose axis coincides with that of the given curve; the distance Aa between the vertices being $= p$ the semi-parameter.

2. The ellipse

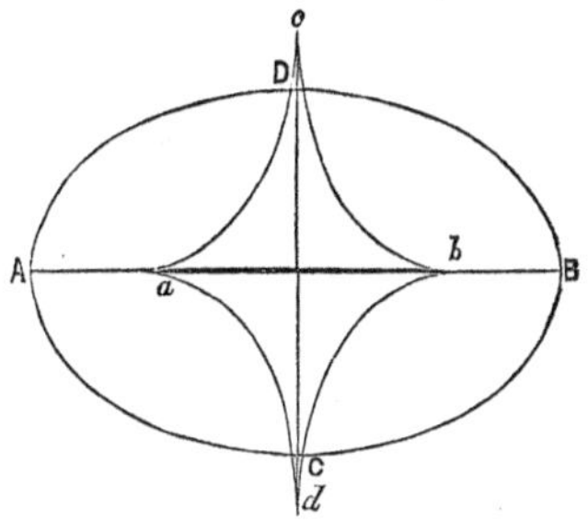

$$A^2y^2 + B^2x^2 = A^2B^2 \; . \; . \; . \; . \; (1).$$

$$p' = -\frac{B^2x}{A^2y}, \quad p'' = -\frac{B^4}{A^2y^3},$$

$$1 + p'^2 = \frac{A^4y^2 + B^4x^2}{A^4y^2},$$

$$\frac{p'}{p''} = \frac{xy^2}{B^2}, \quad \therefore a = x - \frac{x(A^4y^2 + B^4x^2)}{A^4B^2},$$

or $\quad a = \dfrac{xA^2(A^2B^2 - A^2y^2) - B^4x^3}{A^4B^2} = \dfrac{x^3f^2}{A^4}$ where $f^2 = A^2 - B^2$,

and $\quad b = y - \dfrac{y(A^4y^2 + B^4x^2)}{A^2B^4} = \dfrac{A^4y^3 - B^2y(A^2B^2 - B^2x^2)}{-A^2B^4} = -\dfrac{y^3f^2}{B^4}.$

$$\therefore x^2 = \frac{a^{\frac{2}{3}} A^{\frac{8}{3}}}{f^{\frac{4}{3}}}, \quad y^2 = \frac{b^{\frac{2}{3}} B^{\frac{8}{3}}}{f^{\frac{4}{3}}},$$

which values substituted in (1) give

$$A^2 \frac{b^{\frac{2}{3}} B^{\frac{8}{3}}}{f^{\frac{4}{3}}} + B^2 \frac{a^{\frac{2}{3}} A^{\frac{8}{3}}}{f^{\frac{4}{3}}} = A^2B^2,$$

or $$A^{\frac{2}{3}}a^{\frac{2}{3}} + B^{\frac{2}{3}}b^{\frac{2}{3}} = (A^2 - B^2)^{\frac{2}{3}},$$

the equation of the required evolute.

When $a = 0,\ b = \pm\dfrac{f^2}{B}$; and when $b = 0,\ a = \pm\dfrac{f^2}{A}$.

The curve consists of four branches presenting their convexities towards the axis, and tangent to each other as shown in the diagram.

The equilateral hyperbola referred to its asymptotes.

$$xy = c^2 \ldots\ldots (1).$$

$$p' = -\frac{c^2}{x^2},\quad p'' = \frac{2c^2}{x^3},\quad 1 + p'^2 = \frac{c^4 + x^4}{x^4},\quad \frac{p'}{p''} = -\frac{x}{2}.$$

$$\therefore a = x + \frac{c^4 + x^4}{2x^3},\quad b = y + \frac{c^4 + x^4}{2c^2x}.$$

$$\therefore a + b = \frac{c}{2}\left[\frac{c}{x} + \frac{x}{c}\right]^3 \text{ and } a - b = \frac{c}{2}\left[\frac{c}{x} - \frac{x}{c}\right]^3.$$

$$\therefore \sqrt[3]{a + b} + \sqrt[3]{a - b} = \frac{2c}{x}\sqrt[3]{\frac{c}{2}},$$

and $$\sqrt[3]{a + b} - \sqrt[3]{a - b} = \frac{2x}{c}\sqrt[3]{\frac{c}{2}}.$$

Hence by multiplication $(a + b)^{\frac{2}{3}} - (a - b)^{\frac{2}{3}} = (4c)^{\frac{2}{3}}$.

139. *Prop.* Normals to the involute are tangents to the evolute.

From the equation of the osculatory circle $(x-a)^2+(y-b)^2=r^2$, we get by differentiation

$$x - a + p'(y - b) = 0 \ldots\ldots (1),$$

a relation alike applicable to the circle and the given curve, since x, y and p' are the same in both.

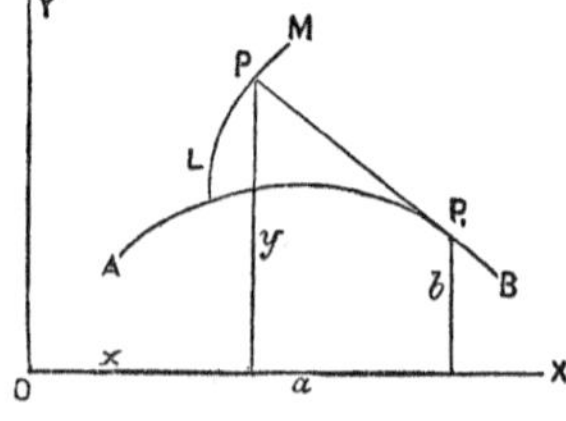

Now when we pass from a point x,y to another point on the circle, the quantities x, y and p' must be considered variable, but a and b constant; but when we pass to a point on the curve, x, y, p', a, and b will all vary, and in both cases p'' will be the same.

The first supposition gives, by differentiating (1) with respect to x,

$$1 + p'^2 + p''(y - b) = 0 \quad \ldots . \quad (2),$$

and the second gives

$$1 - \frac{da}{dx} + p'^2 - p'\frac{db}{dx} + p''(y - b) = 0 \quad \ldots . \quad (3).$$

Whence by combining (2) and (3)

$$\frac{da}{dx} + p'\frac{db}{dx} = 0.$$

$$\therefore \frac{\frac{db}{dx}}{\frac{da}{dx}} = \frac{db}{da} = -\frac{1}{p'}.$$

Now $\frac{db}{da}$ represents the tangent of the angle formed by the axis of x, with the tangent to the evolute AB at the point P_1, and $-\frac{1}{p'}$ = tangent of the angle formed by the same axis with the normal PP_1 to the involute LM at the point x,y; which normal passes through the point P_1. Hence this normal not only passes through the point a,b, but it also coincides in direction with the tangent to the evolute at that point.

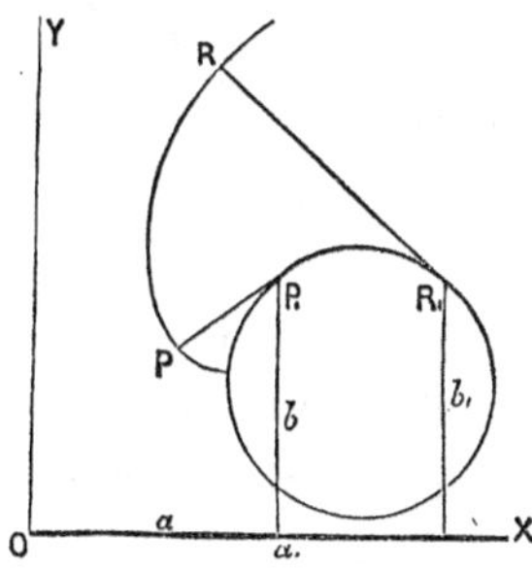

140. *Prop.* The difference of any two radii of curvature is equal to the arc of the evolute intercepted between those radii.

Resuming the equation

$$(x - a)^2 + (y - b)^2 = r^2,$$

and differentiating with respect to a, as an independent variable, we obtain

$$(x-a)\left(\frac{dx}{da}-1\right)+(y-b)\left(\frac{dy}{da}-\frac{db}{da}\right)=r\frac{dr}{da}.$$

But $\quad (x-a)\frac{dx}{da}+(y-b)\frac{dy}{da}=\frac{dx}{da}\left[x-a+(y-b)\frac{dy}{dx}\right]=0.$

$$\therefore x-a+(y-b)\frac{db}{da}=-r\frac{dr}{da};$$

or since $\frac{db}{da}=-\frac{1}{p'}=\frac{y-b}{x-a}$, $\quad \therefore (x-a)\left(1+\frac{db^2}{da^2}\right)=-r\frac{dr}{da}\cdot\cdot(1).$

Also, $\quad (x-a)^2+(y-b)^2=(x-a)^2\left(1+\frac{db^2}{da^2}\right)=r^2.$

or, $\quad (x-a)\left(1+\frac{db^2}{da^2}\right)^{\frac{1}{2}}=\pm r.\ .\ .\ .\ (2).$

Dividing (1) by (2), there results,

$$\left(1+\frac{db^2}{da^2}\right)^{\frac{1}{2}}=\mp\frac{dr}{da}.$$

But $\left(1+\frac{db^2}{da^2}\right)^{\frac{1}{2}}=\frac{ds}{da}$, where s is the arc of the evolute which terminates at the point a, b.

$$\therefore \frac{ds}{da}=\mp\frac{dr}{da}, \quad \text{and} \quad ds=\mp dr.$$

Thus it appears that the increment of s is always numerically equal to the increment of r.

Hence s must always differ from r by a constant quantity, or we must have $s=c\mp r$; and similarly for the arc s_1, which terminates at the point a_1,b_1, $\;s_1=c\mp r_1$, $\;\therefore s_1-s=r-r_1$, which result agrees with the enunciation.

141. In finding the evolutes of polar curves, it is usually most convenient to employ the relation between r and p, the radius vector and the perpendicular on the tangent; thus, let r = radius vector of the given curve, p = the perpendicular on the tangent, r_1 = radius vector of the evolute, p_1 = the perpendicular on its tangent.

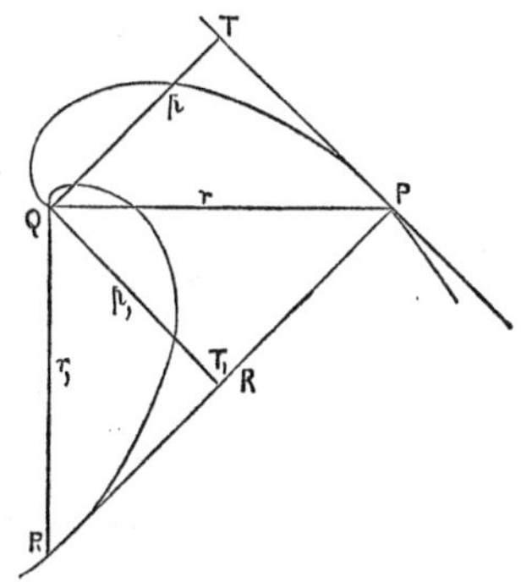

Then since the radius of curvature $PP_1 = R$, at the point P, is tangent to the evolute at P_1, the perpendicular QT_1, is parallel to the tangent TP.

Also QT is parallel to PP_1.

$$\therefore PT_1 = QT = p,$$

and $$PT = QT_1 = p_1.$$

$$\therefore r_1^2 = R^2 + r^2 - 2Rp, \quad . \; . \; . \; (1).$$

$$r^2 = p^2 + p_1^2. \quad . \; . \; . \; . \; (2).$$

Also $R = r\dfrac{dr}{dp} \cdots (3)$ (Art. 135). And $r = Fp, \cdots\cdot (4)$, the equation of the given curve.

By eliminating r, p, and R, between (1), (2), (3), and (4), there will result a relation between r_1 and p_1 which will be the equation of the required evolute.

Ex. The logarithmic or equiangular spiral $r = cp. \; . \; . \; . \; . \; (4).$

$$\therefore \frac{dr}{dp} = c, \quad \text{and} \quad R = cr, \; . \; . \; . \; . \; (3). \quad r^2 = p^2 + p_1^2, \; . \; . \; . \; . \; (2).$$

$$r_1^2 = R^2 + r^2 - 2Rp. \; . \; . \; . \; . \; (1).$$

From (1) and (3), $r_1^2 = c^2r^2 + r^2 - 2crp$, which combined with (4) gives

$$r_1^2 = c^2p^2(1 + c^2) - 2c^2p^2 = c^2p^2(c^2 - 1). \; . \; . \; . \; (5).$$

From (2) and (4), $c^2p^2 = p^2 + p_1^2$, or, $p^2(c^2-1) = p_1^2. \; . \; . \; (6).$

Then from (5) and (6), $r_1^2 = c^2p_1^2$, or, $r_1 = cp_1$, the equation of a similar and equal spiral.

CHAPTER IV.

CONSECUTIVE LINES AND CURVES.

142. If different values be successively assigned to the constants or parameters which enter into the equation of any curve, the several relations thus produced will represent as many distinct curves, differing from each other in form, or in position, or in both these particulars, but all belonging to the same class or family of curves. When the parameters are supposed to vary by indefinitely small increments, the curves are said to be *consecutive*.

Thus let $F(x, y, a) = 0$, ... (1), be the equation of a curve, and let the parameter a take an increment h, converting (1), into $F(x, y, a + h) = 0$, ... (2), then if h be supposed indefinitely small, the curves (1) and (2) will be consecutive. Moreover, the curves (1) and (2) will usually intersect, and the positions of the points of intersection will vary with the value of h, becoming fixed and determinate when the curves are consecutive.

143. *Prop.* To determine the points of intersection of consecutive lines or curves.

To effect this object, we must combine the equations

$$F(x, y, a) = 0, \ldots . (1). \quad \text{and} \quad F(x, y, a + h) = 0, \ldots . (2).$$

and then make $h = 0$, in the result.

Expanding (2) as a function of $a + h$ by Taylor's Theorem, and observing that x, and y, being the same in (1) and (2), (since they refer to the points of intersection,) are to be considered constant in this development, we obtain

$$F(x, y, a+h) = F(x, y, a) + \frac{dF(x, y, a)}{da} \cdot \frac{h}{1}$$

$$+ \frac{d^2F(x, y, a)}{da^2} \cdot \frac{h^2}{1 \cdot 2} + \&c. = 0.$$

But $$F(x, y, a) = 0,$$

$$\therefore \frac{dF(x, y, a)}{da} \cdot \frac{h}{1} + \frac{d^2F(x, y, a)}{da^2} \cdot \frac{h^2}{1 \cdot 2} + \&c. = 0.$$

or, dividing by h, $$\frac{dF(x, y, a)}{da} + \frac{d^2F(x, y, a)}{da^2} \cdot \frac{h}{1 \cdot 2} + \&c. = 0.$$

And when $h=0$ this reduces to $$\frac{dF(x, y, a)}{da} = 0. \quad . \; . \; . \; . \quad (3).$$

The two conditions (1) and (3), serve to determine the co-ordinates x and y, of the required points of intersection.

144. *Ex.* To determine the points of intersection of consecutive normals to any plane curve.

The general equation of a normal is

$$(y - y_1)p' + x - x_1 = 0, \quad . \; . \; . \; . \quad (1).$$

in which x_1, y_1, and p', are parameters, all of which vary together.

Differentiating (1) with respect to x_1, and observing that y_1 and p' are functions of x_1, and that x and y are to be considered constant, we get

$$(y - y_1)p'' - p'^2 - 1 = 0, \quad . \; . \; . \; . \quad (2).$$

or, $$y = y_1 + \frac{1 + p'^2}{p''}, \; . \; . \; (3). \text{ and } \therefore x = x_1 - \frac{p'(1+p'^2)}{p''} \; . \; . \; . \; (4).$$

The values (3) and (4) being identical with those of the co-ordinates of the centre of the osculatory circle, it follows that consecutive normals intersect at the centre of curvature. This principle is sometimes employed in determining the value of the radius of curvature.

145. *Prop.* The curve which is the locus of all the points of intersection of a series of consecutive curves touches each curve in the series.

If we eliminate the parameter a between the two equations

$$F(x, y, a) = 0 \ldots (1) \quad \text{and} \quad \frac{dF(x, y, a)}{da} = 0 \ldots (2),$$

the resulting equation will be a relation between the general co-ordinates x and y of the points of intersection, independent of the particular curve whose parameter is a, or, in other words, the equation of the locus.

Resolving (2) with respect to a the result may be written

$$a = \varphi(x, y),$$

and this substituted in (1) gives

$$F[x, y, \varphi(x,y)] = 0 \ldots . (3),$$

which will be the equation of the locus.

Now if the differential coefficient $\frac{dy}{dx}$ be the same whether derived from (1) or (3), the two curves will have a common tangent at the point x,y, and therefore will be tangent to each other.

Differentiating with respect to x, we obtain from (1)

$$\frac{dF(x, y, a)}{dx} + \frac{dF(x, y, a)}{dy} \cdot \frac{dy}{dx} = 0 \ldots . (4),$$

and from (3), $\frac{dF[x,y,\varphi(x,y)]}{dx} + \frac{dF[x, y, \varphi(x,y)]}{dy} \cdot \frac{dy}{dx}$

$$+ \frac{dF[x, y, \varphi(x,y)]}{d\varphi(x,y)} \cdot \left[\frac{d\varphi(x,y)}{dx}\right] = 0 \ldots (5).$$

But the first and second terms of (4) and (5) are identical, and the third term of (5) is equal to zero by (2).

Hence the values of $\frac{dy}{dx}$ given by (4) and (5), and by (1) and (3), are the same, and consequently the two curves (1) and (3) are tangent to each other.

146. The curve (3) which touches each curve of the series, is called the *envelope* of the series.

147. 1. To determine the envelope of a series of equal circles whose centres lie in the same straight line.

Assuming the line of centres as the axis of x, the equation of one of these circles will be of the form

$$(x - a)^2 + y^2 - r^2 = 0 \ldots. (1),$$

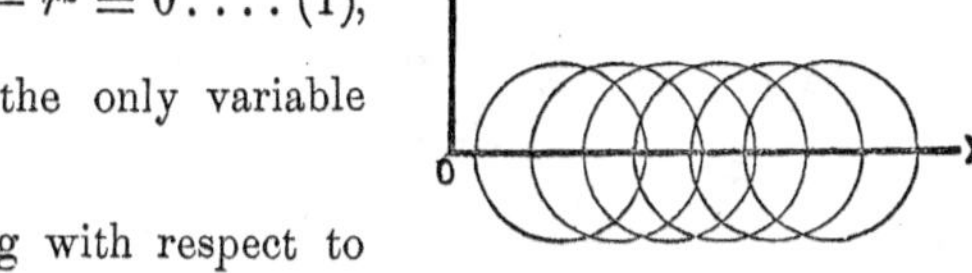

in which a is the only variable parameter.

Differentiating with respect to a, we get

$$-2x + 2a = 0 \ldots. (2)$$

From (2) $a = x$, and this substituted in (1) gives

$$y^2 - r^2 = 0. \quad \therefore y = \pm r.$$

This is the equation of two straight lines parallel to and equidistant from the axis of x, a result easily foreseen.

2. The envelope of a series of equal circles whose centres lie in the circumference of a given circle.

Let $$x_1^2 + y_1^2 - r_1^2 = 0 \ldots. (1)$$

be the equation of the fixed circle.

$$(x - x_1)^2 + (y - y_1)^2 - r^2 = 0 \ldots. (2)$$

that of one of the moveable circles.

The variable parameters are x_1 and y_1, the latter being a function of the former.

From (2) we have $$-2(x - x_1) - 2(y - y_1)\frac{dy_1}{dx_1} = 0 \ldots (3).$$

But from (1) $x_1 + y_1\dfrac{dy_1}{dx_1} = 0$ or $\dfrac{dy_1}{dx_1} = -\dfrac{x_1}{y_1} = -\dfrac{x_1}{\sqrt{r_1^2 - x_1^2}}$.

This value in (3) gives

$$-(x-x_1)+(y-\sqrt{r_1^2-x_1^2})\frac{x_1}{\sqrt{r_1^2-x_1^2}}=0.$$

or $$-x+\frac{yx_1}{\sqrt{r_1^2-x_1^2}}=0, \quad \text{and} \quad \therefore x_1=\frac{r_1x}{\sqrt{x^2+y^2}} \quad \cdots\cdot (4).$$

Now combining (1), (2), and (4), so as to eliminate x_1 and y_1, we get

$$\left(x-\frac{r_1x}{\sqrt{x^2+y^2}}\right)^2+\left(y-\sqrt{r_1^2-\frac{r_1^2x^2}{x^2+y^2}}\right)^2-r^2=0,$$

or $$x^2+y^2-\frac{2r_1(x^2+y^2)}{\sqrt{x^2+y^2}}+r_1^2=r^2.$$

$$\therefore \sqrt{x^2+y^2}-r_1=\pm r. \qquad \therefore x^2+y^2=(r_1\pm r)^2.$$

This is the equation of two concentric circles whose radii are r_1+r and r_1-r respectively.

3. The curve which touches every chord connecting the extremities of conjugate diameters of an ellipse.

Let Q_1P_1 and Q_2P_2 be conjugate diameters of the ellipse $ACBD$, x_1 and y_1 the co-ordinates of P_1, x_2 and y_2 those of P_2.

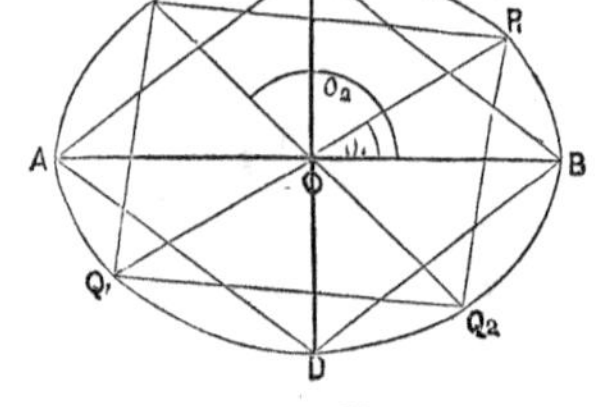

Put $AO=a$, $OC=b$,

$$\tan P_1OB=\tan\theta_1=\frac{y_1}{x_1}=t_1,$$

$$\tan P_2OB=\tan\theta_2=\frac{y_2}{x_2}=t_2.$$

Then, since by the property of the ellipse, $t_1t_2=-\frac{b^2}{a^2}$.

$$\therefore -b^2x_1x_2=a^2y_1y_2 \quad \text{and} \quad x_2=-\frac{a^2y_1y_2}{b^2x_1}.$$

Also $$a^2y_1^2+b^2x_1^2=a^2y_2^2+b^2x_2^2=a^2y_2^2+\frac{a^4y_1^2y_2^2}{b^2x_1^2}$$

$$=\frac{a^2y_2^2}{b^2x_1^2}(a^2y_1^2+b^2x_1^2).$$

$$\therefore \frac{a^2 y_2^2}{b^2 x_1^2} = 1 \quad \text{and} \quad \therefore y_2 = \frac{bx_1}{a} \quad \text{and} \quad x_2 = -\frac{a^2 y_1 y_2}{b^2 x_1} = -\frac{ay_1}{b}.$$

$\therefore$ The equation of the line P_1P_2 is

$$y - y_1 = \frac{y_1 - y_2}{x_1 - x_2}(x - x_1) = \frac{y_1 - \dfrac{bx_1}{a}}{x_1 + \dfrac{ay_1}{b}}(x - x_1)$$

or
$$x_1\left(y + \frac{bx}{a}\right) - y_1\left(x - \frac{ay}{b}\right) - ab = 0 \;.\;.\;.\;.\; (1).$$

Differentiating (1) with respect to x_1 we get

$$y + \frac{bx}{a} - \left(x - \frac{ay}{b}\right)\frac{dy_1}{dx_1} = 0 \;.\;.\;.\;.\; (2).$$

But $a^2y_1^2 + b^2x_1^2 = a^2b^2 \;.\;.\;.\;.\; (3)$, and $\therefore \dfrac{dy_1}{dx_1} = -\dfrac{b^2x_1}{a^2y_1}$

Hence (2) can be reduced to

$$a^2y_1\left(y + \frac{bx}{a}\right) + b^2x_1\left(x - \frac{ay}{b}\right) = 0 \;.\;.\;.\;.\; (4).$$

Combining (1) and (4) we have $\quad x_1 = \dfrac{ba^2(bx + ay)}{2(a^2y^2 + b^2x^2)}$

and
$$\therefore y_1 = \frac{-bx_1(bx - ay)}{a(bx + ay)} = -\frac{ab^2(bx - ay)}{2(a^2y^2 + b^2x^2)}$$

These values reduce (3) to the form

$$\frac{a^2b^2[(bx + ay)^2 + (bx - ay)^2]}{4(a^2y^2 + b^2x^2)^2} = 1, \quad \text{or} \quad \frac{x^2}{\frac{1}{2}a^2} + \frac{y^2}{\frac{1}{2}b^2} = 1,$$

the equation of an ellipse whose semi-axes are $a\sqrt{\frac{1}{2}}$ and $b\sqrt{\frac{1}{2}}$, and which is, therefore, similar to the original ellipse.

4. The envelope of a series of lines drawn from every point in a parabola, and forming with the tangent angles equal to those included between the tangent and the axis.

Let PD be one of the lines.

Put $\quad DPT = PTD = \theta,\ AE = x_1\ EP = y_1.$

Then $PDE = 2\theta$, and the equation of the line PD is

$$y - y_1 = \text{tang}\, 2\theta\, (x - x_1) \ldots\ldots (1).$$

But $\text{tang}\, 2\theta = \dfrac{2\tan\theta}{1 - \tan^2\theta} = \dfrac{2\dfrac{dy_1}{dx_1}}{1 - \dfrac{dy_1^2}{dx_1^2}}$; and since $y_1^2 = 2px_1$, $\dfrac{dy_1}{dx_1} = \dfrac{p}{y_1}$,

$$\therefore \tan 2\theta = \frac{2\dfrac{p}{y_1}}{1 - \dfrac{p_2}{y_1^2}} = \frac{2py_1}{y_1^2 - p^2}$$

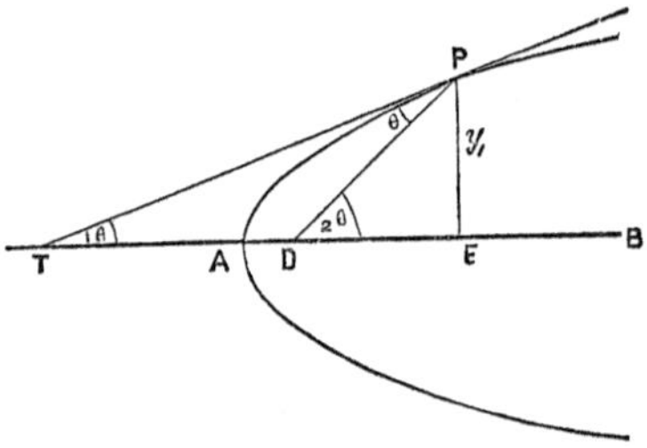

$$\therefore y - y_1 = \frac{2py_1}{y_1^2 - p^2}(x - x_1)$$

$$= \frac{2py_1}{y_1^2 - p^2}\left(x - \frac{y_1^2}{2p}\right)$$

Figure labels: P, y_1, θ, 2θ, θ, T, A, D, E, B

or $\quad yy_1^2 - p^2y + p^2y_1 - 2pxy_1 = 0. \ldots\ldots (2).$

Differentiating (2) with respect to y_1, we find

$$2yy_1 + p^2 - 2px = 0, \quad \text{and} \quad y_1 = \frac{2px - p^2}{2y},$$

This value, substituted in (2), gives

$$\frac{4p^2x^2 - 4p^3x + p^4}{4y} - p^2y + p^2.\frac{2px - p^2}{2y} - 2px\frac{2px - p^2}{2y} = 0,$$

or by reduction $\quad (2x - p)^2 + (2y)^2 = 0 \ldots (3).$

This can be satisfied only by making $2x - p = 0$ and $y = 0$, $\therefore$ (3) represents a point whose co-ordinates are $x = \frac{1}{2}p$ and $y = 0$. Thus the lines will all pass through the focus; as might have been foreseen from the well-known property of the parabola.

5. From every point in the circumference of a circle, pairs of tangents are drawn to another circle; to find the curve which touches every chord connecting corresponding points of contact.

Let P_1 be a point on the first circle P_1P_2 and P_1P_3 a pair of tangents, P_2P_3 one of the chords, O the origin at the centre of the second circle, x_1y_1 the co-ordinates of P_1, x_2y_2 those of P_2, x_3y_3 those of P_3. $OP_3 = r$, $CP_1 = r_1$, $OC = a$.

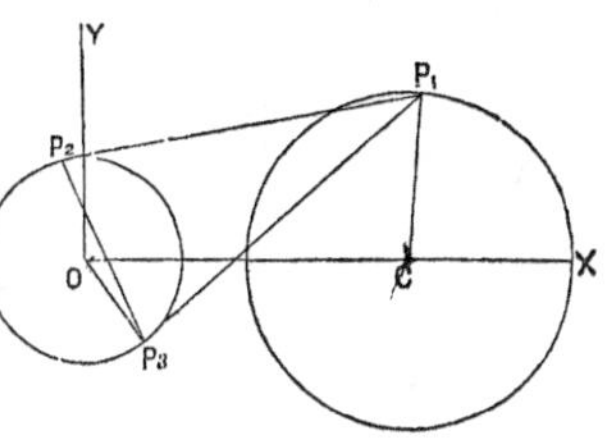

Then $y - y_2 = \frac{y_2 - y_3}{x_2 - x_3}(x - x_2)$ ..... (1) is the equation of the chord P_2P_3.

Also $y_1y_2 + x_1x_2 = r^2$ ..... (2) the equation of the tangent P_2P_1 applied to the point P_1.

$y_1y_3 + x_1x_3 = r^2$ ... (3) the equation of the tangent P_3P_1 applied to the point P_1.

Then $y_1(y_2 - y_3) + x_1(x_2 - x_3) = 0$, and $\therefore \frac{y_2 - y_3}{x_2 - x_3} = -\frac{x_1}{y_1}$,

which reduces (1) to $y - y_2 = -\frac{x_1}{y_1}(x - x_2)$

or $$yy_1 + xx_1 = y_1y_2 + x_1x_2 = r^2 \text{ ..... (4).}$$

Now differentiating (4) with respect to x_1, we get

$$y\frac{dy_1}{dx_1} + x = 0. \text{ But } y_1^2 + (x_1 - a)^2 = r_1^2 \text{ ..... (5).}$$

$$\therefore y_1\frac{dy_1}{dx_1} + x_1 - a = 0 \text{ and } \therefore y\frac{a - x_1}{y_1} + x = 0,$$

or $$yx_1 - xy_1 = ay \text{ .... (6).}$$

Combining (4) and (6) we have

$$y_1 = \frac{r^2y - ayx}{x^2 + y^2} \text{ and } x_1 = \frac{r^2x + ay^2}{x^2 + y^2}.$$

These values substituted in (5) give

$$\frac{(x^2+y^2)(r^4 - 2ar^2x + a^2x^2)}{(x^2+y^2)^2} = r_1^2 \text{ or } r_1^2y^2 + (r_1^2 - a^2)x^2 + 2ar^2x = r^4.$$

Hence the curve required is always a conic section. It is a circle when $a = 0$, an ellipse when $a < r_1$, a parabola when $a = r_1$ and a hyperbola when $a > r_1$.

CHAPTER V.

SINGULAR POINTS OF CURVES.

148. Those points of a curve which enjoy some property not common to the other points, are called *singular* points. Such are *multiple* points, or those through which several branches of the curve pass; *conjugate*, or isolated points; *cusps*, or points at which two tangential branches terminate; *points of inflexion*, &c. These will be examined successively.

Multiple Points.

149. These are of two kinds, viz.: 1st. When two or more branches intersect in passing through a point, their several tangents at that point being inclined to each other; and 2d. When the branches are tangent to each other, their rectilinear tangents being coincident.

150. *Prop.* To determine whether a given curve has multiple points of the first species.

At such a point, there must be as many rectilinear tangents, and therefore as many different values of the differential coefficient $\frac{dy}{dx}$ as there are intersecting branches.

Let $F(x, y) = 0 = u, \ldots\ldots (1)$, be the equation of the given curve, freed from radicals.

Then since $p' = \frac{dy}{dx} = -\frac{\frac{du}{dx}}{\frac{du}{dy}}$, and since differentiation never introduces radicals where they do not exist in the expression differentiated, the value of p' above given cannot contain radicals, and therefore cannot be susceptible of several values, unless it assumes the indeterminate form $\frac{0}{0}$.

Hence the condition $p' = \frac{0}{0}$ will characterise the points sought. To discover whether such points exist, and if so, to find their positions, we form the partial differential coefficients $\frac{du}{dx}$ and $\frac{du}{dy}$ from the equation of the curve, then place their values equal to zero, and determine the corresponding values of x and y.

If these values prove real, and satisfy (1), they *may* belong to a multiple point. We then determine the value of p' by the method applicable to functions which assume the indeterminate form $\frac{0}{0}$, and if there be several real and unequal values of p', they will correspond to as many intersecting branches of the curve, passing through the point examined.

EXAMPLES.

151. 1. To determine whether the curve $x^4 + 2ax^2y - ay^3 = 0$, has multiple points of the first species.

$$u = x^4 + 2ax^2y - ay^3 = 0, \; \ldots \; (1).$$

$$\frac{du}{dx} = 4x^3 + 4axy, \; \ldots \; (2). \qquad \frac{du}{dy} = 2ax^2 - 3ay^2, \; \ldots \; (3).$$

$$\therefore \; p' = \frac{4x^3 + 4axy}{3ay^2 - 2ax^2}. \; \ldots \; (4).$$

Placing (2) and (3) equal to zero, we get

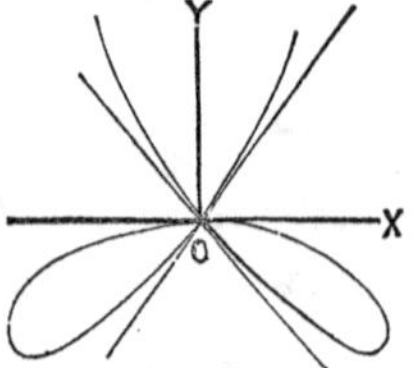

$$x(x^2 + ay) = 0, \ldots . (5).$$

and, $\quad 2x^2 - 3y^2 = 0. \ldots . (6).$

Combining (5) and (6) we have three pairs of values for x and y, viz.:

$$x = 0, \text{ and } y = 0,$$

or, $x = +\frac{1}{3}a\sqrt{6}$, and $y = -\frac{2}{3}a$, or, $x = -\frac{1}{3}a\sqrt{6}$, and $y = -\frac{2}{3}a$.

The first pair of values will alone satisfy (1), and therefore the origin is the only point to be examined.

Placing $x = 0$, and $y = 0$, in (4), there results

$$p' = \frac{12x^2 + 4ay + 4axp'}{6ayp' - 4ax} = \frac{0}{0} \text{ when } \begin{cases} x = 0 \\ y = 0, \end{cases}$$

or by substituting for numerator and denominator their differential coefficients,

$$p' = \frac{24x + 8ap' + 4axp''}{6ap'^2 + 6ayp'' - 4a} = \frac{8ap'}{6ap'^2 - 4a} \text{ when } \begin{cases} x = 0 \\ y = 0. \end{cases}$$

$$\therefore \ p'(6ap'^2 - 4a) = 8ap', \text{ and consequently}$$

$$p' = 0, \text{ or, } p' = +\sqrt{2}, \text{ or, } p' = -\sqrt{2}.$$

Hence the origin is a triple point, the branches being inclined to the axis in angles whose tangents are 0, $+\sqrt{2}$, and $-\sqrt{2}$, respectively.

The form of the curve is shown in the diagram.

2. The curve $ay^3 - x^3y - ax^3 = 0 = u. \ldots . (1).$

$$\frac{du}{dx} = -3x^2y - 3ax^2 = 0, \ldots . (2). \quad \frac{du}{dy} = 3ay^2 - x^3 = 0. \ldots (3).$$

From (2) and (3), $x = 0$, and $y = 0$, or, $x = a\sqrt[3]{3}$, and $y = -a$.

The first pair of values satisfies (1), but the second does not. Therefore the origin is the point to be examined.

Hence $p' = \frac{3x^2y + 3ax^2}{3ay^2 - x^3} = \frac{6xy + 3x^2p' + 6ax}{6ayp' - 3x^2} = \frac{0}{0}$ when $\begin{cases} x = 0 \\ y = 0. \end{cases}$

$$= \frac{6y + 12xp' + 3x^2p'' + 6a}{6ap'^2 + 6ayp'' - 6x} = \frac{6a}{6ap'^2} \text{ when } \begin{cases} x = 0 \\ y = 0. \end{cases}$$

$$\therefore \; p'^3 = 1, \text{ and } p' = 1.$$

This being the only real value of p', there is but one branch passing through the origin, and therefore the curve has no multiple points.

3. The curve $x^4 - 2ay^3 - 3a^2y^2 - 2a^2x^2 + a^4 = 0 = u. \dots (1).$

$\frac{du}{dx} = 4(x^3 - a^2x) = 0. \dots (2).$ $\frac{du}{dy} = -6(ay^2 + a^2y) = 0. \dots (3).$

From (2) and (3) we get six pairs of values, viz.:

$x = 0$, and $y = 0$, or, $x = 0$, and $y = -a$,

or, $x = a$, and $y = 0$, or, $x = -a$, and $y = 0$,

or, $x = a$, and $y = -a$, or, $x = -a$, and $y = -a$.

But of these six pairs of values, the 2d, 3d, and 4th, are the only ones which satisfy (1), and therefore there are but three points to be examined.

$$p' = \frac{2x^3 - 2a^2x}{3ay^2 + 3a^2y} = \frac{6x^2 - 2a^2}{(6ay + 3a^2)p'} = \frac{4}{3p'} \text{ when } \begin{cases} x = \pm a \\ y = 0 \end{cases}$$

and $p' = \frac{2}{3p'}$ when $\begin{cases} x = 0 \\ y = -a. \end{cases}$

$\therefore \; p' = \pm\left(\frac{4}{3}\right)^{\frac{1}{2}}$ at the point where $x = a$ and $y = 0$

$p' = \pm\left(\frac{4}{3}\right)^{\frac{1}{2}}$ " " $x = -a$ and $y = 0$

$p' = \pm\left(\frac{2}{3}\right)^{\frac{1}{2}}$ " " $x = 0$ and $y = -a$.

Thus the curve has three double points.

152. *Prop.* To determine whether a given curve has multiple points of the second species.

Here the mode of proceeding is similar to that in the last proposition, but the resulting values of p' prove equal although given by an equation of the second or higher degree.

Ex. The curve $x^4 + x^2y^2 - 6ax^2y + a^2y^2 = 0 = u \ldots . (1).$

$$\frac{du}{dx} = 4x^3 + 2xy^2 - 12axy = 0 \,..\, (2), \quad \frac{du}{dy} = 2x^2y - 6ax^2 + 2a^2y = 0 \,..\, (3).$$

From (2) and (3) $x = 0$ and $y = 0$, and this is the only pair of values which will satisfy (1). Hence the origin is the only point to be examined.

$$p' = \frac{12axy - 2xy^2 - 4x^3}{2x^2y - 6ax^2 + 2a^2y} = \frac{12ay + 12axp' - 2y^2 - 4xyp' - 12x^2}{4xy + 2x^2p' - 12ax + 2a^2p'} = \frac{0}{2a^2p'},$$

when $x = 0$ and $y = 0$. $\therefore p'^2 = \dfrac{0}{2a^2} = 0$ and $p' = \pm 0$.

And the origin is a double point of the 2d species.

153. We may prove directly that at a double point of the 2d kind, the condition $p' = \dfrac{0}{0}$ is always fulfilled.

Thus suppose the two branches to have contact of the nth order. Then the first n differential coefficients will be the same for the two branches, but the $(n + 1)th$ differential coefficient wiil be different at the double point.

Let $P\dfrac{dy}{dx} + Q = 0 \ldots . (1)$ be the result obtained by differentiating the given equation once, in which P and Q are functions of x and y, the original equation having been freed from radicals.

By repeating the differentiation n times, we get

$$P\frac{d^{n+1}y}{dx^n} + Q_1 = 0,$$

in which P is the same as in (1), and Q_1 is a function of x, y, and the differential coefficients of the several orders less than $(n+1)$.

Now the $(n+1)th$ differential coefficient has, by supposition, two different values a and b for the same values of P and Q_1.

$$\therefore Pa + Q_1 = 0, \quad \text{and} \quad Pb + Q_1 = 0,$$

and by subtraction $P(a-b) = 0$. $\therefore P = 0$ since a and b are unequal.

This value of P substituted in (1) gives $Q = 0$.

$$\therefore \frac{dy}{dx} = -\frac{Q}{P} = \frac{0}{0}.$$

Multiple points of the 2d species are characterized by having but one value (or rather two or more equal values) for $\frac{dy}{dx}$, but several unequal values for $\frac{d^2y}{dx^2}$ or some higher coefficient.

Conjugate or Isolated Points.

154. These are points whose co-ordinates satisfy the equation of a curve, but from which no branches proceed. When p' assumes the imaginary form for real values of x and y, the corresponding point will be isolated, as the curve will then have no direction; and since imaginary values occur only where radicals are introduced, the condition $p' = \frac{0}{0}$ will also hold true in such cases.

The converse proposition, viz.: that at a conjugate point p' will be imaginary, is not always true; for if in the development

$$y_1 = F(x \pm h) = y \pm \frac{dy}{dx} \cdot \frac{h}{1} + \frac{d^2y}{dx^2} \cdot \frac{h^2}{1.2} \pm \frac{d^3y}{dx^3} \cdot \frac{h^3}{1.2.3} + \&c.$$

any one of the differential coefficients should prove imaginary, y_1 would be imaginary also.

To determine with certainty whether a point (a,b) is isolated, substitute successively $a+h$ and $a-h$ for x, and if both values of y_1 prove imaginary (h being small), the point will be imaginary; otherwise it will not.

155. If the coefficient $p' = \frac{dy}{dx}$ be found to have multiple values, some being real and some imaginary, we may regard the result as indicating the indefinitely near approach of a conjugate point to a real branch of the curve.

EXAMPLES.

156. 1. To determine whether the curve

$$ay^2 - x^3 + 4ax^2 - 5a^2x + 2a^3 = 0 = u \dots \dots (1)$$

has conjugate points.

$$\frac{du}{dx} = -3x^2 + 8ax - 5a^2 = 0 \dots\dots (2), \quad \frac{du}{dy} = 2ay = 0 \dots\dots (3).$$

From (2) and (3), $x = a$ and $y = 0$, or $x = \frac{5}{3}a$ and $y = 0$.

The first pair of values satisfies (1), and therefore the point $(a,0)$ must be examined.

$$p' = \frac{3x^2 - 8ax + 5a^2}{2ay} = \frac{6x - 8a}{2ap'} = -\frac{1}{p'} \quad \text{when} \quad \begin{cases} x = a \\ y = 0. \end{cases}$$

$$\therefore p'^2 = -1, \quad p' = \pm\sqrt{-1}.$$

This result being imaginary, we conclude that the point examined is isolated.

2. The curve $(c^2y - x^3)^2 = (x-a)^5(x-b)^6$, in which $a > b$.

$$u = (c^2y - x^3)^2 - (x-a)^5(x-b)^6 = 0 \dots\dots (1),$$

$$\frac{du}{dy} = 2c^2(c^2y - x^3) = 0 \dots\dots (3),$$

$$\frac{du}{dx} = -6x^2(c^2y - x^3) - 5(x-a)^4(x-b)^6 - 6(x-a)^5(x-b)^5 = 0 \dots (2).$$

The equations (2) and (3) give

$$x = a \text{ and } y = \frac{a^3}{c^2}, \text{ or } x = b \text{ and } y = \frac{b^3}{c^2};$$

both of which pairs of values satisfy (1), and therefore both require examination.

$$p' = \frac{6x^2(c^2y - x^3) + 5(x-a)^4(x-b)^6 + 6(x-a)^5(x-b)^5}{2c^2(c^2y - x^3)}$$

$$= \frac{6x^2(c^2y - x^3)}{2c^2(c^2y - x^3)} + \frac{5(x-a)^4(x-b)^6 + 6(x-a)^5(x-b)^5}{2c^2(x-a)^{\frac{5}{2}}(x-b)^3}$$

$$= \frac{3x^2}{c^2} + \frac{5(x-a)^{\frac{3}{2}}(x-b)^3 + 6(x-a)^{\frac{5}{2}}(x-b)^2}{2c^2} = \frac{3b^2}{c^2} \text{ when } x = b,$$

$$= \frac{3a^2}{c^2} \text{ when } x = a.$$

Thus p' is real at both points. But if we substitute $b \pm h$ for x in (1), and solve with respect to y, we get

$$y = \frac{(b \pm h)^3}{c^2} + \frac{1}{c^2}(b \pm h - a)^{\frac{5}{2}}(\pm h)^3,$$

both of which values of y are imaginary when h is taken less than $a - b$; so that the point where $x = b$ and $y = \frac{b^3}{c^2}$ is a conjugate point, although p' is real.

This result is confirmed by forming the succeeding differential coefficients; thus

$$p'' = \frac{1}{c^2}\left[6x + \frac{3}{2}\cdot\frac{5}{2}(x-a)^{\frac{1}{2}}(x-b)^3 + 15(x-a)^{\frac{3}{2}}(x-b)^2 \right.$$

$$\left. + 6(x-a)^{\frac{5}{2}}(x-b)\right] = \frac{6b}{c^2}, \text{ when } x = b.$$

This is a real value also.

But the next coefficient will contain the term $6(x-a)^{\frac{5}{2}} = 6(b-a)^{\frac{5}{2}}$ which is imaginary, since $a > b$.

The value $x = a$ does not belong to a conjugate point, as is seen by substituting $a \pm h$ for x in (1), and solving with respect to y, thus,

$$y = \frac{(a \pm h)^3}{c^2} + \frac{1}{c^2}(\pm h)^{\frac{5}{2}}(a - b \pm h)^3$$

which is real when $h > 0$, but imaginary when $h < 0$.

Cusps.

157. A cusp is that peculiar kind of double point of the second species at which two tangential branches terminate without passing through the point.

Cusps are of two kinds, viz.:

1st. That in which the two branches lie on different sides of the tangent, as in Fig. 1.

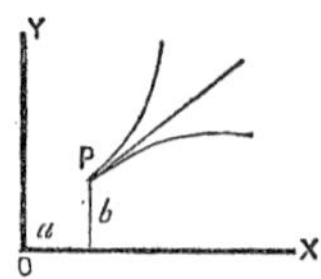

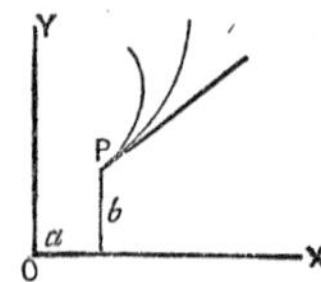

2d. That in which they lie on the same side of the tangent, as in Fig. 2.

The test of a cusp is that $\frac{dy}{dx}$ shall have two real and equal values at some point (a,b), and that when we substitute $a+h$ and $a-h$ for x, we shall find, in one case, two real and unequal values of y, and in the other two imaginary values. The only exception to this is that offered by the case shown in Fig. 3, where a cusp of the first kind occurs at a point P, with the tangent parallel to the axis of y. It will then be more convenient to form the value of $\frac{dx}{dy}$, which should be ± 0, and to try whether the successive substitution of $b + h$ and $b - h$ for y will render x, in one case, real and double, and in the other imaginary. The condition $p' = \frac{0}{0}$ serves as a guide in selecting the points to be examined.

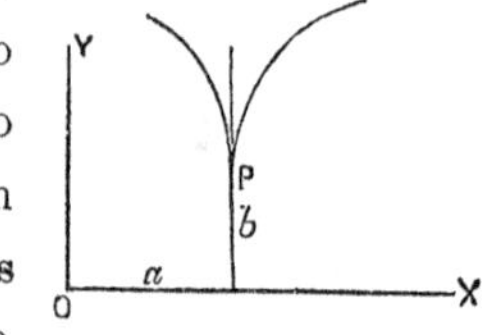

EXAMPLES.

158. 1. To determine whether the curve $(by - cx)^2 = (x - a)^5$ has a cusp, and if so, of which kind.

$$u = (by - cx)^2 - (x - a)^5 = 0 \ldots\ldots (1),$$

$$\frac{du}{dx} = -2c(by - cx) - 5(x - a)^4 = 0 \ldots\ldots (2).$$

$$\frac{du}{dy} = 2b(by - cx) = 0 \ldots\ldots (3).$$

From (2) and (3) we obtain $x = a$, and $y = \frac{ac}{b}$,

and as these values satisfy (1), we must examine the point $\left(a, \frac{ac}{b}\right)$

$$p' = \frac{2c(by - cx) + 5(x - a)^4}{2b(by - cx)} = \frac{0}{0} \text{ when } \left\{ \begin{array}{l} x = a \\ y = \frac{ac}{b} \end{array} \right.$$

$$= \frac{2bcp' - 2c^2 + 20(x - a)^3}{2b^2p' - 2bc} = \frac{2bcp' - 2c^2}{2b^2p' - 2bc}, \text{ when } x = a,$$

$$\therefore b^2p'^2 - 2bcp' = -c^2, \quad p'^2 - 2\frac{c}{b}p' = -\frac{c^2}{b^2} \text{ and } p' = \frac{c}{b} \pm 0.$$

$\therefore$ There are two equal values of p', and consequently two tangential branches proceed from the point, $a, \frac{ac}{b}$.

Now put successively $x = a + h$, and $x = a - h$, and solve with respect to y.

when $x = a + h$, $y = \frac{ca + ch \pm \sqrt{(+h)^5}}{b}$, two real and unequal values,

when $x = a - h$, $y = \frac{ca - ch \pm \sqrt{(-h)^5}}{b}$ two imaginary values.

Hence there is a cusp at the point $a, \frac{ac}{b}$, and the tangent at that point is inclined to the axes of x and y.

Again, the ordinate Y of the tangent corresponding to the abscissa $a + h$, is $\frac{ac}{b} + p'h = \frac{ac + ch}{b}$ which is greater than one of the corresponding values of y, and less than the other. Therefore the branches lie on different sides of the tangent, and the cusp is of the first kind.

Remark. The kind of cusp can usually be found very easily by examining the values of the second differential coefficient; for the deflection of the curve from the tangent is controlled by the sign of $\frac{d^2y}{dx^2}$. Hence, when the two values of this coefficient have contrary signs, the cusp will be of the first kind, but when the signs are alike, it will be of the second kind.

2. The semi-cubical parabola $cy^2 = x^3$.

$$u = cy^2 - x^3 \ldots (1), \quad \frac{du}{dx} = -3x^2 = 0 \ldots (2), \quad \frac{du}{dy} = 2cy = 0 \ldots (3).$$

$\therefore x = 0$, and $y = 0$, and as these satisfy (1) there may be a cusp at the origin.

$$p' = \frac{3x^2}{2cy} = \frac{6x}{2cp'} = \frac{0}{2cp'} \text{ when } x = 0.$$

$$\therefore p'^2 = \frac{0}{2c} = 0 \text{ and } p' = \pm 0,$$

two real and equal values.

Now put $0 \pm h$ for x in (1), and there will result,

when $x = 0 + h$, $y = \pm\sqrt{\frac{h^3}{c}}$ two real and unequal values,

" $x = 0 - h$, $y = \pm\sqrt{-\frac{h^3}{c}}$ two imaginary values.

$\therefore$ There is a cusp at the origin. Also the ordinate Y of the tangent corresponding to the abscissa $0 + h$, is $0 + p'h = 0$, which being intermediate in value between the two corresponding values of y, the cusp is of the first kind.

159. Sometimes it is more convenient to solve the equation with respect to y before differentiating.

Ex. $$(y - b - cx^2)^2 = (x - a)^5$$

$$y = b + cx^2 \pm (x - a)^{\frac{5}{2}}, \quad p' = 2cx \pm \frac{5}{2}(x - a)^{\frac{3}{2}}.$$

Now y has but one value $b + ca^2$, or to speak more correctly, it has two equal values $(b + ca^2 \pm 0)$ when $x = a$, and $p' = 2ca \pm 0$ has then two equal values also.

When $x=a+h$, $y=b+c(a+h)^2\pm(+h)^{\frac{5}{2}}$ two real and unequal values.

" $x=a-h$, $y=b+c(a-h)^2\pm(-h)^{\frac{5}{2}}$ two imaginary values.

Hence there is a cusp at the point

$$(a, b + ca^2).$$

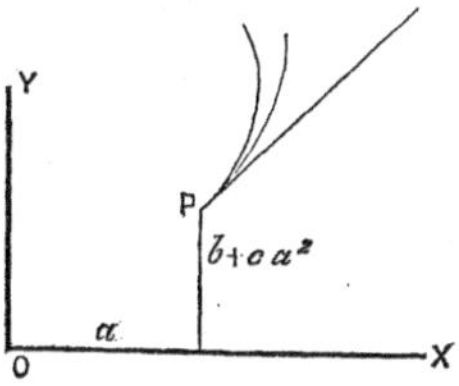

Also $p'' = 2c \pm \frac{3}{2}\cdot\frac{5}{2}(x - a)^{\frac{1}{2}} = 2c \pm 0$ when $x = a$.

And since the two values of p'' have the same sign, the cusp at the point $(a, b + ca^2)$ is of the second kind. The kind of cusp would also appear by comparing the ordinate Y of the tangent with the two values of y.

For when $x = a + h$, $Y = b + ca^2 + p'h = b + ca^2 + 2cah$, which is less than either value of y, when h is small.

Points of Inflexion.

160. Points of inflexion or contrary flexure are those at which the curve changes the direction of its curvature, being successively convex and concave towards a fixed line as the axis of x.

It has already been remarked that a curve is convex towards the axis of x when $\frac{d^2y}{dx^2}$ is positive and concave when $\frac{d^2y}{dx^2}$ is negative. Hence a point of inflexion will be characterized by having the second differential coefficient affected with contrary signs, at points situated near to, but on different sides of the point in question. But since a variable quantity changes its sign only when its value passes through zero or infinity, the condition $\frac{d^2y}{dx^2} = 0$ or $\frac{d^2y}{dx^2} = \infty$ will belong to a point of inflexion. But the converse is not necessarily true, for the sign of $\frac{d^2y}{dx^2}$ does not always change after its value has reached 0 or ∞. We must therefore see whether a change in the sign of $\frac{d^2y}{dx^2}$ will or will not occur.

We may also recognize a point of inflexion by the consideration that at such a point the tangent intersects the curve, and therefore the ordinate of the tangent will, on one side of the point be greater, and on the other less than the corresponding ordinate of the curve.

EXAMPLES.

161. 1. The cubical parabola $a^2y = x^3$.

$$y = \frac{x^3}{a^2}, \quad p' = \frac{3x^2}{a^2}, \quad p'' = \frac{6x}{a^2} = 0 \quad \text{when} \quad x = 0.$$

$\therefore$ The origin is a point to be examined.

Put $x = 0 + h$, and $y = y_1$,

$x = 0 - h$, and $y = y_2$.

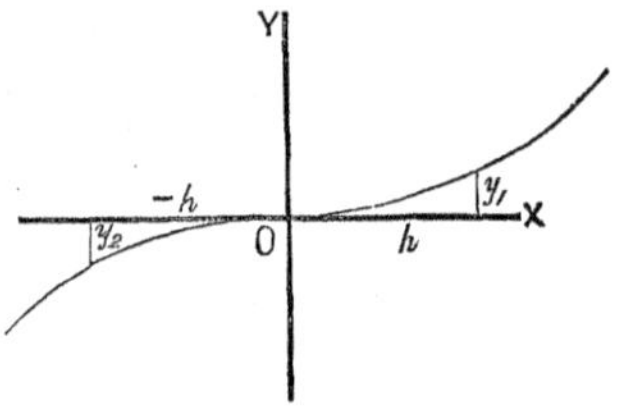

Then $$\frac{d^2y_1}{dx^2} = \frac{6h}{a^2} > 0,$$

$$\frac{d^2y_2}{dx^2} = -\frac{6h}{a^2} < 0.$$

Hence the origin is a point of inflexion. The condition $p'' = \infty$ gives $x = \infty$, and therefore is not applicable.

162. Sometimes it happens that two of the peculiarities which characterize singular points occur at the same point of a curve.

Ex. $\quad a^3y^2 - 2abx^2y - x^5 = 0 = u \; . \; . \; . \; . \; (1),$

$\dfrac{du}{dx} = -4abxy - 5x^4 = 0 \; . \; . \; . \; . \; (2),$

$\dfrac{du}{dy} = 2a^3y - 2abx^2 = 0 \; . \; . \; . \; . \; (3).$

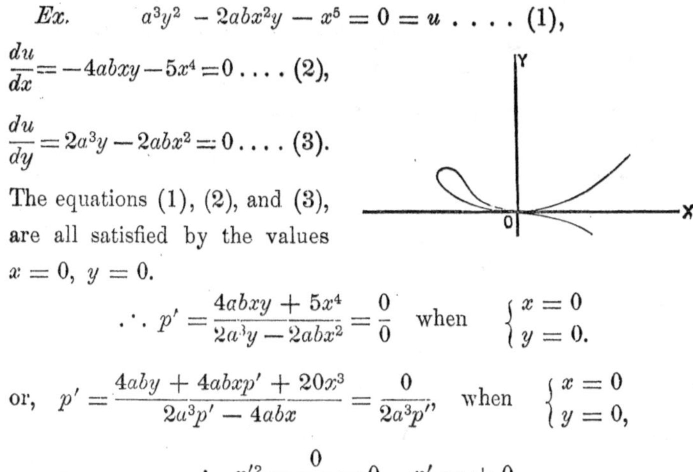

The equations (1), (2), and (3), are all satisfied by the values $x = 0$, $y = 0$.

$$\therefore \; p' = \frac{4abxy + 5x^4}{2a^3y - 2abx^2} = \frac{0}{0} \quad \text{when} \quad \begin{cases} x = 0 \\ y = 0. \end{cases}$$

$$\text{or,} \quad p' = \frac{4aby + 4abxp' + 20x^3}{2a^3p' - 4abx} = \frac{0}{2a^3p''}, \quad \text{when} \quad \begin{cases} x = 0 \\ y = 0, \end{cases}$$

$$\therefore \; p'^2 = \frac{0}{2a^3} = 0, \quad p' = \pm\, 0,$$

and there is either a cusp or a double point at the origin, the axis of x being tangent to the curve.

If $x = 0 + h$, $\; y = \dfrac{bh^2}{a^2} \pm \sqrt{\dfrac{b^2h^4 + ah^5}{a^4}}$, two real values, one greater and the other less than the ordinate (0) of the tangent.

If $x = 0 - h$, $\; y = \dfrac{bh^2}{a^2} \pm \sqrt{\dfrac{b^2h^4 - ah^5}{a^4}}$, two real values when h is small, but both greater than 0.

Hence there is a double point of the second species at the origin, and one branch of the curve has an inflexion at that point.

163. In addition to the singular points already described, two other classes may be noticed, viz.: *Stop Points*, or those at which a single branch terminates abruptly; and *Shooting Points*, at which two or more branches terminate without being tangent to each other. Both are of rare occurrence, but the following are examples of curves belonging to these classes.

1. $y = x \, . \log x$. This curve has a stop point at the origin. For, y has but one value, and that is real when $x > 0$; but the

value of y is impossible when $x < 0$, since negative quantities can not properly be regarded as having any logarithms.

2. $y = x \tan^{-1} \frac{1}{x}$, or, $y = x \cot^{-1} x$.

This curve has a shooting point at the origin, for

$$\frac{dy}{dx} = \tan^{-1}\frac{1}{x} - \frac{x}{1 + x^2} = \tan^{-1}(+\infty)$$

$$= \frac{1}{2}\pi = 1.5708 \quad \text{when} \quad x = +0$$

$$= \tan^{-1}(-\infty) = -\frac{1}{2}\pi = -1.5708 \quad \text{when} \quad x = -0,$$

and whether x be positive or negative, y will have but one value.

164. When a curve has the spiral form, and is therefore more conveniently referred to polar co-ordinates, we may distinguish the existence of a point of contrary flexure by the condition that $\frac{dp}{dr} = 0$ at that point, and that it shall have contrary signs on different sides of that point. This we proceed to show.

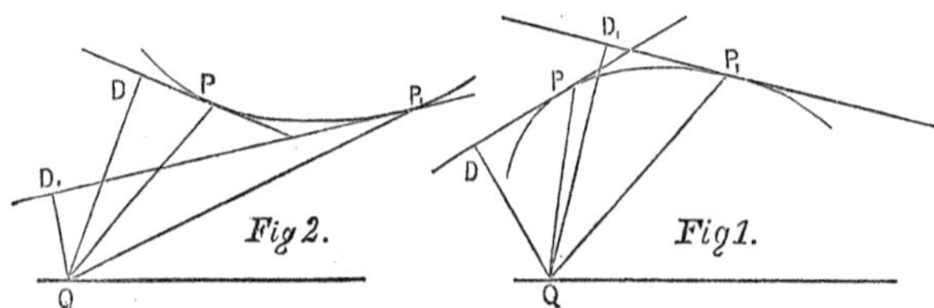

In Fig. 1, the curve is concave to the pole Q; and in Fig. 2, it is convex.

In the first case r and p increase together, and therefore $\frac{dp}{dr}$ is positive. In the second case, p diminishes as r increases, and therefore $\frac{dp}{dr}$ is negative. Hence, in passing through a point of contrary flexure, $\frac{dp}{dr}$ will change its sign, becoming equal to zero at that point, for $\frac{dp}{dr}$ plainly could not become infinite, since p cannot exceed r.

CHAPTER VI.

CURVILINEAR ASYMPTOTES.

165. When two curves continually approach each other, and meet only at an infinite distance, each is said to be an asymptote to the other.

166. *Prop.* To determine the conditions necessary to render two curves asymptotes to each other.

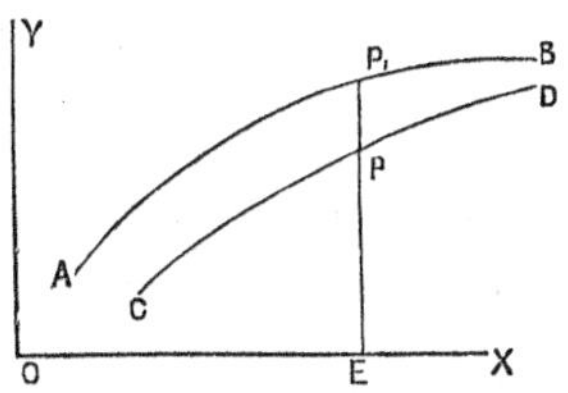

Let the curves be referred to rectangular axes, and let the ordinates EP and EP', corresponding to the same abscissa $OE = x$, be expressed by means of the equations of the curves in terms of x. The difference $PP' = y_1 - y$ can then be expressed in terms of x, and if this difference be reduced to zero by making $x = \infty$, (being finite for all other values of x,) the curves will be asymptotes to each other.

This condition is fulfilled only when the difference (expanded into a series, contains none but negative powers of x, without an absolute term, for in such cases only will the difference $y_1 - y$ become zero when $x = \infty$.

Hence we must be able to express $y_1 - y$ in the form

$$y_1 - y = Ax^{-a} + Bx^{-b} + Cx^{-c} + \&c.,$$

or the difference $x_1 - x$ of the two abscissæ, corresponding to the same ordinate, must admit of being expressed in the form

$$x_1 - x = A_1 y^{-a_1} + B_1 y^{-b_1} + C_1 y^{-c_1} + \&c.$$

167. *Cor.* If there be three curves, (A), (B), and (C), and if the difference of the corresponding ordinates of (A) and (B), and that of the ordinates of (A) and (C), be thus expressed.

$$y_2 - y_1 = Ax^{-a} + Bx^{-(a+1)} + Cx^{-(a+2)}, \dots . (1).$$

$$y_3 - y_1 = B_1x^{-(a+1)} + C_1x^{-(a+2)} + \&c, \dots . (2).$$

the three curves will be asymptotes to each other, and, moreover, the curve (C) will lie nearer to (A) than (B) does. For, by making x sufficiently large, the term Ax^{-a}, or $\frac{A}{x^a}$ may be rendered greater than the sum of the succeeding terms of (1), or greater than the sum of those terms increased by the series (2).

168. *Cor.* The curve whose equation can be written in the form

$$y = D + Ax^a + Bx^b + Cx^c + A_1x^{-a_1} + B_1x^{-b_1} + C_1x^{-c_1} + \&c.,$$

can have an infinite number of curvilinear asymptotes.

For by taking any curve whose equation is of the form

$$y_1 = D + Ax^a + Bx^b + Cx^c + A_2x^{-a_2} + B_2x^{-b_2} + \&c.$$

in which the absolute term D, and the terms involving the positive powers of x, are the same as in the given equation, the difference $y_1 - y$ will reduce to zero when $x = \infty$.

169. *Prop.* To find the general form of the expanded value of the ordinate in such curves as admit of a rectilinear asymptote.

Since the equation of the rectilinear asymptote has the form $y = A_1x + B_1$, the equation of the desired curve must take the form

$$y = A_1x + B_1 + Ax^{-a} + Bx^{-b} + Cx^{-c} + \&c.$$

170. 1. The common hyperbola $a^2y^2 - b^2x^2 = - a^2b^2$.

$$y = \pm \frac{b}{a}(x^2 - a^2)^{\frac{1}{2}} = \pm \frac{b}{a}\left(x - \frac{1}{2}a^2x^{-1} - \frac{1}{8}a^4x^{-3} - \&c.\right)$$

But $y = \pm \frac{b}{a}x$ is the equation of two straight lines passing through the origin and equally inclined to the axis of x. Hence these lines are asymptotes to the hyperbola.

2. To determine whether the curve $y = b(x^2 - a^2)^{-\frac{1}{2}}$ has either rectilinear or curvilinear asymptotes.

By expansion

$$y = b(x^{-1} + \frac{1}{2}a^2x^{-3} + \&c.) = bx^{-1} + \frac{1}{2}ba^2x^{-3} + \&c.$$

But $y = 0$ is the equation of the axis of x. Hence that axis is a rectilinear asymptote to the curve.

To discover whether there is an asymptote parallel to the axis of y, let the equation be solved with respect to x; thus

$$x = \pm (a^2 + b^2y^{-2})^{\frac{1}{2}} = \pm (a + \frac{1}{2}b^2a^{-1}y^{-2} - \&c.)$$

Here it is evident that two lines parallel to the axis of y, and at distances therefrom equal to $+a$ and $-a$ respectively, will be asymptotes to the curve, their equations being

$$x = +a \quad \text{and} \quad x = -a.$$

The hyperbola whose equation (referred to its asymptotes) is $xy = b$ will be a curvilinear asymptote, and there may be found any number of other curvilinear asymptotes.

CHAPTER VII.

TRACING OF CURVES.

171. In this chapter it is proposed to give such general directions as are necessary in tracing a curve from its given equation, and in discovering the chief peculiarities which characterize it.

The following steps will be found useful:

1st. Having resolved the equation, if possible with respect to y, let different positive values be assigned to x from $x = 0$ to $x = \infty$, and let those points be noticed particularly where $y = 0$, $y = \infty$, or $y =$ an imaginary value. The first indicates an intersection with the axis of x, the second shows the existence of an infinite branch, and the third gives the limits of the curve in the direction of x positive.

2d. Assign to x all negative values from $x = 0$ to $x = -\infty$, and observe the same peculiarities with respect to y as when x was positive. In both cases the negative as well as the positive values of y must be examined so as to include the branches below as well as those above the axis of x.

3d. Determine whether the curve has asymptotes, and determine their position.

4th. Find the value of the differential coefficient $\frac{dy}{dx}$ and determine from thence the angles at which the curve cuts the axes, as well as the points at which the tangent is parallel to either axis.

5th. From the value of $\frac{d^2y}{dx^2}$ ascertain the direction of the curvature and the positions of the points of contrary flexure when they exist.

6th. Determine the positions and character of the other singular points, if there be such.

EXAMPLES.

172. 1. Let the equation of the proposed curve be

$$y^2 = \frac{x^3 - a^3}{x + b}.$$

Resolving with respect to y we have

$$y = \pm \sqrt{\frac{x^3 - a^3}{x + b}},$$

and since each value of x gives two values of y numerically equal but having contrary signs, the curve must be divided symmetrically by the axis of x.

If x be positive and numerically less than a, y will be imaginary, and there will be no point of the curve between the axis of y and a parallel thereto at a distance equal to a on the right of the origin.

When $x = a$, $y = 0$, when $x > a$, y is real, and continues so for all greater values of x, becoming infinite when $x = \infty$.

If x be negative and numerically less than b, y is imaginary, and there is no point between the axis of y and a parallel thereto at the distance $= b$, on the left of the origin.

When $x = -b$, y becomes infinite; and when $x < -b$, that is, negative and numerically greater that b, y becomes real and continues to increase without limit as the numerical value of x increases, being infinite when $x = -\infty$.

Thus it appears that the curve has six infinite branches.

Again, since $x = -b$ makes y infinite, there is an asymptote parallel to the axis of y, and at a distance therefrom equal to $-b$.

Also by resolving the given equation with respect to y, and expanding, we get

$$y = \pm \frac{(x^3 - a^3)^{\frac{1}{2}}}{(x+b)^{\frac{1}{2}}} = \pm x\left(1 - \frac{a^3}{x^3}\right)^{\frac{1}{2}}\left(1 + \frac{b}{x}\right)^{-\frac{1}{2}}$$

$$= \pm x\left(1 - \frac{1}{2}\frac{b}{x} + \frac{3}{8}\frac{b^2}{x^2}, \text{ \&c.}\right) = \pm\,(x - \frac{1}{2}b + \text{ terms involv-}$$

ing powers of x).

Hence $y = \pm(x - \frac{1}{2}b)$ is the equation of two straight lines, which are asymptotes to the curve, and are inclined to the axis of x at angles of 45° and 135° respectively.

If we combine this equation of these asymptotes with that of the curve, we shall find that each of the asymptotes intersects that branch of the curve which lies on the right of the axis of y.

Forming the value of $\frac{dy}{dx}$ from the equation of the curve, we have

$$\frac{dy}{dx} = \frac{2x^3 + 3bx^2 + a^3}{2(x^3 - a^3)^{\frac{1}{2}}(x+b)^{\frac{3}{2}}}$$

which, placed equal to zero, gives the cubic equation

$$x^3 + \frac{3}{2}bx^2 + \frac{1}{2}a^3 = 0,$$

in which there must be one real and negative root, since the absolute term is positive. The other two roots are imaginary, as is easily seen from the form of the equation. Thus there are two points corresponding to the same negative abscissa, one above and the other equally below the axis of x, at which the tangent is parallel to the axis of x.

By making $\frac{dy}{dx} = \infty$, we get $x = a$ or $x = -b$. The first corresponds to a point at which the curve intersects the axis of x perpendicularly. The second belongs to the point of contact of one of the asymptotes as before seen.

By forming the value of $\frac{d^2y}{dx^2}$, we should find that the curve is concave to the axis of x when x is positive, and convex when x is negative.

The curve has neither multiple points, cusps, conjugate points, nor inflexions.

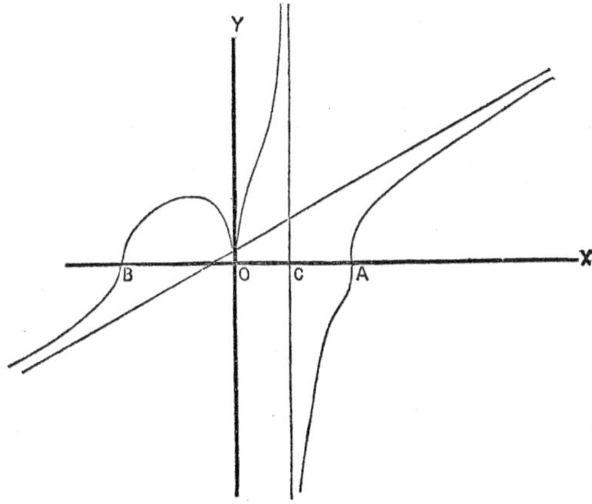

2. The curve whose equation is $y^3 = \frac{x^4 - a^2x^2}{2x - a}$.

When $x = 0$, $y = 0$, and therefore the curve passes through the origin.

When $x = \frac{a}{2}$, $y = \pm \infty$, when $x = + \infty$, $y = + \infty$, and when $x = - \infty$, $y = - \infty$.

Thus the curve has four infinite branches.

When $x = a$, or $x = - a$, $y = 0$ corresponding to two intersections with the axis of x.

Since $x = \frac{a}{2}$ renders $y = \pm \infty$, there is one asymptote whose equation is

$$x = \frac{a}{2}.$$

Also, by resolving with respect to y, and expanding, we get

$$y = x\frac{\left(1-\frac{a^2}{x^2}\right)^{\frac{1}{3}}}{2^{\frac{1}{3}}\left(1-\frac{a}{2x}\right)^{\frac{1}{3}}} = \frac{x}{2^{\frac{1}{3}}}\left(1-\frac{a^2}{x^2}\right)^{\frac{1}{3}}\left(1-\frac{a}{2x}\right)^{-\frac{1}{3}}$$

$$= \frac{x}{2^{\frac{1}{3}}}\left(1 - \frac{1}{3}\frac{a^2}{x^2} - \&c.\right)\left(1 + \frac{1 . a}{6x} + \&c.\right).$$

$$= \frac{1}{2^{\frac{1}{3}}}\left(x + \frac{a}{6} + \text{terms involving negative powers of } x\right)$$

$\therefore y = \dfrac{1}{2^{\frac{1}{3}}}\left(x + \dfrac{a}{6}\right)$ is the equation of a second asymptote.

Forming the value of the differential coefficient $\frac{dy}{dx}$, we have

$$\frac{dy}{dx} = \frac{6x^4 - 2a^2x^2 - 4ax^3 + 2a^3x}{3(2x-a)^{\frac{4}{3}}(x^4 - a^2x^2)^{\frac{2}{3}}}.$$

This expression becomes infinite when $x = \frac{a}{2}$, when $x = \pm a$, and when $x = 0$.

Hence the curve cuts the axis of x perpendicularly at the origin, and at distances therefrom $= + a$ and $- a$ respectively. The value of $\frac{dy}{dx}$ becomes zero when $6x^4 - 4ax^3 - 2a^2x^2 + 2a^3x = 0$, which corresponds to a value of x between 0 and $-a$. The corresponding value of y is a maximum.

There are inflexions at the points where $x = a$ and $x = - a$, as will readily appear by substituting for x values alternately a little greater and somewhat less than a, and similarly for values greater and less than $-a$. For if x be rather greater than a in the equation $y^3 = \frac{x^4 - a^2x^2}{2x - a}$, y will be positive; but if x be somewhat less than a, y will become negative. Thus the curve will cross the tangent at the point where it meets the axis. The same will be true when $x = -a$.

There will be a third inflexion between $x = 0$ and $x = \frac{1}{2}a$, for the curve touches the axis of y at the origin, and a parallel asymptote at the distance $\frac{1}{2}a$ from that axis, and, therefore, must change the direction of its curvature between those two parallels.

Finally by making the value of $\frac{dy}{dx} = \frac{0}{0}$ we shall find that there is a cusp of the first kind at the origin. The form of the curve is represented in the diagram

PART III.

THEORY OF CURVED SURFACES.

CHAPTER I.

TANGENT AND NORMAL PLANES AND LINES.

173. The consideration of surfaces affords an application of the theory of functions of two independent variables. Thus if x, y, and z, be the co-ordinates of any point in the surface, and $z = \varphi(x,y)$ the equation of the surface, the values of x and y may be assumed arbitrarily, and that of z will become determinate.

174. *Prop.* To determine the general differential equation of a plane drawn tangent to any curved surface at a given point (x_1, y_1, z_1) situated in the surface.

Let the surface and plane be intersected by planes respectively parallel to xz and yz, and passing through the point (x_1, y_1, z_1).

The equations of the line cut from the tangent plane by the plane parallel to xz will be of the forms

$$x - x_1 = t(z - z_1) \dots (1), \quad \text{and} \quad y = y_1 \dots (2),$$

and those of the intersection parallel to yz will be of the forms

$$y - y_1 = s(z - z_1) \dots (3) \quad \text{and} \quad x = x_1 \dots (4).$$

Also the equation of the tangent plane, which contains these lines, will have the form

$$A(x - x_1) + B(y - y_1) + C(z - z_1) = 0 \dots (5).$$

The equation of its trace on xz is $A(x-x_1) = -C(z-z_1) + By_1 \dots (6)$.

" " " " yz " $B(y-y_1) = -C(z-z_1) + Ax_1 \dots (7)$.

But the trace (6) is parallel to the intersection (1) (2), and the trace (7) is parallel to the intersection (3) (4).

$$\therefore t = -\frac{C}{A} \quad \text{and} \quad s = -\frac{C}{B}.$$

which values reduce (5) to the form

$$z - z_1 = \frac{1}{t}(x - x_1) + \frac{1}{s}(y - y_1) \ldots . (8).$$

Now since the intersections (1) (2) and (3) (4) are respectively tangent to the corresponding curves cut from the surface, we must have $t = \frac{dx_1}{dz_1}$ and $s = \frac{dy_1}{dz_1}$ or $\frac{1}{t} = \frac{dz_1}{dx_1}$ and $\frac{1}{s} = \frac{dz_1}{dy_1}$.

Hence (8) reduces to

$$z - z_1 = \frac{dz_1}{dx_1}(x - x_1) + \frac{dz_1}{dy_1}(y - y_1) \ldots (9), \text{ the desired equation.}$$

The expressions $\frac{dz_1}{dx_1}$ and $\frac{dz_1}{dy_1}$ are the partial differential coefficients derived from the equation of the surface, and they will have the same values at the point (x_1, y_1, z_1), as the similar coefficients derived from the equation of the plane, tangent at that point.

175. *Cor.* If the equation of the surface be given under the form

$$u = \varphi(x, y, z,) = 0,$$

the equation of the tangent plane will take a more symmetrical form. For we then have (Art. 57)

$$\left[\frac{du}{dx}\right] = \frac{du}{dx} + \frac{du}{dz} \cdot \frac{dz}{dx} = 0, \quad \text{and} \quad \left[\frac{du}{dy}\right] = \frac{du}{dy} + \frac{du}{dz} \cdot \frac{dz}{dy} = 0.$$

Hence

$$\frac{dz_1}{dx_1} = -\frac{\frac{du}{dx_1}}{\frac{du}{dz_1}}, \quad \frac{dz_1}{dy_1} = -\frac{\frac{du}{dy_1}}{\frac{du}{dz_1}},$$

and by substitution in (9) and reduction, we obtain the more symmetrical form

$$(x - x_1)\frac{du}{dx_1} + (y - y_1)\frac{du}{dy_1} + (z - z_1)\frac{du}{dz_1} = 0.$$

176. *Prop.* To determine the equations of a line normal to a curved surface at a given point (x_1, y_1, z_1).

The equations of a line passing through the point (x_1, y_1, z_1), have the forms

$$x - x_1 = t(z - z_1), \qquad y - y_1 = s(z - z_1);$$

and since the normal line is perpendicular to the tangent plane, we have by the conditions of perpendicularity of a line and plane ($A = Ct$ and $B = Cs$), the following relations:

$$t = \frac{A}{C} = -\frac{dz_1}{dx_1} = \frac{\frac{du}{dx_1}}{\frac{du}{dz_1}}, \quad s = \frac{B}{C} = -\frac{dz_1}{dy_1} = \frac{\frac{du}{dy_1}}{\frac{du}{dz_1}}.$$

These conditions give for the equations of the normal line

$$\left.\begin{aligned} x - x_1 + \frac{dz}{dx_1}(z - z_1) = 0 \\ y - y_1 + \frac{dz}{dy_1}(z - z_1) = 0 \end{aligned}\right\} \text{ or } \left\{\begin{aligned} \frac{du}{dz_1}(x - x_1) = \frac{du}{dx_1}(z - z_1). \\ \frac{du}{dz_1}(y - y_1) = \frac{du}{dy_1}(z - z_1). \end{aligned}\right.$$

177. *Cor.* If θ_1, θ_2, θ_3, be the angles formed by the normal with the axes of x, y, and z, respectively, or those formed by the tangent plane with the planes of yz, xz, and xy, we shall have

$$\cos\theta_1 = \frac{A}{\sqrt{A^2 + B^2 + C^2}} = -\frac{-\frac{dz_1}{dx_1}}{\sqrt{\frac{dz_1^2}{dx_1^2} + \frac{dz_1^2}{dy_1^2} + 1}} = \frac{\frac{du}{dx_1}}{\sqrt{\frac{du^2}{dx_1^2} + \frac{du^2}{dy_1^2} + \frac{du^2}{dz_1^2}}},$$

$$\cos\theta_2 = -\frac{-\frac{dz_1}{dy_1}}{\sqrt{\frac{dz_1^2}{dx_1^2} + \frac{dz_1^2}{dy_1^2} + 1}} = \frac{\frac{du}{dy_1}}{\sqrt{\frac{du^2}{dx_1^2} + \frac{du^2}{dy_1^2} + \frac{du^2}{dz_1^2}}}.$$

$$\cos\theta_3 = \frac{1}{\sqrt{\frac{dz_1^2}{dx_1^2} + \frac{dz_1^2}{dy_1^2} + 1}} = \frac{\frac{du}{dz_1}}{\sqrt{\frac{du^2}{dx_1^2} + \frac{du^2}{dy_1^2} + \frac{du^2}{dz_1^2}}}.$$

178. *Prop.* To determine the equations of a line drawn tangent to a curve of double curvature, at a given point (x_1,y_1,z_1), on the curve.

The curve will be given by the equations of its projections on two of the co-ordinate planes, as xz, and yz; thus

$$F(x,z)=0,\ \ldots.\ (1).\quad \text{and} \quad \varphi(y,z)=0.\ \ldots.\ (2).$$

The equations of the required tangent will have the forms

$$x-x_1=t(z-z_1),\ \ldots.\ (3).\quad \text{and} \quad y-y_1=s(z-z_1),\ \ldots.\ (4);$$

and since the projections of the tangent are tangent to the projections of the curve, (3) and (4) will take the forms

$$z-z_1=\frac{dz_1}{dx_1}(x-x_1),\ ..\ (5).\quad \text{and} \quad z-z_1=\frac{dz_1}{dy_1}(y-y_1),\ ..\ (6).$$

in which equations the values of $\frac{dz_1}{dx_1}$ and $\frac{dz_1}{dy_1}$ are to be derived from (1) and (2), the equations of the given curve.

179. *Prop.* To determine the equation of a plane drawn through a given point of a curve of double curvature, and normal to the curve at that point.

The equation of a plane passing through the point (x_1,y_1,z_1), is of the form

$$A(x-x_1)+B(y-y_1)+C(z-z_1)=0.\ \ldots\ (1).$$

But, since the plane is to be perpendicular to the tangent line, we must have the conditions

$$A=Ct=C\frac{dx_1}{dz_1},\quad \text{and} \quad B=Cs=C\frac{dy_1}{dz_1},$$

which values reduce (1) to the form

$$(x-x_1)\frac{dx_1}{dz_1}+(y-y_1)\frac{dy_1}{dz_1}+(z-z_1)=0,$$

the required equation.

EXAMPLES OF TANGENT PLANES TO SURFACES.

180. 1. The tangent plane to the sphere whose equation is

$$u = x^2 + y^2 + z^2 - r^2 = 0.$$

Here $$\frac{du}{dx} = 2x, \quad \frac{du}{dy} = 2y, \quad \frac{du}{dz} = 2z.$$

Therefore by substitution in the general differential equation of a tangent plane to a curved surface, we get

$$(x-x_1)\frac{du}{dx_1}+(y-y_1)\frac{du}{dy_1}+(z-z_1)\frac{du}{dz_1} = 2x_1(x-x_1)+2y_1(y-y_1) + 2z_1(z-z_1) = 0.$$

$\therefore\ xx_1 + yy_1 + zz_1 = x_1^2 + y_1^2 + z_1^2 = r^2$, the required equation.

2. The ellipsoid $u = \frac{x^2}{a^2} + \frac{y^2}{b^2} + \frac{z^2}{c^2} - 1 = 0.$

$$\frac{du}{dx} = \frac{2x}{a^2}, \quad \frac{du}{dy} = \frac{2y}{b^2}, \quad \frac{du}{dz} = \frac{2z}{c^2}.$$

$$\therefore\ \frac{2x_1}{a^2}(x - x_1) + \frac{2y_1}{b^2}(y - y_1) + \frac{2z_1}{c^2}(z - z_1) = 0.$$

or, $\frac{xx_1}{a^2} + \frac{yy_1}{b^2} + \frac{zz_1}{c^2} = 1$, the required equation of the tangent plane.

3. The hyperboloid of one sheet $u = \frac{x^2}{a^2} + \frac{y^2}{b^2} - \frac{z^2}{c^2} - 1 = 0.$

$$\frac{du}{dx} = \frac{2x}{a^2}, \quad \frac{du}{dy} = \frac{2y}{b^2}, \quad \frac{du}{dz} = -\frac{2z}{c^2}.$$

$$\therefore\ \frac{2x_1}{a^2}(x - x_1) + \frac{2y_1}{b^2}(y - y_1) - \frac{2z_1}{c^2}(z - z_1) = 0.$$

$\therefore\ \frac{xx_1}{a^2} + \frac{yy_1}{b^2} - \frac{zz_1}{c^2} - 1 = 0$, the equation of the tangent plane.

4. The conoid $u = c^2x^2 + y^2z^2 - r^2z^2 = 0$.

$$\frac{du}{dx} = 2c^2x, \quad \frac{du}{dy} = 2z^2y, \quad \frac{du}{dz} = 2y^2z - 2r^2z.$$

$$\therefore\ 2c^2x_1(x - x_1) + 2z_1^2y_1(y - y_1) + 2z_1(y_1^2 - r^2)(z - z_1) = 0.$$

or, $c^2xx_1 + z_1^2yy_1 + (y_1^2 - r^2)zz_1 = y_1^2z_1^2$, the equation of the tangent plane.

CHAPTER II.

CYLINDRICAL SURFACES, CONICAL SURFACES. AND SURFACES OF REVOLUTION.

181. *Prop.* To determine the general differential equation of all cylindrical surfaces.

These surfaces are generated by the motion of a straight line, which touches a fixed curve, and remains parallel to a fixed line in every position.

Let the equations of the fixed curve or directrix be

$$F(x, z) = 0, \ldots . (1). \qquad F_1(y, z) = 0, \ldots . (2),$$

those of the generatrix, in one of its positions, being

$$x = tz + a, \ldots . (3). \qquad y = sz + b, \ldots . (4).$$

Since the generatrix continues parallel to a fixed line, the values of t and s will continue constant for all positions of the generatrix, but a and b will vary with its position.

Eliminating x between (1) and (3), and y between (2) and (4), we get one relation between z and a, and a second between z and b. Then combining these equations to eliminate z, we obtain a relation between a and b, which may be written

$$b = \varphi a, \ldots . (5).$$

But from (3) and (4), $a = x - tz$, and $b = y - sz$.

$\therefore$ (5) becomes $y - sz = \varphi(x - tz)$, ... (6).

This is a general equation of all cylindrical surfaces, but it contains the unknown function φ. To eliminate this function, differentiate (6) with respect to x and y successively, and divide the first result by the second; thus

$$-s\frac{dz}{dx} = \frac{d\varphi(x - tz)}{d(x - tz)} \times \frac{d(x - tz)}{dx}$$

and
$$1 - s\frac{dz}{dy} = \frac{d\varphi(x - tz)}{d(x - tz)} \times \frac{d(x - tz)}{dy}.$$

$$\frac{-s\dfrac{dz}{dx}}{1 - s\dfrac{dz}{dy}} = \frac{1 - t\dfrac{dz}{dx}}{-t\dfrac{dz}{dy}},$$

whence $t\dfrac{dz}{dx} + s\dfrac{dz}{dy} = 1$ .... (7), the required equation.

182. *Cor.* If we denote the primitive or integrated equation of a cylindrical surface by $f(x, y, z) = u = 0$ the differential equation (7) may be reduced to a more symmetrical form. For since

$$\frac{dz}{dx} = -\frac{\dfrac{du}{dx}}{\dfrac{du}{dz}} \quad \text{and} \quad \frac{dz}{dy} = -\frac{\dfrac{du}{dy}}{\dfrac{du}{dz}}.$$

we obtain by substitution in (7) and reduction

$$t\frac{du}{dx} + s\frac{du}{dy} + \frac{du}{dz} = 0 \text{ .... (8)},$$

a form often more convenient than (7).

183. *Prop.* To determine the equation of the cylindrical surface which envelops a given surface, and whose axis is parallel to a given line.

The enveloping and enveloped surfaces being tangent to each

other, will have a common tangent plane at every point in the curve of contact, and the equation of one of these planes will be

$$z - z_1 = (x - x_1)\frac{dz_1}{dx_1} + (y - y_1)\frac{dz_1}{dy_1}$$

or $$(x - x_1)\frac{du}{dx_1} + (y - y_1)\frac{du}{dy_1} + (z - z_1)\frac{du}{dz_1} = 0,$$

in which $x_1\, y_1\, z_1$ refer to a point of contact. Moreover the differential coefficients $\frac{dz_1}{dx_1}, \frac{dz_1}{dy_1}$ or $\frac{du}{dx_1}, \frac{du}{dy_1}, \frac{du}{dz_1}$ are the same whether derived from the equation of the cylinder or from that of the enveloped surface. Hence, if we form the differential coefficients from the equation of the given surface, and substitute their values in the differential equation of the cylinder, the result will characterize the points of contact, being the equation of a surface containing those points. This equation, when combined with that of the enveloped surface, will give the equations of the curve of contact, and thence the cylinder can be determined.

184. *Ex.* A sphere $u = x^2 + y^2 + z^2 - r^2 = 0$ is enveloped by a cylinder whose axis is parallel to the axis of z; to find the curve of contact.

Here we have $x = a$ the equation of the projection of the generatrix on xz, and $y = b$ the equation of the projection of the generatrix on yz.

$$\therefore\ t = 0, \quad s = 0.$$

Also $$\frac{du}{dx} = 2x, \quad \frac{du}{dy} = 2y, \quad \frac{du}{dz} = 2z.$$

Hence by substitution in (8),

$$0 \,.\, 2x + 0 \,.\, 2y + 2z = 0 \quad \text{or} \quad z = 0,$$

and the points of contact all lie in the plane of xy.

Combining the equations $x^2 + y^2 + z^2 - r^2 = 0$ and $z = 0$, there results

$$x^2 + y^2 - r^2 = 0.$$

$\therefore$ The curve of contact is a great circle of the sphere, as might have been foreseen.

185. *Prop.* If any surface of the second order be enveloped by a cylinder, the curve of contact will be an ellipse, hyperbola or parabola, or a variety of one of those curves.

The general equation of surfaces of the second order is

$$Az^2+Bzy+Cy^2+Dzx+Ex^2+Fxy+Gz+Hy+Ix+K=0=u \,..\, (1).$$

$$\therefore \frac{du}{dx} = Dz + 2Ex + Fy + I, \quad \frac{du}{dy} = Bz + 2Cy + Fx + H,$$

$$\frac{du}{dz} = By + 2Az + Dx + G.$$

$$\therefore t\frac{du}{dx} + s\frac{du}{dy} + \frac{du}{dz} = t(Dz+2Ex+Fy+I)+s(Bz+2Cy+Fx+H) + (By + 2Az + Dx + G) = 0,$$

which is the equation of a plane.

Hence the points of contact are confined to one plane. But any section, by a plane, of the surface represented by the equation (1), will necessarily be a line of the second order, and therefore the truth of the proposition is apparent.

Conical Surfaces.

186. *Prop.* To determine the general differential equation of all conical surfaces.

These surfaces are generated by the motion of a straight line which touches constantly a fixed curve and passes through a fixed point.

Let the equations of the directrix be

$$F(x,z) = 0 \, .\,.\,.\,. \, (1), \quad F_1(y,z) = 0 \, .\,.\,.\,. \, (2);$$

those of the generatrix in one of its positions being

$$x - a = t\,(z - c) \, .\,.\,.\,. \, (3), \quad \text{and} \quad y - b = s(z - c) \, .\,.\,.\,. \, (4),$$

where a, b, and c, denote the co-ordinates of the fixed point or vertex.

The quantities t and s vary with the position of the generatrix, but a, b, and c, are constant.

Eliminating x between (1) and (3), and y between (2) and (4), we get one relation between z and t, and a second between z and s. Then combining these equations to eliminate z, we obtain a relation between t and s, which may be written

$$s = \varphi t \;.\;.\;.\;.\;.\;.\;.\; (5).$$

But from (3) and (4), $t = \dfrac{x-a}{z-c}$, and $s = \dfrac{y-b}{z-c}$.

$\therefore$ (5) becomes $\qquad \dfrac{y-b}{z-c} = \varphi\left[\dfrac{x-a}{z-c}\right] \;.\;.\;.\;.\;.\; (6).$

This is an equation of conical surfaces, but it contains the unknown function φ. To eliminate this function, differentiate (6) with respect to x and y successively, and divide the first result by the second; thus

$$-\frac{y-b}{(z-c)^2}\cdot\frac{dz}{dx} = \frac{d\varphi[\;]}{d[\;]} \times \frac{d[\;]}{dx} = \frac{d\varphi[\;]}{d[\;]} \times \left[\frac{1}{z-c} - \frac{x-a}{(z-c)^2}\cdot\frac{dz}{dx}\right]$$

and

$$\frac{1}{z-c} - \frac{y-b}{(z-c)^2}\cdot\frac{dz}{dy} = \frac{d\varphi[\;]}{d[\;]} \times \frac{d[\;]}{dy} = \frac{d\varphi[\;]}{d[\;]} \times \left[-\frac{x-a}{(z-c)^2}\cdot\frac{dz}{dy}\right],$$

in which expressions the [ ] is used to signify $\left[\dfrac{x-a}{z-c}\right]$.

Now by division

$$\frac{-(y-b)\dfrac{dz}{dx}}{z-c-(y-b)\dfrac{dz}{dy}} = \frac{z-c-(x-a)\dfrac{dz}{dx}}{-(x-a)\dfrac{dz}{dy}}.$$

$\therefore z - c = (x-a)\dfrac{dz}{dx} + (y-b)\dfrac{dz}{dy} \;.\;.\;.\; (7)$ the required equation.

187. *Cor.* If we denote the primitive or integrated equation of a conical surface by $f(x, y, z) = u = 0$, the differential equation (7) may be reduced to a more symmetrical form.

For since $\dfrac{dz}{dx} = -\dfrac{\frac{du}{dx}}{\frac{du}{dz}}$, and $\dfrac{dz}{dy} = -\dfrac{\frac{du}{dy}}{\frac{du}{dz}}$,

we obtain by substitution in (7) and reduction

$(x - a)\dfrac{du}{dx} + (y - b)\dfrac{du}{dy} + (z - c)\dfrac{du}{dz} = 0 \ldots\ldots (8)$, a form often more convenient than (7).

188. *Prop.* To determine the equation of the conical surface which envelopes a given surface, and whose vertex is situated at a given point.

If we form the differential coefficients $\dfrac{dz}{dx}$, and $\dfrac{dz}{dy}$ or $\dfrac{du}{dx}, \dfrac{du}{dy}$ and $\dfrac{du}{dz}$ from the equation of the given surface, and substitute their values in (7) or (8), the differential equation of the conical surface, the resulting relation will characterize the points of contact, being the equation of a surface which contains those points. This equation, combined with that of the enveloped surface, will give the equations of the curve of contact, and thence the cone can be determined.

Ex. A sphere $x^2 + y^2 + z^2 - r^2 = 0 = u$ is enveloped by a cone whose vertex is situated on the axis of y, at a distance b from the origin; to find the curve of contact.

Here we have the co-ordinates of the vertex $a = 0$, $b = b$, $c = 0$.

Also, $$\frac{du}{dx} = 2x, \ \frac{du}{dy} = 2y, \ \frac{du}{dz} = 2z.$$

$\therefore$ By substitution in the equation of conical surfaces

$$(x - 0)\,2x + (y - b)\,2y + (z - 0)\,2z = 0;$$

or, $$x^2 + y^2 + z^2 - by = 0.$$

This being the equation of a sphere having a radius $= \frac{1}{2}b$, and its centre on the axis of y at a distance $\frac{1}{2}b$ from the origin, the points of contact must lie in the surface of such a sphere.

By combining the equations of the two spheres, we get

$$by = r^2 \text{ or } y = \frac{r^2}{b} \text{ and } x^2 + z^2 = \frac{r^2}{b^2}(b^2 - r^2).$$

Hence the curve of contact is a circle perpendicular to the axis of y, and at a distance $\frac{r^2}{b}$ from the origin.

189. *Prop.* If any surface of the second order be enveloped by a cone, the curve of contact will be an ellipse, hyperbola, or parabola, or a variety of one of these curves.

The general equation of surfaces of the second order is

$$Az^2 + Bzy + Cy^2 + Dzx + Ex^2 + Fxy + Gz + Hy + Ix + K = 0 = u \;..(1).$$

$$\therefore \frac{du}{dx} = Dz + Fy + 2Ex + I, \quad \frac{du}{dy} = Bz + Fx + 2Cy + H,$$

$$\frac{du}{dz} = By + Dx + 2Az + G.$$

$$\therefore (x-a)\frac{du}{dx} + (y-b)\frac{du}{dy} + (z-c)\frac{du}{dz}$$

$$= [Dz + Fy + 2Ex + I](x-a) + [Bz + Fx + 2Cy + H](y-b)$$
$$+ [By + Dx + 2Az + G](z-c) = 0,$$

or,
$$2[Az^2 + Cy^2 + Ex^2] + 2[Bzy + Dzx + Fxy]$$
$$+ [G - Da - Bb - 2Ac]z + [H - Fa - Bc - 2Cb]y$$
$$+ [I - Fb - Dc - 2Ea]x - [Gc + Hb + Ia] = 0 \ldots (2).$$

By combining (1) and (2), we get

$$[G + Da + Bb + 2Ac]z + [H + Fa + Bc + 2Cb]y$$
$$+ [I + Fb + Dc + 2Ea]x + 2K + Gc + Hb + Ia = 0.$$

This is the equation of a plane, and therefore the curve of contact is the intersection of the given surface by a plane, and consequently an ellipse, hyperbola, or parabola.

190. *Prop.* To determine the general differential equation of all surfaces of revolution.

Let $\left.\begin{array}{l} x = tz + a \ldots . (1) \\ y = sz + b \ldots . (2) \end{array}\right\}$ be the equations of the axis.

$$F(x,z) = 0 \ldots . (3), \quad \text{and} \quad F_1(y,z) = 0 \ldots . (4),$$

those of the generatrix.

The characteristic property of this surface is, that every plane section perpendicular to the axis is a circle. Now the equation of a plane perpendicular to the line (1) (2) is

$$z + tx + sy = c,$$

and the circle cut from the surface by this plane may be supposed situated on the surface of a sphere whose centre may be assumed at any point on the axis, and whose radius will be determined by the value of c, when the centre has been chosen.

Take the centre of the sphere at the point $(a, b, 0)$, where the axis pierces the plane of xy, and the equation of the sphere will be

$$(x - a)^2 + (y - b)^2 + z^2 = r^2.$$

But r and c are mutually dependent upon each other, which fact may be indicated by the equation $c = \varphi(r^2)$. Hence

$$z + tx + sy = \varphi\left[(x - a)^2 + (y - b)^2 + z^2\right] \ldots . (5).$$

To eliminate the unknown function φ, differentiate (5) with respect to y and x successively, and divide the first result by the second.

$$\therefore \frac{dz}{dy} + s = \frac{d\varphi[\;]}{d[\;]} \times \frac{d[\;]}{dy} \quad \text{and} \quad \frac{dz}{dx} + t = \frac{d\varphi[\;]}{d[\;]} \times \frac{d[\;]}{dx}.$$

$$\therefore \frac{\frac{dz}{dy} + s}{\frac{dz}{dx} + t} = \frac{y - b + z\frac{dz}{dy}}{x - a + z\frac{dz}{dx}},$$

or $$(x-a-tz)\frac{dz}{dy} - (y-b-sz)\frac{dz}{dx} + (x-a)s - (y-b)t = 0 \ldots (6),$$

which is the required equation of surfaces of revolution.

Cor. When the axis of revolution coincides with that of z, we have

$$t = 0 \quad \text{and} \quad s = 0, \quad a = 0 \quad \text{and} \quad b = 0.$$

$\therefore$ (6) reduces to $$x\frac{dz}{dy} - y\frac{dz}{dx} = 0 \ldots . (7).$$

191. *Prop.* A given curved surface revolves about a fixed axis; to determine the surface which touches and envelopes the moveable surface in every position.

The required surface will obviously be a surface of revolution, whose generatrix will be the curve of contact of that surface with one of the moveable surfaces.

Hence if we determine the values of the differential coefficients $\frac{dz}{dx}$ and $\frac{dz}{dy}$ from the given surface, and substitute them in the general differential equation of all surfaces of revolution, the result will characterize the points of contact, being the equation of a surface containing those points. This equation, combined with that of the given surface, will give the equations of the curve of contact or the required generatrix.

192. 1. A right cone with a circular base, whose vertex is at the origin, and whose axis coincides originally with the axis of x, is caused to revolve about the axis of z: to determine the form of the enveloping surface.

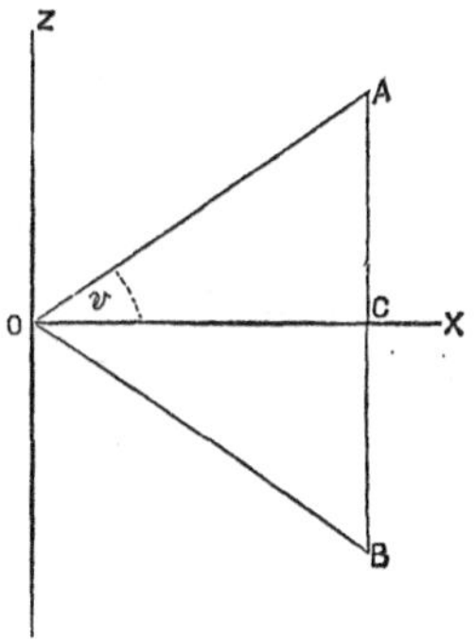

Put the semi-angle AOC of the cone $= v$, and $\tan v = t$.

Then the equation of the cone, in the position AOB will be

$$z^2 + y^2 = t^2x^2, \quad \text{or} \quad z^2 = t^2x^2 - y^2 \ldots . (1)$$

$$\therefore \frac{dz}{dx} = \frac{t^2x}{z} \quad \text{and} \quad \frac{dz}{dy} = -\frac{y}{z},$$

which values substituted in the differential equation of surfaces of revolution, viz.

$$x\frac{dz}{dy} - y\frac{dz}{dx} = 0, \quad \text{gives} \quad \frac{xy}{z} + \frac{t^2xy}{z} = 0.$$

$$\therefore x = 0 \quad \text{or} \quad y = 0.$$

Combining the first of these results, $x = 0$, with the equation of the cone, we get

$$z^2 + y^2 = 0. \quad \therefore z = 0 \quad \text{and} \quad y = 0,$$

which conditions apply to the origin exclusively; but the second result $y = 0$, gives by combination with (1)

$$z^2 = t^2x^2 \quad \text{or} \quad z = \pm\, tx \quad \text{and} \quad y = 0,$$

which are the equations of the lines OA and OB.

Hence the required envelope is a double cone generated by the revolution of the lines OA and OB about OZ.

2. A sphere $(x - a)^2 + (y - b)^2 + z^2 = r^2$, revolves about the axis of z; to find the enveloping surface. Here we have

$$\frac{dz}{dx} = -\frac{x - a}{z} \quad \text{and} \quad \frac{dz}{dy} = -\frac{y - b}{z}.$$

$$\therefore x\frac{dz}{dy} - y\frac{dz}{dx} = -\frac{xy - bx}{z} + \frac{xy - ay}{z} = 0.$$

$\therefore bx - ay = 0$, the equation of a plane passing through the axis of z, and the centre (a,b) of the sphere.

This plane intersects the sphere in a great circle, whose equation, in its own plane, is

$$(r_1 - a_1)^2 + z^2 = r^2,$$

in which $\quad r_1{}^2 = x^2 + y^2, \quad \text{and} \quad a_1{}^2 = a^2 + b^2.$

$$\therefore \left[(x^2 + y^2)^{\frac{1}{2}} - (a^2 + b^2)^{\frac{1}{2}}\right]^2 + z^2 = r^2.$$

or, $$x^2 + y^2 + z^2 - 2(a^2 + b^2)^{\frac{1}{2}}(x^2 + y^2)^{\frac{1}{2}} = r^2 - a^2 - b^2.$$

the equation of the required surface.

When $a^2 + b^2 = r^2$, this reduces to

$$x^2 + y^2 + z^2 - 2r(x^2 + y^2)^{\frac{1}{2}} = 0;$$

and when $a = 0, \quad b = 0, \qquad x^2 + y^2 + z^2 = r^2,$

the equation of the sphere.

3. An ellipsoid $\frac{x^2}{a^2} + \frac{y^2}{b^2} + \frac{z^2}{c^2} = 1$, revolves about the axis of y; to determine the enveloping surface.

The differential equation of the surface is, in this case

$$z\frac{dy}{dx} - x\frac{dy}{dz} = 0.$$

Also, $$\frac{dy}{dx} = -\frac{b^2x}{a^2y}, \quad \frac{dy}{dz} = -\frac{b^2z}{c^2y}.$$

$$\therefore \; -z\frac{b^2x}{a^2y} + x\frac{b^2z}{c^2y} = 0, \quad \therefore \; xz = 0,$$

and consequently $\quad x = 0, \quad$ or, $\quad z = 0.$

But when $x = 0$, $\frac{y^2}{b^2} + \frac{z^2}{c^2} = 1$, an ellipse in the plane of yz.

And when $z = 0$, $\frac{x^2}{a^2} + \frac{y^2}{b^2} = 1$, an ellipse in the plane of xy.

Hence the required envelope consists of two ellipsoids of revolution, whose equations are

$$\frac{y^2}{b^2} + \frac{x^2 + z^2}{c^2} = 1 \qquad \text{and} \qquad \frac{y^2}{b^2} + \frac{x^2 + z^2}{a^2} = 1.$$

CHAPTER III.

CONSECUTIVE SURFACES AND ENVELOPES.

193. In the last chapter we have presented some examples of surfaces enveloping a series of other surfaces, but in the only case considered, the enveloped surface was supposed to be of invariable form, and its change of position was effected only by a revolution around a fixed axis. In that case, the enveloping surface was necessarily a surface of revolution.

It is now proposed to consider the envelopes to any series of consecutive surfaces.

194. If different values be successively assigned to the constants or parameters which enter in the equation of any surface, the several relations thus produced, will represent as many distinct surfaces, differing from each other in form, or in position, or in both these particulars, but all belonging to the same class or family of surfaces. When the parameters are supposed to vary by infinitely small increments, the surfaces are said to be consecutive.

Thus let $F(x, y, z, a) = 0, \ldots$ (1), be the equation of a surface, and let the parameter a, take an increment h, converting (1), into $F(x, y, z, a + h) = 0, \ldots$ (2); then if h be supposed indefinitely small, the surfaces (1) and (2) will be consecutive. Moreover, the surfaces (1) and (2) will usually intersect, and their intersection will vary with the value of h, becoming fixed and determinate when the snrfaces are consecutive.

195. *Prop.* To determine the equations of the intersection of consecutive surfaces.

To effect this object, we must combine the equations

$$F(x, y, z, a) = 0, \ldots . (1), \quad \text{and} \quad F(x, y, z, a + h) = 0, \ldots . (2),$$

and then make $h = 0$.

By reasoning precisely as in the case of consecutive curves, (Art 143) we prove that the two conditions

$$F(x, y, z, a) = 0, \ldots . (1), \quad \text{and} \quad \frac{dF(x, y, z, a)}{da} = 0, \ldots . (3),$$

must be satisfied at the same time.

By combining these equations, so as to eliminate first y, and then x, we shall have the equations of the projections of the required intersection on xz, and yz.

196. *Prop.* The surface which is the locus of all the intersections of a series of consecutive surfaces, touches each surface in the series.

If we eliminate the parameter a between the two equations

$$F(x, y, z, a) = 0, \ldots . (1), \quad \text{and} \quad \frac{dF(x, y, z, a)}{da} = 0, \ldots . (2),$$

the resulting equation will be a relation between the general co-ordinates x, y, z, of the points of the various intersections, independent of the particular curve whose parameter is a, or in other words, the equation of the locus.

Resolving (2) with respect to a, the result may be written

$$a = \varphi(x, y, z),$$

and this substituted in (1) gives

$$F[x, y, z, \varphi(x, y, z)] = 0, \ldots . (3),$$

which will be the equation of the locus.

Now differentiating both (1) and (3) first with respect to x, and then with respect to y, we readily prove, precisely as in the case of consecutive curves, that the values of $\frac{dz}{dx}$ and $\frac{dz}{dy}$ are the same whether derived from (1) or (3). Hence the two surfaces (1) and (3) will have a common tangent plane, and will therefore be mutually tangent to each other at all points common to those surfaces.

197. The surface (3), which touches each surface of the series, is called the envelope of the series.

198. *Ex.* To determine the envelope of a series of equal spheres whose centres lie in the same straight line.

Assuming the line of centres as the axis of x, the equation of one of these spheres will be of the form

$$(x - a)^2 + y^2 + z^2 - r^2 = 0 \dots . (1),$$

in which a is the only variable parameter.

Differentiating with respect to a we get

$$-2x + 2a = 0 \dots . (2).$$

From (2) $a = x$, and this substituted in (1) gives

$$y^2 + z^2 - r^2 = 0.$$

This is the equation of a right cylinder with a circular base, the axis of which coincides with that of x.

199. When the equation of the proposed surface contains two parameters a, b, independent of each other, we must have the three conditions

$$F(x, y, z, a, b) = 0 \dots (1), \quad \frac{dF(x, y, z, a, b)}{da} = 0 \dots . (2).$$

and
$$\frac{dF(x, y, z, a, b)}{db} = 0 \dots . (3).$$

And by eliminating a and b between (1), (2), and (3), the equation of the required envelope will be obtained. Also, if the proposed equation should contain three or more parameters a, b, c, &c., two of which, a and b, are arbitrary, and the others connected with them by given relations, such relations will enable us to eliminate the additional parameters and to obtain a final equation between x, y, and z.

200. 1. A plane whose equation is $\frac{x}{a}+\frac{y}{b}+\frac{z}{c}=1$, is touched in every position by a surface, the variable parameters a, b, and c being connected by the relation $abc = m^3$: to determine the equation of the surface or envelope.

From $\frac{x}{a}+\frac{y}{b}+\frac{z}{c}-1=0 \ldots.$ (1) we obtain by differentiation, regarding a and b as independent, and c dependent upon them,

$$-\frac{x}{a^2}-\frac{z}{c^2}\cdot\frac{dc}{da}=0 \ldots. (2), \quad \text{and} \quad -\frac{y}{b^2}-\frac{z}{c^2}\cdot\frac{dc}{db}=0 \ldots. (3).$$

But the condition $abc = m^3 \ldots.$ (4) gives by differentiation

$$bc + ab\frac{dc}{da} = 0, \quad \text{and} \quad ac + ab\frac{dc}{db} = 0.$$

$$\therefore \frac{dc}{da} = -\frac{c}{a}, \quad \text{and} \quad \frac{dc}{db} = -\frac{c}{b},$$

which values substituted in (2) and (3) reduce them to the forms

$$-\frac{x}{a^2}+\frac{z}{c^2}\cdot\frac{c}{a}=0, \quad \text{and} \quad -\frac{y}{b^2}+\frac{z}{c^2}\cdot\frac{c}{b}=0,$$

whence
$$\frac{x}{a}=\frac{z}{c} \quad \text{and} \quad \frac{y}{b}=\frac{z}{c}.$$

These values in (1) give $\frac{z}{c}+\frac{z}{c}+\frac{z}{c}=1,$

or
$$\frac{3z}{c}=1. \quad \therefore c = 3z.$$

And similarly $\qquad b = 3y, \quad a = 3x.$

Finally by replacing a, b, and c, in (4), by their values just found, we obtain $xyz = \frac{m^3}{27}$ as the equation of the enveloping surface.

2. To find the envelope of all the spheres whose centres lie in the

same plane, and whose radii are proportional to the distances of their centres from a fixed point in that plane.

Assuming the plane of the centres as that of xy, and the origin at the fixed point, the equation of one of the spheres will take the form

$$(x - a)^2 + (y - b)^2 + z^2 - r^2 = 0 \;.\;.\;.\;. \;(1),$$

in which a, b, and r, are variable parameters, a and b being independent, and r connected with them by the relation

$$r^2 = t^2(a^2 + b^2) \;.\;.\;.\;. \;(2) \quad \text{where } t \text{ is a constant.}$$

Eliminating r between (1) and (2) we have

$$(x - a)^2 + (y - b)^2 + z^2 - t^2(a^2 + b^2) = 0 \;.\;.\;.\;. \;(3).$$

Differentiating with respect to a and b successively,

$$-(x - a) - t^2a = 0 \;.\;.\;. \;(4), \quad \text{and} \quad -(y - b) - t^2b = 0 \;.\;.\;. \;(5).$$

$$\therefore a = \frac{x}{1 - t^2}, \quad \text{and} \quad b = \frac{y}{1 - t^2}; \quad \text{which values in (3) give}$$

$$\left(x - \frac{x}{1 - t^2}\right)^2 + \left(y - \frac{y}{1 - t^2}\right)^2 + z^2 - t^2\frac{x^2 + y^2}{(1 - t^2)^2} = 0$$

$$\therefore (x^2 + y^2)(t^2 - t^4) = z^2\,(1 - t^2)^2 \quad \text{or} \quad x^2 + y^2 = \frac{1 - t^2}{t^2}\,z^2.$$

This is the equation of a right cone with a circular base, its axis being coincident with that of z, and its vertex at the origin.

CHAPTER IV.

CURVATURE OF SURFACES.

201. Two surfaces are said to be tangent to each other when they have a common point, $(x, y, z,)$ and a common tangent plane at that point.

Let the equations of the two surfaces be

$$F(X, Y, Z,) = 0 \ldots . (1), \quad \text{and} \quad \varphi(x, y, z) = 0 \ldots . (2).$$

The analytical conditions necessary for a simple contact, or contact of the *first order*, are

$$X = x, \ Y = y, \ Z = z, \ \frac{dZ}{dX} = \frac{dz}{dx}, \ \frac{dZ}{dY} = \frac{dz}{dy}.$$

If the second differential coefficients, derived from the equations of the two surfaces be also equal, viz.:

$$\frac{d^2Z}{dX^2} = \frac{d^2z}{dx^2}, \ \frac{d^2Z}{dY^2} = \frac{d^2z}{dy^2} \quad \text{and} \quad \frac{d^2Z}{dXdY} = \frac{d^2z}{dxdy},$$

the contact is said to be of the *second* order. If the third differential coefficients be also equal, the contact is of the third order, &c.

202. In order to show that the contact will be more intimate as the number of equal differential coefficients becomes greater, let the arbitrary increments h and k be given to the independent variables, $X = x$ and $Y = y$, converting Z and z into Z_1 and z_1, we shall then have (Art. 82)

$$Z_1 = Z + \frac{dZ}{dX} \cdot \frac{h}{1} + \frac{dZ}{dY} \cdot \frac{k}{1} + \frac{d^2Z}{dX^2} \cdot \frac{h^2}{1.2} + \frac{d^2Z}{dXdY} \cdot \frac{hk}{1} + \frac{d^2Z}{dY^2} \cdot \frac{k^2}{1.2} + \&c.$$

$$z_1 = z + \frac{dz}{dx}\cdot\frac{h}{1} + \frac{dz}{dy}\cdot\frac{k}{1} + \frac{d^2z}{dx^2}\cdot\frac{h^2}{1.2} + \frac{d^2z}{dxdy}\cdot\frac{hk}{1} + \frac{d^2z}{dy^2}\cdot\frac{k^2}{1.2} + \&c.$$

and when $Z = z$.

$$Z_1 - z_1 = \left[\frac{dZ}{dX} - \frac{dz}{dx}\right]\frac{h}{1} + \left[\frac{dZ}{dY} - \frac{dz}{dy}\right]\frac{k}{1} + \left[\frac{d^2Z}{dX^2} - \frac{d^2z}{dx^2}\right]\frac{h^2}{1.2} + \&c.$$

Now the value of this difference will depend (when h and k are very small), chiefly on the terms containing the lowest powers of h and k. If, therefore, the first differential coefficients, derived from the equations (A), (B), and (C), of three surfaces, at a common point, be equal, and if the second differential coefficients, derived from (A) and (B), be also equal, but those of (A) and (C) unequal, the surfaces (A) and (B) will separate more slowly, in departing from the common point than will the surfaces (A) and (C).

203. The order of closest possible contact between one surface entirely given, and another given only in species, will depend on the number of arbitrary parameters contained in the equation of the second surface.

Thus a contact of the first order requires three conditions, and therefore there must be three arbitrary parameters. A contact of the second order requires six parameters; one of the third order, ten parameters, &c. Hence the plane, whose equation has three parameters, may have contact of the first order. The sphere cannot, except at particular points, have contact of the second order, since its equation has but four parameters; but of two tangent spheres, one may have closer contact than the other.

The ellipsoid, hyperboloid, and paraboloid, can each have contact of the second order.

204. *Prop.* To determine the radius of curvature of a normal section of a given surface at a given point.

Assume the tangent plane at the given point as that of xy; the normal coinciding with the axis of z.

Let OX_1 be the trace of the secant plane on that of xy, forming with OX an angle θ. AOB the normal section, and P a point in that section. Put

$OE = x$, $ED = y$, $DP = z$, $OD = x_1$

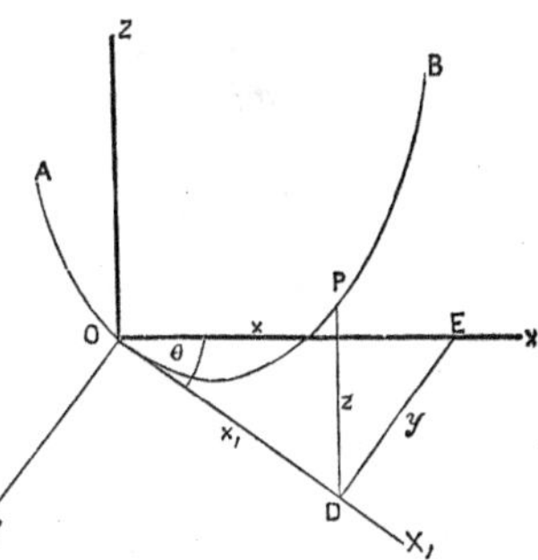

The co-ordinates of the curve AOB, estimated in its own plane, are x_1 and z; and the general value of the radius of curvature of a plane curve where x_1 and z are the co-ordinates, and any quantity t the independent variable, is (Art. 131.)

$$R = \frac{\left[\frac{ds}{dt}\right]^3}{\frac{d^2z}{dt^2}\cdot\frac{dx_1}{dt} - \frac{d^2x_1}{dt^2}\cdot\frac{dz}{dt}}$$

which, applied to the present case, making $t = x$, and observing that at O $\frac{ds}{dx} = \frac{dx_1}{dx}$ and $\frac{dz}{dx} = 0$, reduces to

$$R = \frac{\left[\frac{dx_1}{dx}\right]^2}{\frac{d^2z}{dx^2}} \quad \ldots\ldots (1).$$

In this expression, the coefficient $\frac{d^2z}{dx^2}$ has reference to those points of the surface which lie in the curve AOB, and therefore it differs from the partial differential coefficient $\frac{d^2z}{dx^2}$ derived from the equation of the surface, which latter refers to the change in z produced by a change in x only, while y is constant.

Let $z = \varphi(x,y)$ be the equation of the surface; then (Art. 55)

$\left[\frac{dz}{dx}\right] = \frac{dz}{dx} + \frac{dz}{dy}\cdot\frac{dy}{dx} = \frac{dz}{dx} + \frac{dz}{dy}\tan\theta$, since $\frac{dy}{dx} = \tan\theta$ in the present case.

$$\therefore \left[\frac{d^2z}{dx^2}\right] = \frac{d^2z}{dx^2} + 2\frac{d^2z}{dxdy} \cdot \tan\theta + \frac{d^2z}{dy^2}\tan^2\theta.$$

Also $\frac{dx_1}{dx} = \frac{1}{\cos\theta}$. Hence by substitution in (1) and reduction,

$$R = \frac{1}{\frac{d^2z}{dx^2}\cos^2\theta + 2\frac{d^2z}{dxdy}\cos\theta \,.\, \sin\theta + \frac{d^2z}{dy^2} \cdot \sin^2\theta} \quad \ldots [P].$$

205. *Prop.* The sum of the curvatures of any two normal sections of a curved surface, drawn through the same point of the surface, and perpendicular to each other, is constant, those curvatures being measured by the reciprocals of the radii of curvature.

Let θ and θ_1 be the inclinations of the secant planes to the plane of xz; R and R_1 the radii of curvature of the two sections at their common point. Then, since the sections are perpendicular to each other,

$$\theta_1 = \frac{1}{2}\pi + \theta, \quad \text{and} \quad \therefore \cos\theta = \sin\theta_1, \quad \sin\theta = -\cos\theta_1,$$

and by formula $[P]$

$$\frac{1}{R} = \frac{d^2z}{dx^2} \cdot \cos^2\theta + 2\frac{d^2z}{dxdy} \cdot \cos\theta\sin\theta + \frac{d^2z}{dy^2} \cdot \sin^2\theta.$$

$$\frac{1}{R_1} = \frac{d^2z}{dx^2}\sin^2\theta - 2\frac{d^2z}{dxdy} \cdot \sin\theta\cos\theta + \frac{d^2z}{dy^2} \cdot \cos^2\theta.$$

Hence by addition and reduction

$$\frac{1}{R} + \frac{1}{R_1} = \frac{d^2z}{dx^2} + \frac{d^2z}{dy^2} = \text{a constant for the same point.}$$

Cor. The normal sections of greatest and least curvature at any point of a curved surface, are perpendicular to each other.

For since $\frac{1}{R} + \frac{1}{R_1}$ is constant, $\frac{1}{R}$ will be greatest when $\frac{1}{R_1}$ is least, and it will be least when $\frac{1}{R_1}$ is greatest.

206. The sections of greatest and least curvature are called *principal sections*, and the corresponding radii are called *principal radii*.

207. *Prop.* To determine the principal radii of curvature at a given point of a curved surface.

By differentiating $\frac{1}{R}$ with respect to θ, as an independent variable, and placing the differential coefficient equal to zero, we get

$$\frac{d\frac{1}{R}}{d\theta} = -2\frac{d^2z}{dx^2}\cdot\cos\theta\sin\theta + 2\frac{d^2z}{dxdy}(\cos^2\theta - \sin^2\theta)$$
$$+ 2\frac{d^2z}{dy^2}\cdot\sin\theta\cos\theta = 0.$$

$$\therefore\ \cot^2\theta\,\frac{d^2z}{dxdy} + \cot\theta\left[\frac{d^2z}{dy^2} - \frac{d^2z}{dx^2}\right] = \frac{d^2z}{dxdy}\cdot\ \ldots\ (Q).$$

From which we obtain two values of $\cot\theta$, viz.:

$$\cot\theta = \frac{\frac{d^2z}{dx^2} - \frac{d^2z}{dy^2} \pm \sqrt{\left[\frac{d^2z}{dy^2} - \frac{d^2z}{dx^2}\right]^2 + 4\left(\frac{d^2z}{dxdy}\right)^2}}{2\frac{d^2z}{dxdy}}.$$

Substituting this value in the formula (P), which may be written thus

$$R = \frac{1 + \cot^2\theta}{\frac{d^2z}{dx^2}\cot^2\theta + 2\frac{d^2z}{dxdy}\cot\theta + \frac{d^2z}{dy^2}},$$

and denoting by R_1 and R_2 the least and greatest radii of curvature, there results

$$R_1 = \frac{2}{\frac{d^2z}{dx^2} + \frac{d^2z}{dy^2} + \sqrt{\left[\frac{d^2z}{dy^2} - \frac{d^2z}{dx^2}\right]^2 + 4\left(\frac{d^2z}{dxdy}\right)^2}}\ \ldots\ (R).$$

$$R_2 = \frac{2}{\frac{d^2z}{dx^2} + \frac{d^2z}{dy^2} - \sqrt{\left[\frac{d^2z}{dy^2} - \frac{d^2z}{dx^2}\right]^2 + 4\left(\frac{d^2z}{dxdy}\right)^2}}\ \ldots\ (S).$$

208. *Prop.* To express the radius of curvature of any normal section in terms of the principal radii R_1 and R_2, and the angle φ formed by that section with the principal section of greatest curvature.

If we make successively $\theta = 0$, and $\theta = \frac{1}{2}\pi$ in [P] we obtain

$$R = \frac{1}{\frac{d^2z}{dx^2}}, \quad \text{and} \quad R = \frac{1}{\frac{d^2z}{dy^2}},$$

and these will be the values of R_1 and R_2, if the planes of xz and yz be supposed to coincide with those of greatest and least curvature. Thus we shall have, upon this supposition,

$$\frac{d^2z}{dx^2} = \frac{1}{R_1}, \quad \text{and} \quad \frac{d^2z}{dy^2} = \frac{1}{R_2}.$$

The same supposition renders $\frac{d^2z}{dxdy} = 0$, as appears when we put $\theta = 0$ in (Q).

These conditions reduce (P), when θ is replaced by φ, to the form

$$R = \frac{R_1R_2}{R_2\cos^2\varphi + R_1\sin^2\varphi}. \quad \ldots\ldots [T].$$

the desired formula.

209. *Prop.* If the two principal sections of a curved surface, at any point, have their concavities turned in the same direction, then every normal section through that point will be concave in the same direction.

In the formula (T), the signs of R_1 and R_2 depend upon those of $\frac{d^2z}{dx^2}$ and $\frac{d^2z}{dy^2}$; and the signs of these coefficients indicate the directions of the curvature of the principal sections.

In the case under consideration, the signs of R_1 and R_2 must be alike, and therefore if both be +, the sign of R will be + also; but if both be −, then the sign of R will likewise be negative.

From which the truth of the proposition is apparent.

210. *Cor.* If R_1 and R_2 be also equal, then $R = R_1 = R_2$ for every value of φ, and every normal section, through the same point, will have the same curvature. This occurs at the vertices of surfaces of revolution.

211. *Prop.* If one principal section of a surface be concave, and the other convex, it will be possible to select a value φ_1 for φ, which shall render R infinite, or the section a straight line; also, between the values $\varphi = -\varphi_1$ and $\varphi = +\varphi_1$, the signs of R and R_1 will be alike; but from $\varphi = \varphi_1$ to $\varphi = \pi - \varphi_1$, the signs of R and R_2 will be alike.

In the formula [T], suppose R_1 negative, and it will become

$$R = \frac{-R_1 R_2}{R_2 \cos^2\varphi - R_1 \sin^2\varphi}$$

in which transformed expression, the quantities R_1 and R_2 are to be considered essentially positive.

Now suppose φ so taken that $R_2 \cos^2\varphi - R_1 \sin^2\varphi = 0$, a condition that will be fulfilled when

$$\varphi = \varphi_1 = \tan^{-1}\left[\frac{R_2}{R_1}\right]^{\frac{1}{2}} \quad \text{or,} \quad \varphi_1 = \tan^{-1} - \left[\frac{R_2}{R_1}\right]^{\frac{1}{2}}.$$

Then $$R = \frac{-R_1 R_2}{0} = \infty.$$

Thus there are two sections corresponding to the angles φ_1 and $-\varphi_1$ which give straight lines. Also, if $\varphi > -\varphi_1$ and $\varphi < \varphi_1$; then $R_2 \cos^2\varphi - R_1 \sin^2\varphi > 0$, and $\therefore R < 0$.

But if $\varphi > \varphi_1$ and $\varphi < \pi - \varphi_1$, then $R_2 \cos^2\varphi - R_1 \sin^2\varphi < 0$, and $R > 0$.

Hence the surface may be divided into four parts by two planes, and if the first of these parts be supposed concave the second will be convex, the third concave and the fourth convex.

212. *Prop.* To determine whether the principal radii at any point have the same or contrary signs, the co-ordinate planes not being coincident with the principal sections.

The general values of R_1 and R_2 may be reduced to the forms

$$R_1 = \frac{2}{p'' + q'' + \sqrt{(p''+q'')^2 - 4(p''q'' - s''^2)}},$$

$$R_2 = \frac{2}{p'' + q'' - \sqrt{(p''+q'')^2 - 4(p''q'' - s''^2)}},$$

in which $\quad p'' = \dfrac{d^2z}{dx^2}, \quad q'' = \dfrac{d^2z}{dy^2}, \quad$ and $\quad s'' = \dfrac{d^2z}{dxdy},$

and these values will have the same sign when $p''q'' - s''^2 > 0$, and contrary signs when $p''q'' - s''^2 < 0$.

213. *Prop.* At every point of a curved surface, a paraboloid (either elliptical or hyperbolic) can be applied, with its vertex at that point, which shall have contact of the second order with the given surface.

Assume the point of contact as the origin, the normal being taken as the axis of z, and the planes of xz and yz coincident with the principal sections of the surface.

Take the normal as the axis of the paraboloid, its vertex being at the point of contact, and turn the paraboloid about its axis until *its* principal sections coincide with xz and yz. The equation of the paraboloid when in this position will be $Ax^2 \pm By^2 = Cz$,

which may be written $\quad z = \dfrac{x^2}{2P} \pm \dfrac{y^2}{2P_1},$

where $2P = \dfrac{C}{A}$ and $2P_1 = \dfrac{C}{B}$, which represent the parameters of the principal sections, are entirely arbitrary.

Take $\quad P = R_1, \quad$ and $\quad P_1 = R_2. \quad$ Then $\quad z = \dfrac{x^2}{2R_1} \pm \dfrac{y^2}{2R_2}.$

Hence $\quad \dfrac{d^2z}{dx^2} = \dfrac{1}{R_1} \quad$ and $\quad \dfrac{d^2z}{dy^2} = \pm \dfrac{1}{R_2},$

and therefore R_1 and R_2 are the principal radii of curvature of the paraboloid also. Then, for any other normal section of the paraboloid, we shall have $R = \dfrac{\pm R_1 R_2}{R_1 \sin^2\varphi \pm R_2 \cos^2\varphi}$, the same value as that of the radius of curvature of the corresponding normal section of the surface. (Art. 208).

Cor. It appears that when the principal sections of two tangent surfaces have contact of the second order, every other normal sections made by the same plane drawn through the same point will likewise have contact of the second order.

214. *Prop.* To determine the radius of curvature of an oblique section of a curved surface.

Take the point of contact as the origin, and the tangent plane as that of xy.

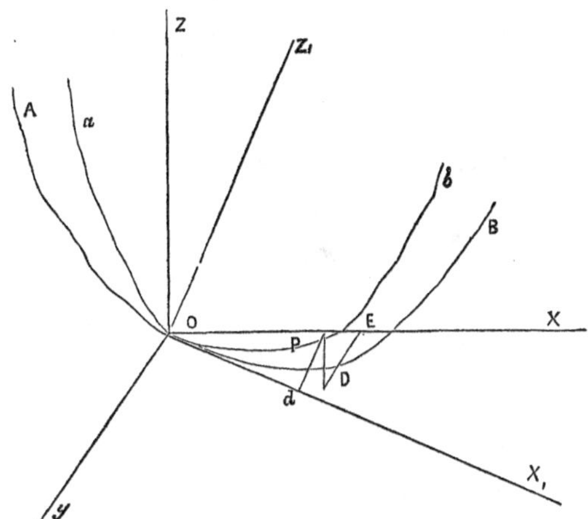

Let OX_1 be the trace of the secant plane on xy, aOb the section of the surface by that plane, AOB the normal section by the plane ZOX, R the radius of curvature of AOB at O, r the radius of curvature of aOb at O. Draw OZ_1 perpendicular to OX_1, in the plane aOb, and refer that section to the rectangular axes OX_1 and OZ_1.

Put $Od = x_1$, $dp = z_1$, $\lambda =$ angle between aOb and AOB, $pD = z$, $DE = y$, $OE = x$.

Then at the point O we shall have

$$r = \frac{\left[\frac{ds_1}{dx}\right]^2}{\frac{d^2z_1}{dx^2}}, \quad R = \frac{\left[\frac{ds}{dx}\right]^2}{\frac{d^2z}{dx^2}}.$$

But $z = z_1 \cos\lambda$. $\therefore \frac{d^2z}{dx^2} = \frac{d^2z_1}{dx^2} \cdot \cos\lambda$. Also $\frac{ds_1}{dx} = \frac{dx_1}{dx} = \frac{ds}{dx}$.

$$\therefore r = R \cdot \cos\lambda,$$

and consequently radius of the oblique section = projection of the radius of the normal section, on the plane of the oblique section. This result is known as *Meusnier's Theorem.*

Cor. If a sphere be described whose radius shall be identical with that of the normal section, and if through the tangent to that section any plane be drawn intersecting the sphere and the given surface, then will the small circle cut from the sphere be osculatory to the curve cut from the surface.

Lines of Curvature.

215. If, through the consecutive points of any curve traced upon a given surface, normals to that surface be drawn, such consecutive normals will not usually lie in the same plane, and therefore will not intersect; but when the consecutive normals do intersect, the corresponding curves (which enjoy peculiar properties) are called *lines of curvature.*

216. *Prop.* To determine the lines of curvature passing through any point on a curved surface.

Let the equations of the normals passing through any point (x_1, y_1, z_1), be

$x - x_1 + t(z - z_1) = 0 = P \ldots (1)$ and $y - y_1 + s(z - z_1) = 0 = Q \ldots (2)$,

and suppose the independent variables x and y to receive the increments h and k.

Then the equations of the normal in the new position will be

$$P+\frac{dP}{dx_1}\cdot\frac{h}{1}+\frac{dP}{dy_1}\frac{k}{1}+\text{\&c.}=0\ldots\ldots(3),$$

and
$$Q+\frac{dQ}{dx_1}\cdot\frac{h}{1}+\frac{dQ}{dy_1}\cdot\frac{k}{1}+\text{\&c.}=0\ldots\ldots(4).$$

If these two normals intersect, the equations (1), (2), (3), and (4), will apply to the point of intersection; and if the co-ordinates x, y, and z of that point be eliminated between the four equations, the result will be a relation between the increments h and k and constants, it being observed that $t=\frac{dz_1}{dx_1}$ and $s=\frac{dz_1}{dy_1}$, are constant for the same point, and the same is true of $\frac{dP}{dx_1}$, $\frac{dP}{dy_1}$, &c.

This relation between h and k implies a necessary relation between the new values of x and y, in order that an intersection of the normals may be possible; and when the normals are consecutive, $h=0$, and $k=0$, and $\frac{k}{h}=\frac{dy_1}{dx_1}$. Thus by omitting P and Q (each of which is equal to zero) in (3) and (4), then dividing by h, and finally making $h=0$, those equations become

$$\frac{dP}{dx_1}+\frac{dP}{dy_1}\cdot\frac{dy_1}{dx_1}=0\ldots.(5),\quad\text{and}\quad\frac{dQ}{dx_1}+\frac{dQ}{dy_1}\cdot\frac{dy_1}{dx_1}=0\ldots(6),$$

or, by forming the values of the partial differential coefficients,

$$\frac{dP}{dx_1},\ \frac{dP}{dy_1},\ \frac{dQ}{dx_1},\ \text{and}\ \frac{dQ}{dy_1},\ \text{from (1) and (2)},$$

$$\left.\begin{aligned}&-1+(z-z_1)\frac{d^2z_1}{dx_1^2}-\frac{dz_1^2}{dx_1^2}+(z-z_1)\frac{d^2z_1}{dx_1dy_1}\cdot\frac{dy_1}{dx_1}-\frac{dz_1}{dx_1}\cdot\frac{dz_1}{dy_1}\cdot\frac{dy_1}{dx_1}=0,\\ &\frac{d^2z_1}{dx_1dy_1}(z-z_1)-\frac{dz_1}{dy_1}\cdot\frac{dz_1}{dx_1}-\frac{dy_1}{dx_1}+(z-z_1)\frac{d^2z_1}{dy_1^2}\cdot\frac{dy_1}{dx_1}-\frac{dz_1^2}{dy_1^2}\cdot\frac{dy_1}{dx_1}=0;\end{aligned}\right\}(7),$$

and by eliminating $z-z$, putting

$$\frac{dz_1}{dx_1}=p',\ \frac{dz_1}{dy_1}=q',\ \frac{d^2z_1}{dx_1^2}=p'',\ \frac{d^2z_1}{dy_1^2}=q'',\ \text{and}\ \frac{d^2z_1}{dx_1dy_1}=s'',$$

we obtain

$$\frac{dy_1^2}{dx_1^2}[s''(1+q'^2)-p'q'q''^2]$$

$$+\frac{dy_1}{dx_1}[p''(1+q'^2)-q''(1+p'^2)]-s''(1+p'^2)+p'q'p''=0\ldots(U).$$

This is a quadratic equation, giving two values of $\frac{dy_1}{dx_1}$, the tangent of the angle between the axis of x and the projection of the tangent to the line of curvature passing through $(x_1 y_1 z_1)$, upon the plane of xy. Hence there will be two lines of curvature passing through each point of the surface; and if p', q', &c., be replaced in (U) by their general values derived from the equation of the surface, the result will be the differential equation of the projection of every pair of lines of curvature upon the plane of xy.

217. *Prop.* The lines of curvature at any point of a curved surface intersect each other at right angles, and they are respectively tangent to the sections of greatest and least curvature.

If we suppose the plane of xy, (which in the last proposition was assumed arbitrarily) to coincide with the tangent plane at the point under consideration, we shall have

$$p'=\frac{dz_1}{dx_1}=0, \quad \text{and} \quad q'=\frac{dz_1}{dy_1}=0.$$

Hence the equation (U) may be reduced to the form

$$\frac{dy_1^2}{dx_1^2}+\frac{p''-q''}{s''}\cdot\frac{dy_1}{dx_1}-1=0 \;.\;.\;.\;.\;.\; (V).$$

Hence if θ_1 and θ_2 denote two angles determined by the condition that $\tan\theta_1$ and $\tan\theta_2$ shall be the roots of this equation, we shall have, by the theory of equations,

$$\tan\theta_1 \tan\theta_2 = -1, \quad \text{or} \quad 1+\tan\theta_1\tan\theta_2 = 0,$$

which is the condition of perpendicularity of two lines in the plane of xy forming angles θ_1 and θ_2 with the axis of x. Thus the tangents to the two lines of curvature intersect at right angles.

218. Again, if we divide equation (V) by $\frac{dy_1^2}{dx_1^2} = \tan^2\theta$ and replace $\frac{1}{\tan\theta}$ by $\cot\theta$, the result will become identical in form with equation (Q), which serves to determine the two angles formed by the principal sections with the plane of xz, and hence the directions of the lines of curvature are tangent to the curves of principal section.

219. *Prop.* The consecutive normals to a surface drawn through points in the lines of curvature, intersect at the same points as the consecutive normals to the principal sections to which the lines of curvature are tangent.

Regarding the tangent plane at the given point of the surface as still coincident with that of xy, we shall have

$z_1 = 0$, $\frac{dz_1}{dx_1} = 0$ and $\frac{dz_1}{dy_1} = 0$ and the equation (7), gives

$$z = \frac{1}{\frac{d^2z_1}{dx_1^2} + \frac{d^2z_1}{dx_1dy_1}\tan\theta}, \quad \text{or} \quad \frac{\tan\theta}{\frac{d^2z_1}{dx_1dy_1} + \frac{d^2z_1}{dy_1^2}\cdot\tan\theta}.$$

Now if the plane of xz be supposed coincident with a principal section, these expressions will be still further simplified, since $\frac{d^2z_1}{dx_1dy_1}$ will then be $= 0$; thus,

$$z = \frac{1}{\frac{d^2z_1}{dx_1^2}} \quad \text{or} \quad z = \frac{1}{\frac{d^2z_1}{dy_1^2}}.$$

But these expressions are precisely the same as those previously found for the radii of curvature of the principal sections, and hence the centres of curvature of the principal sections must coincide with the points of intersection of consecutive normals to the surface through points in the lines of curvature.

INTEGRAL CALCULUS. PART I.

CHAPTER I.

FIRST PRINCIPLES.

1. The object of the Integral Calculus is to determine the function from which any proposed differential has been obtained. The process by which this is effected is called *integration*, and is indicated by the sign $\int$, the result being called the *integral* of the proposed differential.

2. Whenever the given differential can be reduced to a known form, we may return to the function by simply reversing the rules for differentiation.

3. Since $d(a \,.\, Fx) = a \,.\, d(Fx) = aF_1x \,.\, dx$, we infer that

$$\int aF_1x \,.\, dx = a \int F_1x \,.\, dx,$$

that is, we may remove any constant factor from under the sign of integration, placing it as a factor exterior to that sign.

4. Again $\int F_1x \,.\, dx = \int \frac{a}{a} F_1x \,.\, dx = \frac{1}{a} \int a \,.\, F_1x \,.\, dx.$

Therefore we may introduce a constant factor under the integral sign, provided we write its reciprocal, as a factor, exterior to that sign.

5. To differentiate the algebraic sum of several functions, we differentiate each function separately, and take the algebraic sum of the several differentials. Hence, in order to integrate the algebraic sum of several differentials, we have only to integrate the several terms successively.

Thus $\int(adx + bdy - cdz + edv) = \int adx + \int bdy - \int cdz + \int edv$
$$= ax + by - cz + ev.$$

6. Again, since differentiation causes all constants connected with the variables by the signs $+$ and $-$ to disappear, it follows, that in effecting an integration, we should always add a constant, in order to provide for that which may have disappeared by differentiation: thus we write

$$\int adx = ax + c,$$

in which the value of c will be arbitrary, unless fixed by other conditions.

Suppose, for example, that the general value of the integral is X, so that

$$X = ax + c;$$

and that for a particular value x_1 of x, the integral assumes a known value X_1: then

$$X_1 = ax_1 + c, \text{ and } \therefore\ c = X_1 - ax_1.$$

And this value substituted in the general integral, gives

$$X = a(x - x_1) + X_1.$$

Integration of the Form (Fx)ⁿdFx.

7. *Prop.* To integrate the form $(Fx)^n dFx$.

Here we have $\int (Fx)^n dFx = \frac{1}{n+1} \int (n+1)(Fx)^n dFx$

$$= \frac{1}{n+1} \int d(Fx)^{n+1} = \frac{(Fx)^{n+1}}{n+1} + c.$$

The same process can obviously be applied, whenever the quantity exterior to the parenthesis, can be rendered the exact differential of that within, by the introduction or suppression of a constant.

Hence we have the following rule for the integration of this form, viz.:

Divide the given expression by the differential of the quantity within the ( ), then increase the exponent of the ( ) by unity, and finally, divide by the exponent thus increased.

EXAMPLES.

8. 1. To integrate ax^3dx.

$$\int ax^3dx = a\int x^3dx = \frac{a}{4}\int 4x^3dx = \frac{ax^4}{4} + c.$$

2. To integrate $\sqrt{b^2 + x^4}\,.\,3cx^3dx.$

$$\int (b^2 + x^4)^{\frac{1}{2}}.\,3cx^3dx = \frac{3c}{4}\cdot\frac{2}{3}\int\frac{3}{2}(b^2 + x^4)^{\frac{1}{2}}.\,4x^3dx = \frac{c}{2}(b^2 + x^4)^{\frac{3}{2}} + c.$$

3. To integrate $dy = (2a + 3bx)^3dx.$

This may be integrated in two ways; thus

$$y = \int(2a + 3bx)^3dx = \int(8a^3 + 36a^2bx + 54ab^2x^2 + 27b^3x^3)dx$$

$$= \int 8a^3dx + \int 36a^2bxdx + \int 54ab^2x^2dx + \int 27b^3x^3dx$$

$$= 8a^3x + 18a^2bx^2 + 18ab^2x^3 + \frac{27}{4}b^3x^4 + c.\ \ .\ .\ .\ .\ (1).$$

Again

$$y = \int(2a + 3bx)^3dx = \frac{1}{12b}\int 4(2a+3bx)^3.\,3bdx = \frac{1}{12b}(2a+3bx)^4 + c_1$$

$$= \frac{4a^4}{3b} + 8a^3x + 18a^2bx^2 + 18ab^2x^3 + \frac{27}{4}b^3x^4 + c_1.\ \ .\ .\ .\ .\ (1).$$

The formulæ (1) and (2) are identical. For if y_1 denote the particular value of y when $x = 0$, we shall have from (1) $y_1 = c$; and from (2) $y_1 = \dfrac{4a^4}{3b} + c_1$, $\therefore\ c = \dfrac{4a^4}{3b} + c_1.$

4. To integrate $dy = 3(4bx^2 - 2cx^3)^{\frac{1}{3}}(4bx - 3cx^2)dx$

$$y = \frac{3}{2}\int(4bx^2 - 2cx^3)^{\frac{1}{3}}(8bx - 6cx^2)dx = \frac{9}{8}(4bx^2 - 2cx^3)^{\frac{4}{3}} + c.$$

9. In each of the preceding examples the proposed differential has been brought to the required form, viz.: that in which the part exterior to the ( ) is the exact differential of that within, by introducing a constant factor. To ascertain when this is possible, take the last example, and denote by A the required unknown factor: then

$$y = \frac{1}{A}\int(4bx^2 - 2cx^3)^{\frac{1}{3}}(12Abx - 9Acx^2)dx,$$

and if this be of the required form, we must have

$$d(4bx^2 - 2cx^3) = (12Abx - 9Acx^2)dx$$

or
$$8bx - 6cx^2 = 12Abx - 9Acx^2,$$

and since this condition must be satisfied without reference to the value of x, we must have, by the principle of indeterminate coefficients, the two separate conditions

$$8b = 12Ab \ldots. (1) \quad \text{and} \quad -6c = -9Ac \ldots. (2).$$

From (1) $A = \frac{8b}{12b} = \frac{2}{3}$, and from (2) $A = \frac{6c}{9c} = \frac{2}{3}$.

The values of A derived from (1) and (2) being identical, the proposed reduction is possible.

The next example will illustrate the contrary case.

1. $$dy = (4b^2x + 3ax^2)^{\frac{1}{4}}(2b^2 + 8ax)dx.$$

If possible, let A be the required factor. Then

$$y = \frac{1}{A}\int(4b^2x + 3ax^2)^{\frac{1}{4}}(2b^2A + 8aAx)dx,$$

and
$$\therefore d(4b^2x + 3ax^2) = (2b^2A + 8aAx)dx,$$

or
$$4b^2 + 6ax = 2b^2A + 8aAx,$$

which gives the two separate conditions

$$4b^2 = 2b^2A \dots (1) \qquad \text{and} \qquad 6a = 8aA \dots (2).$$

From (1) $A = \frac{4b^2}{2b^2} = 2$, and from (2) $A = \frac{6a}{8a} = \frac{3}{4}$.

These values of A being different, the desired reduction is impossible.

2. To integrate $\qquad dy = \frac{adx}{7x^4}$.

$$y = \frac{a}{7}\int x^{-4}dx = -\frac{a}{21}\int -3x^{-4}dx = -\frac{ax^{-3}}{21} + c = -\frac{a}{21x^3} + c.$$

3. $$dy = \frac{adx}{x\sqrt{3bx + 4c^2x^2}}.$$

$$y = a\int x^{-1}(3bx + 4c^2x^2)^{-\frac{1}{2}}dx = \frac{a}{3b}\int (3bx^{-1} + 4c^2)^{-\frac{1}{2}}.\,3bx^{-2}dx$$

$$= -\frac{2a}{3b}(3bx^{-1} + 4c^2)^{\frac{1}{2}} + c = -\frac{2a(3bx + 4c^2x^2)^{\frac{1}{2}}}{3bx} + c.$$

4. $$dy = \frac{axdx}{(2bx + x^2)^{\frac{3}{2}}}.$$

$$y = a\int (2bx + x^2)^{-\frac{3}{2}}.\,xdx = a\int (2bx^{-1} + 1)^{-\frac{3}{2}}(x^2)^{-\frac{3}{2}}.\,xdx$$

$$= \frac{a}{2b}\int (2bx^{-1} + 1)^{-\frac{3}{2}}.\,x^{-2}.\,2bdx = \frac{a}{b}(2bx^{-1} + 1)^{-\frac{1}{2}} + c$$

$$= \frac{a}{b}\left[\frac{2bx + x^2}{x^2}\right]^{-\frac{1}{2}} + c = \frac{ax}{b\sqrt{2bx + x^2}} + c.$$

5. $$dy = \frac{3x^4(x^3 - a^3)}{x - a}dx.$$

$$y = 3\int x^4(x^2 + ax + a^2)dx = 3\int (x^6 + ax^5 + a^2x^4)dx$$

$$= 3\left(\frac{x^7}{7} + \frac{ax^6}{6} + \frac{a^2x^5}{5}\right) + c.$$

CHAPTER II.

ELEMENTARY TRANSCENDENTAL FORMS.

Logarithmic Forms.

10. *Prop.* To integrate the forms $\frac{adx}{x}$ and $\frac{ad(Fx)}{Fx}$.

Since $$d(a \log x) = \frac{adx}{x}. \quad \therefore \int \frac{adx}{x} = a \log x + c.$$

Also since $$d(a \,.\log Fx) = \frac{a \,.\, dFx}{Fx}. \quad \therefore \int \frac{a \,.\, dFx}{Fx} = a \,.\log Fx + c.$$

EXAMPLES.

11. 1. To integrate $$dy = \frac{adx}{b + cx}.$$

$$y = \frac{a}{c}\int \frac{cdx}{b + cx} = \frac{a}{c} \log (b + cx) + C = \log \left[(b + cx)^{\frac{a}{c}}\right] + C.$$

2. To integrate $$dy = \frac{8x^3dx}{a + 2x^4}.$$

$$y = \int \frac{8x^3dx}{a+2x^4} = \log(a+2x^4) + C = \log(a+2x^4) + \log c = \log[c(a+2x^4)].$$

In this example the constant introduced by the integration is put into the form of a logarithm (which is always admissible) for the purpose of simplifying the form to which the integral is finally reduced.

3. To integrate $$dy = \frac{7xdx}{8a - 3x^2}.$$

$$y = \int - \frac{-7xdx}{8a - 3x^2} = -\frac{7}{6}\int \frac{-6xdx}{8a - 3x^2} = -\log(8a - 3x^2)^{\frac{7}{6}} + C$$

$$= \log c - \log(8a - 3x^2)^{\frac{7}{6}} = \log \frac{c}{(8a - 3x^2)^{\frac{7}{6}}}.$$

4. To integrate $$dy = \frac{b(3x - a^2)^4 dx}{cx^3}.$$

$$y = \frac{b}{c}\int \frac{(81x^4 - 108x^3a^2 + 54x^2a^4 - 12xa^6 + a^8)dx}{x^3},$$

or, $$y = \frac{b}{c}\int \left[81x - 108a^2 + \frac{54a^4}{x} - \frac{12a^6}{x^2} + \frac{a^8}{x^3}\right] dx$$

$$= \frac{b}{c}\left[\frac{81}{2}x^2 - 108a^2x + 54a^4 \log x + \frac{12a^6}{x} - \frac{a^8}{2x^2}\right] + C.$$

Circular Forms.

12. *Prop.* To integrate the form $dy = \pm \frac{dx}{\sqrt{a^2 - b^2x^2}}$.

Taking the upper sign, we have

$$y = \int \frac{+dx}{\sqrt{a^2 - b^2x^2}} = \int \frac{\frac{1}{a}dx}{\sqrt{1 - \frac{b^2x^2}{a^2}}} = \frac{1}{b}\int \frac{\frac{b}{a}dx}{\sqrt{1 - \frac{b^2x^2}{a^2}}}.$$

Let the quantity under the sign of integration be compared with the well known form $d(\sin^{-1}z) = \frac{dz}{\sqrt{1 - z^2}}$, and it will be found identical therewith, provided we make $\frac{b}{a}x = z$.

But $\int \frac{dz}{\sqrt{1 - z^2}} = \sin^{-1}z + c$, $\therefore \int \frac{\frac{b}{a}dx}{\sqrt{1 - \frac{b^2x^2}{a^2}}} = \sin^{-1}\frac{bx}{a} + c.$

$$\therefore y = \frac{1}{b}\sin^{-1}\frac{bx}{a} + c.$$

Similarly, since $\int \frac{-dz}{\sqrt{1-z^2}} = \cos^{-1} z + c.$

$$\therefore\ y = \int \frac{-dx}{\sqrt{a^2 - b^2x^2}} = \frac{1}{b} \cdot \cos^{-1} \frac{bx}{a} + c.$$

13. *Prop.* To integrate the form $dy = \pm \frac{dx}{a^2 + b^2x^2}$.

Taking the upper sign, we have

$$y = \int \frac{+dx}{a^2 + b^2x^2} = \int \frac{\frac{1}{a^2} dx}{1 + \frac{b^2x^2}{a^2}} = \frac{1}{ab} \int \frac{\frac{b}{a} dx}{1 + \frac{b^2x^2}{a^2}}.$$

Comparing the expression under the sign of integration with the well known form $d(\tan^{-1} z) = \frac{dz}{1 + z^2}$, they become identical by making $\frac{bx}{a} = z$.

But $\int \frac{dz}{1 + z^2} = \tan^{-1} z + c.\ \therefore \int \frac{\frac{b}{a} dx}{1 + \frac{b^2x^2}{a^2}} = \tan^{-1} \frac{b}{a} x + c.$

$$\therefore\ y = \frac{1}{ab} \tan^{-1} \frac{bx}{a} + c.$$

And similarly, since $\int \frac{-dz}{1 + z^2} = \cot^{-1} z + c.$

$$\therefore\ y = \int \frac{-dx}{a^2 + b^2x^2} = \frac{1}{ab} \cot^{-1} \frac{bx}{a} + c.$$

14. *Prop.* To integrate the form $dy = \pm \frac{dx}{x\sqrt{b^2x^2 - a^2}}$.

Taking the upper sign, we have

$$y = \int \frac{+dx}{x\sqrt{b^2x^2 - a^2}} = \int \frac{\frac{1}{a} dx}{x\sqrt{\frac{b^2x^2}{a^2} - 1}} = \frac{1}{a} \int \frac{\frac{b}{a} dx}{\frac{bx}{a}\sqrt{\frac{b^2x^2}{a^2} - 1}}.$$

Comparing the expression under the sign of integration with the known form $d(\sec^{-1}z)=\dfrac{dz}{z\sqrt{z^2-1}}$, they become identical by making $\dfrac{bx}{a}=z$.

But $\displaystyle\int\frac{dz}{z\sqrt{z^2-1}}=\sec^{-1}z+c.$ $\therefore$ $\displaystyle\int\frac{\frac{b}{a}dx}{\frac{bx}{a}\sqrt{\frac{b^2x^2}{a^2}-1}}=\sec^{-1}\frac{bx}{a}+c.$

$$\therefore\ y=\frac{1}{a}\sec^{-1}\frac{bx}{a}+c.$$

And similarly, since $\displaystyle\int\frac{-dz}{z\sqrt{z^2-1}}=\text{cosec}^{-1}z+c.$

$$\therefore\ y=\int\frac{-dx}{x\sqrt{b^2x^2-a^2}}=\frac{1}{a}\text{cosec}^{-1}\frac{bx}{a}+c.$$

15. *Prop.* To integrate the form $dy=\pm\dfrac{dx}{\sqrt{a^2x-b^2x^2}}$.

Taking the upper sign, we have

$$y=\int\frac{+dx}{\sqrt{a^2x-b^2x^2}}=\int\frac{\frac{2b}{a^2}dx}{\sqrt{\frac{4b^2x}{a^2}-\frac{4b^4x^2}{a^4}}}=\frac{1}{b}\int\frac{\frac{2b^2}{a^2}dx}{\sqrt{\frac{4b^2x}{a^2}-\frac{4b^4x^2}{a^4}}}$$

$$=\frac{1}{b}\int\frac{\frac{2b^2}{a^2}dx}{\sqrt{2\left(\frac{2b^2x}{a^2}\right)-\left(\frac{2b^2x}{a^2}\right)^2}}$$

Comparing the expression under the sign of integration with the known form $d(\text{versin}^{-1}z)=\dfrac{dz}{\sqrt{2z-z^2}}$, they become identical by making $\dfrac{2b^2x}{a^2}=z$.

But $$\int\frac{dz}{\sqrt{2z-z^2}}=\text{versin}^{-1}z+c,$$

$$\therefore\ \int\frac{\frac{2b^2}{a^2}dx}{\sqrt{2\left(\frac{2b^2x}{a^2}\right)-\left(\frac{2b^2x}{a^2}\right)^2}}=\text{versin}^{-1}\frac{2b^2x}{a^2}+c.$$

$$\therefore\ y = \frac{1}{b}\,\text{versin}^{-1}\frac{2b^2x}{a^2} + c.$$

And similarly, since $\int \frac{-dz}{\sqrt{2z - z^2}} = \text{coversin}^{-1}z.$

$$\therefore\ y = \int \frac{-dx}{x\sqrt{a^2x - b^2x^2}} = \frac{1}{b}\,\text{coversin}^{-1}\frac{2b^2x}{a^2} + c.$$

EXAMPLES.

16. 1. To integrate $dy = \frac{xdx}{\sqrt{a^2 - b^2x^4}}.$

$$y = \frac{1}{2b}\int \frac{\frac{2bx}{a}dx}{\sqrt{1 - \frac{b^2x^4}{a^2}}} = \frac{1}{2b}\sin^{-1}\frac{bx^2}{a} + c.$$

2. To integrate $dy = \frac{x^2dx}{1 + x^6}.$

$$y = \frac{1}{3}\int \frac{3x^2dx}{1 + x^6} = \frac{1}{3}\tan^{-1}(x^3) + c.$$

3. To integrate $dy = \frac{8x^{-\frac{2}{3}}dx}{\sqrt{2x^{\frac{1}{3}} - 6x^{\frac{2}{3}}}}.$

$$y = \sqrt{6}\int \frac{8x^{-\frac{2}{3}}dx}{\sqrt{2\,.\,6x^{\frac{1}{3}} - 6\,.\,6x^{\frac{2}{3}}}} = 4\sqrt{6}\int \frac{2x^{-\frac{2}{3}}dx}{\sqrt{2\,.\,6x^{\frac{1}{3}} - 6\,.\,6x^{\frac{2}{3}}}}$$

$$= 4\sqrt{6}\,.\,\text{versin}^{-1}\left(6x^{\frac{1}{3}}\right) + c.$$

17. Since each of the trigonometrical functions can be expressed in terms of any other, all the circular forms must apply, whenever one is applicable. To illustrate this, take the example

$$dy = \frac{x^{\frac{1}{2}}dx}{\sqrt{2 - 4x^3}}$$

$$y=\int\frac{\frac{1}{\sqrt{2}}x^{\frac{1}{2}}dx}{\sqrt{1-2x^3}}=\frac{1}{2}\int\frac{\sqrt{2}\,x^{\frac{1}{2}}dx}{\sqrt{1-2x^3}}=\frac{1}{3}\int\frac{\frac{3}{2}\sqrt{2}\,.\,x^{\frac{1}{2}}dx}{\sqrt{1-2x^3}}=\frac{1}{3}\sin^{-1}\sqrt{2x^3}+c.$$

$$\text{or } y=-\frac{1}{3}\int\frac{-\frac{3}{2}\sqrt{2}\,.\,x^{\frac{1}{2}}dx}{\sqrt{1-2x^3}}=-\frac{1}{3}\cos^{-1}\sqrt{2}x^3+c_1.$$

Again,

$$y=\int\frac{x^2dx}{\sqrt{2x^3-4x^6}}=\frac{1}{6}\int\frac{12x^2dx}{\sqrt{2\,.\,4x^3-(4x^3)^2}}=\frac{1}{6}\,\text{versin}^{-1}(4x^3)+c_2,$$

$$\text{or, } y=-\frac{1}{6}\int\frac{-12x^2dx}{\sqrt{2\,.\,4x^3-(4x^3)^2}}=-\frac{1}{6}\,\text{coversin}^{-1}(4x^3)+c_3.$$

$$\text{Again, } y=\int\frac{x^{\frac{1}{2}}x^{-\frac{3}{2}}dx}{\sqrt{2x^{-3}-4}}=\int\frac{\frac{1}{2}\sqrt{\frac{1}{2}}x^{-\frac{5}{2}}dx}{\sqrt{\frac{1}{2}}\cdot x^{-\frac{3}{2}}\sqrt{\frac{1}{2}x^{-3}-1}}$$

$$=-\frac{1}{3}\int\frac{-\frac{3}{2}\sqrt{\frac{1}{2}}\,.\,x^{-\frac{5}{2}}dx}{\sqrt{\frac{1}{2}}\cdot x^{-\frac{3}{2}}\sqrt{\frac{1}{2}x^{-3}-1}}=-\frac{1}{3}\sec^{-1}\sqrt{\frac{1}{2}x^{-3}}+c_4,$$

or

$$y=\frac{1}{3}\,\text{cosec}^{-1}\sqrt{\frac{1}{2}x^{-3}}+c_5.$$

$$\text{Finally, } y=\int\frac{1}{2}x^{\frac{1}{2}}x^{-\frac{3}{2}}\left(\frac{1}{2}x^{-3}-1\right)^{-\frac{1}{2}}dx$$

$$=-\frac{1}{3}\int\frac{-\frac{3}{4}x^{-4}\left(\frac{1}{2}x^{-3}-1\right)^{-\frac{1}{2}}dx}{1+\left(\frac{1}{2}x^{-3}-1\right)}$$

$$=-\frac{1}{3}\tan^{-1}\sqrt{\frac{1}{2}x^{-3}-1}+c_6,$$

or

$$y=\frac{1}{3}\,\text{cotan}^{-1}\sqrt{\frac{1}{2}x^{-3}-1}+c_7.$$

Trigonometrical Forms.

18. *Prop.* To integrate the forms $\sin x dx$, $\cos x\, dx$, $\sec^2 x dx$, $\operatorname{cosec}^2 x dx$, $\sec x \tan x dx$, and $\operatorname{cosec} x \cot x dx$.

Since $d(\cos x) = -\sin x\, dx$, $\therefore \int \sin x dx = -\int -\sin x\, dx = -\cos x + c.$

" $d(\sin x) = \cos x dx$, $\therefore \int \cos x dx = \sin x + c.$

" $d(\tan x) = \sec^2 x dx$, $\therefore \int \sec^2 x dx = \tan x + c.$

" $d(\cot x) = -\operatorname{cosec}^2 x dx$, $\therefore \int \operatorname{cosec}^2 x dx = -\cot x + c.$

" $d(\sec x) = \sec x \tan x dx$, $\therefore \int \sec x \tan x dx = \sec x + c.$

" $d(\operatorname{cosec} x) = -\operatorname{cosec} x \cot x dx$, $\therefore \int \operatorname{cosec} x \cot x dx = -\operatorname{cosec} x + c.$

EXAMPLES.

19. 1. To integrate $dy = 2\cos 3x \,.\, dx.$

$$y = \int 2\cos 3x \,.\, dx = \frac{2}{3}\int \cos 3x \,.\, d(3x) = \frac{2}{3}\sin 3x + c.$$

2. $$dy = 5\sec^2(x^3) \,.\, x^2 dx.$$

$$y = \int 5\sec^2(x^3) \,.\, x^2 dx = \frac{5}{3}\int \sec^2(x^3)\, 3x^2 dx = \frac{5}{3}\int \sec^2(x^3)\, d(x^3)$$

$$= \frac{5}{3}\tan(x^3) + c.$$

3. $$dy = 6\sec 4x \,.\, \tan 4x \,.\, dx$$

$$y = \frac{6}{4}\int \sec 4x \,.\, \tan 4x \,.\, d(4x) = \frac{3}{2}\sec 4x + c.$$

4. $$dy = 2\sin(a + 3x)dx,$$

$$y = \frac{2}{3}\int \sin(a + 3x)\, 3dx = \frac{2}{3}\int \sin(a + 3x) \,.\, d(a + 3x)$$

$$= -\frac{2}{3}\cos(a + 3x) + c.$$

5. $$dy = \frac{3}{2}\operatorname{cosec}^2(\sqrt{2x}) \,.\, x^{-\frac{1}{2}} dx.$$

$$y = \frac{3}{\sqrt{2}}\int \operatorname{cosec}^2(\sqrt{2x}) \cdot \frac{1}{2}\sqrt{2} \,.\, x^{-\frac{1}{2}} dx = -\frac{3}{\sqrt{2}}\cot\sqrt{2x} + c.$$

6. $$dy = 2 \operatorname{cosec}(nx) \, . \cot(nx) \, . \, dx.$$

$$y = \frac{2}{n}\int \operatorname{cosec}(nx) \cot(nx) \, . \, d(nx) = -\frac{2}{n} \operatorname{cosec}(nx) + c.$$

Exponential Forms.

20. *Prop.* To integrate the form $dy = a^x dx$.

Since $da^x = \log a \, . \, a^x dx$, $\therefore \int a^x dx = \frac{1}{\log a}\int \log a \, . \, a^x dx$

$$= \frac{a^x}{\log a} + c.$$

EXAMPLES.

21. 1. To integrate $dy = 3e^x dx$, where e is the Naperian base.

$$y = 3\int e^x dx = \frac{3e^x}{\log e} + c = 3e^x + c.$$

2. $$dy = ba^{3x} dx,$$

$$y = b\int a^{3x} dx = \frac{b}{3 \log a} \cdot \int \log a \, . \, a^{3x} d(3x) = \frac{ba^{3x}}{3 \log a} + c.$$

3. $$dy = me^{nx} dx.$$

$$y = \frac{m}{n}\int e^{nx} d(nx) = \frac{m}{n} e^{nx} + c.$$

The cases which have now been considered include all the elementary forms.

CHAPTER III.

RATIONAL FRACTIONS.

22. Having disposed of the simple and elementary forms, or such as admit of being brought to such by some very obvious process, we shall proceed to the consideration of more complicated expressions, endeavoring in each case to resolve them by a systematic process into one or more of the elementary forms.

23. The first form, in point of simplicity, which we shall have occasion to consider, is that of a rational algebraic fraction, and in such expressions we may always regard the highest exponent of the variable in the numerator as less than the corresponding exponent in the denominator, since the fraction, when not given originally in that form, may be reduced by actual division, to a series of monomial terms and a fraction of the desired form.

24. *Prop.* To integrate the form

$$dy = \frac{bx^{n-1} + cx^{n-2} \ldots . + lx + k}{a_1x^n + b_1x^{n-1} + c_1x^{n-2} \ldots . + l_1x + k_1} dx.$$

1*st Case.* When the denominator of the proposed fraction can be resolved into real and unequal factors of the first degree.

To illustrate this case, take the example

$$dy = \frac{ax + c}{x^2 + bx} dx = \frac{ax + c}{x(x + b)} dx.$$

Assume $\dfrac{ax + c}{x^2 + bx} = \dfrac{A}{x} + \dfrac{B}{x + b}$, where A and B are unknown constants whose values are to be determined by the condition that this assumed equality shall be verified.

Reducing the terms of the second member to a common denominator, we have

$$\frac{ax + c}{x^2 + bx} = \frac{A(x + b)}{x^2 + bx} + \frac{Bx}{x^2 + bx} = \frac{Ax + Ab + Bx}{x^2 + bx}.$$

Hence $$ax + c = Ax + Ab + Bx;$$

and since this condition is to be fulfilled without reference to the value of x, the principle of indeterminate coefficients will furnish the separate equations

$$c = Ab, \quad \text{and} \quad a = A + B.$$

Thus we shall have two equations with which to determine the values of the two constants A and B. Resolving them, we find

$$A = \frac{c}{b} \quad \text{and} \quad B = a - A = a - \frac{c}{b} = \frac{ab - c}{b}.$$

Hence by substitution

$$y = \int \frac{ax + c}{x^2 + bx}\,dx = \int \frac{A}{x}\,dx + \int \frac{B}{x + b}\,dx = \frac{c}{b}\int \frac{dx}{x} + \frac{ab-c}{b}\int \frac{dx}{x+b}$$

$$= \frac{c}{b}\log x + \frac{ab - c}{b}\log(x + b) + C.$$

As a second illustration take the following example

$$dy = \frac{a}{x^2 + bx}\,dx.$$

Assume $$\frac{a}{x^2 + bx} = \frac{A}{x} + \frac{B}{x + b}.$$

Then $$\frac{a}{x^2 + bx} = \frac{A(x + b)}{x^2 + bx} + \frac{Bx}{x^2 + bx} = \frac{Ax + Ab + Bx}{x^2 + bx}.$$

$\therefore a = Ax + Ab + Bx$, and consequently by the principle of indeterminate coefficients

$a = Ab$ and $0 = A + B$, whence $A = \frac{a}{b}$ and $B = -A = -\frac{a}{b}$.

And by substitution

$$y=\int\frac{adx}{bx}-\int\frac{adx}{b(x+b)}=\frac{a}{b}\int\frac{dx}{x}-\frac{a}{b}\int\frac{dx}{x+b}$$

$$=\frac{a}{b}\log x-\frac{a}{b}\log(x+b)+\log c$$

$$=\log\left(x^{\frac{a}{b}}\right)-\log\left[(x+b)^{\frac{a}{b}}\right]+\log c$$

$$=\log\left[c\left(\frac{x}{x+b}\right)^{\frac{a}{b}}\right].$$

Ex. To integrate $dy=\dfrac{(2+3x-4x^2)dx}{4x-x^3}$.

Here the factors of the denominator are x, $2+x$, and $2-x$, and we therefore assume

$$\frac{2+3x-4x^2}{4x-x^3}=\frac{A}{x}+\frac{B}{2+x}+\frac{C}{2-x}$$

$$=\frac{4A-Ax^2+2Bx-Bx^2+2Cx+Cx^2}{4x-x^3}.$$

$$\therefore 2+3x-4x^2=4A-Ax^2+2Bx-Bx^2+2Cx+Cx^2,$$

and by comparing the coefficients of the like powers of x, we have

$$2=4A,\quad 3=2B+2C,\quad -4=-A-B+C.$$

These conditions give

$$A=\frac{1}{2},\quad B+C=\frac{3}{2},\quad \text{and}\quad B-C=4-A=\frac{7}{2}.$$

$$\therefore A=\frac{1}{2},\quad B=\frac{5}{2},\quad C=-1.$$

$$\therefore y=\frac{1}{2}\int\frac{dx}{x}+\frac{5}{2}\int\frac{dx}{2+x}+\int\frac{-dx}{2-x}$$

$$=\frac{1}{2}\log x+\frac{5}{2}\log(2+x)+\log(2-x)+c.$$

25. A similar decomposition into partial fractions, each integrable by the logarithmic form, will be possible whenever the denominator can be resolved into simple and unequal factors. For if the number of such factors be n, each constant numerator, as A, B, C, &c., will be multiplied (in the reduction to a common denominator) by all the denominators except its own; and since each denominator contains only the first power of the variable x, it follows that there will appear in the numerator of the sum of the reduced fractions every power of x to the $(n-1)th$ power inclusive, and also an absolute term. Hence the number of equations formed by placing the absolute terms, and the coefficients of the like powers of x equal to each other, will be n, and therefore just sufficient to determine the n constants A, B, C, &c.

26. When the factors of the denominator are not immediately apparent, we may place the denominator equal to zero, determine the roots x_1, x_2, &c., of the equation so formed, if practicable, and employ the factors $x - x_1$, $x - x_2$, &c.

Ex.
$$dy = \frac{(4+7x)dx}{2x^2 - 4x - 10}.$$

Put
$$2x^2 - 4x - 10 = 0 \quad \text{or} \quad x^2 - 2x - 5 = 0.$$

Then $x = 1 \pm \sqrt{6}$, and the factors of the denominator are

$$x - 1 + \sqrt{6} \quad \text{and} \quad x - 1 - \sqrt{6}.$$

$$\therefore y = \frac{1}{2}\int \frac{(4+7x)dx}{x^2 - 2x - 5} = \frac{1}{2}\int \frac{(4+7x)dx}{(x - 1 + \sqrt{6})(x - 1 - \sqrt{6})}$$

$$= \frac{1}{2}\int \frac{Adx}{x - 1 + \sqrt{6}} + \frac{1}{2}\int \frac{Bdx}{x - 1 - \sqrt{6}}.$$

$$\therefore 4 + 7x = Ax - A - A\sqrt{6} + Bx - B + B\sqrt{6},$$

whence $4 = -A - A\sqrt{6} - B + B\sqrt{6}$ and $7 = A + B$,

from which we deduce

$$A = \frac{7\sqrt{6} - 11}{2\sqrt{6}} \quad \text{and} \quad B = \frac{7\sqrt{6} + 11}{2\sqrt{6}}.$$

$$\therefore y = \frac{7\sqrt{6}-11}{4\sqrt{6}}\int\frac{dx}{x-1+\sqrt{6}} + \frac{7\sqrt{6}+11}{4\sqrt{6}}\int\frac{dx}{x-1-\sqrt{6}}$$

$$= \frac{7\sqrt{6}-11}{4\sqrt{6}}\log(x-1+\sqrt{6}) + \frac{7\sqrt{6}+11}{4\sqrt{6}}\log(x-1-\sqrt{6})+c.$$

27. 2*d Case.* When the denominator of the proposed fraction contains equal factors of the first degree.

To illustrate, take the example $dy = \frac{a+bx+cx^2}{(x+h)^3}dx.$

If we attempt, as in the first case, to resolve this into three fractions having denominators of the first degree, by assuming

$$\frac{a+bx+cx^2}{(x+h)^3} = \frac{A}{x+h} + \frac{B}{x+h} + \frac{C}{x+h},$$

there will result

$$a+bx+cx^2 = (A+B+C)(x+h)^2,$$

and $\therefore a=(A+B+C)h^2$, $b=(A+B+C)2h$, and $c=(A+B+C)$,

whence
$$c = \frac{b}{2h} = \frac{a}{h^2}.$$

Thus the assumed condition would establish a necessary relation between the constants a, b, c, and h, where none such should exist, those constants being entirely arbitrary.

It is easily seen that such a result might have been anticipated: for since $\frac{A}{x+h} + \frac{B}{x+h} + \frac{C}{x+h} = \frac{A+B+C}{x+h}$, the proposed expression $\frac{a+bx+cx^2}{(x+h)^3}$ can only be reduced to this form when the numerator is divisible by $(x+h)^2$. Hence the decomposition of the proposed expression into three fractions of this form is not usually possible, and when possible it is not necessary because the form of the fraction can be modified by reducing it to simpler terms.

But if we put $x+h=z$, we shall have $dx=dz$, and by substitution

$$\frac{(a+bx+cx^2)dx}{(x+h)^3}=\frac{[a+b(z-h)+c(z^2-2zh+h^2)]dz}{z^3}$$

$$=\left[\frac{a-bh+ch^2}{z^3}+\frac{b-2ch}{z^2}+\frac{c}{z}\right]dz$$

$$=\left[\frac{a-bh+ch^2}{(x+h)^3}+\frac{b-2ch}{(x+h)^2}+\frac{c}{x+h}\right]dx.$$

Hence the proposed fraction can be resolved into three fractions having the forms

$$\frac{A}{(x+h)^3},\quad \frac{B}{(x+h)^2},\quad \text{and}\quad \frac{C}{x+h},$$

and since the same reasoning would apply if the number of equal factors were greater, we may in general assume

$$\frac{a+bx+cx^2\ldots\ldots+ix^{n-1}}{(x+h)^n}=\frac{A}{(x+h)^n}+\frac{B}{(x+h)^{n-1}}\cdots\cdots+\frac{I}{x+h},$$

the number of such fractions being n.

EXAMPLES.

28. 1. To integrate $\quad dy=\dfrac{2-3x}{(x-a)^2}dx.$

Assume $\quad \dfrac{2-3x}{(x-a)^2}=\dfrac{A}{(x-a)^2}+\dfrac{B}{x-a}.$

$$\therefore \frac{2-3x}{(x-a)^2}=\frac{A}{(x-a)^2}+\frac{B(x-a)}{(x-a)^2}=\frac{A+Bx-Ba}{(x-a)^2}.$$

$\therefore\ 2-3x=A+Bx-Ba$, whence $2=A-Ba$, and $-3=B$.

$\therefore\ B=-3$ and $A=2+Ba=2-3a$, and consequently

$$y=(2-3a)\int\frac{dx}{(x-a)^2}-3\int\frac{dx}{x-a}=(2-3a)\frac{1}{a-x}-3\log(x-a)+c.$$

When the denominator contains both equal and unequal factors of the first degree, the two methods must be combined.

2. $$dy = \frac{x^2 - 4x + 3}{x^3 - 6x^2 + 9x} dx.$$

Since $x^3 - 6x^2 + 9x = x(x^2 - 6x + 9) = x(x-3)^2$ we assume

$$\frac{x^2 - 4x + 3}{x^3 - 6x^2 + 9x} = \frac{A}{x} + \frac{B}{(x-3)^2} + \frac{C}{(x-3)} = \frac{A(x-3)^2 + Bx + Cx(x-3)}{x^3 - 6x^2 + 9x}.$$

$$\therefore x^2 - 4x + 3 = A(x^2 - 6x + 9) + Bx + C(x^2 - 3x),$$

whence $3 = 9A$, $-4 = -6A + B - 3C$, and $1 = A + C$.

$$\therefore A = \frac{1}{3}, \quad C = \frac{2}{3}, \quad \text{and} \quad B = 0.$$

$$\therefore y = \frac{1}{3}\int\frac{dx}{x} + \frac{2}{3}\int\frac{dx}{x-3} = \frac{1}{3}\log x + \frac{2}{3}\log(x-3) + C$$

$$= \frac{1}{3}\log x + \frac{1}{3}\log(x-3)^2 + \frac{1}{3}\log c = \frac{1}{3}\log\left[cx(x-3)^2\right]$$

$$= \log\left[cx(x-3)^2\right]^{\frac{1}{3}}.$$

3. $$dy = \frac{dx}{(x-2)^2(x+3)^2}.$$

Assume $$\frac{1}{(x-2)^2(x+3)^2} = \frac{A}{(x-2)^2} + \frac{B}{x-2} + \frac{C}{(x+3)^2} + \frac{D}{(x+3)}.$$

$$\therefore 1 = A(x+3)^2 + B(x-2)(x+3)^2 + C(x-2)^2 + D(x-2)^2(x+3),$$

or $$1 = A(x^2 + 6x + 9) + B(x^3 + 4x^2 - 3x - 18)$$
$$+ C(x^2 - 4x + 4) + D(x^3 - x^2 - 8x + 12).$$

$$\therefore 0 = B + D, \quad 0 = A + 4B + C - D, \quad 0 = 6A - 3B - 4C - 8D,$$

and $$1 = 9A - 18B + 4C + 12D.$$

These equations give, by elimination,

$$A = \frac{1}{25}, \quad B = -\frac{2}{125}, \quad C = \frac{1}{25}, \quad \text{and} \quad D = \frac{2}{125}.$$

$$\therefore y = \frac{1}{25}\int\frac{dx}{(x-2)^2} - \frac{2}{125}\int\frac{dx}{x-2} + \frac{1}{25}\int\frac{dx}{(x+3)^2} + \frac{2}{125}\int\frac{dx}{x+3}$$

$$= -\frac{1}{25(x-2)} - \frac{2}{125}\log(x-2) - \frac{1}{25(x+3)} + \frac{2}{125}\log(x+3) + c.$$

29. *Case 3d.* When the simple factors of the denominator are imaginary.

These factors, which correspond to the imaginary roots of an equation, enter in pairs, and are of the forms

$$x \pm a + b\sqrt{-1}, \quad \text{and} \quad x \pm a - b\sqrt{-1}.$$

and their product gives the real quadratic factor

$$x^2 \pm 2ax + a^2 + b^2 = (x \pm a)^2 + b^2$$

Hence, if there be but one pair of simple imaginary factors, or a single quadratic factor, in the denominator, the corresponding partial fraction will be of the form $\frac{Ax + B}{(x \pm a)^2 + b^2}$, in which the numerator must consist of two terms, one containing the first power of x, and the other an absolute term, because the denominator now contains the second power of x; and, therefore, if we introduced a constant only into the numerator, we should not provide for having the exponent of the highest power of x, in the numerator, only *one* less than the corresponding power in the denominator,

But when there are several equal quadratic factors, the denominator being of the form

$$[(x \pm a)^2 + b^2]^n,$$

the partial fractions will be of the forms

$$\frac{Ax + B}{[(x \pm a)^2 + b^2]^n} + \frac{Cx + D}{[(x \pm a)^2 + b^2]^{n-1}} \cdots\cdots\cdots + \frac{Ex + F}{(x \pm a)^2 + b^2}$$

the number of such fractions being n.

That such a decomposition is possible in all cases, will appear more clearly by the following illustration. Let the proposed fraction be

$$\frac{cx^5 + ex^4 + fx^3 + gx^2 + hx + i}{[(x \pm a)^2 + b^2]^3}.$$

Put $\qquad x \pm a = y, \qquad \text{and} \qquad y^2 + b^2 = z^2.$

Then the fraction can be reduced successively to the following forms

$$\frac{c(y \mp a)^5 + e(y \mp a)^4 + f(y \mp a)^3 + g(y \mp a)^2 + h(y \mp a) + i}{z^6}$$

$$= \frac{cy^5 + e_1y^4 + f_1y^3 + g_1y^2 + h_1y + i_1}{z^6}$$

$$= \frac{(cy + e_1)(z^2 - b^2)^2 + (f_1y + g_1)(z^2 - b^2) + h_1y + i_1}{z^6}$$

$$= \frac{[c(x \pm a) + e_1](z^4 - 2z^2b^2 + b^4) + [f_1(x \pm a) + g_1](z^2 - b^2) + h_1(x \pm a) + i_1}{z^6}$$

$$= \frac{(cb^4 - f_1b^2 + h_1)x + b^4(e_2 \pm a)c - f_1b^2(g_2 \pm a) + i_1 \pm h_1a}{[(x \pm a)^2 + b^2]^3}$$

$$+ \frac{(-2c\, b^2 + f_1)x - 2b^2c\, (e_2 \pm a) + f_1(g_2 \pm a)}{[(x \pm a)^2 + b^2]^2} + \frac{cx + c(e_2 \pm a)}{(x \pm a)^2 + b^2},$$

which is of the form

$$\frac{Ax + B}{[(x \pm a)^2 + b^2]^3} + \frac{Cx + D}{[(x \pm a)^2 + b^2]^2} + \frac{Ex + F}{(x \pm a)^2 + b^2}.$$

And a similar decomposition would evidently be possible, if there were n equal quadratic factors in the denominator.

30. It appears therefore, that when the denominator contains simple imaginary factors, the general form presented for integration, will be

$$dy = \frac{(Ax + B)\,dx}{[(x \pm a)^2 + b^2]^n}, \text{ where } n \text{ may be any integer.}$$

Put $\qquad x \pm a = z, \quad$ then

$$dy = \frac{(Az \mp Aa + B)\,dz}{(z^2 + b^2)^n}.$$

$$\therefore y = \int \frac{Az\,dz}{(z^2 + b^2)^n} + \int \frac{(B \mp Aa)\,dz}{(z^2 + b^2)^n} = -\frac{A}{2(n - 1)(z^2 + b^2)^{n-1}}$$

$$+ \int \frac{A_1\,dz}{(z^2 + b^2)^n}$$

by making $B \mp Aa = A_1$. Thus the proposed integral is found to depend on the more simple form, $\int \frac{A_1 dz}{(z^2+b^2)^n}$

It will now be shown that this latter can be caused to depend on the form $\int \frac{dz}{(z^2+b^2)^{n-1}}$, in which the exponent of the parenthesis is diminished by unity. Thus we have

$$\frac{dz}{(z^2+b^2)^{n-1}} = \frac{(z^2+b^2)dz}{(z^2+b^2)^n} = \frac{z^2dz}{(z^2+b^2)^n} + \frac{b^2dz}{(z^2+b^2)^n}$$

$$\therefore \int \frac{dz}{(z^2+b^2)^n} = \frac{1}{b^2}\int \frac{dz}{(z^2+b^2)^{n-1}} - \frac{1}{b^2}\int \frac{z^2dz}{(z^2+b^2)^n} \quad \cdots\cdots (1).$$

But $$d\left(\frac{z}{(z^2+b^2)^{n-1}}\right) = \frac{dz}{(z^2+b^2)^{n-1}} - \frac{2(n-1)z^2dz}{(z^2+b^2)^n},$$

$$\therefore \int \frac{z^2dz}{(z^2+b^2)^n} = \frac{1}{2(n-1)}\int \frac{dz}{(z^2+b^2)^{n-1}} - \frac{1}{2(n-1)} \cdot \frac{z}{(z^2+b^2)^{n-1}},$$

which value, substituted in (1), reduces it to the form

$$\int \frac{dz}{(z^2+b^2)^n} = \frac{1}{b^2}\int \frac{dz}{(z^2+b^2)^{n-1}} - \frac{1}{2b^2(n-1)}\int \frac{dz}{(z^2+b^2)^{n-1}}$$
$$+ \frac{z}{2b^2(n-1)(z^2+b^2)^{n-1}}$$
$$= \frac{z}{2b^2(n-1)(z^2+b^2)^{n-1}} + \frac{2n-3}{2b^2(n-1)}\int \frac{dz}{(z^2+b^2)^{n-1}}.$$

Similarly, $\int \frac{dz}{(z^2+b^2)^{n-1}}$ can be rendered dependent upon $\int \frac{dz}{(z^2+b^2)^{n-2}}$, &c., so that eventually, the original integral will depend on the form $\int \frac{dz}{z^2+b^2} = \frac{1}{b}\tan^{-1}\frac{z}{b}$.

31. We infer, therefore, that the integration of a rational fraction can be effected whenever its denominator can be resolved into simple or quadratic factors, and that the integral will be expressed in the form of logarithms, powers, or circular arcs.

GENERAL EXAMPLES.

32. 1. $$dy = \frac{dx}{x^3 - 1}.$$

Since $(x^3 - 1) = (x^2 + x + 1)(x - 1)$, and $x^2 + x + 1$

$$= \left(x + \frac{1}{2} + \frac{1}{2}\sqrt{-3}\right)\left(x + \frac{1}{2} - \frac{1}{2}\sqrt{-3}\right),$$

we assume $$\frac{1}{x^3 - 1} = \frac{Ax + B}{x^2 + x + 1} + \frac{C}{x - 1},$$

$$\therefore\ 1 = Ax^2 + Bx - Ax - B + Cx^2 + Cx + C.$$

whence $0 = A + C$, $0 = B + C - A$ and $1 = C - B$.

$$\therefore\ A = -\frac{1}{3},\ B = -\frac{2}{3} \text{ and } C = \frac{1}{3}.$$

$$\therefore\ dy = \frac{\left(-\frac{1}{3}x - \frac{2}{3}\right)dx}{x^2 + x + 1} + \frac{\frac{1}{3}dx}{x - 1},$$

and if we put $x + \frac{1}{2} = z$, or $x^2 + x + \frac{1}{4} = z^2$, $x = z - \frac{1}{2}$ and $dx = dz$, there will result

$$y = \frac{1}{3}\int\frac{dx}{x - 1} - \frac{1}{3}\int\frac{(x + 2)dx}{\left(x + \frac{1}{2}\right)^2 + \frac{3}{4}} = \frac{1}{3}\log(x - 1) - \frac{1}{3}\int\frac{\left(z + \frac{3}{2}\right)dz}{z^2 + \frac{3}{4}}$$

$$= \frac{1}{3}\log(x - 1) - \frac{1}{6}\int\frac{2z\,dz}{z^2 + \frac{3}{4}} - \frac{1}{\sqrt{3}}\int\frac{\frac{2dz}{\sqrt{3}}}{\frac{4z^2}{3} + 1}$$

$$= \frac{1}{3}\log(x - 1) - \frac{1}{6}\log\left(z^2 + \frac{3}{4}\right) - \frac{1}{\sqrt{3}}\tan^{-1}\frac{2z}{\sqrt{3}} + c$$

$$= \frac{1}{3}\log(x - 1) - \frac{1}{6}\log(x^2 + x + 1) - \frac{1}{\sqrt{3}}\cdot\tan^{-1}\frac{2x + 1}{\sqrt{3}} + c.$$

2. $$dy = \frac{dx}{x(a + bx^2)^2}.$$

Assume $\dfrac{1}{x(a+bx^2)^2} = \dfrac{A}{x} + \dfrac{Bx+C}{(a+bx^2)^2} + \dfrac{Dx+E}{a+bx^2}.$

$\therefore\ 1 = A(a^2+2abx^2+b^2x^4)+Bx^2+Cx+D(ax^2+bx^4)+E(ax+bx^3),$

$\therefore\ 1 = Aa^2,\ 0 = 2Aab+B+Da,\ 0 = Ab^2+Db,\ 0 = C+Ea,\ 0 = Eb,$

$$\therefore\ A = \frac{1}{a^2},\ B = -\frac{b}{a},\ C = 0,\ D = -\frac{b}{a^2},\ E = 0.$$

Hence, $y = \dfrac{1}{a^2}\displaystyle\int\frac{dx}{x} - \frac{b}{a}\int\frac{xdx}{(a+bx^2)^2} - \frac{b}{a^2}\int\frac{xdx}{a+bx^2}$

$$= \frac{1}{a^2}\log x + \frac{1}{2a}\cdot\frac{1}{a+bx^2} - \frac{1}{2a^2}\log(a+bx^2) + c.$$

$$= \frac{1}{2a(a+bx^2)} + \frac{1}{2a^2}\log\frac{x^2}{a+bx^2} + c.$$

3. $$dy = \frac{x^2dx}{x^4+x^2-2}$$

Put $x^4 + x^2 - 2 = 0$, and resolve with respect to x^2.

$$\therefore\ x^2 = -\frac{1}{2} \pm \frac{3}{2} = 1, \text{ or } x^2 = -2,$$

$\therefore\ (x^4+x^2-2) = (x^2+2)(x^2-1) = (x^2+2)(x+1)(x-1),$
and we may assume,

$$\frac{x^2}{x^4+x^2+2} = \frac{A}{x+1} + \frac{B}{x-1} + \frac{Cx+D}{x^2+2}.\quad \text{Then}$$

$x^2 = A(x^3-x^2+2x-2)+B(x^3+x^2+2x+2)+C(x^3-x)+D(x^2-1).$

$\therefore\ 0 = A+B+C,\ 1 = -A+B+D,\ 0 = 2A+2B-C,$
$0 = -2A+2B-D.$

$$\therefore\ A = -\frac{1}{6},\ B = \frac{1}{6},\ C = 0,\ D = \frac{2}{3}.$$

$$\therefore\ y = -\frac{1}{6}\int\frac{dx}{x+1} + \frac{1}{6}\int\frac{dx}{x-1} + \frac{2}{3}\int\frac{dx}{x^2+2}$$

$$= -\log(x+1)^{\frac{1}{6}} + \log(x-1)^{\frac{1}{6}} + \frac{\sqrt{2}}{3}\tan^{-1}\frac{x}{\sqrt{2}} + c.$$

4. $$dy = \frac{dx}{1-x^6}.\quad \text{Since}$$

$1-x^6=(1-x^3)(1+x^3)=(1-x)(1+x+x^2)(1+x)(1-x+x^2)$, put

$$\frac{1}{1-x^6} = \frac{A}{1+x} + \frac{B}{1-x} + \frac{Cx+D}{1+x+x^2} + \frac{Ex+F}{1-x+x^2}$$

$$\begin{aligned}\therefore 1 = {} & A(1-x+x^2-x^3+x^4-x^5)+B(1+x+x^2+x^3+x^4+x^5)\\ & +C(x-x^2+x^4-x^5)+D(1-x+x^3-x^4)+E(x+x^2-x^4-x^5)\\ & +F(1+x-x^3-x^4).\end{aligned}$$

$$\begin{aligned}\therefore\ & 1 = A+B+D+F,\quad 0 = -A+B+C-D+E+F,\\ & 0 = A+B-C+E,\quad 0 = -A+B+D-F,\\ & 0 = A+B+C-D-E-F,\quad 0 = -A+B-C-E.\end{aligned}$$

$$\therefore A = \frac{1}{6},\ B = \frac{1}{6},\ C = \frac{1}{6},\ D = \frac{1}{3},\ E = -\frac{1}{6},\ F = \frac{1}{3}.$$

$$\therefore y = \frac{1}{6}\int\frac{dx}{1+x} + \frac{1}{6}\int\frac{dx}{1-x} + \frac{1}{6}\int\frac{(x+2)dx}{1+x+x^2} - \frac{1}{6}\int\frac{(x-2)dx}{1-x+x^2}$$

$$\begin{aligned}= {} & \frac{1}{6}\log(1+x) - \frac{1}{6}\log(1-x) + \frac{1}{12}\int\frac{(2x+1)dx}{1+x+x^2} + \frac{1}{12}\int\frac{3dx}{\left(x+\frac{1}{2}\right)^2+\frac{3}{4}}\\ & - \frac{1}{12}\int\frac{(2x-1)dx}{1-x+x^2} + \frac{1}{12}\int\frac{3dx}{\left(x-\frac{1}{2}\right)^2+\frac{3}{4}}\end{aligned}$$

$$\begin{aligned}= {} & \log(1+x)^{\frac{1}{6}} - \log(1-x)^{\frac{1}{6}} + \frac{1}{12}\log(1+x+x^2) - \frac{1}{12}\log(1-x+x^2)\\ & + \frac{1}{2\sqrt{3}}\int\frac{\frac{2dx}{\sqrt{3}}}{\frac{4\left(x+\frac{1}{2}\right)^2}{3}+1} + \frac{1}{2\sqrt{3}}\int\frac{\frac{2dx}{\sqrt{3}}}{\frac{4\left(x-\frac{1}{2}\right)^2}{3}+1}\end{aligned}$$

$$\begin{aligned}= {} & \log\left(\frac{1+x}{1-x}\right)^{\frac{1}{6}} + \log\left(\frac{1+x+x^2}{1-x+x^2}\right)^{\frac{1}{12}} + \frac{1}{2\sqrt{3}}\left(\tan^{-1}\frac{2x+1}{\sqrt{3}}\right.\\ & \left. + \tan^{-1}\frac{2x-1}{\sqrt{3}}\right) + c.\end{aligned}$$

CHAPTER IV.

IRRATIONAL FRACTIONS.

33. The differential form next to be considered is that which is still algebraic, but which involves irrational or surd quantities. As the general mode of treating such expressions is the same in principle, whether presented in the entire or fractional form, they will be considered in the latter, which is of very frequent occurrence, and which presents some difficulties peculiar to itself.

34. When an irrational fraction, which does not belong to one of the known elementary forms, is presented for integration, we endeavor to *rationalize* it, that is, to transform it into a rational form by suitable substitutions. The following are the principal cases in which this is possible.

35. *Case* 1*st.* When the fraction contains none but monomial terms. As an example to illustrate this case, suppose

$$dy = \frac{ax^{\frac{n_1}{n}} + bx^{\frac{m_1}{m}}}{a_1x^{\frac{c_1}{c}} + b_1x^{\frac{e_1}{e}}}\,dx.$$

Put $x = z^{nmce}$ or $x = z^l$, where l is any common multiple of the denominators n, m, c, and e.

Then $x^{\frac{n_1}{n}} = z^{n_1mce}$, or $x^{\frac{n_1}{n}} = z^{\frac{n_1l}{n}}$ where $\frac{n_1l}{n}$ is an integer since l is a multiple of n.

Similarly $x^{\frac{m_1}{m}} = z^{m_1nce}$, $x^{\frac{c_1}{c}} = z^{c_1nme}$, and $x^{\frac{e_1}{e}} = z^{e_1nmc}$.

Also $$dx = nmce \,.\, z^{nmce-1} dz.$$

Hence $$dy = \frac{a z^{n_1 mce} + b z^{m_1 nce}}{a_1 z^{c_1 mne} + b_1 z^{e_1 nmc}} (nmce \,.\, z^{nmce-1} \, dz),$$

which is a fraction entirely rational.

Ex. To integrate $$dy = \frac{2x^{\frac{1}{2}} - 3x^{\frac{1}{6}}}{3x^{\frac{2}{3}} + x^{\frac{3}{4}}} dx.$$

Assume $x = z^{12}$: then $x^{\frac{1}{2}} = z^6$, $x^{\frac{1}{6}} = z^2$, $x^{\frac{2}{3}} = z^8$, $x^{\frac{3}{4}} = z^9$,

and $$dx = 12z^{11} \,.\, dz.$$

$$\therefore \; dy = \frac{2z^6 - 3z^2}{3z^8 + z^9} 12z^{11} \,.\, dz = \frac{24z^9 - 36z^5}{z + 3} dz$$

$$= (24z^8 - 72z^7 + 216z^6 - 648z^5) dz$$

$$+ 1908\left(z^4 - 3z^3 + 9z^2 - 27z + 81 - \frac{243}{z + 3}\right) dz.$$

$$\therefore \; y = 12 \int (2z^8 - 6z^7 + 18z^6 - 54z^5) dz$$

$$+ 1908 \int \left(z^4 - 3z^3 + 9z^2 - 27z + 81 - \frac{243}{z + 3}\right) dz$$

$$= 12\left(\frac{2}{9} z^9 - \frac{3}{4} z^8 + \frac{18}{7} z^7 - 9z^6\right)$$

$$+ 1908 \left[\frac{1}{5} z^5 - \frac{3}{4} z^4 + 3z^3 - \frac{27}{2} z^2 + 81z - 243 \log(z + 3)\right] + c$$

$$= 12\left(\frac{2}{9} x^{\frac{3}{4}} - \frac{3}{4} x^{\frac{2}{3}} + \frac{18}{7} x^{\frac{7}{12}} - 9x^{\frac{1}{2}}\right)$$

$$+ 1908 \left[\frac{1}{5} x^{\frac{5}{12}} - \frac{3}{4} x^{\frac{1}{3}} + 3x^{\frac{1}{4}} - \frac{27}{2} x^{\frac{1}{6}} + 81x^{\frac{1}{12}} - 243 \log(x^{\frac{1}{12}} + 3)\right] + c.$$

36. 2*d Case.* When the surds which enter the given expression contain no quantity within the ( ) but one of the form $(a + bx)$.

As an example, take

$$dy = \frac{(a + bx)^{\frac{n_1}{n}} + (a + bx)^{\frac{m_1}{m}}}{(a + bx)^{\frac{c_1}{c}} + h} dx.$$

Put $a + bx = z^{nmc}$ where the exponent of z is a multiple of all the denominators n, m, and c.

Then $(a+bx)^{\frac{n_1}{n}} = z^{n_1mc}$, $(a+bx)^{\frac{m_1}{m}} = z^{m_1nc}$, $(a+bx)^{\frac{c_1}{c}} = z^{c_1nm}$,

and $$dx = \frac{nmc}{b} \cdot z^{nmc-1}dz,$$
and by substitution
$$dy = \frac{z^{n_1mc} + z^{m_1nc}}{b(z^{c_1nm} + h)} \cdot nmc \,.\, z^{nmc-1}dz,$$

which form is entirely rational.

37. 1. To integrate $dy = \dfrac{x^3dx}{(1+4x)^{\frac{5}{2}}}$.

Assume $1 + 4x = z^2$. Then

$x = \dfrac{z^2-1}{4}$, $dx = \dfrac{zdz}{2}$, $x^3 = \dfrac{1}{64}(z^6 - 3z^4 + 3z^2 - 1)$, and $(1+4x)^{\frac{5}{2}} = z^5$.

$$\therefore dy = \frac{1}{128}\left(\frac{z^6 - 3z^4 + 3z^2 - 1}{z^5}\right)zdz = \frac{1}{128}\left(z^2 - 3 + \frac{3}{z^2} - \frac{1}{z^4}\right)dz.$$

$$\therefore y = \frac{1}{128}\left(\frac{z^3}{3} - 3z - \frac{3}{z} + \frac{1}{3z^3}\right) + c,$$

or $$y = \frac{1}{128}\left[\frac{(1+4x)^{\frac{3}{2}}}{3} - 3(1+4x)^{\frac{1}{2}} - \frac{3}{(1+4x)^{\frac{1}{2}}} + \frac{1}{3(1+4x)^{\frac{3}{2}}}\right] + c.$$

2. $$dy = \frac{dx}{x\sqrt{1+x}}.$$

Put $1 + x = z^2$. Then $x = z^2 - 1$, $dx = 2zdz$, and $\sqrt{1+x} = z$

$$\therefore dy = \frac{2zdz}{(z^2-1)z} = \frac{2dz}{z^2-1} = \frac{dz}{z-1} - \frac{dz}{z+1}.$$

$$\therefore y = \log(z-1) - \log(z+1) + c,$$

or $$y = \log\frac{z-1}{z+1} + c = \log\frac{\sqrt{1+x} - 1}{\sqrt{1+x} + 1} + c.$$

38. *Case 3d.* When the proposed fraction contains no surd except one of the form

$$\sqrt{a + bx \pm c^2x^2} = c\sqrt{\frac{a}{c^2} + \frac{b}{c^2}x \pm x^2}.$$

When the last term is positive, assume

$$\sqrt{\frac{a}{c^2} + \frac{b}{c^2}x + x^2} = z + x; \quad \text{then} \quad \frac{a}{c^2} + \frac{b}{c^2}x + x^2 = z^2 + 2zx + x^2.$$

$$\therefore x = \frac{a - c^2z^2}{2c^2z - b}, \quad dx = -\frac{2c^2(a - bz + c^2z^2)}{(2c^2z - b)^2}dz,$$

and $$\sqrt{a+bx+c^2x^2} = c(z+x) = c\left(z + \frac{a - c^2z^2}{2c^2z - b}\right) = \frac{c(a - bz + c^2z^2)}{2c^2z - b}.$$

The values of x, $\sqrt{a + bx + c^2x^2}$, and dx, being all expressed rationally in terms of z, the proposed differential when transformed will also be rational.

Again, if the term involving x^2 in the surd be negative, denote by x_1 and x_2 the roots of the equation

$$x^2 - \frac{bx}{c^2} - \frac{a}{c^2} = 0; \quad \text{then} \quad x^2 - \frac{bx}{c^2} - \frac{a^2}{c^2} = (x - x_1)(x - x_2),$$

and therefore $$\frac{a^2}{c^2} + \frac{bx}{c^2} - x^2 = (x_2 - x)(x - x_1).$$

Now assume $\sqrt{(x_2 - x)(x - x_1)} = (x - x_1) . z.$

$$\therefore x_2 - x = (x - x_1)z^2, \quad \text{whence} \quad x = \frac{x_2 + x_1z^2}{1 + z^2}$$

$$dx = \frac{2(x_1 - x_2)zdz}{(1 + z^2)^2} \quad \text{and} \quad \sqrt{a + bx - c^2x^2} = cz(x - x_1) = \frac{cz(x_2 - x_1)}{1 + z^2}.$$

Hence the several expressions which enter into the proposed differential will be rational, and therefore that differential will become entirely rational.

39. 1. To integrate $dy = \dfrac{dx}{\sqrt{1 + x + x^2}}.$

Assume $\sqrt{1+x+x^2} = z + x$; then $1 + x + x^2 = z^2 + 2zx + x^2$

$$x = \frac{1-z^2}{2z-1}, \quad dx = -\frac{2(1-z+z^2)}{(2z-1)^2}dz, \text{ and } \sqrt{1+x+x^2} = \frac{1-z+z^2}{2z-1}.$$

$$\therefore y = -\int \frac{2z-1}{1-z+z^2} \times \frac{2(1-z+z^2)}{(2z-1)^2}dz = -\int \frac{2dz}{2z-1} = -\log(2z-1)+c$$

$$= \log\frac{1}{2z-1} + c = \log\frac{1}{2\sqrt{1+x+x^2}-(2x+1)} + c$$

$$= \log[2\sqrt{1+x+x^2} + (2x+1)] - \log 3 + c$$

$$= \log[2\sqrt{1+x+x^2} + 2x + 1] + c_1.$$

2. $$dy = \frac{dx}{\sqrt{1+x-x^2}}.$$

Put $1 + x - x^2 = 0$, and find the roots x_1 and x_2, thus

$$x_1 = \frac{1}{2} + \frac{1}{2}\sqrt{5} \quad \text{and} \quad x_2 = \frac{1}{2} - \frac{1}{2}\sqrt{5}.$$ Now assume

$$\sqrt{1+x-x^2} \text{ or } \sqrt{(x-x_1)(x_2-x)} = \sqrt{x-\frac{1}{2}-\frac{1}{2}\sqrt{5}}\,.\sqrt{\frac{1}{2}-\frac{1}{2}\sqrt{5}-x}$$

$$= (x - \frac{1}{2} - \frac{1}{2}\sqrt{5})\, z.$$

$$\therefore \frac{1}{2} - \frac{1}{2}\sqrt{5} - x = (x - \frac{1}{2} - \frac{1}{2}\sqrt{5})\, z^2 \text{ and } x = \frac{\frac{1}{2}-\frac{1}{2}\sqrt{5}+\left(\frac{1}{2}+\frac{1}{2}\sqrt{5}\right)z^2}{1+z^2}$$

$$dx = \frac{2\sqrt{5}\,.\,zdz}{(1+z^2)^2} \quad \text{and} \quad \sqrt{1+x-x^2} = -\frac{\sqrt{5}\,.\,z}{1+z^2}.$$

$$\therefore y = -\int \frac{1+z^2}{\sqrt{5}\,.\,z} \times \frac{2\sqrt{5}\,.\,zdz}{(1+z^2)^2} = -2\int \frac{dz}{1+z^2} = -2\tan^{-1}z + c.$$

$$= -2\tan^{-1}\frac{\sqrt{1+x-x^2}}{x-\frac{1}{2}-\frac{1}{2}\sqrt{5}} + c = -2\tan^{-1}\sqrt{\frac{\frac{1}{2}-\frac{1}{2}\sqrt{5}-x}{x-\frac{1}{2}-\frac{1}{2}\sqrt{5}}} + c.$$

3. $$dy = \frac{dx}{x\sqrt{a^2 + b^2x^2}}.$$

Assume $\sqrt{\frac{a^2}{b^2} + x^2} = z + x$; then $\frac{a^2}{b^2} + x^2 = z^2 + 2zx + x^2$

$$x = \frac{a^2 - b^2z^2}{2b^2z}, \quad dx = -\frac{2b^2(a^2 + b^2z^2)}{(2b^2z)^2}\,dz, \text{ and } \sqrt{a^2 + b^2x^2} = \frac{a^2 + b^2z^2}{2bz}.$$

$$\therefore y = -\int \frac{2b^2z}{a^2 - b^2z^2} \times \frac{2bz}{a^2 + b^2z^2} \times \frac{2b^2(a^2 + b^2z^2)}{(2b^2z)^2}\,dz = -\int \frac{2b\,dz}{a^2 - b^2z^2}$$

$$= -\frac{1}{a}\int \frac{b\,dz}{a + bz} + \frac{1}{a}\int \frac{-b\,dz}{a - bz} = \frac{1}{a}\log\frac{a - bz}{a + bz} + c.$$

$$= \frac{1}{a}\log\frac{a + bx - \sqrt{a^2 + b^2x^2}}{a - bx + \sqrt{a^2 + b^2x^2}} + c = \frac{1}{a}\log\frac{\sqrt{a^2 + b^2x^2} - a}{bx} + c.$$

40. The other irrational forms which admit of being rationalized, are chiefly those belonging to the binomial class, which it is proposed to consider carefully in the next chapter.

CHAPTER V.

BINOMIAL DIFFERENTIALS.

41. *Prop.* To determine the conditions under which the general form

$$dy = x^{\frac{m_1}{m}}\left(a + bx^{\frac{n_1}{n}}\right)^{\frac{r_1}{r}} dx,\ \text{can be rendered rational.}$$

If we put $x = z^{mn}$, there will result $x^{\frac{m_1}{m}} = z^{m_1 n}$, $x^{\frac{n_1}{n}} = z^{n_1 m}$ and

$$dx = mn \,.\, z^{mn-1} dz.$$

$$\therefore\ dy = z^{m_1 n}\left(a + bz^{n_1 m}\right)^{\frac{r_1}{r}} nmz^{nm-1} dz,$$

so that the form will be equally general if written thus

$$dy = x^m (a + bx^n)^p dx \ldots\ldots (1),$$

in which p is the only fractional exponent.

Assume $\quad a + bx^n = z$; then $x = \left(\dfrac{z-a}{b}\right)^{\frac{1}{n}}$

$$x^{m+1} = \left(\frac{z-a}{b}\right)^{\frac{m+1}{n}} \text{ and } \quad x^m dx = \frac{1}{nb^{\frac{m+1}{n}}}(z-a)^{\frac{m+1}{n}-1} . dz,$$

and by substitution in (1),

$$dy = \frac{1}{nb^{\frac{m+1}{n}}}(z-a)^{\frac{m+1}{n}-1} z^p dz.$$

Hence, if $\dfrac{m+1}{n}$ be a positive or negative integer, or zero, the quantity $(z-a)^{\frac{m+1}{n}-1}$ can be developed in the form of a series of monomials (with a limited number of terms), a rational fraction, or unity, and thus the value of dy can be rendered entirely rational. For, although p is a fractional exponent, the expression can be so transformed as to remove the fraction, by the method explained in the first case of irrational fractions.

Again, since $x^m(a + bx^n)^p = x^{m+np}(ax^{-n} + b)^p$, if we put
$ax^{-n} + b = z$, there will result,

$$x = \left(\frac{z-b}{a}\right)^{-\frac{1}{n}}, \quad x^{m+np+1} = \left(\frac{z-b}{a}\right)^{\frac{m+np+1}{-n}}$$

$$\therefore\ x^{m+np}dx = -\frac{a^{\frac{m+np+1}{n}}}{n}(z-b)^{-\frac{m+1}{n}-p-1}dz.$$

$$\therefore\ dy = -\frac{a^{\frac{m+np+1}{n}}}{n}(z-b)^{-\frac{m+1}{n}-p-1}.z^p.dz.$$

And this can be readily rationalized when $\frac{m+1}{n} + p$ is a positive or negative integer or zero.

We conclude, therefore, that there are two cases in which it will be possible to rationalize the general binomial differential, viz.:

1st. When the exponent m of x exterior to the parenthesis, increased by unity, is exactly divisible by the exponent n of x within the ( ); or

2d. When the fraction thus formed, increased by the exponent p of the ( ) is an integer or zero.

42. These two relations are called the *conditions of integrability* of binomial differentials.

43. 1. To integrate $dy = x^5(a + x^2)^{\frac{1}{3}}dx$.

Here $m = 5$, $n = 2$, $\therefore\ \frac{m+1}{n} = 3$, an integer,

and the expression can be rendered rational.

Put $\quad a + x^2 = z$, $\therefore\ x = (z - a)^{\frac{1}{2}}$, $x^6 = (z - a)^3$

$x^5dx = \frac{1}{2}(z-a)^2dz$, and $dy = \frac{1}{2}(z-a)^2 . z^{\frac{1}{3}}dz = \frac{1}{2}(z^{\frac{7}{3}} - 2az^{\frac{4}{3}} + a^2z^{\frac{1}{3}})dz$

$$\therefore y = \frac{1}{2}\int (z^{\frac{7}{3}} - 2az^{\frac{4}{3}} + a^2z^{\frac{1}{3}})dz = \frac{1}{2}\left(\frac{3}{10}z^{\frac{10}{3}} - \frac{6}{7}az^{\frac{7}{3}} + \frac{3}{4}a^2z^{\frac{4}{3}}\right) + c,$$

$$= \frac{3}{20}(a + x^2)^{\frac{10}{3}} - \frac{3}{7}(a + x^2)^{\frac{7}{3}}.\,a + \frac{3}{8}(a + x^2)^{\frac{4}{3}}.\,a^2 + c.$$

2. $$dy = \frac{dx}{x^2(1 + x^2)^{\frac{3}{2}}} = x^{-2}(1 + x^2)^{-\frac{3}{2}}dx.$$

Here $m = -2,\quad n = 2,\quad$ and $\quad p = -\frac{3}{2}.$

$$\therefore \frac{m+1}{n} = -\frac{1}{2}, \text{ a fraction;}$$

but $$\frac{m+1}{n} + p = -\frac{1}{2} - \frac{3}{2} = -2, \text{ a negative integer,}$$

and the expression can therefore be rendered rational by the second transformation.

Put $x^{-2} + 1 = z,\quad \therefore x = (z - 1)^{-\frac{1}{2}},\quad x^{-4} = (z - 1)^2$

$$x^{-5}.\,dx = -\frac{1}{2}(z - 1)dz, \quad \text{and} \quad dy = -\frac{1}{2}(z - 1)z^{-\frac{3}{2}}dz.$$

or, $$dy = -\frac{1}{2}z^{-\frac{1}{2}}dz + \frac{1}{2}z^{-\frac{3}{2}}dz.$$

$$\therefore y = -\frac{1}{2}\int z^{-\frac{1}{2}}dz + \frac{1}{2}\int z^{-\frac{3}{2}}dz = -(z^{\frac{1}{2}} + z^{-\frac{1}{2}}) + c.$$

$$= -\frac{z+1}{z^{\frac{1}{2}}} + c = -\frac{x^{-2}+2}{\sqrt{x^{-2}+1}} + c = -\frac{1}{\sqrt{1+x^2}}\left(\frac{1}{x} + 2x\right) + c.$$

CHAPTER VI.

FORMULÆ OF REDUCTION.

44. When a binomial differential satisfies either of the conditions of integrability, it is possible to transform it into a rational expression; but, instead of applying this process of rationalization directly, it is often more convenient to employ certain *formulæ of reduction*, which render the proposed integral dependent upon others of simpler form, or such as have been previously integrated.

45. Such formulæ are deduced by employing another known as the formula for *integration by parts*.

Thus, since $d(uv) = udv + vdu$, we have

$$\int udv = uv - \int vdu. \quad . \quad . \quad . \quad . \quad (1).$$

By this formula, $\int udv$, is made to depend upon $\int vdu$, which latter integral may be more simple.

46. *Prop.* To obtain a formula for diminishing the exponent m of x, exterior to the ( ), in the general binomial form

$$y = \int x^m (a + bx^n)^p \, dx.$$

Put $(a + bx^n)^p x^{n-1} dx = dv$, and $x^{m-n+1} = u$.

Then $v = \dfrac{(a + bx^n)^{p+1}}{nb(p+1)}$, and $du = (m - n + 1)x^{m-n}dx$.

But $y = \int x^{m-n+1}(a + bx^n)^p x^{n-1} dx = \int udv = uv - \int vdu$.

$$\therefore y = \frac{x^{m-n+1}(a + bx^n)^{p+1}}{nb(p+1)} - \frac{m-n+1}{nb(p+1)} \int x^{m-n}(a + bx^n)^{p+1} dx \quad (2).$$

The formula (2), effects the object of diminishing the exponent m of x, but it increases by unity the exponent p of the ( ), and as this would often be an inconvenience, we must endeavor to modify (2).

Now $\int x^{m-n}(a+bx^n)^{p+1}dx = \int x^{m-n}(a+bx^n)^p(a+bx^n)dx$

$$= a\int x^{m-n}(a+bx^n)^p dx + b\int x^m(a+bx^n)^p dx.$$

$$\therefore\ y = \int x^m(a+bx^n)^p dx = \frac{x^{m-n+1}(a+bx^n)^{p+1}}{nb(p+1)}$$

$$-\frac{(m-n+1)a}{nb(p+1)}\int x^{m-n}(a+bx^n)^p dx - \frac{m-n+1}{n(p+1)}\int x^m(a+bx^n)^p dx.$$

Transposing the last term to the first member and reducing, we have

$$\frac{np+m+1}{n(p+1)}\int x^m(a+bx^n)^p dx = \frac{x^{m-n+1}(a+bx^n)^{p+1}}{nb(p+1)}$$

$$-\frac{(m-n+1)a}{nb(p+1)}\int x^{m-n}(a+bx^n)^p dx.$$

Hence $y = \int x^m(a+bx^n)^p dx$

$$= \frac{x^{m-n+1}(a+bx^n)^{p+1} - (m-n+1)a\int x^{m-n}(a+bx^n)^p dx}{b(np+m+1)} \quad \ldots (A).$$

47. By this formula, the proposed integral is made to depend upon another of a similar form, but having the exponent $m-n$ of x, exterior to the ( ), less than the original exponent m, by n, the exponent of x within the ( ).

48. *Prop.* To obtain a formula for diminishing the exponent p of the ( ), in the general integral

$$y = \int x^m(a+bx^n)^p dx.$$

Since $\int x^m(a+bx^n)^p dx = \int x^m(a+bx^n)^{p-1}(a+bx^n)dx$

$$= a\int x^m(a+bx^n)^{p-1}dx + b\int x^{m+n}(a+bx^n)^{p-1}dx;$$

and since by applying formula (A) to the last integral, replacing m by $(m+n)$, and p by $(p-1)$, we get

$$\int x^{m+n}(a+bx^n)^{p-1}dx = \frac{x^{m+1}(a+bx^n)^p - (m+1)a\int x^m(a+bx^n)^{p-1}dx}{b(np+m+1)}.$$

$$\therefore \quad y = \int x^m(a+bx^n)^p dx = a\int x^m(a+bx^n)^{p-1}dx + \frac{x^{m+1}(a+bx^n)^p}{np+m+1}$$
$$- \frac{(m+1)a}{np+m+1}\int x^m(a+bx^n)^{p-1}dx$$
$$= \frac{x^{m+1}(a+bx^n)^p + pna\int x^m(a+bx^n)^{p-1}dx}{np+m+1} \quad \ldots\ldots (B).$$

49. By the use of this formula, the proposed integral is made to depend upon a similar integral, but having the exponent of the ( ) diminished by unity.

50. When the exponents m and p are negative, and numerically large, it is generally convenient to increase them, so as to bring their values nearer to zero, and hence we require two additional formulæ, one for increasing the exponent of the variable exterior to the ( ), and the other, for increasing the exponent of the ( ).

51. *Prop.* To obtain a formula for increasing the exponent $-m$, of the parenthesis in the general integral

$$y = \int x^{-m}(a+bx^n)^p dx.$$

From formula (A), we obtain, by transposition and reduction,

$$\int x^{m-n}(a+bx^n)^p dx = \frac{x^{m-n+1}(a+bx^n)^{p+1} - b(np+m+1)\int x^m(a+bx^n)^p dx}{a(m-n+1)}.$$

Now making $m-n = -m_1$, or $m = n - m_1$, there results

$$\int x^{-m_1}(a+bx^n)^p dx$$
$$= \frac{x^{-m_1+1}(a+bx^n)^{p+1} - b(np+n-m_1+1)\int x^{-m_1+n}(a+bx^n)^p dx}{a(-m_1+1)},$$

or by omitting the subscript accents and reducing,

$$y = \int x^{-m}(a+bx^n)^p dx$$
$$= \frac{x^{-m+1}(a+bx^n)^{p+1} + b(m-np-n-1)\int x^{-m+n}(a+bx^n)^p dx}{-a(m-1)} \quad \ldots (C).$$

52. By the use of this formula the exponent $-m$ of x exterior to the ( ) is increased by n the exponent of x within the ( ).

53. *Prop.* To obtain a formula for increasing the exponent $-p$ of the ( ) in the general integral

$$y = \int x^m (a + bx^n)^{-p} dx.$$

From formula (B) we obtain, by transposition and reduction,

$$\int x^m (a+bx^n)^{p-1} dx = \frac{x^{m+1}(a+bx^n)^p - (np+m+1)\int x^m (a+bx^n)^p dx}{-pna}.$$

Now making $p - 1 = -p_1$ or $p = 1 - p_1$, there results

$$\int x^m (a + bx^n)^{-p_1} dx$$

$$= \frac{x^{m+1}(a + bx^n)^{-p_1+1} + (np_1 - n - m - 1)\int x^m (a + bx^n)^{-p_1+1} dx}{-na(-p_1 + 1)},$$

or by omitting accents and reducing

$$y = \int x^m (a + bx^n)^{-p} dx$$

$$= \frac{x^{m+1}(a+bx^n)^{-p+1} + (np-n-m-1)\int x^m (a+bx^n)^{-p+1} dx}{na(p - 1)} \cdots\cdot (D).$$

54. By the use of this formula, the exponent $-p$ of the ( ) is increased by unity.

Applications of Formulæ (A), (B), (C), *and* (D).

55. 1. To integrate $dy = \dfrac{x^m dx}{\sqrt{1 - x^2}}$ where m is an odd integer.

Put m successively equal to 1, 3, 5, 7, &c., and apply (A). Thus,

$$\int \frac{x dx}{\sqrt{1 - x^2}} = -\sqrt{1 - x^2} + C_1 \text{ by the rule for powers.}$$

$$\int \frac{x^3 dx}{\sqrt{1 - x^2}} = -\frac{1}{3} x^2 \sqrt{1 - x^2} + \frac{2}{3} \int \frac{x dx}{\sqrt{1 - x^2}}$$ by formula (A), in which $m = 3$, $n = 2$, and $p = -\frac{1}{2}$.

$$\int \frac{x^5 dx}{\sqrt{1-x^2}} = -\frac{1}{5}x^4\sqrt{1-x^2} + \frac{4}{5}\int \frac{x^3 dx}{\sqrt{1-x^2}}$$ by formula (A), in which $m = 5$, $n = 2$, and $p = -\frac{1}{2}$.

$$\int \frac{x^7 dx}{\sqrt{1-x^2}} = -\frac{1}{7}x^6\sqrt{1-x^2} + \frac{6}{7}\int \frac{x^5 dx}{\sqrt{1-x^2}};$$ and generally

$$\int \frac{x^m dx}{\sqrt{1-x^2}} = -\frac{1}{m}x^{m-1}\sqrt{1-x^2} + \frac{m-1}{m}\int \frac{x^{m-2} dx}{\sqrt{1-x^2}}.$$

Hence by substitution,

$$\int \frac{x^3 dx}{\sqrt{1-x^2}} = -\left(\frac{1}{3}x^2 + \frac{1.2}{1.3}\right)\sqrt{1-x^2} + C_3,$$

$$\int \frac{x^5 dx}{\sqrt{1-x^2}} = -\left(\frac{1}{5}x^4 + \frac{1.4}{3.5}x^2 + \frac{1.2.4}{1.3.5}\right)\sqrt{1-x^2} + C_5,$$

$$\int \frac{x^7 dx}{\sqrt{1-x^2}} = -\left(\frac{1}{7}x^6 + \frac{1.6}{5.7}x^4 + \frac{1.4.6}{3.5.7}x^2 + \frac{1.2.4.6}{1.3.5.7}\right)\sqrt{1-x^2} + C_7,$$

and generally

$$\int \frac{x^m dx}{\sqrt{1-x^2}} = -\left[\frac{1}{m}x^{m-1} + \frac{1.(m-1)}{(m-2).m}x^{m-3} + \frac{1.(m-3)(m-1)}{(m-4)(m-2)m}x^{m-5} + \&c.\right.$$

$$\left. \ldots + \frac{1.2.4.6\ldots(m-3)(m-1)}{1.3.5.7\ldots(m-2)(m)}\right]\sqrt{1-x^2} + C_m.$$

2. To integrate $dy = \frac{x^m dx}{\sqrt{1-x^2}}$, where m is an even integer.

Put $m = 0, 2, 4, 6$, &c., and apply (A) thus

$$\int \frac{dx}{\sqrt{1-x^2}} = \sin^{-1}x + C_0$$ by one of the circular forms.

$$\int \frac{x^2 dx}{\sqrt{1-x^2}} = -\frac{1}{2}x\sqrt{1-x^2} + \frac{1}{2}\int \frac{dx}{\sqrt{1-x^2}}$$ by formula (A), in which $m = 2$, $n = 2$, and $p = -\frac{1}{2}$.

$\int \frac{x^4 dx}{\sqrt{1-x^2}} = -\frac{1}{4} x^3 \sqrt{1-x^2} + \frac{3}{4} \int \frac{x^2 dx}{\sqrt{1-x^2}}$ by formula (A), in which $m = 4$, $n = 2$, and $p = -\frac{1}{2}$.

$\int \frac{x^6 dx}{\sqrt{1-x^2}} = -\frac{1}{6} x^5 \sqrt{1-x^2} + \frac{5}{6} \int \frac{x^4 dx}{\sqrt{1-x^2}}$; and generally

$$\int \frac{x^m dx}{\sqrt{1-x^2}} = -\frac{1}{m} x^{m-1} \sqrt{1-x^2} + \frac{m-1}{m} \int \frac{x^{m-2} dx}{\sqrt{1-x^2}}.$$

Hence by substitution,

$$\int \frac{x^2 dx}{\sqrt{1-x^2}} = -\frac{1}{2} x \sqrt{1-x^2} + \frac{1}{2} \sin^{-1} x + C_2.$$

$$\int \frac{x^4 dx}{\sqrt{1-x^2}} = -\left(\frac{1}{4} x^3 + \frac{1.3}{2.4} x\right) \sqrt{1-x^2} + \frac{1.3}{2.4} \sin^{-1} x + C_4.$$

$$\int \frac{x^6 dx}{\sqrt{1-x^2}} = -\left(\frac{1}{6} x^5 + \frac{1.5}{4.6} x^3 + \frac{1.3.5}{2.4.6} x\right) \sqrt{1-x^2} + \frac{1.3.5}{2.4.6} \sin^{-1} x + C_3,$$

and generally

$$\int \frac{x^m dx}{\sqrt{1-x^2}} = -\left(\frac{1}{m} x^{m-1} + \frac{1.(m-1)}{(m-2)m} x^{m-3} + \frac{1.(m-3)(m-1)}{(m-4)(m-2)m} x^{m-5} + \&c. \right.$$

$$\left. \dots + \frac{1.3.5.7 \dots (m-3)(m-1)}{2.4.6.8 \dots (m-2)m}\right) \sqrt{1-x^2}$$

$$+ \frac{1.3.5.7 \dots (m-3)(m-1)}{2.4.6.8 \dots (m-2).m} \sin^{-1} x + C_m.$$

3. $$dy = \frac{dx}{x^2 \sqrt{1+x}} = x^{-2}(1+x)^{-\frac{1}{2}} dx.$$

Make $m = +2$, $p = -\frac{1}{2}$, and $n = 1$, and apply (C): then

$$\int \frac{dx}{x^2 \sqrt{1+x}} = -\frac{x^{-1}(1+x)^{\frac{1}{2}}}{1.(2-1)} - \frac{\frac{1}{2} \int x^{-1}(1+x)^{-\frac{1}{2}} dx}{1.(2-1)}$$

$$= -\frac{\sqrt{1+x}}{x} - \frac{1}{2} \int \frac{dx}{x\sqrt{1+x}}.$$

Now put $1 + x = z^2$, then $x = z^2 - 1$, $dx = 2zdz$ and $\sqrt{1+x} = z$.

$$\therefore \int \frac{dx}{x\sqrt{1+x}} = \int \frac{2zdz}{(z^2-1)z} = \int \frac{2dz}{z^2-1} = \int \frac{dz}{z-1} - \int \frac{dz}{z+1}$$

$$= \log \frac{z-1}{z+1} + C = \log \frac{\sqrt{1+x}-1}{\sqrt{1+x}+1} + C.$$

$$\therefore y = -\frac{\sqrt{1+x}}{x} - \frac{1}{2} \log \frac{\sqrt{1+x}-1}{\sqrt{1+x}+1} + C_1.$$

4. $$dy = \frac{(a+bx)^{\frac{3}{2}}}{x} dx = x^{-1}(a+bx)^{\frac{3}{2}} dx.$$

Put $m = -1$, $n = 1$, and $p = \frac{3}{2}$, and apply (B); then

$$\int \frac{(a+bx)^{\frac{3}{2}} dx}{x} = \frac{x^0 (a+bx)^{\frac{3}{2}}}{\frac{3}{2} + 1 - 1} + \frac{\frac{3}{2}a}{\frac{3}{2}} \int x^{-1}(a+bx)^{\frac{1}{2}} dx.$$

Now put $m = -1$, $n = 1$, and $p = \frac{1}{2}$, and apply (B) to the last integral; thus

$$\int x^{-1}(a+bx)^{\frac{1}{2}} dx = \frac{x^0(a+bx)^{\frac{1}{2}}}{\frac{1}{2} + 1 - 1} + \frac{\frac{1}{2}a}{\frac{1}{2}} \int x^{-1}(a+bx)^{-\frac{1}{2}} dx.$$

Now put $a + bx = z^2$; then $x = \frac{z^2 - a}{b}$, $dx = \frac{2zdz}{b}$, and $\sqrt{a+bx} = z$.

$$\therefore \int x^{-1}(a+bx)^{-\frac{1}{2}} dx = \int \frac{2dz}{z^2 - a} = \int \frac{\frac{1}{\sqrt{a}} dz}{z - \sqrt{a}} - \int \frac{\frac{1}{\sqrt{a}} dz}{z + \sqrt{a}}$$

$$= \frac{1}{\sqrt{a}} \log \frac{z - \sqrt{a}}{z + \sqrt{a}} + C = \frac{1}{\sqrt{a}} \log \frac{\sqrt{a+bx} - \sqrt{a}}{\sqrt{a+bx} + \sqrt{a}} + C.$$

and by substitution,

$$\therefore\ y = \frac{2}{3}(a + bx)^{\frac{3}{2}} + 2a\,(a + bx)^{\frac{1}{2}} + a^{\frac{3}{2}} \log \frac{\sqrt{a + bx} - \sqrt{a}}{\sqrt{a + bx} + \sqrt{a}} + C_1.$$

5.
$$dy = \frac{dx}{x(1 + 2x)^{\frac{3}{2}}} = x^{-1}(1 + 2x)^{-\frac{3}{2}}dx.$$

Put $m = -1$, $n = 1$, $p = \frac{3}{2}$, and apply (D), then

$$\int \frac{dx}{x(1 + 2x)^{\frac{3}{2}}} = \frac{x^0(1 + 2x)^{-\frac{1}{2}}}{1\,.\left(\frac{3}{2} - 1\right)} - \frac{1 - 1 - \frac{3}{2} + 1}{\frac{1}{2}} \int x^{-1}(1 + 2x)^{-\frac{1}{2}}dx.$$

But $\int x^{-1}(1 + 2x)^{-\frac{1}{2}}dx = \log \frac{\sqrt{1 + 2x} - 1}{\sqrt{1 + 2x} + 1} + C$, by the last example.

$$\therefore\ y = \frac{2}{\sqrt{1 + 2x}} + \log \frac{\sqrt{1 + 2x} - 1}{\sqrt{1 + 2x} + 1} + C_1.$$

CHAPTER VII.

LOGARITHMIC AND EXPONENTIAL FUNCTIONS.

56. We shall now proceed to the integration of those forms which involve transcendental functions, beginning with the case of logarithmic functions.

57. Of the logarithmic forms, only a very limited number can be integrated, except by methods of approximation. The principal integrable forms will be examined.

58. *Prop.* To integrate the form $dy = X \,.\, \log^n x \,.\, dx$, in which X is a given algebraic function of x.

Put $Xdx = dv$, and $\log^n x = u$, and apply the formula for integration by parts. Thus

$$v = \int Xdx, \text{ and } du = n \,.\, \log^{n-1} x \,.\, \frac{dx}{x}, \text{ and since } \int udv = uv - \int vdu,$$

$$\therefore \int X \,.\, \log^n x \,.\, dx = \log^n x \,.\, \int Xdx - \int \left[n \,.\, \log^{n-1} x \,.\, \int (Xdx) \,.\, \frac{dx}{x} \right]$$

or, by making $\int Xdx = X_1$

$$y = X_1 \log^n x - n \int \frac{X_1}{x} \cdot \log^{n-1} x \,.\, dx.$$

If, therefore, it be possible to integrate the algebraic form Xdx, the proposed integral will depend upon another of the same general form, but having the exponent of the logarithm less by unity.

Now put $\int \frac{X_1}{x} dx = X_2$, and by a similar process, there will result

$$\int \frac{X_1}{x} \log^{n-1} x \,.\, dx = X_2 \log^{n-1} x - (n-1) \int \frac{X_2}{x} \log^{n-2} x \,.\, dx.$$

If n be a positive integer, the repeated application of this formula will cause the proposed integral to depend ultimately upon the algebraic form $\int \frac{X_n}{x} dx$, provided we can integrate Xdx, $\frac{X_1}{x} dx$, $\frac{X_2}{x} dx$, &c., obtaining in each an integral in the algebraic form.

Ex. To integrate $$dy = \frac{\log x \,.\, dx}{(1+x)^2}.$$

Here $$X = \frac{1}{(1+x)^2} = (1+x)^{-2}.$$

$$\therefore \int Xdx = \int (1+x)^{-2} dx = -\frac{1}{1+x} = X_1.$$

Also, $n = 1$, and $\therefore\ y = -\frac{\log x}{1+x} + \int \frac{dx}{x+x^2}.$

But $\int \frac{dx}{x+x^2} = \int \frac{dx}{x} - \int \frac{dx}{1+x} = \log x - \log(1+x) + C.$

$$\therefore y = -\frac{\log x}{1+x} + \log x - \log(1+x) + C = \frac{x}{1+x}\log x - \log(1+x) + C.$$

59. *Prop.* To integrate the form $dy = x^m \,.\, \log^n x \,.\, dx$, in which n is a positive integer.

Put $x^m = X$; then $X_1 = \int Xdx = \int x^m dx = \frac{x^{m+1}}{m+1}.$

And, therefore, by the last proposition,

$$y = \int x^m \,.\, \log^n x \,.\, dx = \frac{x^{m+1}}{m+1} \log^n x - \frac{n}{m+1} \int x^m \,.\, \log^{n-1} x \,.\, dx :$$

and similarly,

$$\int x^m \log^{n-1} x dx = \frac{x^{m+1}}{m+1} \log^{n-1} x - \frac{n-1}{m+1} \int x^m \log^{n-2} x dx.$$

$$\int x^m \log^{n-2} x dx = \frac{x^{m+1}}{m+1} \log^{n-2} x - \frac{n-2}{m+1} \int x^m \log^{n-3} x . dx.$$

&c. &c. &c.

Hence by successive substitutions,

$$y = \int x^m \log^n x dx = \frac{x^{m+1}}{m+1}\left[\log^n x - \frac{n}{m+1}\log^{n-1}x + \frac{n(n-1)}{(m+1)^2}\log^{n-2}x\right.$$
$$- \frac{n(n-1)(n-2)}{(m+1)^3}\log^{n-3}x + \&c.$$
$$\left. \ldots\ldots \pm \frac{n(n-1)(n-2)(n-3)\ldots . 3.2.1}{(m+1)^n}\right] + C.$$

Cor. This formula ceases to be applicable when $m = -1$, as the terms become infinite; but we then have

$$\int x^{-1}\log^n x \,.\, dx = \int \log^n x \cdot \frac{dx}{x} = \int \log^n x \,.\, d(\log x) = \frac{\log^{n+1}x}{n+1} + C.$$

Ex. To integrate $\quad dy = x^3 \,.\, \log^3 x \,.\, dx.$

Here $m = 3, \quad n = 3, \quad m + 1 = 4, \quad n - 1 = 2, \quad n - 2 = 1.$

$$\therefore y = \int x^3 . \log^3 x . dx = \frac{x^4}{4}\left[\log^3 x - \frac{3}{4}\log^2 x + \frac{3.2}{4^2}\log x - \frac{3.2.1}{4^3}\right] + C.$$

2. $$dy = \frac{\log^5 x}{x^{\frac{3}{2}}} \cdot dx = x^{-\frac{3}{2}} \log^5 x \,.\, dx.$$

Here $\quad m = -\frac{3}{2} \quad$ and $\quad n = 5.$

$$\therefore m + 1 = -\frac{1}{2}, \; n - 1 = 4, \; n - 2 = 3, \; n - 3 = 2, \; n - 4 = 1.$$

$$\therefore y = -\frac{2}{x^{\frac{1}{2}}}[\log^5 x + 5 \,.\, 2\log^4 x + 5 \,.\, 4 \,.\, 2^2\log^3 x$$
$$+ 5 \,.\, 4 \,.\, 3 \,.\, 2^3\log^2 x + 5 \,.\, 4 \,.\, 3 \,.\, 2 \,.\, 2^4\log x + 5 \,.\, 4 \,.\, 3 \,.\, 2 \,.\, 1 \,.\, 2^5] + C.$$

Remark. If we suppose n to be a positive fraction, the same formula will apply, but the series will not terminate.

60. *Prop*. To integrate the form $dy = \frac{x^m dx}{\log^n x}$, in which n is a positive integer.

Put $\quad x^{m+1} = u \quad$ and $\quad \frac{1}{\log^n x} \cdot \frac{dx}{x} = dv$, then

$$du = (m+1)x^m dx, \quad \text{and} \quad v = \frac{-1}{(n-1)\log^{n-1}x}.$$

Applying the formula $\int udv = uv - \int vdu$, we obtain

$$\int \frac{x^m dx}{\log^n x} = -\frac{x^{m+1}}{(n-1)\log^{n-1}x} + \frac{m+1}{n-1}\int \frac{x^m dx}{\log^{n-1}x}, \text{ and similarly}$$

$$\int \frac{x^m dx}{\log^{n-1}x} = -\frac{x^{m+1}}{(n-2)\log^{n-2}x} + \frac{m+1}{n-2}\int \frac{x^m dx}{\log^{n-2}x}.$$

$$\int \frac{x^m dx}{\log^{n-2}x} = -\frac{x^{m+1}}{(n-3)\log^{n-3}x} + \frac{m+1}{n-3}\int \frac{x^m dx}{\log^{n-3}x}. \ \&c. \&c. \&c.$$

$$\therefore y = \int \frac{x^m dx}{\log^n x} = -\frac{x^{m+1}}{n-1}\Big[\frac{1}{\log^{n-1}x} + \frac{m+1}{n-2}\cdot\frac{1}{\log^{n-2}x}$$

$$+ \frac{(m+1)^2}{(n-2)(n-3)}\cdot\frac{1}{\log^{n-3}x}, \ \&c.$$

$$\ldots . + \frac{(m+1)^{n-2}}{(n-2)(n-3)(n-4)\ldots 3.2.1}\cdot\frac{1}{\log x}\Big]$$

$$+ \frac{(m+1)^{n-1}}{(n-1)(n-2)(n-3)\ldots 3.2.1}\int \frac{x^m dx}{\log x}.$$

The last integral admits of only an approximate determination, but its form may be simplified; thus,

put $z = x^{m+1}$, then $dz = (m+1)x^m dx$, and $(m+1)\log x = \log z$.

$$\therefore \int \frac{x^m dx}{\log x} = \int \frac{dz}{\log z}.$$

This, also, can only be integrated approximately by expanding the expression under the sign of integration into a series, and then integrating the terms separately, a method which will be considered more at length in a future chapter.

3. To integrate approximately $dy = \dfrac{x^4 dx}{\log^3 x}$.

Here $m = 4$ and $n = 3$, $\therefore m + 1 = 5$, $n - 1 = 2$, $n - 2 = 1$.

$$\therefore y = \int \frac{x^4 dx}{\log^3 x} = -\frac{x^5}{2}\Big[\frac{1}{\log^2 x} + \frac{5}{1}\cdot\frac{1}{\log x}\Big] + \frac{25}{2}\int \frac{x^4 dx}{\log x}.$$

Now put $x^5 = z$, then $\displaystyle\int \frac{x^4 dx}{\log x} = \int \frac{dz}{\log z}$;

and, making $\log z = t$, we have $z = e^t$, $dz = e^t . dt$.

$$\therefore \int \frac{dz}{\log x} = \int \frac{e^t dt}{t} = \int \left(1 + \frac{t}{1} + \frac{t^2}{1.2} + \frac{t^3}{1.2.3} + \&c.\right)\frac{dt}{t}$$

$$= \log t + t + \frac{t^2}{1.2^2} + \frac{t^3}{1.2.3^2} + \&c.$$

$$= \log [\log x^5] + \log x^5 + \frac{1}{1.2^2}\log^2 x^5 + \frac{1}{1.2.3^2}\log^3 x^5 + \&c.$$

$$\therefore y = -\frac{x^5}{2\log^2 x} - \frac{5x^5}{2\log x} + \frac{25}{2}[\log(\log x^5) + \log x^5 + \frac{1}{1.2^2}\log^2 x^5$$

$$+ \frac{1}{1.2.3^2}\log^3 x^5 + \&c.].$$

Exponential Functions.

61. To integrate the form $dy = a^x . x^m . dx$, when m is a positive integer.

Put $a^x dx = dv$, and $x^m = u$; then $v = \frac{1}{\log a} \cdot a^x$ and $du = mx^{m-1}dx$.

$$\therefore \int a^x . x^m . dx = \frac{a^x . x^m}{\log a} - \frac{m}{\log a}\int a^x . x^{m-1}dx, \text{ and similarly}$$

$$\int a^x . x^{m-1}dx = \frac{a^x . x^{m-1}}{\log a} - \frac{m-1}{\log a}\int a^x . x^{m-2}dx$$

$$\int a^x . x^{m-2}dx = \frac{a^x x^{m-2}}{\log a} - \frac{m-2}{\log a}\int a^x . x^{m-3}dx, \ \&c.\ \&c.\ \&c.$$

Hence, by substitution,

$$y = \int a^x . x^m . dx = \frac{a^x}{\log a}\left[x^m - \frac{mx^{m-1}}{\log a}\right.$$

$$+ \frac{m(m-1)x^{m-2}}{\log^2 a} - \frac{m(m-1)(m-2)x^{m-3}}{\log^3 a} + \&c.$$

$$\left. \ldots\ldots \pm \frac{m(m-1)\ldots 2.1}{\log^m a}\right] + C.$$

62. *Prop.* To integrate $dy = \frac{a^x}{x^m}dx$, when m is a positive integer.

Put $x^{-m}dx = dv$ and $a^x = u$; then $v = -\frac{x^{-m+1}}{m-1}$ and $du = \log a \,.\, a^x dx$.

$$\therefore \int \frac{a^x dx}{x^m} = -\frac{a^x}{(m-1)x^{m-1}} + \frac{\log a}{m-1}\int \frac{a^x dx}{x^{m-1}}, \text{ and similarly,}$$

$$\int \frac{a^x dx}{x^{m-1}} = -\frac{a^x}{(m-2)x^{m-2}} + \frac{\log a}{m-2} \cdot \int \frac{a^x dx}{x^{m-2}}$$

$$\int \frac{a^x dx}{x^{m-2}} = -\frac{a^x}{(m-3)x^{m-3}} + \frac{\log a}{m-3} \cdot \int \frac{a^x dx}{x^{m-3}}, \text{ \&c. \&c. \&c.}$$

Hence, by substitution,

$$y = \int \frac{a^x dx}{x^m} = -\frac{a^x}{(m-1)x^{m-1}}\left[1 + \frac{\log a}{m-2}x + \frac{\log^2 a}{(m-2)(m-3)}x^2 + \text{\&c.}\right.$$

$$\left. \ldots + \frac{\log^{m-2} a}{(m-2)(m-3)\ldots 2 \,.\, 1}x^{m-2}\right]$$

$$+ \frac{\log^{m-1} a}{(m-1)(m-2)\ldots 2 \,.\, 1}\int \frac{a^x dx}{x}.$$

The last integral can only be found approximately.

1. To integrate $dy = a^x \,.\, x^3 dx.$

Here $m = 3, \quad m - 1 = 2, \quad m - 2 = 1.$ Hence

$$y = \frac{a^x}{\log a}\left[x^3 - \frac{3x^2}{\log a} + \frac{6x}{\log^2 a} - \frac{6}{\log^3 a}\right] + C$$

2. To integrate $dy = e^x \,.\, x^4 \,.\, dx.$

Here $m = 4, \quad m - 1 = 3, \quad m - 2 = 2, \quad m - 3 = 1, \log e = 1.$

$$\therefore y = e^x(x^4 - 4x^3 + 12x^2 - 24x + 24) + C.$$

3. $dy = e^{-x}x^2 dx = e^{-x}(-x)^2 dx = -e^{-x}(-x)^2 d(-x).$

Here $m = 2$, $m - 1 = 1$, $\log e = 1$, and x in the general formula is to be replaced by $-x$.

$$\therefore y = \int e^{-x} \,.\, x^2 \,.\, dx = -e^{-x}(x^2 + 2x + 2) + C.$$

4. $$dy = \frac{a^x}{x^4}\,dx.$$

Here $$m = 4, \quad m - 1 = 3, \quad \&c.$$

$$\therefore y = \int \frac{a^x}{x^4}\,dx = -\frac{a^x}{3x^3}\left[1 + \frac{\log a}{2}x + \frac{\log^2 a}{2}x^2\right] + \frac{\log^3 a}{6}\int \frac{a^x}{x}\,dx,$$

the last integral being found, approximately, as in a previous example, by expanding a^x.

5. To integrate $$dy = \frac{1 + x^2}{(1 + x)^2}\,e^x\,.\,dx.$$

Put $$1 + x = z.$$

$$\therefore x = z - 1, \quad 1 + x^2 = 1 + z^2 - 2z + 1 = z^2 - 2z + 2, \quad dx = dz, \quad e^x = e^{z-1}.$$

$$\therefore y = \int \frac{(z^2 - 2z + 2)e^{z-1}dz}{z^2} = \frac{1}{e}\int\left(1 - \frac{2}{z} + \frac{2}{z^2}\right)e^z\,.\,dz$$

$$= \frac{1}{e}\left[e^z - 2\int \frac{e^z}{z}\,dz + 2\int \frac{e^z}{z^2}\,dz\right],$$

or by integrating the last term by parts

$$y = \frac{1}{e}\left[e^z - 2\int \frac{e^z}{z}\,dz - 2\frac{e^z}{z} + 2\int \frac{e^z}{z}\,dz\right] = e^{z-1} - 2\frac{e^{z-1}}{z} + C$$

$$= e^x\left(1 - \frac{2}{1 + x}\right) + C = e^x\left(\frac{x - 1}{1 + x}\right) + C.$$

CHAPTER VIII.

TRIGONOMETRICAL AND CIRCULAR FUNCTIONS.

63. Since the tangent, cotangent, secant, cosecant. versed-sine. and coversed-sine, can all be expressed rationally in terms of the sine and cosine, it will only be necessary to investigate formulæ for the integration of expressions involving sines and cosines.

64. *Prop.* To obtain a formula for diminishing the exponent m of $\sin x$, in the general integral

$$y = \int \sin^m x \,.\, \cos^n x \,.\, dx, \quad \text{when } m \text{ is an integer.}$$

Put $\qquad \cos^n x \,.\, \sin x \,.\, dx = dv, \quad \text{and} \quad \sin^{m-1} x = u;$

then $\quad v = -\dfrac{\cos^{n+1} x}{n+1}, \quad \text{and} \quad du = (m-1)\sin^{m-2} x \,.\, \cos x \,.\, dx.$

and by the formula for integration by parts

$$y = \int \sin^m x \,.\, \cos^n x \,.\, dx = -\frac{\sin^{m-1} x \,.\, \cos^{n+1} x}{n+1} + \frac{m-1}{n+1}\int \sin^{m-2} x \,.\, \cos^{n+2} x \,.\, dx.$$

But $\qquad \cos^{n+2} x = \cos^2 x \,.\, \cos^n x = (1 - \sin^2 x)\cos^n x.$

$$\therefore y = -\frac{\sin^{m-1} x \,.\, \cos^{n+1} x}{n+1}$$

$$+ \frac{m-1}{n+1}\int \sin^{m-2} x \,.\, \cos^n x \,.\, dx - \frac{m-1}{n+1}\int \sin^m x \,.\, \cos^n x \,.\, dx.$$

Transposing the last term and reducing, we obtain

$$\int \sin^m x \,.\, \cos^n x \,.\, dx = -\frac{\sin^{m-1} x \,.\, \cos^{n+1} x}{m+n} + \frac{m-1}{m+n}\int \sin^{m-2} x \,.\, \cos^n x \,.\, dx.$$

And similarly,

$$\int \sin^{m-2}x \, . \cos^n x \, . \, dx = -\frac{\sin^{m-3}x \, . \cos^{n+1}x}{m+n-2} + \frac{m-3}{m+n-2}\int \sin^{m-4}x . \cos^n x . dx,$$

$$\int \sin^{m-4}x \, . \cos^n x \, . \, dx = -\frac{\sin^{m-5}x \, . \cos^{n+1}x}{m+n-4} + \frac{m-5}{m+n-4}\int \sin^{m-6}x . \cos^n x \, dx,$$

&c. &c. &c.

Hence by successive substitutions,

$$y = \int \sin^m x \, . \cos^n x \, . \, dx = -\frac{\cos^{n+1}x}{m+n}\Big[\sin^{m-1}x + \frac{m-1}{m+n-2}\sin^{m-3}x$$

$$+ \frac{(m-1)(m-3)}{(m+n-2)(m+n-4)}\sin^{m-5}x + \text{\&c.}$$

$$\ldots\ldots + \frac{(m-1)(m-3)(m-5)\ldots\ldots 4 \text{ or } 3}{(m+n-2)(m+n-4)(m+n-6)\ldots(n+3) \text{ or } (n+2)}$$

$$\times \sin^2 x \quad \text{or} \quad \sin x\Big]$$

$$+ \frac{(m-1)(m-3)(m-5)\ldots\ldots 2 \text{ or } 1}{(m+n)(m+n-2)(m+n-4)\ldots(n+3) \text{ or } (n+2)}$$

$$\times \int \sin x \, . \cos^n x \, . \, dx \quad \text{or} \quad \int \cos^n x \, . \, dx \quad \ldots\ldots (E).$$

65. This formula renders the proposed integral dependent upon that of the form

$$\sin x \, . \cos^n x \, . \, dx \quad \text{or} \quad \cos^n x \, . \, dx,$$

according as m is odd or even, the effect of the formula first obtained being to diminish by 2 the exponent m of $\sin x$, at each application.

Also the first of these two final forms is immediately integrable by the rule for powers: for

$$\int \cos^n x \, . \sin x \, . \, dx = -\int \cos^n x \, . \, d(\cos x) = -\frac{\cos^{n+1}x}{n+1} + C.$$

Hence we have only to obtain a formula for the integration of the form $\cos^n x \, . \, dx$, in order to effect the complete integration of the proposed differential.

66. *Prop.* To integrate the form $dy = \cos^n x \,.\, dx$ where n is an integer.

Put $\cos x \,.\, dx = dv,$ and $\cos^{n-1} x = u$;

then $v = \sin x$ and $du = -(n-1)\cos^{n-2} x \sin x \, dx.$

Hence by substitution in the formula $\int u dv = uv - \int v du$, we obtain

$$\int \cos^n x \,.\, dx = \sin x \,.\, \cos^{n-1} x + (n-1)\int \cos^{n-2} x \,.\, \sin^2 x \,.\, dx$$

$$= \sin x \,.\, \cos^{n-1} x + (n-1)\int \cos^{n-2} x (1-\cos^2 x) dx$$

$$= \sin x \,.\, \cos^{n-1} x + (n-1)\int \cos^{n-2} x \,.\, dx - (n-1)\int \cos^n x \,.\, dx.$$

Transposing the last term and reducing, we get

$$\int \cos^n x \,.\, dx = \frac{\sin x \,.\, \cos^{n-1} x}{n} + \frac{n-1}{n}\int \cos^{n-2} x \,.\, dx, \text{ and similarly}$$

$$\int \cos^{n-2} x \,.\, dx = \frac{\sin x \,.\, \cos^{n-3} x}{n-2} + \frac{n-3}{n-2}\int \cos^{n-4} x . dx.$$

$$\int \cos^{n-4} x \,.\, dx = \frac{\sin x \,.\, \cos^{n-5} x}{n-4} + \frac{n-5}{n-4}\int \cos^{n-6} x . dx.$$

&c. &c. &c.

Hence by successive substitutions,

$$y = \int \cos^n x \,.\, dx = \frac{\sin x}{n}\Big[\cos^{n-1} x + \frac{n-1}{n-2}\cos^{n-3} x$$

$$+ \frac{(n-1)(n-3)}{(n-2)(n-4)}\cos^{n-5} x + \&c.$$

$$\ldots\ldots + \frac{(n-1)(n-3)(n-5)\ldots 4 \text{ or } 3}{(n-2)(n-4)(n-6)\ldots 3 \text{ or } 2}\cos^2 x \text{ or } \cos x\Big]$$

$$+ \frac{(n-1)(n-3)(n-5)\ldots 2 \text{ or } 1}{n(n-2)(n-4)\ldots\ldots 3 \text{ or } 2}\int \cos x \,.\, dx \text{ or } \int dx. \;. \;(F).$$

This formula renders the proposed form dependent upon one of two known forms, viz.;

$$\int \cos x \,.\, dx = \sin x + C, \quad \text{when } n \text{ is odd,}$$

or, $$\int dx = x + C, \quad \text{when } n \text{ is even.}$$

67. The two propositions just given effect the complete integration of $\sin^m x \, . \cos^n x \, . \, dx$, when m and n are integers, by first diminishing the exponent m of the sine, and then the exponent n of the cosine. But it is often preferable to reduce n first, and for this purpose we require the following proposition.

68. *Prop.* To integrate $dy = \sin^m x \, . \cos^n x \, . \, dx$, by first diminishing the exponent n of the cosine.

If in the formula (E), we make $x = \frac{1}{2}\pi - x_1$, $m = n_1$, and $n = m_1$, then

$$\sin x = \cos x_1, \quad \cos x = \sin x_1, \quad dx = -dx_1,$$

and by substitution we shall obtain

$$\int \sin^m x \, . \cos^n x \, . \, dx = -\int \cos^{n_1} x_1 \, . \sin^{m_1} x_1 dx_1$$

$$= -\frac{\sin^{m_1+1} x_1}{n_1 + m_1} [\cos^{n_1-1} x_1 + \&c.],$$

or by omitting the accents and changing signs,

$$\int \sin^m x \, . \cos^n x \, . \, dx = \frac{\sin^{m+1} x}{n+m} [\cos^{n-1} x + \frac{n-1}{n+m-2} \cos^{n-3} x$$

$$+ \frac{(n-1)(n-3)}{(n+m-2)(n+m-4)} \cos^{n-5} + \&c.$$

$$\ldots\ldots + \frac{(n-1)(n-3)(n-5) \ldots\ldots . 4 \text{ or } 3}{(n+m-2)(n+m-4)(n+m-6) \ldots (m+3) \text{ or } (m+2)}$$

$$\times \cos^2 x \text{ or } \cos x]$$

$$+ \frac{(n-1)(n-3)(n-5) \ldots\ldots 2 \text{ or } 1}{(n+m)(n+m-2)(n+m-4) \ldots (m+3) \text{ or } (m+2)}$$

$$\times \int \cos x \, . \sin^m x \, . \, dx \text{ or } \int \sin^m x \, . \, dx. \; \ldots \; (G).$$

But $\int \cos x \, . \sin^m x \, . \, dx = \int \sin^m x \, . \, d(\sin x) = \frac{\sin^{m+1} x}{m+1} + C.$

which will be the required form when n is odd.

We have therefore only to provide a formula for the integration of the form $\sin^m x \, . \, dx$, which will be necessary when n is even.

This may be readily effected by substituting in formula (F), m for n, and $\frac{1}{2}\pi - x$ for x, and changing the signs. Thus

$$\int \sin^m x \, . \, dx = -\frac{\cos x}{m}\left[\sin^{m-1}x + \frac{m-1}{m-2}\sin^{m-3}x \right.$$

$$+ \frac{(m-1)(m-3)}{(m-2)(m-4)}\sin^{m-5}x + \&c.$$

$$\left. \ldots + \frac{(m-1)(m-3)(m-5)\ldots 4 \text{ or } 3}{(m-2)(m-4)(m-6)\ldots 3 \text{ or } 2}\sin^2 x \text{ or } \sin x\right]$$

$$+ \frac{(m-1)(m-3)(m-5)\ldots 2 \text{ or } 1}{m(m-2)(m-4)\ldots 3 \text{ or } 2}\int \sin x dx \text{ or } \int dx. \ldots (H).$$

69. The formulæ (G) and (H) effect the same object as (E) and (F), reducing the integral $\int \sin^m x \, . \cos^n x \, . \, dx$ to one of the known forms

$$\int dx = x + C, \quad \int \cos x \, . \, dx = \sin x + C, \text{ or, } \int \sin x \, . \, dx = -\cos x + C,$$

the exponent m or n which is first reduced being an even integer, and the other exponent an even or odd integer.

But if m be odd, (E) alone will effect the integration, whether n be an integer or fraction; and similarly, if n be odd, (G), alone will suffice.

70. *Prop.* To integrate the forms $\frac{\sin^m x}{\cos^n x}dx$, and $\frac{\cos^n x}{\sin^m x}dx$, where m and n are integers.

By the formula (E) the first of these forms may be reduced to $\frac{dx}{\cos^n x}$, or $\frac{\sin x \, . \, dx}{\cos^n x}$, and by (G), the second may be reduced to $\frac{dx}{\sin^m x}$, or $\frac{\cos x \, . \, dx}{\sin^m x}$.

But $\int \frac{\sin x \, . \, dx}{\cos^n x} = \frac{\cos^{-n+1}x}{n-1} + C$, and $\int \frac{\cos x \, . \, dx}{\sin^m x} = -\frac{\sin^{-m+1}x}{m-1} + C$.

Hence there will remain to be integrated the forms

$$\cos^{-n}x \, . \, dx \ldots\ldots (1), \quad \text{and} \quad \sin^{-m}x \, . \, dx \ldots\ldots (2).$$

Put in (1), $\cos x \,.\, dx = dv$, and $\cos^{-n-1} x = u$,

then $v = \sin x$, and $du = (n+1)\cos^{-n-2} x \,.\, \sin x \,.\, dx$,

and by substitution in the formula for integration by parts,

$$\int \cos^{-n} x \,.\, dx = \sin x \,.\, \cos^{-n-1} x - (n+1) \int \sin^2 x \,.\, \cos^{-n-2} x \,.\, dx$$

$$= \sin x \,.\, \cos^{-n-1} x - (n+1) \int \cos^{-n-2} x \,.\, dx + (n+1) \int \cos^{-n} x \,.\, dx.$$

Transposing and reducing, we get

$$\int \frac{dx}{\cos^{n+2} x} = \frac{\sin x}{(n+1)\cos^{n+1} x} + \frac{n}{n+1} \int \frac{dx}{\cos^n x};$$ and by analogy

$$\int \frac{dx}{\cos^n x} = \frac{\sin x}{(n-1)\cos^{n-1} x} + \frac{n-2}{n-1} \int \frac{dx}{\cos^{n-2} x},$$ and similarly

$$\int \frac{dx}{\cos^{n-2} x} = \frac{\sin x}{(n-3)\cos^{n-3} x} + \frac{n-4}{n-3} \int \frac{dx}{\cos^{n-4} x}.$$

&c. &c. &c. Hence by substitution

$$y = \int \frac{dx}{\cos^n x} = \frac{\sin x}{n-1}\left[\frac{1}{\cos^{n-1} x} + \frac{n-2}{(n-3)\cos^{n-3} x} + \frac{(n-2)(n-4)}{(n-3)(n-5)\cos^{n-5} x} + \text{\&c.} \right.$$

$$\left. \ldots\ldots + \frac{(n-2)(n-4)(n-6)\ldots\ldots 3 \text{ or } 2}{(n-3)(n-5)(n-7)\ldots\ldots 2 \text{ or } 1 \,.\, \cos^2 x \text{ or } \cos x}\right]$$

$$+ \frac{(n-2)(n-4)(n-6)\ldots. 1 \text{ or } 0}{(n-1)(n-3)(n-5)\ldots. 2 \text{ or } 1} \int \frac{dx}{\cos x} \text{ or } \int dx. \ldots (I).$$

The second of these integrals, $\int dx = x + C$, will never be required, because its coefficient is zero, and therefore we stop at the preceding term. For the first we have

$$\int \frac{dx}{\cos x} = \int \frac{\cos x \,.\, dx}{\cos^2 x} = \int \frac{\cos x \,.\, dx}{1 - \sin^2 x} = \frac{1}{2} \int \frac{\cos x \,.\, dx}{1 + \sin x} + \frac{1}{2} \int \frac{\cos x \,.\, dx}{1 - \sin x}$$

$$= \frac{1}{2} \log(1 + \sin x) - \frac{1}{2} \log(1 - \sin x) + C = \log \left[\frac{1 + \sin x}{1 - \sin x}\right]^{\frac{1}{2}} + C$$

$$= \log \left[\frac{2 \sin\left(\frac{1}{4}\pi + \frac{1}{2}x\right)\cos\left(\frac{1}{4}\pi - \frac{1}{2}x\right)}{2 \sin\left(\frac{1}{4}\pi - \frac{1}{2}x\right)\cos\left(\frac{1}{4}\pi + \frac{1}{2}x\right)}\right]^{\frac{1}{2}} + C = \log \tan\left(\frac{1}{4}\pi + \frac{1}{2}x\right) + C \qquad (I_1).$$

71. To obtain a formula for $z = \int \frac{dx}{\sin^m x}$, replace in (I), n by m, x by $\frac{1}{2}\pi - x$, and y by z. Then

$$z = \int \frac{dx}{\sin^m x} = -\frac{\cos x}{m-1}\left[\frac{1}{\sin^{m-1}x} + \frac{m-2}{(m-3)\sin^{m-3}x} + \frac{(m-2)(m-4)}{(m-3)(m-5)\sin^{m-5}x} + \text{\&c.} \right.$$

$$\left. \ldots + \frac{(m-2)(m-4)(m-6)\ldots 3 \text{ or } 2}{(m-3)(m-5)(m-7)\ldots 2 \text{ or } 1 \,.\, \sin^2 x \text{ or } \sin x}\right]$$

$$+ \frac{(m-2)(m-4)(m-6)\ldots 1 \text{ or } 0}{(m-1)(m-3)(m-5)\ldots 2 \text{ or } 1}\int \frac{dx}{\sin x} \text{ or } \int dx \ldots . (K).$$

The second integral has a coefficient equal to zero, and therefore will never be used. For the first we have, by replacing x by $\frac{1}{2}\pi - x$ in (I_1), and changing signs

$$\int \frac{dx}{\sin x} = -\log \cot \frac{1}{2}x = \log \frac{1}{\cot \frac{1}{2}x} = \log \tan \frac{1}{2}x + C.$$

72. *Prop.* To integrate $dy = \frac{dx}{\sin^m x \,.\, \cos^n x}$ where m and n are integers.

Since $$\sin^2 x + \cos^2 x = 1.$$

$$y = \int \frac{dx}{\sin^m x \,.\, \cos^n x} = \int \frac{(\sin^2 x + \cos^2 x)dx}{\sin^m x \,.\, \cos^n x}$$

$$= \int \frac{dx}{\sin^{m-2}x \,.\, \cos^n x} + \int \frac{dx}{\sin^m x \,.\, \cos^{n-2}x}$$

$$= \int \frac{(\sin^2 x + \cos^2 x)dx}{\sin^{m-2}x \,.\, \cos^n x} + \int \frac{(\sin^2 x + \cos^2 x)dx}{\sin^m x \,.\, \cos^{n-2}x}$$

$$= \int \frac{dx}{\sin^{m-4}x \,.\, \cos^n x} + \int \frac{2dx}{\sin^{m-2}x \,.\, \cos^{n-2}x} + \int \frac{dx}{\sin^m x \,.\, \cos^{n-4}x};$$

and by continuing to introduce the factor

$$\sin^2 x + \cos^2 x = 1,$$

we obtain finally one or more of the following known forms

$$\int \frac{dx}{\sin^m x}, \int \frac{dx}{\cos^n x}, \int \frac{\sin x . dx}{\cos^n x}, \int \frac{\cos x . dx}{\sin^m x}, \int \frac{\sin x . dx}{\cos x}, \int \frac{\cos x . dx}{\sin x}.$$

Applications of Formulæ (E), (F), (G), (H), (I,) *and* (K).

73. 1. To integrate $dy = \sin^5 x . \cos^5 x . dx.$

Here $m = 5$, and $n = 5$, and since both are odd we may apply (E) or (G) with equal advantage. Employing (E) we have

$$y = -\frac{\cos^6 x}{10}[\sin^4 x + \frac{4}{8}\sin^2 x] + \frac{4.2}{10.8}\int \sin x . \cos^5 x . dx$$

$$= -\frac{\cos^6 x}{10}[\sin^4 x + \frac{1}{2}\sin^2 x] - \frac{1}{10.6}\cos^6 x + C$$

$$= -\frac{\cos^6 x}{10}\left[\sin^4 x + \frac{1}{2}\sin^2 x + \frac{1}{6}\right] + C.$$

2. $$dy = \sin^6 x . \cos^3 x . dx.$$

Here $m = 6$, $n = 3$, and since n is odd we apply (G).

$$\therefore y = \frac{\sin^7 x}{9}[\cos^2 x] + \frac{2}{9}\int \cos x . \sin^6 x . dx = \frac{\sin^7 x}{9}\left(\cos^2 x + \frac{2}{7}\right) + C.$$

3. $$dy = \sin^6 x . dx.$$

In (H) make $m = 6$.

$$\therefore y = -\frac{\cos x}{6}[\sin^5 x + \frac{5}{4}\sin^3 x + \frac{5.3}{4.2}\sin x] + \frac{5.3.1}{6.4.2}x + C.$$

4. $$dy = \sin^8 x . \cos^6 x . dx.$$

In (E) make $m = 8$ and $n = 6$.

$$y = -\frac{\cos^7 x}{14}[\sin^7 x + \frac{7}{12}\sin^5 x + \frac{7.5}{12.10}\sin^3 x + \frac{7.5.3}{12.10.8}\sin x]$$
$$+ \frac{7.5.3.1}{14.12.10.8}\int \cos^6 x . dx,$$

and by applying (F) to the last term, we get

$$y = -\frac{\cos^7 x}{14}\left[\sin^7 x + \frac{7}{12}\sin^5 x + \frac{7}{24}\sin^3 x + \frac{7}{64}\sin x\right]$$

$$+\frac{\sin x}{768}\left[\cos^5 x + \frac{5}{4}\cos^3 x + \frac{15}{8}\cos x\right] + \frac{15x}{6144} + C.$$

5. $$dy = \frac{\sin^5 x}{\cos^2 x}\,dx.$$

In (E) make $m = 5$ and $n = -2$. Then

$$y = \frac{-1}{3\cos x}\left[\sin^4 x + \frac{4}{1}\sin^2 x\right] + \frac{4.2}{3.1}\int\frac{\sin x\,.\,dx}{\cos^2 x}$$

$$= \frac{-1}{3\cos x}\left[\sin^4 x + 4\sin^2 x - 8\right] + C.$$

6. $$dy = \frac{dx}{\sin^5 x}.$$

In (K) make $m = 5$. Then

$$y = -\frac{\cos x}{4}\left[\frac{1}{\sin^4 x} + \frac{3}{2\sin^2 x}\right] + \frac{3.1}{4.2}\int\frac{dx}{\sin x}$$

$$= -\frac{\cos x}{4}\left[\frac{1}{\sin^4 x} + \frac{3}{2\sin^2 x}\right] + \frac{3}{8}\log\tan\frac{x}{2} + C.$$

7. $$dy = \frac{dy}{\cos^6 x}.$$

In (I) make $n = 6$. Then

$$y = \frac{\sin x}{5}\left[\frac{1}{\cos^5 x} + \frac{4}{3}\cdot\frac{1}{\cos^3 x} + \frac{4}{3}\cdot\frac{2}{1}\frac{1}{\cos x}\right] + C.$$

8. $$dy = \frac{dx}{\sin^4 x\,.\,\cos^2 x}.$$

Introducing the factor $\sin^2 x + \cos^2 x$, we obtain

$$y = \int\frac{(\sin^2 x + \cos^2 x)\,dx}{\sin^4 x\,.\,\cos^2 x} = \int\frac{dx}{\sin^2 x\,.\,\cos^2 x} + \int\frac{dx}{\sin^4 x}$$

$$= \int\frac{dx}{\cos^2 x} + \int\frac{dx}{\sin^2 x} + \int\frac{dx}{\sin^4 x}$$

$$= \tan x - \cot x - \frac{\cos x}{3}\left[\frac{1}{\sin^3 x} + \frac{2}{\sin x}\right] + C.$$

74. When $m = -n$, formulæ (E) and (G) cease to be applicable, but we then have

$$y = \int \frac{\sin^n x}{\cos^n x} dx = \int \tan^n x \, . \, dx \quad \text{or} \quad y = \int \frac{\cos^n x}{\sin^n x} dx = \cot^n x \, dx.$$

To integrate the first of these expressions, put $\sec^2 - 1$ for $\tan^2$ and in the second put $\operatorname{cosec}^2 - 1$ for $\cot^2$. Thus

$$\int \tan^n x \, . \, dx = \int \tan^2 x \, . \tan^{n-2} x \, . \, dx = \int \sec^2 x \, . \tan^{n-2} x \, . \, dx - \int \tan^{n-2} x \, . dx$$

$$= \frac{1}{n-1} \tan^{n-1} x - \int \tan^{n-2} x \, . \, dx$$

$$= \frac{1}{n-1} \tan^{n-1} x - \int (\sec^2 x - 1) \tan^{n-4} x \, . \, dx$$

$$= \frac{1}{n-1} \tan^{n-1} x - \frac{1}{n-3} \tan^{n-3} x + \int \tan^{n-4} x \, . \, dx$$

$$= \frac{1}{n-1} \tan^{n-1} x - \frac{1}{n-3} \tan^{n-3} x + \frac{1}{n-5} \tan^{n-5} x - \text{\&c.},$$

the last term being

$$\int \tan x \, . \, dx = \int \frac{\sin x \, . \, dx}{\cos x} = -\log \cos x + C = \log \sec x + C$$

when n is odd or $\int dx = x + C$ when n is even.

Similarly, $$\int \frac{\cos^n x dx}{\sin^n x} = \int \cot^n x \, . \, dx$$

$$= -\frac{\cot^{n-1} x}{n-1} + \frac{\cot^{n-3} x}{n-3} - \frac{\cot^{n-5} x}{n-5} + \text{\&c.}$$

The last term being $\int \cot x \, . \, dx = \log \sin x + C$, or $\int dx = x + C$.

75. When the proposed form is $\int \sin^m x \, . \cos^n x dx$, in which m and n are integers, the integration may be conveniently effected by converting the product $\sin^m x \, . \cos^n x$ into a series of terms involving sines or cosines of multiples of x. The integration can then be performed without introducing powers of the sines or cosines.

The proposed transformation can always be accomplished by the repeated application of one or more of the three trigonometrical formulæ.

$$\sin a \cos b = \frac{1}{2}\sin(a+b) + \frac{1}{2}\sin(a-b),$$

$$\sin a \sin b = \frac{1}{2}\cos(a-b) - \frac{1}{2}\cos(a+b),$$

$$\cos a \cos b = \frac{1}{2}\cos(a-b) + \frac{1}{2}\cos(a+b).$$

To illustrate this process take the following example.

$$dy = \sin^3 x \,.\, \cos^2 x dx$$

$$\sin^3 x \,.\, \cos^2 x = \sin x\,(\sin x \,.\, \cos x)^2 = \sin x \left(\frac{\sin 2x}{2}\right)^2$$

$$= \frac{1}{4}\sin x\left(\frac{1-\cos 4x}{2}\right) = \frac{1}{8}\sin x - \frac{1}{8}\sin x \,.\, \cos 4x$$

$$= \frac{1}{8}\sin x + \frac{1}{16}\sin 3x - \frac{1}{16}\sin 5x.$$

$$\therefore\ y = \int\left(\frac{1}{8}\sin x + \frac{1}{16}\sin 3x - \frac{1}{16}\sin 5x\right)dx$$

$$= -\frac{1}{8}\cos x - \frac{1}{48}\cos 3x + \frac{1}{80}\cos 5x + C.$$

76. *Prop.* To integrate the form $dy = b^{ax} \,.\, \sin^n x \,.\, dx$.

Put $\sin x \,.\, dx = dv$, and $b^{ax}\sin^{n-1}x = u$, then $v = -\cos x$,

and $du = (n-1)b^{ax}\sin^{n-2}x \,.\, \cos x dx + a \,.\, \log b \cdot b^{ax}\sin^{n-1}x \,.\, dx.$

$$\therefore\ y = \int b^{ax}\sin^n x \,.\, dx = -b^{ax}\sin^{n-1}x \,.\, \cos x + (n-1)\int b^{ax}\sin^{n-2}x\cos^2 x dx$$
$$+ a \,.\, \log b \,.\int b^{ax}\sin^{n-1}x \,.\, \cos x \,.\, dx.$$

But, by applying the formula $\int u dv = uv - \int v du$ to the last integral, making $\sin^{n-1}x \,.\, \cos x \,.\, dx = dv$ and $b^{ax} = u$, we get

$$\int b^{ax}\sin^{n-1}x \,.\, \cos x \,.\, dx = \frac{1}{n}\sin^n x \,.\, b^{ax} - \frac{1}{n}a\log b\int \sin^n x \,.\, b^{ax} \,.\, dx,$$

and, by replacing $\cos^2 x$ by $1 - \sin^2 x$, we have

$$\int b^{ax}\sin^{n-2}x \,.\, \cos^2 x \,.\, dx = \int b^{ax}\sin^{n-2}x dx - \int b^{ax}\sin^n x \,.\, dx.$$

Hence, by substitution,

$$\int b^{ax} . \sin^n x . dx = - b^{ax}\sin^{n-1}x . \cos x + \frac{a \log b}{n} b^{ax} . \sin^n x$$

$$- \frac{(a \log b)^2}{n} \int \sin^n x . b^{ax} dx + (n-1) \int b^{ax}\sin^{n-2}x . dx$$

$$- (n-1) \int b^{ax}\sin^n x . dx.$$

Transposing, collecting like terms and reducing, we obtain

$$\int b^{ax}\sin^n x . dx = \frac{b^{ax}\sin^{n-1}x}{(a \log b)^2 + n^2} (a \log b . \sin x - n \cos x)$$

$$+ \frac{n(n-1)}{(a \log b)^2 + n^2} \int b^{ax}\sin^{n-2}x . dx \quad . . . (L).$$

By repeated applications of (L) we obtain the final integral.

$\int b^{ax}dx = \dfrac{b^{ax}}{a \log b} + C$, when n is even; and when n is odd,

$\int b^{ax}\sin x . dx$, which is given by (L) without an integration, since the last term then contains the factor $n - 1 = 1 - 1 = 0$, and therefore that term disappears.

77. *Prop.* To integrate the form $dy = b^{ax}\cos^n x . dx$.

Put $x = x_1 - \frac{1}{2}\pi$, then $\cos x = \sin x_1$, $\sin x = -\cos x_1$,

$$b^{ax} = b^{ax_1} . b^{-\frac{1}{2}\pi a}, \quad dx = dx_1.$$

$$\therefore y = b^{-\frac{1}{2}\pi a} \int b^{ax_1}\sin^n x_1 dx_1 = \frac{b^{-\frac{1}{2}\pi a} b^{ax_1}\sin^{n-1}x_1}{(a \log b)^2 + n^2}(a \log b . \sin x_1 - n \cos x_1)$$

$+ \dfrac{n(n-1)b^{-\frac{1}{2}\pi a}}{(a \log b)^2 + n^2} \int b^{ax_1}\sin^{n-2}x_1 dx_1$, and by substitution,

$$\int b^{ax}\cos^n x . dx = \frac{b^{ax}\cos^{n-1}x}{(a \log b)^2 + n^2} (a \log b . \cos x + n \sin x)$$

$$+ \frac{n(n-1)}{(a \log b)^2 + n^2} \int b^{ax} . \cos^{n-2}x . dx \quad . . . . (M).$$

Here the final integral will be $\int b^{ax}dx = \frac{b^{ax}}{a \log b} + C$, when n is even; and when n is odd, $\int b^{ax}\cos x \,.\, dx$, to which (M) applies without an integration.

1. To integrate $\qquad dy = e^{ax} \,.\, \cos x \,.\, dx.$

In (M) make $\quad b = e,\ n = 1,\ \log b = \log e = 1.\quad$ Then

$$y = \frac{e^{ax}}{a^2 + 1}\left[a \cos x + \sin x\right] + C.$$

2. $$dy = e^x \,.\, \sin^3 x \,.\, dx.$$

In (L) make $\ b = e,\ a = 1,\ n = 3,\ \log b = 1.\quad$ Then

$$y = \frac{e^x \,.\, \sin^2 x}{1 + 3^2}\left[\sin x - 3 \cos x\right] + \frac{3 \,.\, 2}{1 + 3^2}\int e^x \sin x \,.\, dx$$

$$= \frac{1}{10} e^x \left[\sin^3 x - 3 \sin^2 x \,.\, \cos x\right] + \frac{6}{10} \cdot \frac{1}{2} e^x(\sin x - \cos x) + C.$$

or, $\ y = \frac{1}{10} e^x\left[\sin^3 x + 3\cos^3 x + 3\sin x - 6\cos x\right] + C.$

3. $$dy = e^{-ax}\sin kx \,.\, dx = \frac{1}{k} e^{-ax}\sin kx \,.\, d(kx).$$

In (L) make $\quad b = e, \quad x = kx, \quad a = -\frac{a}{k},\quad$ Then

$$y = -\frac{e^{-ax}(a \sin kx + k \cos kx)}{k^2 + a^2}.$$

78. *Prop.* To integrate the form $dy = X \,.\, \sin^{-1}x \,.\, dx$, in which X is an algebraic function of x.

Put $\qquad Xdx = dv, \qquad$ and $\qquad \sin^{-1}x = u\,;$

then $\qquad v = \int Xdx = X_1, \qquad$ and $\qquad du = \frac{dx}{\sqrt{1 - x^2}}.$

$$\therefore\ y = X_1\sin^{-1}x - \int \frac{X_1 dx}{\sqrt{1 - x^2}},$$

and the proposed integral is thus caused to depend upon another whose form is algebraic.

79. *Prop.* To integrate the form $dy = X\cos^{-1}x\,.\,dx$, in which X is an algebraic function of x.

Put $\qquad Xdx = dv, \qquad$ and $\qquad \cos^{-1}x = u;$

then $\qquad v = \int Xdx = X_1 \qquad$ and $\qquad du = -\dfrac{dx}{\sqrt{1-x^2}}.$

$$\therefore\; y = X_1\cos^{-1}x + \int\frac{X_1 dx}{\sqrt{1-x^2}}, \text{ an algebraic form.}$$

Cor. The same process will apply to each of the forms

$$X\tan^{-1}xdx, \quad X\cot^{-1}xdx, \quad X\sec^{-1}xdx, \text{ \&c.,}$$

since the differential coefficients of $\tan^{-1}x$, $\cot^{-1}x$, $\sec^{-1}x$, &c., are all algebraic.

1. $$dy = x^2\sin^{-1}x\,.\,dx.$$

Here $X = x^2$, $\qquad \therefore\; X_1 = \int Xdx = x^2dx = \dfrac{x^3}{3}$

and $$\int\frac{X_1dx}{\sqrt{1-x^2}} = \frac{1}{3}\int\frac{x^3dx}{\sqrt{1-x^2}} = -\frac{1}{3}\left(\frac{1}{3}x^2 + \frac{2}{3}\right)\sqrt{1-x^2}.$$

$$\therefore\; y = \frac{x^3}{3}\sin^{-1}x - \frac{1}{3}\left(\frac{1}{3}x^2 + \frac{2}{3}\right)\sqrt{1-x^2} + C.$$

2. $$dy = \frac{x^2dx}{1+x^2}\cdot\tan^{-1}x.$$

Put $dv = \dfrac{x^2dx}{1+x^2} = dx - \dfrac{dx}{1+x^2}, \qquad$ and $\qquad u = \tan^{-1}x.$

$$\therefore\; v = x - \tan^{-1}x, \qquad \text{and} \qquad du = \frac{dx}{1+x^2}.$$

$$\therefore\; y = x\tan^{-1}x - (\tan^{-1}x)^2 - \int\frac{xdx}{1+x^2} + \int\frac{\tan^{-1}x\,.\,dx}{1+x^2}$$

$$= x\tan^{-1}x - (\tan^{-1}x)^2 - \frac{1}{2}\log(1+x^2) + \frac{1}{2}(\tan^{-1}x)^2 + C.$$

$$= \tan^{-1}x\left(x - \frac{1}{2}\tan^{-1}x\right) - \log\sqrt{1+x^2} + C.$$

CHAPTER IX.

APPROXIMATE INTEGRATION.

80. When a given differential cannot be reduced to a form exactly integrable, we may expand the differential coefficient, either by Maclaurin's theorem, by the common binomial theorem, or otherwise; then multiply by dx, and finally integrate the terms successively. If the resulting series be convergent, a limited number of terms will give an approximate value of the integral.

81. This method may also be employed with advantage, when an exact integration would lead to a function of complicated form. And the two methods can be used jointly to discover the form of the developed integral.

EXAMPLES.

82. 1. To integrate $dy = \frac{1}{1+x}dx$, in a series.

Expanding $\frac{1}{1+x}$ by actual division, we have

$$\frac{1}{1+x} = 1 - x + x^2 - x^3 + x^4 - \&c$$

$$\therefore\; y = \int(1 - x + x^2 - x^3 + x^4 - \&c.)dx.$$

$$= x - \frac{1}{2}x^2 + \frac{1}{3}x^3 - \frac{1}{4}x^4 + \frac{1}{5}x^5 - \&c. + C.$$

the required series.

Again $$\int \frac{1}{1+x}\,dx = \log(1+x) + C_1$$

$$\therefore \quad \log(1+x) = x - \frac{1}{2}x^2 + \frac{1}{3}x^3 - \frac{1}{4}x^4 + \frac{1}{5}x^5 - \&c. + c,$$

where $c = C - C_1$.

But when $x = 0$, $\log(1+x) = \log 1 = 0$, $\therefore c = 0$,

$$\therefore \quad \log(1+x) = x - \frac{1}{2}x^2 + \frac{1}{3}x^3 - \frac{1}{4}x^4 + \frac{1}{5}x^5 - \&c.$$

a well known formula.

2. $$dy = x^{\frac{1}{2}}(1-x^2)^{\frac{1}{2}}dx.$$

Expanding $(1-x^2)^{\frac{1}{2}}$ by the binomial theorem,

$$(1-x^2)^{\frac{1}{2}} = 1 - \frac{1}{2}x^2 - \frac{1}{8}x^4 - \frac{1}{16}x^6 - \frac{5}{128}x^8 - \&c.$$

$$\therefore y = \int x^{\frac{1}{2}}(1 - \frac{1}{2}x^2 - \frac{1}{8}x^4 - \frac{1}{16}x^6 - \frac{5}{128}x^8 - \&c.)dx.$$

$$= \frac{2}{3}x^{\frac{3}{2}} - \frac{1}{7}x^{\frac{7}{2}} - \frac{1}{44}x^{\frac{11}{2}} - \frac{1}{120}x^{\frac{15}{2}} - \frac{5}{1216}x^{\frac{19}{2}} - \&c. + C.$$

3. $$dy = \frac{dx}{\sqrt{1+x^2}}.$$

Here $$\frac{1}{\sqrt{1+x^2}} = (1+x^2)^{-\frac{1}{2}} = 1 - \frac{1}{2}x^2 + \frac{1}{2}\cdot\frac{3}{4}x^4 - \frac{1}{2}\cdot\frac{3}{4}\cdot\frac{5}{6}x^6 + \&c.$$

$$\therefore y = \int(1 - \frac{1}{2}x^2 + \frac{1}{2}\cdot\frac{3}{4}x^4 - \frac{1}{2}\cdot\frac{3}{4}\cdot\frac{5}{6}x^6 + \&c.)dx$$

$$= x - \frac{1}{2.3}x^3 + \frac{1.3}{2.4.5}x^5 - \frac{1.3.5}{2.4.6.7}x^7 + \&c. + C.$$

But $$\int \frac{dx}{\sqrt{1+x^2}} = \log(x + \sqrt{1+x^2}) + C_1.$$

$$\therefore \log(x+\sqrt{1+x^2}) = x - \frac{1}{2.3}x^3 + \frac{1.3}{2.4.5}x^5 - \frac{1.3.5}{2.4.6.7}x^7 + \&c. + C - C_1.$$

Now when $x=0$, $\log(x+\sqrt{1+x^2})=\log 1=0$. $\therefore C-C_1=0$.

$$\therefore \log(x+\sqrt{1+x^2})=x-\frac{1}{2.3}x^3+\frac{1.3}{2.4.5}x^5-\frac{1.3.5}{2.4.6.7}x^7+\&c.$$

4. To integrate $dy=\dfrac{dx}{1+x^2}$ both in ascending and descending powers of x.

$$\frac{1}{1+x^2}=1-x^2+x^4-x^6+\&c. \text{ and } \frac{1}{x^2+1}=\frac{1}{x^2}-\frac{1}{x^4}+\frac{1}{x^6}-\frac{1}{x^8}+\&c.$$

$$\therefore y=\int\frac{dx}{1+x^2}=\tan^{-1}x+C=\int(1-x^2+x^4-x^6\,\&c.)\,dx$$

$$=x-\frac{1}{3}x^3+\frac{1}{5}x^5-\frac{1}{7}x^7+\&c.+C.$$

Also
$$y=\int\left(\frac{1}{x^2}-\frac{1}{x^4}+\frac{1}{x^6}-\frac{1}{x^8}+\&c.\right)dx$$

$$=-\frac{1}{x}+\frac{1}{3x^3}-\frac{1}{5x^5}+\frac{1}{7x^7}-\&c.+C_1.$$

The two results become equivalent, by selecting the constants C and C_1 such that $C_1-C=\frac{1}{2}\pi$.

For, the first series $=\tan^{-1}x+C$.

And the second " $=-\tan^{-1}\frac{1}{x}+C_1=-\cot^{-1}x+C_1$.

$\therefore$ In order that the two series may be equal, we must have

$$\tan^{-1}x+C=-\cot^{-1}x+C_1,$$

or
$$\tan^{-1}x+\cot^{-1}x=C_1-C, \text{ or } \frac{1}{2}\pi=C_1-C.$$

5.
$$dy=\frac{\sqrt{1-e^2x^2}}{\sqrt{1-x^2}}dx.$$

Expanding the numerator we have

$$(1-e^2x^2)^{\frac{1}{2}}=1-\frac{1}{2}e^2x^2+\frac{1}{2\,.\,4}e^4x^4-\frac{1\,.\,3}{2\,.\,4\,.\,6}e^6x^6\,\&c.$$

$$\therefore\ y = \int \left(1 - \frac{1}{2}e^2x^2 + \frac{1}{2.4}e^4x^4 - \frac{1.3}{2.4.6}e^6x^6 \text{ \&c.}\right) \frac{dx}{\sqrt{1-x^2}},$$

all the terms of which are of the form $\int \frac{x^n dx}{\sqrt{1-x^2}}$ and have been already integrated in the chapter relating to binomial differentials.

We might also expand $(1-x^2)^{-\frac{1}{2}}$ by the binomial theorem, then perform the multiplication indicated, and finally integrate the terms in succession. Adopting the first course we have

$$y = \sin^{-1}x + \frac{1}{2}e^2\left(\frac{1}{2}x\sqrt{1-x^2} - \frac{1}{2}\sin^{-1}x\right)$$
$$- \frac{1}{2.4}e^4\left[\left(\frac{1}{4}x^3 + \frac{1.3}{2.4}x\right)\sqrt{1-x^2} - \frac{1.3}{2.4}\sin^{-1}x\right] \text{\&c.}$$

83. *Prop.* To obtain a series which shall express the integral of every function of the form Xdx, in terms of X, its differential coefficients, and x.

Put $X = u$, $dx = dv$: then $du = \frac{dX}{dx} \cdot dx$, and $v = x$.

Now substituting in the formula $\int udv = uv - \int vdu$ we get

$$\int Xdx = Xx - \int \frac{dX}{dx} \cdot xdx.$$

Next, put $\frac{dX}{dx} = u$ and $xdx = dv$,

then $du = \frac{d^2X}{dx^2} \cdot dx$ and $v = \frac{1}{1.2}x^2$.

$$\therefore \int \frac{dX}{dx} xdx = \frac{dX}{dx} \cdot \frac{x^2}{1.2} - \int \frac{d^2X}{dx^2} \cdot \frac{x^2}{1.2} dx.$$

Similarly $\int \frac{d^2X}{dx^2} \cdot \frac{x^2}{1.2} dx = \frac{d^2X}{dx^2} \cdot \frac{x^3}{1.2.3} - \int \frac{d^3X}{dx^3} \cdot \frac{x^3}{1.2.3} dx$ &c.&c.

By substiution

$$\int Xdx = Xx - \frac{dX}{dx} \cdot \frac{x^2}{1.2} + \frac{d^2X}{dx^2} \cdot \frac{x^3}{1.2.3} - \frac{d^3X}{dx^3} \cdot \frac{x^4}{1.2.3.4} + \text{\&c.} + C.$$

This formula, called Bernouilli's series, shows the possibility of expressing the integral of every function of a single variable, in terms of that variable, since the several differential coefficients $\frac{dX}{dx}$, $\frac{d^2X}{dx^2}$, &c., can always be formed. But the series is often divergent, and then of no use in giving the value of the integral approximately.

CHAPTER X.

INTEGRATION BETWEEN LIMITS AND SUCCESSIVE INTEGRATION.

84. The integrals determined by the methods hitherto explained are called *indefinite* integrals, because the value of the variable x, and that of the constant C, both of which appear in the integral, remain undetermined. But in applying the Calculus, the nature of the question will always require that the integral should be taken between given limits. Thus, suppose the integral to originate (or its value to reduce to zero) when $x = a$: this condition will fix the value of the constant C. Then, to determine the value of the entire or *definite* integral, we replace x by b, the other extreme value of the variable.

Ex. To integrate $dy = 3x^2dx$, between the limits $x = x_1$ and $x = x_2$.

$$y = \int 3x^2dx = x^3 + C. \quad \text{But when} \quad x = x_1, \quad y = 0.$$

$$\therefore\ 0 = x_1^3 + C \quad \text{and} \quad C = -x_1^3,$$

and by substitution in the indefinite integral

$$y = x^3 - x_1^3.$$

Now make $x = x_2$, and there will result

$$y = x_2^3 - x_1^3,$$

the complete or definite integral.

A slight examination will show that the desired result will always be obtained by substituting in the indefinite integral for the variable x, first the inferior limit x_1 and then the superior limit x_2, and then subtracting the first result from the second. In these substitutions the constant C may be neglected, since it will disappear in the subtraction.

85. The integration of $3x^2dx$ between the limits x_1 and x_2, when x_1 is the inferior limit, or that at which the integral originates, and x_2 the superior limit, is indicated by the notation.

$$\int_{x_1}^{x_2} 3x^2dx.$$

86. The precise signification of this definite integral will, perhaps, be better understood by the aid of the following

Prop. The definite integral $\int_a^b Xdx$, (where X is a function of x, which does not become infinite for any value of x between the limits $x = a$ and $x = b$,) is the limit of the sum of the values assumed by the product Xh, as x is caused to increase by successive equal increments (each $= h$) from $x = a$ to $x = b$; the value of h being continually diminished, and consequently the number of these increments being indefinitely increased.

Thus, if $X_0\ X_1\ X_2\ X_3 \ldots X_{n-1}$ be the values assumed by X, when x takes successively the values a, $a+h$, $a+2h$, $a+3h, \ldots a+(n-1)h$, then will $\int_a^b Xdx$ be the limit to the value $(X_0+X_1+X_2\ldots+X_{n-1})h$, provided $nh = b - a$, and h be diminished indefinitely.

Proof. Let x and $x + h$ be any two successive values of x, and denote by Fx the general or indefinite integral $\int Xdx$.

Then by Taylor's Theorem,

$$F(x+h) = Fx + \frac{dFx}{dx}\frac{h}{1} + \frac{d^2Fx}{dx}\frac{h^2}{1.2} + \frac{d^3Fx}{dx^3}\frac{h^3}{1.2.3} + \&c.,$$

which may be written, $F(x+h) = Fx + Xh + Ph^2, \ldots$ (1), where P is a function of x and h.

Suppose the difference $b - a$ to be divided into n equal parts, each equal to h, so that $b - a = nh$.

Now, putting successively a, $a+h$, $a+2h \ldots a+(n-1)h$ for x in (1), and denoting the corresponding values of P by P_0, P_1, &c., we get

$$F(a+h) = Fa + X_0h + P_0h^2$$

$$F(a+2h) = F[(a+h)+h] = F(a+h) + X_1h + P_1h^2$$

$$F(a+3h) = F[(a+2h)+h] = F(a+2h) + X_2h + P_2h^2$$

&c. &c. &c.

$$F(a+nh) = F[(a+(n-1)h)+h] = F[a+(n-1)h] + X_{n-1}h + P_{n-1}h^2,$$

adding these equations, and omitting the terms common to both members of the sum, there results

$$F(a+nh) = Fa + h(X_0 + X_1 + X_2 \ldots . + X_{n-1})$$
$$+ h^2(P_0 + P_1 + P_2 \ldots . + P_{n-1}).$$

But, since every value of X is finite, none of the values of P will become infinite. If, therefore, we denote the greatest value of P by P, we shall have

$P_0+P_1+P_2 \ldots + P_{n-1} < Pn$, and since $F(a+nh) = Fb$, and $nh = b - a$.

$$\therefore \; Fb - Fa - h(X_0 + X_1 + X_2 \ldots + X_{n-1}) < (b-a)P \, . \, h.$$

But $b - a$ and P are both finite, and therefore by diminishing h, the second member can be rendered less than any assignable quantity. Hence $Fb - Fa$ must approach indefinitely near to equality with $h(X_0 + X_1 + X_2 \ldots . + X_{n-1})$ when h is continually diminished.

Successive Integration.

87. If the second differential coefficient $\frac{d^2y}{dx^2} = X$ be given instead of the first, two successive integrations will be required to determine the original function y in terms of x. Thus, multiplying by dx and integrating, we get

$$\int \frac{d^2y}{dx^2} \cdot dx = \int X dx, \quad \text{or} \quad \frac{dy}{dx} = \int X dx = X_1 + C_1.$$

Multiplying again by dx, and integrating, we get

$$\int \frac{dy}{dx} dx = \int X_1 dx + \int C_1 dx,$$

or

$$y = X_2 + C_1 x + C_2.$$

88. Similarly, if there were given $\frac{d^3y}{dx^3} = X$, three successive integrations would give

$$y = X_3 + \frac{1}{1\,.\,2} C_1 x^2 + C_2 x + C_3.$$

And if there were given $\quad \frac{d^ny}{dx^n} = X, \quad$ then

$$y = X_n + \frac{C_1 x^{n-1}}{1\,.\,2\,.\,3 \ldots (n-1)}$$

$$+ \frac{C_2 x^{n-2}}{1\,.\,2\,.\,3 \ldots (n-2)} + \text{\&c.} \ldots + C_{n-1} x + C_n,$$

the number of arbitrary constants introduced being n.

89. The result obtained by performing the above integrations may be indicated thus

$$\int^n X dx^n = y:$$

it is called the nth integral of $X dx^n$.

90. *Prop.* To develop the nth integral $\int^n X dx^n$ in a series.

Employing Maclaurin's Theorem, we have

$$\int^n X dx^n = \left[\int^n X dx^n\right] + \left[\int^{n-1} X dx^{n-1}\right]\frac{x}{1} + \left[\int^{n-2} X dx^{n-2}\right]\frac{x^2}{1 . 2} + \&c.$$

$$+\left[\int X dx\right]\frac{x^{n-1}}{1 . 2 . 3 .. (n-1)} + [X]\frac{x^n}{1 . 2 . 3 .. n} + \left[\frac{dX}{dx}\right]\frac{x^{n+1}}{1.2.3...(n+1)}$$

$$+\left[\frac{d^2X}{dx^2}\right]\frac{x^{n+2}}{1 . 2 . 3 .. (n+2)} + \left[\frac{d^3X}{dx^3}\right]\frac{x^{n+3}}{1 . 2 . 3 .. (n+3)} + \&c. \dots [R].$$

The terms within the [ ] are the arbitrary constants $C_1\ C_2\ C_3 \dots C_n$, as far as $[\int X dx]$ inclusive, but taken in an inverted order.

91. *Prop.* To deduce the development of $\int^n X dx^n$ from that of X.

By Maclaurin's Theorem, we have

$$X = [X] + \left[\frac{dX}{dx}\right]\frac{x}{1} + \left[\frac{d^2X}{dx^2}\right]\frac{x^2}{1 . 2} + \left[\frac{d^3X}{dx^3}\right]\frac{x^3}{1 . 2 . 3} + \&c.$$

and this may be converted into the series $[R]$ by multiplying each term by x^n, then dividing the successive terms by $1 . 2 . 3 \dots n$, by $2 . 3 . 4 \dots (n+1)$, by $3 . 4 . 5 \dots (n+2)$, &c., and finally annexing terms of the form

$$\frac{C_1 x^{n-1}}{1 . 2 . 3 \dots (n-1)}, \quad \frac{C_2 x^{n-2}}{1 . 2 . 3 \dots (n-2)}, \dots C_n.$$

1. To develop $\displaystyle\int^4 \frac{dx^4}{\sqrt{1-x^2}}$.

Here $X = (1-x^2)^{-\frac{1}{2}} = 1 + \frac{1}{2}x^2 + \frac{1}{2}\cdot\frac{3}{4}x^4 + \frac{1}{2}\cdot\frac{3}{4}\cdot\frac{5}{6}x^6 + \&c.$

Also $n = 4$. Therefore multiplying by x^4 and dividing successively by $1 . 2 . 3 . 4$, by $3 . 4 . 5 . 6$, &c., and finally annexing the terms containing the constants, we get

$$\int^4 \frac{dx^4}{\sqrt{1-x^2}} = C_4 + \frac{C_3 x}{1} + \frac{C_2 x^2}{1 . 2} + \frac{C_1 x^3}{1 . 2 . 3} + \frac{x^4}{1 . 2 . 3 . 4} + \frac{x^6}{2 . 3 . 4 . 5 . 6}$$

$$+ \frac{1 . 3x^8}{2 . 4 . 5 . 6 . 7 . 8} + \frac{1 . 3 . 5x^{10}}{2 . 4 . 6 . 7 . 8 . 9 . 10} + \&c.$$

2. What curves are characterized by the equations $\frac{d^2y}{dx^2} = 0$, and $\frac{d^3y}{dx^3} = 0$, respectively ?

1st. If $\quad \frac{d^2y}{dx^2} = 0, \quad$ then $\quad \int \frac{d^2y}{dx^2} dx = \frac{dy}{dx} = C_1.$

$$\therefore \int \frac{dy}{dx} dx = \int C_1 dx, \quad \text{or} \quad y = C_1 x + C_2, \quad \text{a straight line.}$$

2d. If $\quad \frac{d^3y}{dx^3} = 0, \quad$ then $\quad \int \frac{d^3y}{dx^3} dx = \frac{d^2y}{dx^2} = C_1$

$$\therefore \int \frac{d^2y}{dx^2} dx = \int C_1 dx \quad \text{or} \quad \frac{dy}{dx} = C_1 x + C_2,$$

$$\int \frac{dy}{dx} dx = \int C_1 x dx + \int C_2 dx \quad \text{or} \quad y = \frac{C_1 x^2}{1 \cdot 2} + C_2 x + C_3, \text{ a parabola.}$$

PART II.

RECTIFICATION OF CURVES. QUADRATURE OF AREAS. CUBATURE OF VOLUMES.

CHAPTER I.

RECTIFICATION OF CURVES.

92. To rectify a curve is to determine a straight line whose length shall be equivalent to that of the curve, or simply to obtain an expression for the length of the curve, in terms of the co-ordinates of its two extremities.

93. *Prop.* To obtain a general formula for the length of the arc of a plane curve, when referred to rectangular co-ordinates.

Let AB be the proposed arc, P a point in it, OX and OY the co-ordinate axes.

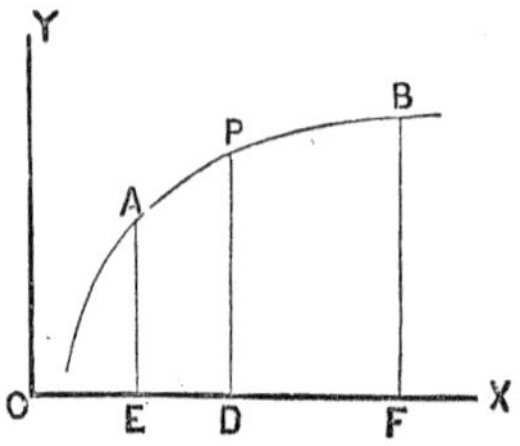

Put $OD = x$, $DP = y$, $AP = s$.

Then since $ds = dx\sqrt{1 + \dfrac{dy^2}{dx^2}}$,

we shall have by integration

$$s = \int\left(1 + \frac{dy^2}{dx^2}\right)^{\frac{1}{2}} dx \;\ldots\ldots\; (S), \quad \text{the required formula.}$$

94. To apply (S) we replace $\dfrac{dy^2}{dx^2}$ by its value, in terms of x, deduced from the equation of the curve, and then integrate between the limits $x = OE$ and $x = OF$.

95. Again if y be taken as the independent variable, we shall have

$$ds = dy\sqrt{1+\frac{dx^2}{dy^2}}, \quad \text{and therefore}$$

$$s = \int\left(1+\frac{dx^2}{dy^2}\right)^{\frac{1}{2}} dy \ldots (S_1), \text{ a second formula.}$$

This will be applied by substituting for $\frac{dx^2}{dy^2}$ its value, in terms of y, derived from the equation of the curve, and then integrating between the proper limits.

EXAMPLES.

96. 1. To find the length of the parabolic arc AB, included between the ordinates b_1 and b_2.

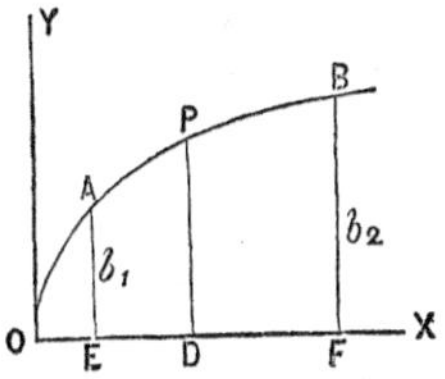

The equation of the curve is $y^2 = 2px$.

$$\therefore \frac{dx}{dy} = \frac{y}{p},$$

which substituted in (S_1) gives

$$s = \int\left[1+\frac{y^2}{p^2}\right]^{\frac{1}{2}} dy = \frac{1}{p}\int(p^2+y^2)^{\frac{1}{2}} dy.$$

But by formula (B),

$$\int(p^2+y^2)^{\frac{1}{2}} dy = \frac{1}{2}(p^2+y^2)^{\frac{1}{2}} y + \frac{1}{2}p^2\int(p^2+y^2)^{-\frac{1}{2}} dy \ldots . (1).$$

To integrate the last term, put $(p^2+y^2)^{\frac{1}{2}} = z + y$.

$$\therefore p^2 + y^2 = z^2 + 2zy + y^2, \quad y = \frac{p^2-z^2}{2z}, \quad dy = -\frac{p^2+z^2}{2z^2}dz.$$

and
$$(p^2+y^2)^{\frac{1}{2}} = z + \frac{p^2-z^2}{2z} = \frac{p^2+z^2}{2z}.$$

$$\therefore \int(p^2+y^2)^{-\frac{1}{2}} dy = -\int\frac{2z}{p^2+z^2}\cdot\frac{p^2+z^2}{2z^2}dz = -\int\frac{dz}{z} = -\log z + C.$$

And by substitution in (1),

$$\int (p^2+y^2)^{\frac{1}{2}}dy = \frac{1}{2}(p^2+y^2)^{\frac{1}{2}}y - \frac{1}{2}p^2.\log[(p^2+y^2)^{\frac{1}{2}} - y] + C_1.$$

$$\therefore s = \frac{(p^2+y^2)^{\frac{1}{2}}.y}{2p} - \frac{1}{2}p.\log[(p^2+y^2)^{\frac{1}{2}} - y] + C_1.$$

To determine the value of C_1, put $y = b_1$ and $s = 0$, since the arc is supposed to commence at the point A.

Thus $$0 = \frac{(p^2+b_1^2)^{\frac{1}{2}}b_1}{2p} - \frac{1}{2}p\log[(p^2+b_1^2)^{\frac{1}{2}} - b_1] + C_1.$$

$$\therefore \quad C_1 = -\frac{(p^2+b_1^2)^{\frac{1}{2}}b_1}{2p} + \frac{1}{2}p\log[(p^2+b_1^2)^{\frac{1}{2}} - b_1],$$ and by substitution

$$s = \frac{(p^2+y^2)^{\frac{1}{2}}y}{2p} - \frac{(p^2+b_1^2)^{\frac{1}{2}}b_1}{2p} - \frac{1}{2}p\log\frac{(p^2+y^2)^{\frac{1}{2}} - y}{(p^2+b_1^2)^{\frac{1}{2}} - b_1},$$ and

when $y = b_2$

$$s = \frac{(p^2+b_2^2)^{\frac{1}{2}}b_2}{2p} - \frac{(p^2+b_1^2)^{\frac{1}{2}}b_1}{2p} - \frac{1}{2}p\log\frac{(p^2+b_2^2)^{\frac{1}{2}} - b_2}{(p^2+b_1^2)^{\frac{1}{2}} - b_1}.$$

If the arc be reckoned from the vertex O, the ordinate $b_1 = 0$,

$$\therefore \quad s = \frac{(p^2+b_2^2)^{\frac{1}{2}}b_2}{2p} - \frac{1}{2}p\log\frac{(p^2+b_2^2)^{\frac{1}{2}} - b_2}{p}$$

$$= \frac{(p^2+b_2^2)^{\frac{1}{2}}b_2}{2p} + \frac{1}{2}p\log\frac{(p^2+b_2^2)^{\frac{1}{2}} + b_2}{p}.$$

2. The cycloid $\quad y = \sqrt{2rx - x^2} + r.\text{versin}^{-1}\frac{x}{r}.$

Here $\frac{dy}{dx} = \sqrt{\frac{2r-x}{x}}$, and

$$\therefore \ 1 + \frac{dy^2}{dx^2} = 1 + \frac{2r-x}{x} = \frac{2r}{x}.$$

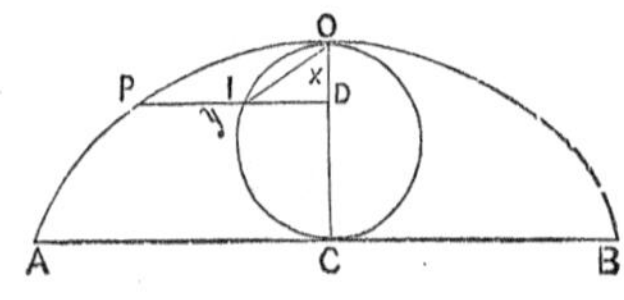

Hence by substitution in formula (S).

$$s = \int \frac{2^{\frac{1}{2}} \cdot r^{\frac{1}{2}}}{x^{\frac{1}{2}}} dx = 2\sqrt{2rx} + C.$$

But when $x = 0$, $s = 0$, $\therefore$ $C = 0$, and hence $s = 2\sqrt{2rx}$,

or, the cycloidal arc $OP = 2$ chord OI of the generating circle.

When $x = 2r$, $s =$ arc $OPA = 2$ diameter OC.

$\therefore$ arc AOB of the entire cycloid $= 4$ diameters of the generating circle.

3. The circle $y^2 = r^2 - x^2$.

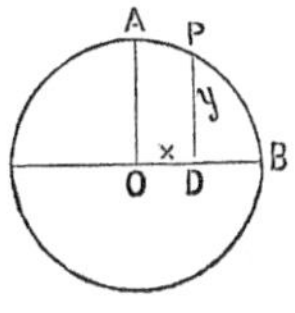

$$\frac{dy}{dx} = -\frac{x}{y}, \quad 1 + \frac{dy^2}{dx^2} = 1 + \frac{x^2}{y^2} = \frac{r^2}{y^2}.$$

$$\therefore \quad s = \int \frac{r}{y} dx = r \int \frac{dx}{\sqrt{r^2 - x^2}} = r \cdot \sin^{-1}\frac{x}{r} + C.$$

This result involves a circular arc, the very quantity we wish to determine, and is therefore inapplicable.

To obtain an approximate result, expand the differential coefficient $(r^2 - x^2)^{-\frac{1}{2}}$ and integrate: thus

$$s = r \int \left(\frac{1}{r} + \frac{1}{2} \cdot \frac{x^2}{r^3} + \frac{1 \cdot 3}{2 \cdot 4} \cdot \frac{x^4}{r^5} + \frac{1 \cdot 3 \cdot 5}{2 \cdot 4 \cdot 6} \cdot \frac{x^6}{r^7} + \text{\&c.} \right) dx$$

$$= r \left[\frac{x}{r} + \frac{1}{2 \cdot 3} \cdot \frac{x^3}{r^3} + \frac{1 \cdot 3}{2 \cdot 4 \cdot 5} \cdot \frac{x^5}{r^5} + \frac{1 \cdot 3 \cdot 5}{2 \cdot 4 \cdot 6 \cdot 7} \cdot \frac{x^7}{r^7} + \text{\&c.} \right] + C.$$

But if $s = 0$ when $x = 0$, then $C = 0$, and $\therefore$ when $x = r$,

$$s = r \left(1 + \frac{1}{2 \cdot 3} + \frac{1 \cdot 3}{2 \cdot 4 \cdot 5} + \frac{1 \cdot 3 \cdot 5}{2 \cdot 4 \cdot 6 \cdot 7} + \text{\&c.} \right)$$

the value of the arc APB of the quadrant.

And if $r = 1$, $s = \frac{1}{2}\pi = 1 + \frac{1}{2 \cdot 3} + \frac{1 \cdot 3}{2 \cdot 4 \cdot 5} + \frac{1 \cdot 3 \cdot 5}{2 \cdot 4 \cdot 6 \cdot 7} + \text{\&c.}$

4. The ellipse $$a^2y^2 + b^2x^2 = a^2b^2.$$

$$\frac{dy^2}{dx^2} = \frac{b^4x^2}{a^4y^2},$$

$$1 + \frac{dy^2}{dx^2} = 1 + \frac{b^4x^2}{a^4y^2} = \frac{a^2(a^2b^2 - b^2x^2) + b^4x^2}{a^2(a^2b^2 - b^2x^2)} = \frac{a^2(a^2 - x^2) + b^2x^2}{a^2(a^2 - x^2)},$$

or, $1 + \dfrac{dy^2}{dx^2} = \dfrac{a^2 - \dfrac{a^2 - b^2}{a^2}x^2}{a^2 - x^2} = \dfrac{a^2 - e^2x^2}{a^2 - x^2}$, where e is the eccentricity.

$$\therefore \quad s = \int \frac{(a^2 - e^2x^2)^{\frac{1}{2}}}{(a^2 - x^2)^{\frac{1}{2}}} dx = \int \frac{(1 - e^2x_1^2)^{\frac{1}{2}}}{(1 - x_1^2)^{\frac{1}{2}}} dx, \text{ by making } \frac{x}{a} = x_1.$$

This expression has already been integrated approximately.

5. To determine what curves of the parabolic class are rectifiable.

The equation of this class of curves is $y^n = ax^m$, in which n and m are positive integers.

$$\frac{dy}{dx} = \frac{m}{n} a^{\frac{1}{n}} x^{\frac{m}{n} - 1}, \quad \therefore \ s = \int \left[1 + \frac{m^2}{n^2} a^{\frac{2}{n}} x^{\frac{2m}{n} - 2}\right]^{\frac{1}{2}} dx,$$

and this can be rationalized, when $\dfrac{1}{2\left(\dfrac{m - n}{n}\right)} = r$, an integer, that is,

when $\dfrac{m}{n} = \dfrac{1 + 2r}{2r}$ (Art. 41).

Hence, if one exponent, n, be even, and the other, m, greater by unity, the curve will be rectifiable; that is, an exact expression for the length of the curve can be obtained in terms of the co-ordinates of its extremities.

The term rectifiable is sometimes restricted to those curves whose lengths can be expressed *algebraically*, or without employing transcendental quantities; and with this restriction, the value of r must be positive, otherwise s would be transcendental.

Now applying the other condition of integrability, we have $\dfrac{1}{2\left(\dfrac{m - n}{n}\right)} + \dfrac{1}{2} = r$, an integer, whence $\dfrac{n}{m} = \dfrac{2r - 1}{2r}$.

Hence, if one of the exponents be an even integer, and the other less by unity, the curve will be rectifiable.

Combining the two results, we find it simply necessary that m and n should differ by unity.

97. *Prop.* To obtain a formula for the rectification of polar curves.

Here we have to express s in terms of r or θ, and for this purpose we must transform the formula $[S]$, by means of the relations

$$\frac{ds^2}{d\theta^2} = \frac{dx^2}{d\theta^2} + \frac{dy^2}{d\theta^2}, \ldots (1). \quad x = r\cos\theta, \ldots (2). \quad y = r\sin\theta \ldots (3),$$

the quantity θ being taken as the independent variable.

Then (2) and (3) give

$$\frac{dx}{d\theta} = -r\sin\theta + \cos\theta\frac{dr}{d\theta}, \quad \text{and} \quad \frac{dy}{d\theta} = r\cos\theta + \sin\theta\frac{dr}{d\theta}.$$

$$\left.\begin{aligned} \therefore \frac{ds^2}{d\theta^2} &= r^2\sin^2\theta - 2r\sin\theta\cos\theta\frac{dr}{d\theta} + \cos^2\theta\frac{dr^2}{d\theta^2} \\ &+ r^2\cos^2\theta + 2r\sin\theta\cos\theta\frac{dr}{d\theta} + \sin^2\theta\frac{dr^2}{d\theta^2} \end{aligned}\right\} = r^2 + \frac{dr^2}{d\theta^2}.$$

$$\therefore s = \int\left[r^2 + \frac{dr^2}{d\theta^2}\right]^{\frac{1}{2}} d\theta \ldots\ldots (T).$$

1. The logarithmic spiral $r = a^\theta$, between the limits $r = r_1$, and $r = r_2$.

$$\frac{dr}{d\theta} = \log a \,.\, a^\theta = \frac{a^\theta}{m}, \text{ where } m \text{ is the modulus.}$$

$$\therefore d\theta = \frac{m}{a^\theta}dr = \frac{m}{r}dr, \text{ and by substitution in } (T),$$

$$\therefore s = \int\left(r^2 + \frac{r^2}{m^2}\right)^{\frac{1}{2}}\frac{m}{r}dr = (m^2+1)^{\frac{1}{2}}\int dr = (m^2+1)^{\frac{1}{2}}r + C.$$

But $s = 0$, when $r = r_1$, $\therefore C = -(m^2+1)^{\frac{1}{2}}r_1$.

$\therefore s = (1+m^2)^{\frac{1}{2}}(r - r_1)$, and when $r = r_2$, $s = (1+m^2)^{\frac{1}{2}}(r_2 - r_1)$.

2. The spiral of Archimedes $r = a\theta$, from the pole to the point $r = r_1$.

$$\frac{dr}{d\theta} = a, \quad d\theta = \frac{dr}{a}. \quad \therefore\; s = \frac{1}{a}\int (r^2 + a^2)^{\frac{1}{2}} dr.$$

This expression is entirely similar to that integrated in rectifying the parabola.

$$\therefore\; s = \frac{r_1(a^2 + r_1{}^2)^{\frac{1}{2}}}{2a} + \frac{1}{2} a \log \frac{r_1 + (a^2 + r_1{}^2)^{\frac{1}{2}}}{a}.$$

3. The lemniscata $r^2 = a^2 \cos 2\theta$,

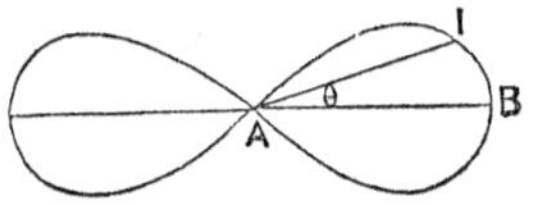

$$\frac{rdr}{d\theta} = -\, a^2 \sin 2\theta, \quad \frac{dr^2}{d\theta^2} = \frac{a^4}{r^2}\left(1 - \frac{r^4}{a^4}\right).$$

$$\therefore\; r^2 + \frac{dr^2}{d\theta^2} = \frac{a^4}{r^2}, \quad d\theta = -\frac{rdr}{a^2\left(1 - \frac{r^4}{a^4}\right)^{\frac{1}{2}}} = -\frac{-\, rdr}{(a^4 - r^4)^{\frac{1}{2}}}.$$

$$\therefore\; s = \int \frac{-\, a^2 dr}{(a^4 - r^4)^{\frac{1}{2}}} = -\, a^2 \int \left[\frac{1}{a^2} + \frac{1}{2}\frac{r^4}{a^6} + \frac{1\,.\,3}{2\,.\,4}\frac{r^8}{a^{10}} + \frac{1\,.\,3\,.\,5}{2\,.\,4\,.\,6}\cdot\frac{r^{12}}{a^{14}} \;\&\text{c.}\right] dr,$$

which, integrated from $r = a$ to $r = 0$, gives for the arc BIA or one-fourth of the entire length of the curve.

$$s = a\left[1 + \frac{1}{2\,.\,5} + \frac{1\,.\,3}{2\,.\,4\,.\,9} + \frac{1\,.\,3\,.\,5}{2\,.\,4\,.\,6\,.\,13}, \;\&\text{c.}\right]$$

98. When the curve is characterized by a relation between the radius vector r and the perpendicular p upon the tangent. To obtain a formula for the rectification in this case, we assume the value of the perpendicular found in the Differen. Calculus, p. 154 viz.:

$$p = \frac{r^2}{\sqrt{r^2 + \frac{dr^2}{d\theta^2}}}; \quad \text{whence} \quad \frac{dr^2}{d\theta^2} = \frac{r^2(r^2 - p^2)}{p^2}, \text{ and}$$

$$\frac{ds^2}{d\theta^2} = \frac{ds^2}{dr^2}\cdot\frac{dr^2}{d\theta^2} = r^2 + \frac{dr^2}{d\theta^2} = \frac{r^4}{p^2}.$$

$$\therefore \frac{ds^2}{dr^2} = \frac{r^4}{p^2} \div \frac{dr^2}{d\theta^2} = \frac{r^4}{p^2} \times \frac{p^2}{r^2(r^2 - p^2)} = \frac{r^2}{r^2 - p^2}.$$

$$\therefore s = \int \frac{rdr}{(r^2 - p^2)^{\frac{1}{2}}} \cdots (U), \text{ the required formula.}$$

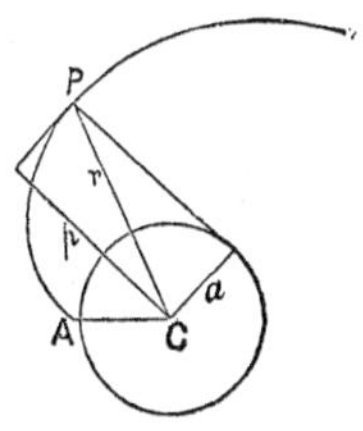

Ex. The involute of the circle from $p = 0$ to $p = 2\pi a$.

Here the equation of the curve is $r^2 = a^2 + p^2$.

$$\therefore s = \int \frac{rdr}{a} = \frac{r^2}{2a} + C = \frac{a^2 + p^2}{2a} + C.$$

But when $p = 0$, $s = 0$. $\therefore C = -\frac{a}{2}$; and when $p = 2\pi a$, $s = 2\pi^2 a$.

CHAPTER II.

QUADRATURE OF PLANE AREAS.

99. The quadrature of a plane curve is the determination of a square equal in area to the space bounded in part or entirely by that curve. The problem is regarded as resolved when an *expression* for the area in terms of known quantities has been obtained, the number of terms being limited.

100. *Prop.* To obtain a general formula for the value of the plane area $ABCD$, included between the curve DC, the axis OX, and the two parallel ordinates AD and BC, the curve being referred to rectangular co-ordinates.

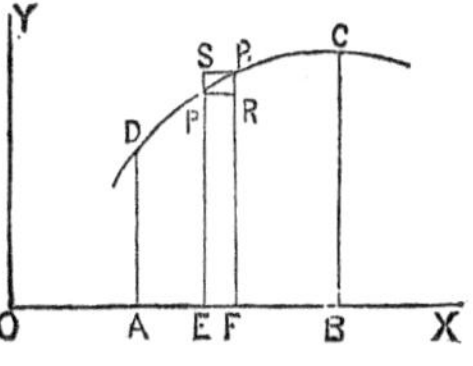

Put $OE = x$, $EP = y$, $EF = h$, $FP_1 = y_1$, and the area $AEPD = A$.

Then when x receives in increment h, the area takes a corresponding increment EPP_1F, intermediate in value between the rectangle FP and the rectangle FS.

$$\text{But } \frac{\square FS}{\square FP} = \frac{y_1 \times h}{y \times h} = \frac{y_1}{y} = \frac{y + \frac{dy}{dx}\cdot\frac{h}{1} + \frac{d^2y}{dx^2}\cdot\frac{h^2}{1.2} + \&c.}{y}$$

$$= 1 + \frac{dy}{dx}\cdot\frac{h}{y} + \frac{d^2y}{dx^2}\cdot\frac{h^2}{1.2.y} + \&c. = 1, \text{ when } h = 0.$$

Hence at the limit, when h is taken indefinitely small, the area EPP_1F, which is always intermediate in value between FP and FS, must become equal to each of these rectangles, or equal to $y \times h$.

$$\therefore\ dA = ydx, \text{ and consequently}$$

$$A = \int ydx \ldots\ldots (V), \text{ the required formula.}$$

101. If the area were included between two curves DC and D_1C_1, we should find by a similar course of reasoning

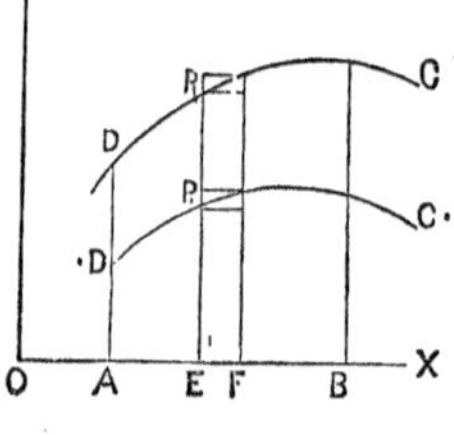

$$A = \int(Y - y)dx \ldots\ldots\ldots (V_1),$$

in which Y and y denote the ordinates EP and EP_1, corresponding to the same abscissa OE.

102. To apply (V) or (V_1), we eliminate y, or y and Y, by employing the equation of one or both curves, and then integrate between the limits $x = OA$ and $x = OB$.

EXAMPLES.

103. 1. The area $ABCD$, included between the parabolic arc DC, the axis of x, and two given ordinates AD and BC.

Put $OA = a_1$, $AD = b_1$, $OB = a_2$, $BC = b_2$, $OE = x$, and $EP = y$.

Then, from the equation of the parabola, we have

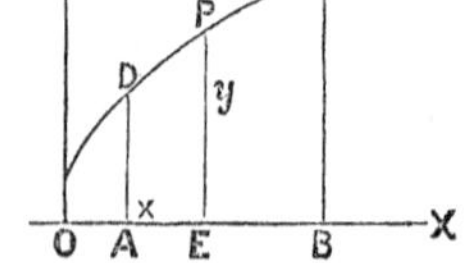

$$y^2 = 2px, \quad \text{or} \quad y = (2p)^{\frac{1}{2}}.x^{\frac{1}{2}}.$$

$\therefore$ And by substitution in formula (V),

$$A = \int(2p)^{\frac{1}{2}}.x^{\frac{1}{2}}dx = \frac{2}{3}(2p)^{\frac{1}{2}}.x^{\frac{3}{2}} + C = \frac{2}{3}(2px)^{\frac{1}{2}}.x + C = \frac{2}{3}xy + C.$$

But $A = O$, when $x = a_1$ and $y = b_1$, $\therefore C = -\frac{2}{3} a_1 b_1$

$$\therefore A = \frac{2}{3}(xy - a_1 b_1) = ADPE; \text{ and when } x = a_2 \text{ and } y = b_2$$

$$A = \frac{2}{3}(a_2 b_2 - a_1 b_1) = ADCB.$$

Cor. If the area $ODCB$ of the semi-parabola were required, we should have

$$a_1 = 0,\ b_1 = 0, \text{ and } \therefore A = \frac{2}{3} a_2 b_2 = \frac{2}{3} \text{ circumscribing } \square;$$

and for the entire area of the parabola

$$2A = \frac{4}{3} a_2 b_2 = \frac{2}{3} a_2 \cdot 2b_2 = \frac{2}{3} \text{ circumscribing } \square.$$

2. The circle $y^2 = r^2 - x^2$, or its segment ACD.

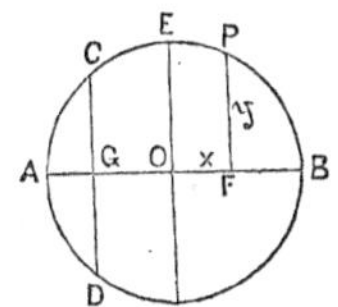

Here $A = \int y dx = \int (r^2 - x^2)^{\frac{1}{2}} dx$,

or by employing formula (B),

$$A = \frac{1}{2} x (r^2 - x^2)^{\frac{1}{2}} + \frac{1}{2} r^2 \int \frac{dx}{(r^2 - x^2)^{\frac{1}{2}}} = \frac{1}{2} x (r^2 - x^2)^{\frac{1}{2}} - \frac{1}{2} r^2 \cos^{-1} \frac{x}{r} + C.$$

Suppose the area to be reckoned from A.

$$\text{area } A = 0 \quad \text{when} \quad x = OA = -r.$$

$$\therefore C = \frac{1}{2} r^2 \cos^{-1}(-1) = \frac{1}{2} \pi r^2.$$

$$\therefore A = \frac{1}{2} \pi r^2 + \frac{1}{2} x (r^2 - x^2)^{\frac{1}{2}} - \frac{1}{2} r^2 \cos^{-1} \frac{x}{r}.$$

And when $x = + r$, $A = \frac{1}{2} \pi r^2 =$ area of semicircle AEB.

$\therefore$ area of entire circle $AEBD = \pi r^2$.

To find the area of the segment ACD, make $x = \mathring{O}G = -a$, then

$$A = ACG = \frac{1}{2}\pi r^2 - \frac{1}{2}a(r^2 - a)^{\frac{1}{2}} - \frac{1}{2}r^2\cos^{-1}\left(-\frac{a}{r}\right)$$

$$= \frac{1}{2}r[\pi r - r \,.\, \cos^{-1}\left(-\frac{a}{r}\right)] - \frac{1}{2}a(r^2 - a^2)^{\frac{1}{2}}$$

$$= \frac{1}{2}r(ACB - CB) - \frac{1}{2}a(r^2 - a^2)^{\frac{1}{2}}$$

$$= \frac{1}{2}r \,.\, AC - \frac{1}{2}a \,.\, CG.$$

$$\therefore \text{ segment } CADC = r \,.\, AC - a \,.\, CG.$$

3. The elliptic segment AC_1D_1.

Here the equation of the curve is

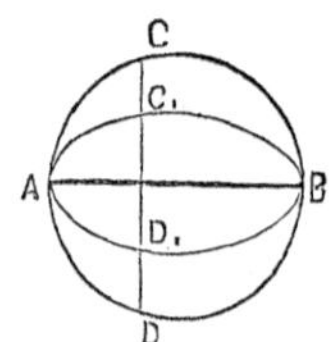

$$y = \frac{b}{a}(a^2 - x^2)^{\frac{1}{2}}.$$

$$\therefore A = \int y dx = \frac{b}{a}\int (a^2 - x^2)^{\frac{1}{2}} dx.$$

$\therefore 2A =$ segment $AC_1D_1 = \frac{b}{a}$. segment ACD of a circle described on AB.

Hence the area of the entire ellipse $= \frac{b}{a} \cdot$ area circle $= \frac{b}{a} \cdot \pi a^2 = \pi ab.$

4. The cycloid $\quad y = (2rx - x^2)^{\frac{1}{2}} + r \,.\, \text{versin}^{-1}\frac{x}{r}$

Put $OD = x, \quad DP = y.$

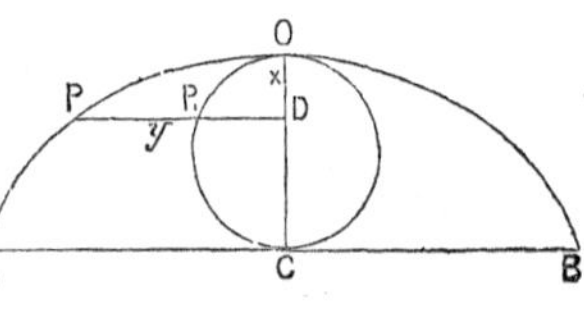

Then the area $OPD = \int y dx.$

But since y is a transcendental function of x, it will be preferable to integrate this expression by parts. Thus

$$A = \int y dx = xy - \int x dy.$$

But from the equation of the curve we have

$$\frac{dy}{dx}=\sqrt{\frac{2r-x}{x}} \quad \text{or} \quad dy=\sqrt{\frac{2r-x}{x}}\,dx.$$

$$\therefore A = xy - \int \sqrt{2rx - x^2}\,.\,dx.$$

Now $\int \sqrt{2rx - x^2}dx = \int y_1 dx$ where y_1 is the ordinate DP_1 of the generating circle, corresponding to the abscissa $OD = x$,

or $$\int \sqrt{2rx - x^2}dx = \text{area } OP_1D.$$

$\therefore$ area $OPD = xy -$ area OP_1D, and when $x = OC = 2r$.

area semi-cycloid $OAC = OC \times CA -$ area semicircle OP_1C

$$= 2r\,.\,\pi r - \frac{1}{2}\pi r^2 = \frac{3}{2}\pi r^2.$$

$\therefore$ area entire cycloid $= 3\pi r^2 = 3$ area generating circle.

104. *Prop.* To determine a general formula for the quadrature of polar curves, their equation having the form $r = F\theta$.

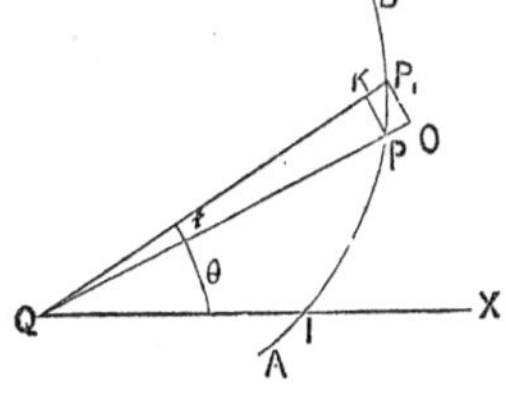

Let QX be the fixed axis, QP the radius vector, forming with QX an angle measured by the arc θ described with radius equal to unity.

Let θ take the increment t, converting r into $r_1 = F(\theta + t)$, and adding the sector QPP_1 to the area $QIP = A$, previously swept over by the radius vector. Now $QPP_1 > QPK$, but $< QP_1O$. Also the ratio

$$\frac{QP_1O}{QPK} = \frac{\frac{1}{2}r_1 \times r_1 t}{\frac{1}{2}r \times rt} = \frac{r_1^2}{r^2} = \frac{\left(r + \frac{dr}{d\theta}\cdot\frac{t}{1} + \frac{d^2r}{d\theta^2}\cdot\frac{t^2}{1\,.\,2} + \&c.\right)^2}{r^2}$$

$$= 1 + 2\frac{dr}{d\theta}\cdot\frac{t}{r} + 2\frac{d^2r}{d\theta^2}\cdot\frac{t^2}{1\,.\,2\,.\,r} + \frac{dr^2}{d\theta^2}\cdot\frac{t^2}{r^2}\ \&c.$$

$$= 1 \quad \text{when} \quad t = 0.$$

Hence at the limit, when t is replaced by $d\theta$, and QPP_1 becomes dA, the value of QPP_1 will be equal to QKP or QP_1O. Thus we shall have $dA = \frac{1}{2} r^2 . d\theta$.

$$\therefore A = \frac{1}{2} \int r^2 d\theta \ldots . (V_2), \text{ the required formula.}$$

1. The spiral of Archimedes $r = a\theta$.

$$A = \frac{1}{2} \int r^2 d\theta = \frac{1}{2} a^2 \int \theta^2 d\theta = \frac{1}{6} a^2 \theta^3 + C = \frac{1}{6} \frac{r^3}{a} + C.$$

If $A = 0$ when $r = r_1$, then $C = - \frac{1}{6} \frac{r_1^3}{a}$.

$$\therefore A = \frac{1}{6} \left(\frac{r^3 - r_1^3}{a} \right); \quad \text{and when} \quad r = r_2, A = \frac{1}{6} \left(\frac{r_2^3 - r_1^3}{a} \right).$$

For the area of one convolution estimated from the pole, we have the limits $r_1 = 0$ and $r_2 = 2\pi a$.

$$\therefore A = \frac{4}{3} \pi^3 a^2.$$

2. The logarithmic spiral from $r = r_1$ to $r = r_2$.

Here $r = a^{\theta}$. $\therefore dr = \log a . a^{\theta} . d\theta$ and $d\theta = \frac{1}{\log a} \cdot \frac{dr}{r}$.

$$\therefore A = \frac{1}{2} \int r^2 d\theta = \frac{1}{2 \log a} \int r dr = \frac{1}{4 \log a} r^2 + C.$$

$= \frac{1}{4} m (r_2^2 - r_1^2)$, between the limits r_1 and r_2: the quantity m denoting the modulus

3. The hyperbolic spiral from $r = r_1$ to $r = r_2$.

Here $r = \frac{a}{\theta}$, $dr = - \frac{a d\theta}{\theta^2}$, $d\theta = - \frac{\theta^2}{a} \cdot dr = - \frac{a}{r^2} dr$.

$\therefore A = - \frac{1}{2} \int a dr = - \frac{1}{2} ar + C = \frac{a(r_1 - r_2)}{2}$ between the limits r_1 and r_2.

4. The lemniscata $r^2 = a^2 \cos 2\theta.$

$$A = \frac{1}{2} \int r^2 d\theta = \frac{1}{2} a^2 \int \cos 2\theta d\theta = \frac{1}{4} a^2 \sin 2\theta + C.$$

Put $A = 0$ when $\theta = 0$; then $C = 0$, and $A = \frac{1}{4} a^2 \sin 2\theta$,

which gives, when $r = 0$, or $\theta = \frac{1}{4}\pi$, $A = \frac{1}{4} a^2$,

$\therefore$ Entire area $= a^2 =$ square described on semi-axis.

105. *Prop.* To find a formula for the quadrature of a plane curve, when its equation is given by a relation between the radius vector, and the perpendicular upon the tangent.

Since $d\theta = \dfrac{pdr}{r(r^2 - p^2)^{\frac{1}{2}}}$, $A = \frac{1}{2} \int r^2 d\theta = \frac{1}{2} \int \dfrac{prdr}{(r^2 - p^2)^{\frac{1}{2}}}$

1. The involute of the circle

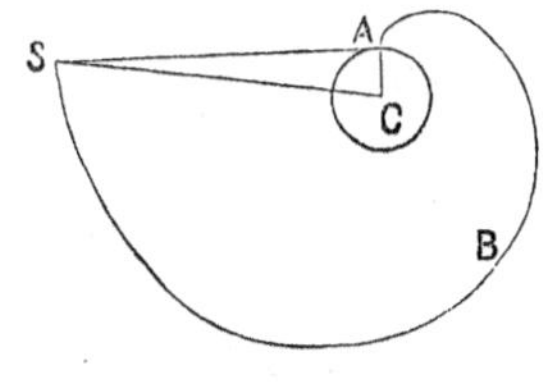

$$r^2 - p^2 = a^2.$$

$$A = \frac{1}{2} \int \frac{prdr}{(r^2 - p^2)^{\frac{1}{2}}} = \frac{1}{2a} \int (r^2 - a^2)^{\frac{1}{2}} rdr$$

$$= \frac{1}{6a}(r^2 - a^2)^{\frac{3}{2}} = \frac{p^3}{6a} + C;$$

and this, between the limits $p = 0$, and $p = 2\pi a$, within which the entire circumference is unwound, gives

$$A = \frac{4}{3} \pi^3 a^2.$$

Cor. The area included between the involute ABS, the circle, and the tangent AS, is equivalent to that swept over by the radius vector, and therefore equal to $\frac{p^3}{6a}$.

2. The epicycloid $p^2 = \dfrac{c^2(r^2 - a^2)}{c^2 - a^2}$, where $c = a + 2b$, a and b being the radii of the fixed and generating circles.

$$A = \frac{1}{2}\int \frac{prdr}{\sqrt{r^2 - p^2}} = \frac{1}{2}\frac{c}{a}\int \frac{(r^2 - a^2)^{\frac{1}{2}}}{(c^2 - r^2)^{\frac{1}{2}}} rdr.$$

Put $(r^2 - a^2)^{\frac{1}{2}} = z, \quad \therefore\ r^2 = z^2 + a^2, \quad rdr = zdz,$

$$(c^2 - r^2)^{\frac{1}{2}} = (c^2 - a^2 - z^2)^{\frac{1}{2}},$$

$$\therefore\ A = \frac{1}{2}\frac{c}{a}\int (c^2 - a^2 - z^2)^{-\frac{1}{2}} z^2 dz$$

$$= -\frac{z(c^2 - a^2 - z^2)^{\frac{1}{2}}c}{4a} + \frac{(c^2 - a^2)c}{4a}\int \frac{dz}{(c^2 - a^2 - z^2)^{\frac{1}{2}}}$$

$$-\frac{z(c^2 - a^2 - z^2)^{\frac{1}{2}}c}{4a} + \frac{(c^2 - a^2)c}{4a} \cdot \sin^{-1}\left[\frac{r^2 - a^2}{c^2 - a^2}\right]^{\frac{1}{2}} + C.$$

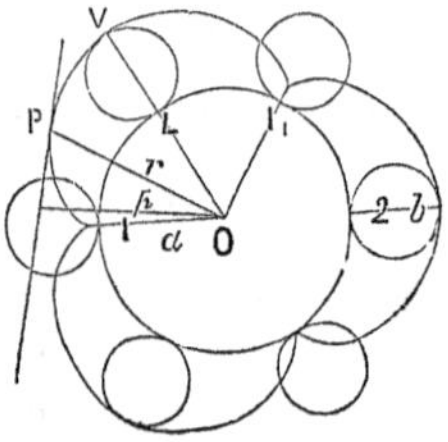

This, between the limits $r = a$ and $r = c$, gives

$$A = \frac{1}{4}\frac{c}{a}(c^2 - a^2)\frac{\pi}{2}$$

$$= \frac{b\pi}{2a}(a^2 + 3ab + 2b^2) = OIPVO.$$

But $\qquad OIL = \frac{1}{2}\pi ab.$

$$\therefore\ IPVL = \frac{1}{2}\text{ epicycloid} = \frac{b^2\pi}{2a}(3a + 2b),$$

and, $IVI_1LI = \dfrac{b^2\pi}{a}(3a + 2b)$, the entire epicycloid.

If $b = \frac{1}{2}a$, then epicycloid $= 4\pi b^2 = \pi a^2 =$ area fixed circle.

If $b = a$, then epicycloid $= 5\pi b^2 = 5\pi a^2 = 5$ area fixed circle.

CHAPTER III.

QUADRATURE OF CURVED SURFACES.

106. *Prop.* To obtain a general formula for the quadrature of a surface of revolution.

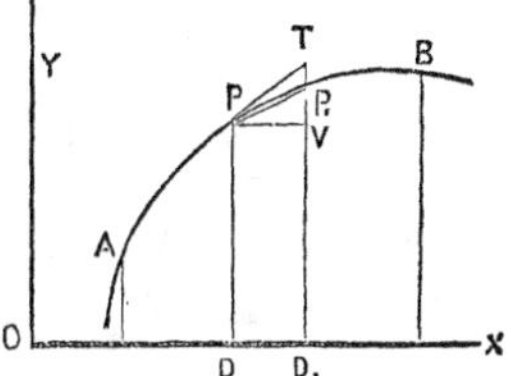

Let AB be the arc of a plane curve which revolves about the axis OX, P and P_1 points taken on the curve so near to each other that the arc PP_1 may present its concavity to OX at every point.

Put $OD = x$, $DP = y$, $DD_1 = h$, $D_1P_1 = y_1$, $AP = s$.

The surface generated by the arc PP_1, is intermediate in magnitude between those generated by the chord PP_1, and the broken line PTP_1. Denoting these surfaces by C and B, we have

$$\frac{B}{C} = \frac{\frac{1}{2}(PD + TD_1)2\pi PT + (TD_1^2 - P_1D_1^2)\pi}{\frac{1}{2}(PD + P_1D_1)PP_1 . 2\pi}$$

$$= \frac{(2PD + VT)PT + (2P_1D_1 + P_1T)P_1T}{(2PD + VP_1)PP_1}$$

$$= \frac{\left(2y+\frac{dy}{dx}\frac{h}{1}\right)\left(1+\frac{dy^2}{dx^2}\right)^{\frac{1}{2}}h+[2y+2\frac{dy}{dx}\frac{h}{1}+\frac{d^2y}{dx^2}\frac{h^2}{1.2}+\&c.][-\frac{d^2y}{dx^2}\frac{h^2}{1.2}\&c.]}{\left[2y+\frac{dy}{dx}\frac{h}{1}+\frac{d^2y}{dx^2}\frac{h^2}{1.2}\&c.\right]\left[h^2+h^2\left(\frac{dy}{dx}+\frac{d^2y}{dx^2}\frac{h}{1.2}\&c.\right)^2\right]^{\frac{1}{2}}}.$$

Dividing numerator and denominator by h, and then passing to the limit, we obtain $\frac{B}{C} = 1$. And hence the limit to the value of the surface C, generated by the chord, will be a proper expression for the elementary surface generated by the arc PP_1, when that arc becomes indefinitely small.

But at the limit, when $h = dx$, $C = 2\pi y\left(1 + \frac{dy^2}{dx^2}\right)^{\frac{1}{2}} dx$.

Hence we have for the differential of the surface,

$$dA = 2\pi y\left(1 + \frac{dy^2}{dx^2}\right)^{\frac{1}{2}} dx, \quad \text{and} \quad \therefore\ A = 2\pi \int y\left(1 + \frac{dy^2}{dx^2}\right)^{\frac{1}{2}} dx, \ldots (W).$$

or,

$$A = 2\pi \int y ds. \ldots . (W_1).$$

107. To apply (W), we eliminate, by means of the equation of the generating curve, y, and $\frac{dy}{dx}$, and then integrate between the given limits. Similarly, we apply (W_1) by expressing y in terms of s, or ds in terms of y and dy.

EXAMPLES.

108. 1. The surface of the sphere.

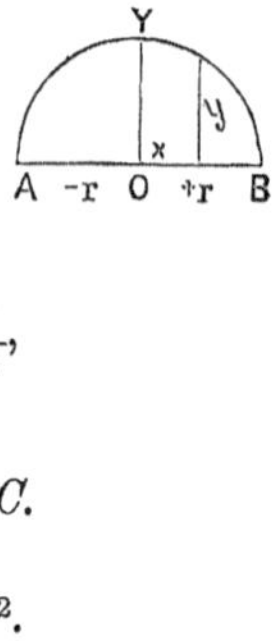

Here the generating curve is a circle whose equation is

$$y^2 = r^2 - x^2.$$

$$\therefore\ \frac{dy}{dx} = -\frac{x}{y}, \quad 1 + \frac{dy^2}{dx^2} = \frac{y^2 + x^2}{y^2} = \frac{r^2}{y^2},$$

$$\therefore\ A = 2\pi \int \frac{rydx}{y} = 2\pi r \int dx = 2\pi r x + C.$$

Put $A = 0$, when $x = -r$; then $C = 2\pi r^2$.

$\therefore\ A = 2\pi r(r + x)$, which, when $x = +r$, gives for the surface of the entire sphere $A = 4\pi r^2 = 4$ great circles.

For the zone whose height is $h = x_2 - x_1$, we have

$$A = 2\pi r(x_2 - x_1) = 2\pi r h.$$

1. The paraboloid of revolution,

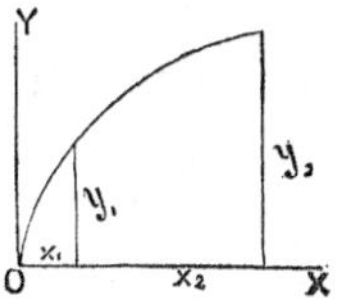

$$y^2=2px,\ \frac{dy}{dx}=\frac{p}{y},\ 1+\frac{dy^2}{dx^2}=1+\frac{p^2}{y^2}=\frac{y^2+p^2}{y^2}.$$

$$\therefore\ A=2\pi\int(y^2+p^2)^{\frac{1}{2}}dx=2\pi\int(2px+p^2)^{\frac{1}{2}}dx$$

$$=\frac{2\pi}{3p}(2px+p^2)^{\frac{3}{2}}+C=\frac{2\pi}{3p}[(2px_2+p^2)^{\frac{3}{2}}-(2px_1+p^2)^{\frac{3}{2}}].$$

If the surface be reckoned from the vertex, we shall have $x_1=0$.

$$\therefore\ A=\frac{2\pi}{3p}[(2px_2+p^2)^{\frac{3}{2}}-p^3].$$

2. The surface generated by the revolution of the Catenary about its axis.

The equation of the curve is $s^2=x^2+2ax$.*

$$\therefore\ ds=\frac{(x+a)dx}{\sqrt{x^2+2ax}}\cdot \text{ and } dy=\sqrt{ds^2-dx^2}=\frac{adx}{\sqrt{x^2+2ax}}=\frac{adx}{s}.$$

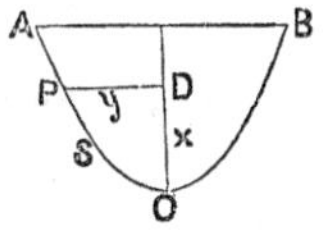

Now, applying formula (W_1), and integrating by parts, we have

$$A=2\pi\int yds=2\pi(ys-\int sdy)=2\pi(ys-a\int dx)$$

$$=2\pi(ys-ax)+C.$$

But when $x=0$, $y=0$ and $s=0$, $\therefore\ C=0$.

$$\therefore\ A=2\pi(y\sqrt{x^2+2ax}-ax).$$

3. The surface generated by the revolution of a semi-cycloid about its axis.

Here $dy=\sqrt{\dfrac{2r-x}{x}}\cdot dx$ and $s=2\sqrt{2rx}=\sqrt{8rx}$.

$$\therefore\ A=2\pi\int yds=2\pi(ys-\int sdy)=2\pi(ys-\int 2\sqrt{2rx}\sqrt{\frac{2r-x}{x}}\cdot dx)$$

$$=2\pi(y\sqrt{8rx}-\sqrt{8r}\int\sqrt{2r-x}.dx)$$

$$=2\pi[y\sqrt{8rx}+\sqrt{8r}.\frac{2}{3}(2r-x)^{\frac{3}{2}}]+C.$$

But when $x=0$, $A=0$, $\therefore\ C=-\dfrac{32}{3}\pi r^2$.

$$\therefore A = 2\pi\left[y\sqrt{8rx} + \sqrt{8r}\,.\,\frac{2}{3}(2r-x)^{\frac{3}{2}} - \frac{16}{3}r^2\right];$$

and when $x = 2r$, $A = 8\pi^2r^2 - \frac{32}{3}\pi r^2$, the entire surface.

4. The surface generated by the revolution of the cycloid about its base.

In the formula $A = 2\pi\int yds$, the quantity y denotes the distance of a point in the revolving curve from the axis of revolution, and must therefore be replaced in the present instance, by $2r - x$.

$$\therefore A = 2\pi\int(2r-x)ds = 2\pi\int(2r-x)\sqrt{\frac{2r}{x}}\,dx$$

$$= 2\pi\sqrt{2r}\left(4rx^{\frac{1}{2}} - \frac{2}{3}x^{\frac{3}{2}}\right) + C.$$

But $A=0$, when $x = 0$. $\therefore C=0$, $\therefore A=2\pi\sqrt{2r}\left(4rx^{\frac{1}{2}}-\frac{2}{3}x^{\frac{3}{2}}\right)$,

and when $x = 2r$.

$$\therefore A = \frac{32}{3}\pi r^2;\ \text{and the entire surface } 2A = \frac{64}{3}\pi r^2.$$

109. *Prop.* To obtain a general formula for the quadrature of any curved surface, whose equation is referred to rectangular co-ordinates.

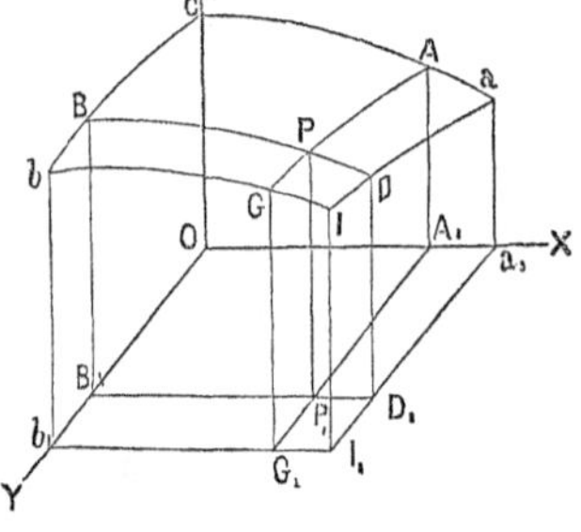

Let $CAPB$ be a portion of the surface included between the planes of xz and yz, and the planes BP_1, AP_1 drawn parallel thereto.

Put $OA_1 = x$, $OB_1 = A_1P_1 = y$, $P_1P = z$, $ACBP = A$, and let $z = F(x,y)\ldots(1)$ be the equation of the surface.

Then, since the value of A will be determined by the assumed values of the independent variables x and y, we shall have $A = \varphi(x,y)$.

Now when x receives an increment $A_1a_1 = h$, the area A takes the increment AD, becoming

$$A_1 = \varphi(x+h,y) = A + \frac{dA}{dx}\cdot\frac{h}{1} + \frac{d^2A}{dx^2}\cdot\frac{h^2}{1.2} + \frac{d^3A}{dx^3}\cdot\frac{h^3}{1.2.3} + \&c.$$

Similarly, when y alone takes an increment $B_1b_1 = k$, A takes the increment BG, becoming

$$A_2 = \varphi(x,y+k) = A + \frac{dA}{dy}\cdot\frac{k}{1} + \frac{d^2A}{dy^2}\cdot\frac{k^2}{1.2} + \frac{d^3A}{dy^3}\cdot\frac{k^3}{1.2.3} + \&c.$$

But, when x and y increase simultaneously, A takes an increment $AD + BG + PI$, becoming

$$A_3 = A + \frac{dA}{dx}\cdot\frac{h}{1} + \frac{dA}{dy}\cdot\frac{k}{1} + \frac{d^2A}{dx^2}\cdot\frac{h^2}{1.2} + \frac{d^2A}{dxdy}\cdot\frac{hk}{1} + \frac{d^2A}{dy^2}\cdot\frac{k^2}{1.2}$$

$$+ \frac{d^3A}{dx^3}\cdot\frac{h^3}{1.2.3} + \frac{d^3A}{dx^2dy}\cdot\frac{h^2k}{1.2} + \frac{d^3A}{dxdy^2}\cdot\frac{hk^2}{1.2} + \frac{d^3A}{dy^3}\cdot\frac{k^3}{1.2.3} + \&c.$$

$$\therefore\ PI = A_3 - A - (A_1 - A) - (A_2 - A)$$

$$= \frac{d^2A}{dxdy}\cdot\frac{hk}{1} + \frac{d^3A}{dx^2dy}\cdot\frac{h^2k}{1.2} + \frac{d^3A}{dxdy^2}\cdot\frac{hk^2}{1.2} + \&c.$$

and
$$\frac{PI}{hk} = \frac{d^2A}{dxdy} + \frac{d^3A}{dx^2dy}\cdot\frac{h}{1.2} + \frac{d^3A}{dx\,.\,dy^2}\cdot\frac{k}{1.2} + \&c.$$

which, at the limit when $h = 0$ and $k = 0$, reduces to

$$\frac{PI}{P_1I_1} = \frac{d^2A}{dxdy} \cdots\cdots\ (1).$$

Now this quotient, which results from dividing the elementary surface PI by its projection P_1I_1 on the plane of xy, is equal to $\frac{1}{\cos v}$, where v denotes the angle formed by the tangent plane at the point P with the plane of xy.

But from the theory of surfaces (Diff. Cal., Art. 177), we have

$$\cos v = \frac{1}{\sqrt{1 + \frac{dz^2}{dx^2} + \frac{dz^2}{dy^2}}}.$$

$$\therefore\ \frac{d^2A}{dxdy} = \sqrt{1 + \frac{dz^2}{dx^2} + \frac{dz^2}{dy^2}}.$$

Now, since the second differential coefficient $\frac{d^2A}{dxdy}$ is obtained by differentiating the function A of x and y, first as though x were alone variable, and then as though y only varied, we shall obtain the value of A by multiplying the value of $\frac{d^2A}{dxdy}$ by $dxdy$, and then performing two successive integrations with respect to x and y, the order of these integrations being immaterial, since that of the differentiations is arbitrary.

This double integration is indicated by the symbol $\int\int$, and the result is called a *double integral.* Thus

$$A = \int\int\left(1 + \frac{dz^2}{dx^2} + \frac{dz^2}{dy^2}\right)^{\frac{1}{2}} dxdy \ldots . (W_2), \text{ the required formula.}$$

The limits of these integrations, in the case represented in the diagram, are $y = 0$ and $y = OB_1 = b$, $x = 0$ and $x = OA_1 = a$. But if the surface were terminated laterally by a cylinder (instead of by planes parallel to xz and yz), the elements of this cylinder being parallel to the axis of z, and its base in the plane of xy represented by the equation $y_1 = fx$, then the superior limit of the first integration would be $y = y_1 = fx$, the inferior limit being still zero. This will be rendered plain by an example.

110. 1. Required the surface of the tri-rectangular triangle ABC.

From the equation of the surface $x^2 + y^2 + z^2 = r^2$, we obtain

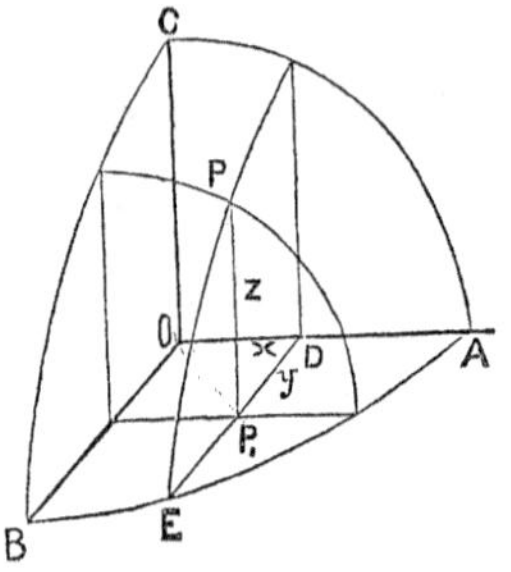

$$\frac{dz}{dx} = -\frac{x}{z} \quad \text{and} \quad \frac{dz}{dy} = -\frac{y}{z}.$$

$$\therefore \sqrt{1 + \frac{dz^2}{dx^2} + \frac{dz^2}{dy^2}} = \frac{r}{z}.$$

$$\therefore A = \int\int\frac{r}{z}dxdy = r\int\int\frac{dxdy}{\sqrt{r^2 - x^2 - y^2}}$$

$$= r\int \sin^{-1}\left(\frac{y}{\sqrt{r^2 - x^2}}\right)dx.$$

The limits of this first integration with respect to y are $y = 0$ and $y = \sqrt{r^2 - x^2} = DE$.

But when $\quad y = 0, \quad \frac{y}{\sqrt{r^2 - x^2}} = 0,$

and when $y = \sqrt{r^2 - x^2}, \quad \frac{y}{\sqrt{r^2 - x^2}} = 1,$ and $\sin^{-1}(1) = \frac{1}{2}\pi.$

$$\therefore A = \frac{1}{2}\pi r \int dx = \frac{1}{2}\pi r x + C = \frac{1}{2}\pi r^2$$

between the limits $x = 0$, and $x = r$.

2. The axes of two equal circular semi-cylinders intersect at right angles, forming the figure called the groin. Required the entire surface intercepted upon the two cylinders.

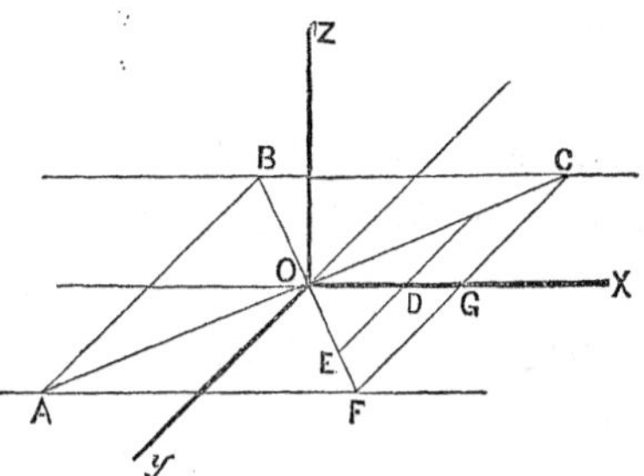

Assuming the axes of the cylinders as those of x and y respectively, the equation of the cylinder whose axis coincides with x will be $y^2 + z^2 = r^2$, and that of the cylinder whose axis coincides with y will be $x^2 + z^2 = r^2$.

The entire surface to be estimated is projected upon xy in the rectangle $ABCF$, and the triangle OGF is the projection of one-eighth of this surface. To compute this portion to which the equation $x^2 + z^2 = r^2$ applies, we have $A = \int\int\left(1 + \frac{dz^2}{dx^2} + \frac{dz^2}{dy^2}\right)^{\frac{1}{2}} dxdy$, in which the limits of integration are $y = 0$ and $y = x$, $x = 0$ and $x = r$.

But from the equation $x^2 + z^2 = r^2$, we get $\frac{dz}{dx} = -\frac{x}{z}, \frac{dz}{dy} = 0.$

$$\therefore A = \int\int\left(1 + \frac{x^2}{z^2}\right)^{\frac{1}{2}} dxdy = \int\int\frac{r}{z}\, dxdy = \int\int\frac{rdxdy}{\sqrt{r^2 - x^2}}$$

$$= r\int\frac{ydx}{\sqrt{r^2 - x^2}} = r\int\frac{xdx}{\sqrt{r^2 - x^2}} \text{ between the given limits.}$$

or $A = r\sqrt{r^2 - x^2} + C = r^2$ between the limits $x = 0$ and $x = r$.

$\therefore 8A = 8r^2$, the entire surface of the groin.

CHAPTER IV.

CUBATURE OF VOLUMES.

111. *Prop* To obtain a general formula for the volume generated by the revolution of a plane figure about a fixed axis.

Let OX, the axis of x, be the axis of revolution, $ABCF$ the generating area. Put

$$OD=x,\ DP=y,\ DD_1=h,\ D_1P_1=y_1,$$

and let $y = Fx$ be the equation of the bounding curve AB.

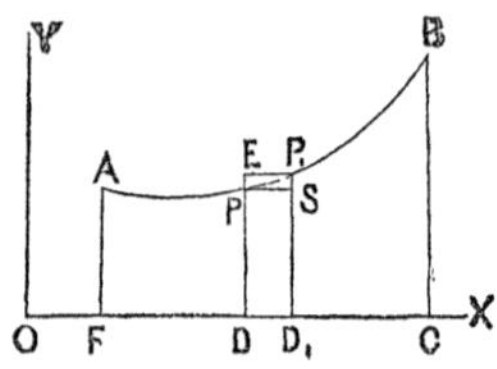

The volume generated by the revolution of the small quadrilateral DPP_1D_1 is intermediate in magnitude between the cylinders generated by the rectangles PD_1 and ED_1. But

$$\frac{\text{cylinder } ED_1}{\text{cylinder } PD_1} = \frac{\pi y_1{}^2 h}{\pi y^2 h} = \frac{y_1{}^2}{y^2} = \frac{\left(y + \dfrac{dy}{dx}\cdot\dfrac{h}{1} + \dfrac{d^2y}{dx^2}\cdot\dfrac{h^2}{1.2} + \&c.\right)^2}{y^2}$$

$$= 1 + 2\frac{dy}{dx}\cdot\frac{h}{y} + 2\frac{d^2y}{dx^2}\cdot\frac{h^2}{1.2.y} + \frac{dy^2}{dx^2}\cdot\frac{h^2}{y^2} + \&c.$$

$$= 1 \quad \text{when} \quad h = 0.$$

Therefore at the limit the volume generated by $DPP_1D_1 =$ cylinder PD_1, or $dV = \pi y^2 dx$, and consequently $V = \pi\int y^2 dx \ldots (X)$ the required formula.

To apply (X), we substitute for y^2 its value in terms of x derived from the equation of the bounding curve AB, and then integrate between the given limits.

112. 1. The sphere.

Here the equation of the circle which bounds the generating area is $x^2 + y^2 = r^2$.

$$\therefore V = \int \pi y^2 dx = \pi \int (r^2 - x^2) dx = \pi (r^2 x - \frac{1}{3} x^3) + C.$$

Put $V = 0$ when $x = -r$,

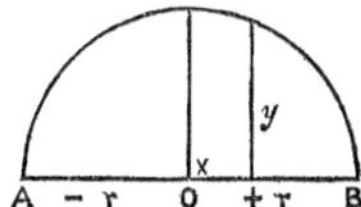

then $C = -\pi\left(\frac{1}{3} r^3 - r^3\right) = \frac{2}{3} \pi r^3.$

$\therefore V = \pi (r^2 x - \frac{1}{3} x^3) + \frac{2}{3} \pi r^3$ and when $x = +r$, $V = \frac{4}{3} \pi r^3$.

2. The ellipsoid of revolution, generated by the revolution of the semi-ellipse about its greater axis $2a$.

Here $y^2 = \frac{b^2}{a^2}(a^2 - x^2)$. $\therefore V = \frac{b^2 \pi}{a^2} \int (a^2 - x^2) dx = \frac{b^2 \pi}{a^2}(a^2 x - \frac{1}{3} x^3) + C$,

which gives between the limits $x = -a$ and $x = +a$.

$$V = \frac{4}{3} \pi b^2 a = \frac{2}{3} (2a \,.\, \pi b^2) = \frac{2}{3} \text{ circumscribing cylinder.}$$

3. The paraboloid of revolution

$$y^2 = 2px. \quad V = 2\pi p \int x dx = \pi p x^2 + C.$$

If $V = 0$ when $x = 0$; then $C = 0$ and $V = \pi p x^2$;

which becomes, when $x = x_1$ and $y = y_1$,

$$V = \pi p x_1^2 = \frac{1}{2} x_1 \,.\, \pi y_1^2 = \frac{1}{2} \text{ circumscribing cylinder.}$$

4. The parabolic spindle generated by the revolution of the parabolic area AQB about the double ordinate AB.

Put $OQ = a$, $OA = b$, $OD = x$, $DP = y$. Then $QC = a - y$.

$x^2 = 2p(a - y)$ and $V = \pi \int \left(a - \frac{x^2}{2p}\right)^2 dx = \frac{\pi}{4p^2} \int (4a^2 p^2 - 4apx^2 + x^4) dx.$

$$\therefore V = \pi \left(a^2 x - \frac{ax^3}{3p} + \frac{x^5}{20p^2}\right) + C.$$

But if $V=0$ when $x=0$, then $C=0$; and when $x=OA=b$.

$$V=\pi\left(a^2b-\frac{ab^3}{3p}+\frac{b^5}{20p^2}\right);$$

or since $$\frac{b^2}{2p}=a.$$

$$V=\pi a^2b\left(1-\frac{2}{3}+\frac{1}{5}\right)=\frac{8}{15}\pi a^2b=\text{volume } AQO.$$

$$\therefore \text{ volume } AQB=\frac{16}{15}\pi a^2b.$$

5. The volume generated by the revolution of the cycloid about its base.

Put $OV=2r$, $OD=x$, $DP=y$, $IV=z=2r-y$.

Then from the equation of the cycloid,

$$\frac{dx}{dz}=\left(\frac{2r-z}{z}\right)^{\frac{1}{2}}, \text{ and since } dz=-dy.$$

$$\therefore \frac{dx}{dy}=-\left(\frac{2r-z}{z}\right)^{\frac{1}{2}}$$

$$=-\left(\frac{y}{2r-y}\right)^{\frac{1}{2}}$$

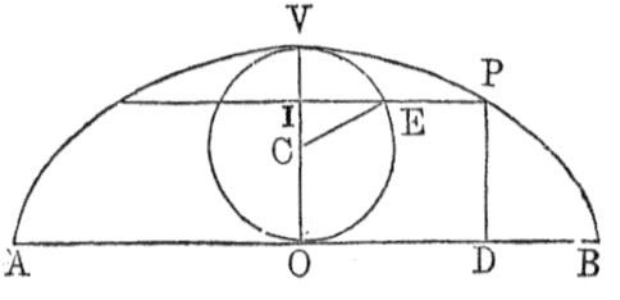

$$\therefore V=\pi\int y^2dx=-\pi\int y^2\left(\frac{y}{2r-y}\right)^{\frac{1}{2}}dy=-\pi\int y^{\frac{5}{2}}(2r-y)^{-\frac{1}{2}}dy.$$

But by formula (A),

$$\int y^{\frac{5}{2}}(2r-y)^{-\frac{1}{2}}dy=-\frac{1}{3}y^{\frac{5}{2}}(2r-y)^{\frac{1}{2}}+\frac{5}{3}r\int y^{\frac{3}{2}}(2r-y)^{-\frac{1}{2}}dy$$

$$\int y^{\frac{3}{2}}(2r-y)^{-\frac{1}{2}}dy=-\frac{1}{2}y^{\frac{3}{2}}(2r-y)^{\frac{1}{2}}+\frac{3}{2}r\int y^{\frac{1}{2}}(2r-y)^{-\frac{1}{2}}dy$$

$$\int y^{\frac{1}{2}}(2r-y)^{-\frac{1}{2}}dy=-y^{\frac{1}{2}}(2r-y)^{\frac{1}{2}}+r\int y^{-\frac{1}{2}}(2r-y)^{-\frac{1}{2}}dy.$$

Also $$\int y^{-\frac{1}{2}}(2r-y)^{-\frac{1}{2}}dy=\int\frac{dy}{\sqrt{2ry-y^2}}=\text{versin}^{-1}\frac{y}{r}.$$

$$\therefore V=\pi(2r-y)^{\frac{1}{2}}\left(\frac{1}{3}y^{\frac{5}{2}}+\frac{5}{6}y^{\frac{3}{2}}r+\frac{5}{2}y^{\frac{1}{2}}r^2\right)-\frac{5}{2}\pi r^3.\text{versin}^{-1}\frac{y}{r}+C.$$

Put $V=0$ and $y=2r$. Then $C=\frac{5}{2}r^3\pi^2$, and when $y=0$.

$$V=\frac{5}{2}r^3\pi^2 = \text{volume } BVO.$$

$\therefore$ volume $BVA = 5r^3\pi^2 = 2\pi r\frac{5}{2}\pi r^2 = \frac{5}{8}$ circumscribing cylinder.

6. The volume generated by the revolution of the cycloid about its axis. (See last Fig.).

Put $VI=x$, $IP=y$, $VO=2r$, and to facilitate the integration, introduce the variable angle $VCE=\theta$.

Then $x=r(1-\cos\theta)$, $y=r(\sin\theta+\theta)$, $dx=r.\sin\theta d\theta$.

$$\therefore V=\pi\int y^2dx = \pi r^3\int(\sin^3\theta + 2\theta.\sin^2\theta + \theta^2.\sin\theta)d\theta.$$

But $\int\sin^3\theta.d\theta = -\frac{1}{3}\sin^2\theta.\cos\theta - \frac{2}{3}\cos\theta = \frac{4}{3}$, from $\theta=0$ to $\theta=\pi$.

$$2\int\theta.\sin^2\theta.d\theta = \int\theta.d\theta - \int\theta.\cos 2\theta d\theta$$

$$=\frac{1}{2}\theta^2 - \frac{1}{2}\theta\sin 2\theta + \frac{1}{2}\int\sin 2\theta.d\theta \text{ by integrating by parts.}$$

$$=\frac{1}{2}\theta^2 - \frac{1}{2}\theta\sin 2\theta - \frac{1}{4}\cos 2\theta = \frac{1}{2}\pi^2, \text{ from } \theta=0 \text{ to } \theta=\pi,$$

and
$$\int\theta^2.\sin\theta.d\theta = -\theta^2\cos\theta + 2\int\theta.\cos\theta d\theta$$
$$= -\theta^2\cos\theta + 2\theta\sin\theta - 2\int\sin\theta d\theta$$
$$= -\theta^2\cos\theta + 2\theta\sin\theta + 2\cos\theta$$
$$= \pi^2 - 4, \text{ from } \theta=0 \text{ to } \theta=\pi.$$

$$\therefore \text{Entire volume} = \pi r^3\left(\frac{3}{2}\pi^2 - \frac{8}{3}\right).$$

113. *Prop.* To obtain a general formula for the volume of all solids which are symmetrical with respect to an axis.

Such solids may be generated by the motion of a plane figure, as $ABCD$, of variable dimensions, and of any form, whose centre G remains upon the axis OX, its plane being always perpendicular to OX, and its variable area X being a function of x, its distance from the origin.

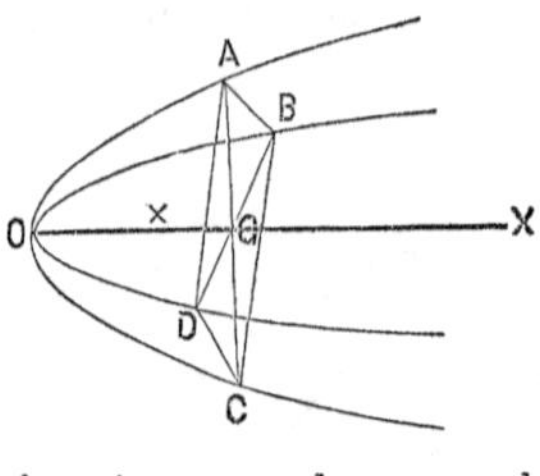

By a method entirely similar to that applied to solids of revolution, we may show that $dV = Xdx$,

and $\therefore V = \int Xdx \ldots . (X_1)$,

the required formula.

To apply (X_1) we must express the value of the area X in terms of x, and then integrate between the proper limits.

Cor. The same formula is applicable to any solid generated by the motion of a section of variable dimensions parallel to a given plane, when the area of the section can be expressed in functions of its distance from the fixed plane.

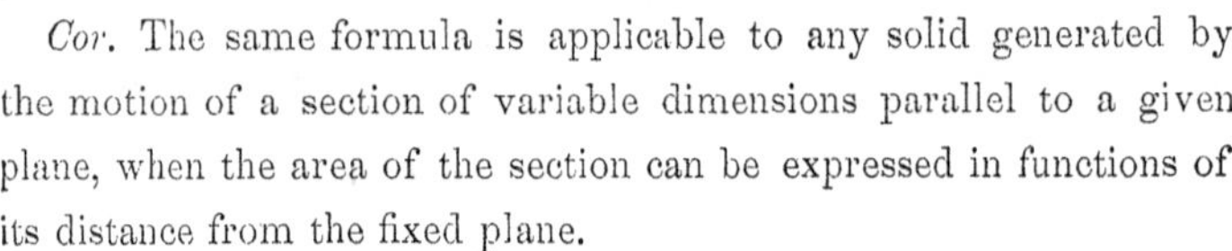

114. 1. The ellipsoid with three unequal axes.

Here we have
$$\frac{x^2}{a^2} + \frac{y^2}{b^2} + \frac{z^2}{c^2} = 1,$$

or
$$b^2c^2x^2 + a^2c^2y^2 + a^2b^2z^2 = a^2b^2c^2.$$

Make $CC_1 = x$, and put successively

$$y = 0 \quad \text{and} \quad z = 0.$$

Then when

$$y = 0, \; z = \frac{c}{a}\sqrt{a^2 - x^2} = B_1C_1;$$

and when

$$z = 0, \; y = \frac{b}{a}\sqrt{a^2 - x^2} = D_1C_1.$$

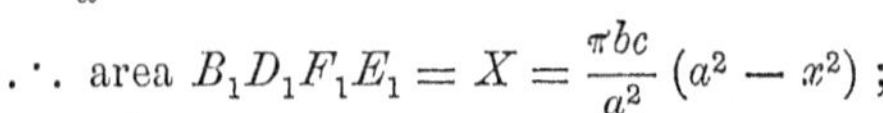

$$\therefore \text{ area } B_1D_1F_1E_1 = X = \frac{\pi bc}{a^2}(a^2 - x^2);$$

and this value substituted in (X_1) gives

$$V = \frac{\pi bc}{a^2}\int(a^2 - x^2)\,dx = \frac{\pi bc}{a^2}\left(a^2x - \frac{1}{3}x^3\right) + C.$$

Put $V = 0$ when $x = -a$; then $C = \frac{\pi bc}{a^2} \cdot \frac{2}{3}a^3$, and when $x = +a$

$$V = \frac{4}{3}\pi bca = \text{entire ellipsoid} = \frac{2}{3} \text{ circumscribing cylinder.}$$

2. The elliptical paraboloid $cz^2 + by^2 = a^2x$.

Put successively $y = 0$ and $z = 0$;

then $CB = a\left(\frac{x}{c}\right)^{\frac{1}{2}}$, and $CD = a\left(\frac{x}{b}\right)^{\frac{1}{2}}$.

Then $X = \frac{\pi a^2 x}{\sqrt{bc}}$. And

$$V = \frac{\pi a^2}{\sqrt{bc}}\int x \,.\, dx = \frac{1}{2}\frac{\pi a^2 x^2}{\sqrt{bc}} + C.$$

If $V = 0$ when $X = 0$, then $C = 0$, and

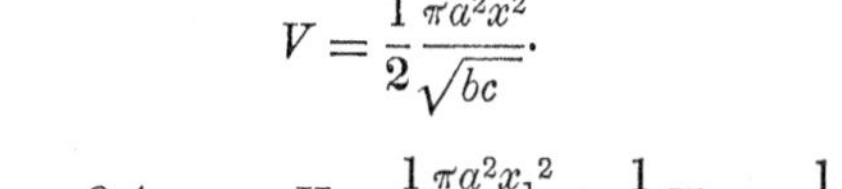

$$V = \frac{1}{2}\frac{\pi a^2 x^2}{\sqrt{bc}}.$$

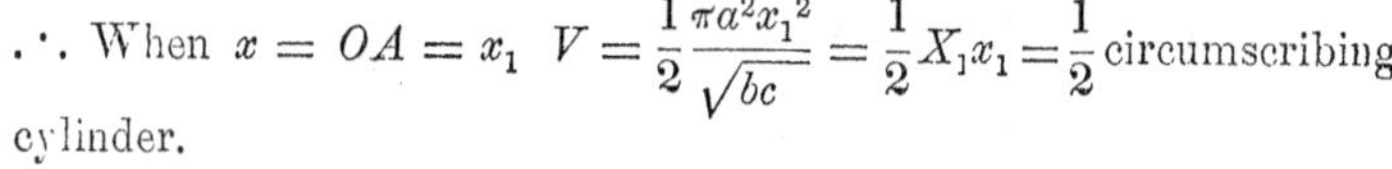

$\therefore$ When $x = OA = x_1$ $V = \frac{1}{2}\frac{\pi a^2 x_1{}^2}{\sqrt{bc}} = \frac{1}{2}X_1 x_1 = \frac{1}{2}$ circumscribing cylinder.

3. The groin or solid formed by the intersection of two cylinders whose axes are perpendicular to each other.

1st. Let the bases of the cylinders be equal semi-circles.

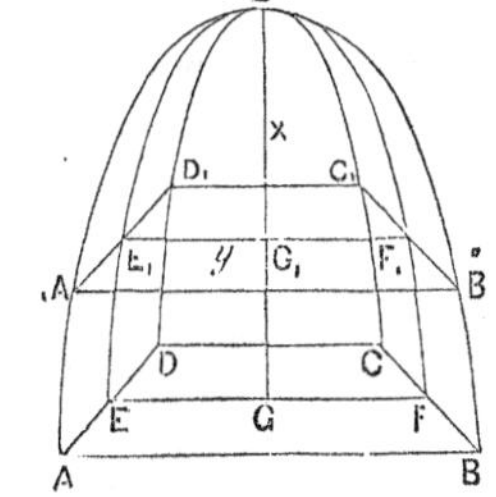

Then the generating section $A_1B_1C_1D_1$ will be a square.

Put $OG = GE = EA = r$, $OG_1 = x$,
$G_1E_1 = y = E_1A_1$.

Then $A_1B_1C_1D_1 = 4y^2$, and from the equation of the circle EOF,

$y^2 = 2rx - x^2$. $\therefore$ $V = \int X dx = \int (8rx - 4x^2)dx = 4rx^2 - \frac{4}{3}x^3 + C.$

But $V = 0$ when $x = 0$, $\therefore$ $C = 0$, and when $x = r$,

$V = \frac{8}{3}r^3 = \frac{2}{3}r \,.\, 2r \,.\, 2r = \frac{2}{3}$ circumscribing parallelopipedon.

2d. Let the bases be unequal parabolas.

Then the generating section will be a rectangle.

Put $OG = a$, $GE = b$, $EA = b_1$, $OG_1 = x$, $G_1E_1 = y$, $E_1A_1 = y_1$

Then $y^2 = 2px$, $y_1{}^2 = 2p_1x$. $\therefore X = 2y \,.\, 2y_1 = 8x\sqrt{pp_1}$.

$V = \int Xdx = 8\sqrt{pp_1}\int xdx = 4x^2\sqrt{pp_1} = 2x \,.\, yy_1$, and when $x = a$.

$V = 2abb_1 = \frac{1}{2}a \,.\, 2b \,.\, 2b_1 = \frac{1}{2}$ circumscribing parallelopipedon.

4. The Conoid, with a circular base.

Put $DA = a$, $DE = 2r$, $DG = x$, $GI = y$.

Then the generating triangle $IFH = X = ay$

$$= a\sqrt{2rx - x^2}.$$

$$\therefore V = \int Xdx = a\int\sqrt{2rx - x^2}\,.\,dx$$

$$= a \,.\, \frac{1}{2} \text{ segment } DGH.$$

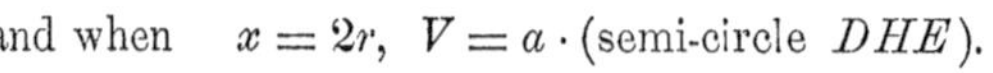

and when $x = 2r$, $V = a \cdot (\text{semi-circle } DHE)$.

or volume conoid $= \frac{1}{2}$ volume circumscribing cylinder.

Cor. A similar result will be obtained if we suppose the base to have any other form, the generating triangle being still perpendicular to the base.

115. *Prop.* To obtain a general formula for the volume of a solid bounded by any curved surface, whose equation is referred to rectangular co-ordinates.

First suppose the volume bounded by the co-ordinate planes of xy, xz, and yz, by planes parallel to xz and yz, respectively, and by the curved surface $CaIb$, whose equation is

$$z_1 = F(x,y).$$

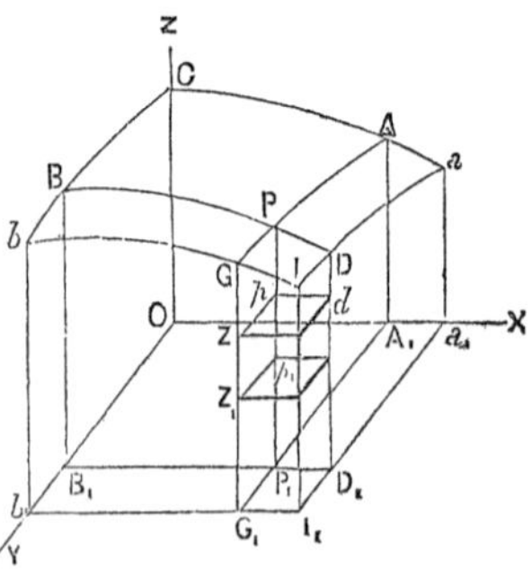

Put $OA_1 = x$, $OB_1 = A_1P_1 = y$,

$P_1p_1 = z$, $P_1P = z_1$,

$A_1a_1 = dx$, $P_1G_1 = dy$, $p_1p = dz$.

Let the volume be intersected by planes AG_1 and aI_1, parallel to yz, and including between them the lamina or slice A_1I: let this lamina be cut by planes bI_1 BD_1, &c.,

dividing it into prisms such as PI_1, &c.; and, finally, let each prism be subdivided into elementary parallelopipedons, such as z_1d by planes parallel to xy, the successive planes being at distances from each other denoted by dx, dy, and dz, respectively. Then the volume of one of these elementary parallelopipedons will be expressed by $dxdydz$; and if this be integrated with respect to z, regarding x and y as constant between the limits $z = 0$ and $z = z_1 = P_1P = F(x,y)$, the result obtained will represent the sum of all the parallelopipedons contained in the prism PI_1. A second integration, with respect to y, between the limits $y=0$ and $y=A_1G_1$, will give the sum of the prisms contained in the lamina AI_1; and a third integration, with respect to x, between the limits $x = 0$ and $x = Oa_1$, will give the sum of the laminæ, which constitute the entire volume.

Hence the required formula is

$$V = \int\int\int dxdydz \dots\dots (1).$$

The symbol $\int\int\int$ denotes three successive integrations, with respect to the variables x, y, and z, and the result is called the *triple integral* of $dxdydz$.

Cor. If the volume were bounded on every side by the curved surface, the same formula (1) would apply, but the limits of integration would be different, those of the first integration being $z = z_1$ and $z = z_2$ where z_1 and z_2 are the two extreme values of z corresponding to the same values of x and y, and derived from the equation of the surface; those of the second integration being $y = y_1$ and $y = y_2$, the extreme values of y corresponding to the same value of x, and derived from the equation of the section perpendicular to OX; and, finally, those of the third integration being x_1 and x_2, the extreme values of x.

116. 1. The tri-rectangular spherical sector.

Here the limits of the integration are $z=0$ and $P_1P=\sqrt{r^2-x^2-y^2}$,

$$y=0 \text{ and } y=D_1E=\sqrt{r^2-x^2},\ x=0, \text{ and } x=OA=r.$$

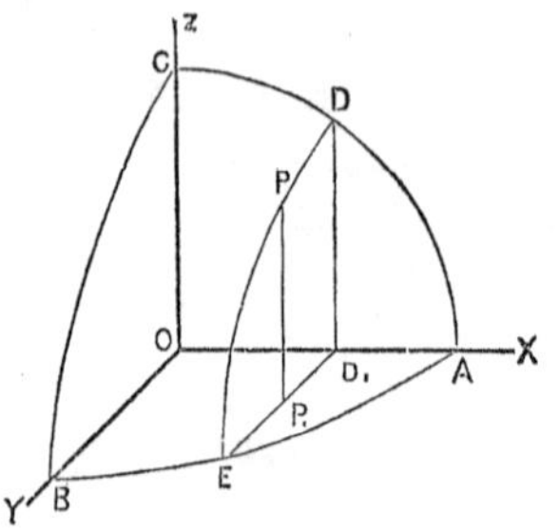

$\therefore\ V = \int\int\int dxdydz = \int\int zdxdy$

$= \int\int\sqrt{r^2 - x^2 - y^2}\,.\,dxdy.$ But

$\int\sqrt{r^2-x^2-y^2}\,.\,dy = \frac{1}{2}y(r^2-x^2-y^2)^{\frac{1}{2}}$

$+\frac{1}{2}(r^2-x^2)\int\frac{dy}{\sqrt{r^2-x^2-y^2}}$

$= \frac{1}{2}y(r^2-x^2-y^2)^{\frac{1}{2}}$

$+\frac{1}{2}(r^2-x^2)\sin^{-1}\frac{y}{\sqrt{r^2-x^2}} = \frac{1}{4}\pi(r^2-x^2)$ between the limits given.

$$\therefore\ V = \frac{1}{4}\pi\int(r^2 - x^2)dx = \pi\left(\frac{1}{4}r^2x - \frac{1}{12}x^3\right) = \frac{1}{6}\pi r^3$$

between the limits.

2. The volume cut from a paraboloid of revolution, the equation of whose generating curve is $y^2 = 2px$, by a right cylinder with a circular base, its axis passing through the focus, and the diameter of its base being equal to p.

The equation of the paraboloid being $y^2 + z^2 = 2px$, and that of the cylinder $y^2 = px - x^2$, the limits of integration in the present case will be

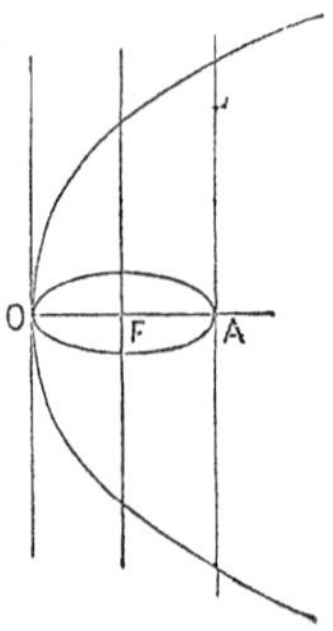

$z = +\sqrt{2px - y^2}$ and $z = -\sqrt{2px - y^2}$,

$y = +\sqrt{px - x^2}$ and $y = -\sqrt{px - x^2}$,

$x = 0$ and $x = p$.

$\therefore\ V = \int\int\int dxdydz = \int\int zdxdy = \int\int 2(2px - y^2)^{\frac{1}{2}}dxdy.$

But $\int(2px-y^2)^{\frac{1}{2}}dy = \frac{1}{2}y(2px-y^2)^{\frac{1}{2}} + px\int\frac{dy}{\sqrt{2px-y^2}}$

$$= \frac{1}{2}y(2px - y^2)^{\frac{1}{2}} + px\,.\,\sin^{-1}\frac{y}{\sqrt{2px}}$$

$= x\sqrt{p^2 - x^2} + 2px\,.\,\sin^{-1}\sqrt{\frac{p-x}{2p}}$ between the limits.

$$\therefore V = 2\int\left[x\sqrt{p^2 - x^2} + 2px \,.\, \sin^{-1}\sqrt{\frac{p-x}{2p}}\right]dx$$

$$= -\frac{2}{3}(p^2 - x^2)^{\frac{3}{2}} + 2px^2 . \sin^{-1}\sqrt{\frac{p-x}{2p}} + p\int\frac{x^2 dx}{\sqrt{p^2 - x^2}}$$

$$= -\frac{2}{3}(p^2 - x^2)^{\frac{3}{2}} + 2px^2 . \sin^{-1}\sqrt{\frac{p-x}{2p}}$$

$$-\frac{1}{2}px(p^2 - x^2)^{\frac{1}{2}} + \frac{1}{2}p^3 \sin^{-1}\frac{x}{p}$$

$$= p^3\left(\frac{2}{3} + \frac{1}{4}\pi\right) \text{ between the limits } x = 0 \text{ and } x = p.$$

117. *Prop.* To obtain a general formula for the volume of a solid bounded by a surface whose equation is referred to polar co-ordinates.

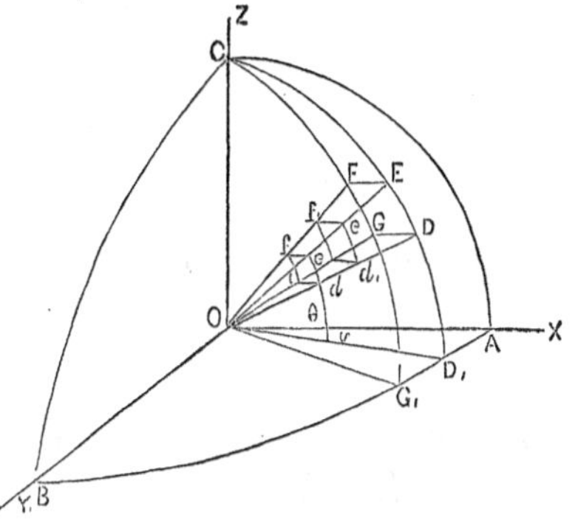

Let the volume be divided into elementary wedges such as G_1D_1CO by planes drawn through the axis OC. Let each wedge be subdivided into elementary pyramids, such as $FGDEO$, by conical surfaces generated by the revolution, about the axis OC, of lines OD, OE, &c., inclined to OC in constant angles. Finally, let each pyramid be subdivided into elementary parallelopipedons, such as fd_1 by concentric spherical surfaces with their centres at the origin O.

The co-ordinates of a point d are $Od = r$, $dOD_1 = \theta$, and $AOD_1 = v$; and the three edges of the elementary parallelopipedon fd_1, are $dd_1 = dr$, $de = rd\theta$, and $dt = r\cos\theta \,.\, dv$; the last expression being obtained by observing that when the line OD revolves around the axis OC, the point d describes a small arc dt whose centre lies upon the axis OC, and whose radius is the perpendicular distance of d from that axis, and therefore expressed by $r . \sin ZOd = r \cos \theta$.

Hence the volume of the parallelopipedon will be expressed by

$$r^2 \cos \theta dv . d\theta dr, \quad \text{and} \quad \therefore V = \int\int\int r^2 \cos \theta dv . d\theta . dr \ldots . (1)$$

will be the required formula for the entire volume.

The first integration, if performed with respect to r, while v and θ remain constant, will give the sum of the parallelopipedons contained in the pyramid $DEFGO$, the limits of the integration being $r = 0$ and $r = OD = F(v, \theta)$.

A second integration with respect to θ, while v remains constant, will give the sum of the pyramids contained in the wedge G_1D_1CO, and the third integration with respect to v will give the sum of the wedges which constitute the entire volume.

118. 1. The hemisphere with radius equal to a.

Here the limits of the integrations are

$$r = 0 \text{ and } r = a, \quad \theta = 0 \text{ and } \theta = \frac{1}{2}\pi, \quad v = 0 \text{ and } v = 2\pi.$$

$$\therefore V = \int\int\int r^2 \cos \theta . dv . d\theta dr = \frac{1}{3}\int\int r^3 \cos \theta . dv d\theta = \frac{1}{3} a^3 \int\int \cos \theta . dv . d\theta$$

$$= \frac{1}{3} a^3 \int \sin \theta . dv = \frac{1}{3} a^3 . \sin \frac{1}{2}\pi \int dv = \frac{1}{3} a^3 v = \frac{1}{3} a^3 . 2\pi = \frac{2}{3} \pi a^3.$$

2. The volume cut from a sphere whose radius is a, by a cylinder with a circular base whose radius $= b$, the centre of the sphere being on the axis of the cylinder.

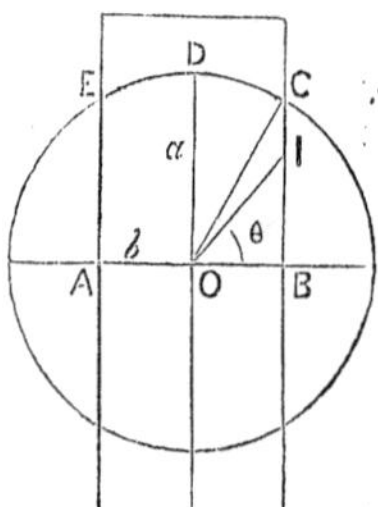

Here we shall have for half the required volume or $ABCDE$,

$$\frac{1}{2} V = \int\int\int r^2 \cos \theta . dv d\theta dr,$$

the limits of integration being

1st. $$r=0 \text{ and } r=OI=b\sec\theta,$$

$$\theta=0 \text{ and } \theta=\cos^{-1}\frac{b}{a}, \quad v=0 \text{ and } v=2\pi.$$

2d. $r=0$ and $r=a$, $\theta=\cos^{-1}\frac{b}{a}$ and $\theta=\frac{1}{2}\pi$, $v=0$ and $v=2\pi$.

The first set of limits give

$$\int\int\int r^2\cos\theta dv.d\theta dr=\frac{1}{3}\int\int r^3\cos\theta dv.d\theta=\frac{1}{3}\int\int b^3\sec^3\theta\cos\theta dv.d\theta$$

$$=\frac{1}{3}b^3\int\int\sec^2\theta.dvd\theta=\frac{1}{3}b^3\int\tan\theta.dv$$

$$=\frac{1}{3}b^3\tan\left(\cos^{-1}\frac{b}{a}\right)\int dv$$

$$=\frac{1}{3}b^3\tan\left(\tan^{-1}\frac{\sqrt{a^2-b^2}}{b}\right)v=\frac{2}{3}\pi b^2\sqrt{a^2-b^2}.$$

And the second set of limits give

$$\int\int\int r^2\cos\theta dv.d\theta dr=\frac{1}{3}\int\int r^3.\cos\theta.dv.d\theta=\frac{1}{3}a^3\int\int\cos\theta dv.d\theta$$

$$=\frac{1}{3}a^3\int\sin\theta.dv$$

$$=\frac{1}{3}a^3\sin\frac{1}{2}\pi\int dv-\frac{1}{3}a^3\sin\left(\sin^{-1}\frac{\sqrt{a^2-b^2}}{a}\right)\int dv$$

$$=\frac{1}{3}a^3\left(1-\frac{\sqrt{a^2-b^2}}{a}\right)v=\frac{2}{3}\pi a^3\left(1-\frac{\sqrt{a^2-b^2}}{a}\right).$$

$$\therefore\frac{1}{2}V=\frac{2}{3}\pi[a^3-(a^2-b^2)\sqrt{a^2-b^2}] \text{ and } V=\frac{4}{3}\pi\left[a^3-(a^2-b^2)^{\frac{3}{2}}\right].$$

PART III.

INTEGRATION OF FUNCTIONS OF TWO OR MORE VARIABLES.

CHAPTER I.

INTEGRATION OF EXPRESSIONS CONTAINING SEVERAL INDEPENDENT VARIABLES.

119. When a differential expression, containing two or more independent variables, can be obtained directly by differentiating some function of those variables, it is said to be an *exact differential.*

Thus $xdy + ydx$ is an exact differential, being equal to $d(xy)$; so also is $3x^2dy - 3ydx + 6xydx - 3xdy$, being equal to $d(3x^2y - 3xy)$; but $x^2dy - 3ydx$ is not an exact differential, there being no expression which, when differentiated, will produce that proposed.

120. If a differential be exact, its integral can be determined in all cases by methods which will be explained, but we shall first establish whereby to distinguish exact differentials.

121. *Prop.* To determine the conditions which indicate that any proposed differential is exact.

Let the proposed expression be $Pdx + Qdy$, in which P and Q may be functions of one or both variables.

If this expression be the exact differential of some function u of x and y, we shall have

$$du = Pdx + Qdy \text{ .... } (1).$$

But by the general process for differentiating a function of two independent variables, we have

$$du = \frac{du}{dx}dx + \frac{du}{dy}dy \ldots\ldots\ldots (2).$$

And since (1) and (2) must, from the nature of the supposition, be identical, the following conditions will exist, viz. :

$$P = \frac{du}{dx} \cdots\cdots (3), \quad Q = \frac{du}{dy} \cdots\cdots (4).$$

Now differentiate (3) with respect to y, and (4) with respect to x, and there will result

$$\frac{dP}{dy} = \frac{d^2u}{dxdy} \quad \text{and} \quad \frac{dQ}{dx} = \frac{d^2u}{dydx}.$$

But it has been shown that the result of differentiating u, with respect to x and y, successively, is the same, without reference to the order of the differentiations, or that

$$\frac{d^2u}{dxdy} = \frac{d^2u}{dydx}. \quad \therefore \frac{dP}{dy} = \frac{dQ}{dx} \cdots\cdots (5).$$

Hence, when the proposed differential $Pdx + Qdy$ is exact, the condition (5) will be fulfilled. The converse is equally true, as will appear fully when we attempt to integrate such expressions, and hence the condition (5) is called the *test of integrability.*

122. Now let the proposed expression be $Pdx + Qdy + Rdz$, involving three independent variables.

If this be an exact differential of some function u of x, y, and z, then

$$du = \frac{du}{dx}dx + \frac{du}{dy}dy + \frac{du}{dz}dz = Pdx + Qdy + Rdz;$$

whence $\quad P = \dfrac{du}{dx}, \quad Q = \dfrac{du}{dy}, \quad R = \dfrac{du}{dz},$ or by differentiation,

$$\frac{dP}{dy} = \frac{d^2u}{dxdy}, \quad \frac{dP}{dz} = \frac{d^2u}{dxdz}, \quad \frac{dQ}{dx} = \frac{d^2u}{dydx}, \quad \frac{dQ}{dz} = \frac{d^2u}{dydz},$$

$$\frac{dR}{dx} = \frac{d^2u}{dzdx}, \quad \frac{dR}{dy} = \frac{d^2u}{dzdy}.$$

But $$\frac{d^2u}{dxdy}=\frac{d^2u}{dydx}, \quad \frac{d^2u}{dxdz}=\frac{d^2u}{dzdx}, \quad \frac{d^2u}{dydz}=\frac{d^2u}{dzdy}.$$

Hence we have three following conditions of integrability,

$$\frac{dP}{dy}=\frac{dQ}{dx}, \quad \frac{dP}{dz}=\frac{dR}{dx}, \quad \frac{dQ}{dz}=\frac{dR}{dy}.$$

Similarly, if the expression were $Pdx + Qdy + Rdz + Sds +$ &c., involving n independent variables, there would be $\frac{1}{2}n(n-1)$, conditions of the forms

$$\frac{dP}{dy}=\frac{dQ}{dx}, \quad \frac{dP}{dz}=\frac{dR}{dx}. \quad \frac{dP}{ds}=\frac{dS}{dx}, \text{ \&c. } \frac{dQ}{dz}=\frac{dR}{dy}, \quad \frac{dQ}{ds}=\frac{dS}{dy} \text{ \&c.}$$

123. 1. Is $a^2ydx + x^3dx + b^3dy + a^2xdy$ an exact differential?

Here $\qquad P = a^2y + x^3$ and $Q = b^3 + a^2x$.

$$\therefore \frac{dP}{dy}=a^2, \quad \frac{dQ}{dx}=a^2, \quad \therefore \frac{dP}{dy}=\frac{dQ}{dx}$$ and the expression is integrable.

2. Is $\dfrac{dx}{(x^2+y^2)^{\frac{1}{2}}}+\dfrac{dy}{y}-\dfrac{xdy}{y(x^2+y^2)^{\frac{1}{2}}}$ an exact differential?

Here $\qquad P=(x^2+y^2)^{-\frac{1}{2}}$ and $Q=y^{-1}[1-(x^2+y^2)^{-\frac{1}{2}}x]$.

$$\frac{dP}{dy}=-y(x^2+y^2)^{-\frac{3}{2}}, \quad \frac{dQ}{dx}=y^{-1}[-(x^2+y^2)^{-\frac{1}{2}}+x^2(x^2+y^2)^{-\frac{3}{2}}],$$

or $\dfrac{dQ}{dx}=-y(x^2+y^2)^{-\frac{3}{2}}=\dfrac{dP}{dy}$, and the expression is integrable

3. Is $3xdy - 4y^2dx$ an exact differential?

$$P=-4y^2, \quad Q=3x, \quad \frac{dP}{dy}=-8y, \quad \frac{dQ}{dx}=3,$$

and since $8y$ and 3 are not equal, the expression is not integrable.

124. *Prop.* To oblain a general formula for the integration of the form $du = Pdx + Qdy$, when the condition of integrability is satisfied.

Since the term Pdx has resulted from the differentiation of the function u, with respect to x only, y being regarded as invariable, it follows that u will be obtained by integrating Pdx with reference to x alone; but as u may have contained terms involving y alone, which terms necessarily disappear in a differentiation with reference to x, we must complete the integration of Pdx, not as usual by adding a constant C, but by adding a quantity Y, which is some unknown function of y and constant, and we thus provide for the reappearance of such terms as may have disappeared in the first differentiation. Thus we get

$$u = \int Pdx + Y, \quad . \ . \ . \ . \ (1),$$

in which the value of Y remains to be determined.

Differentiating (1) with respect to y, there results

$$\frac{du}{dy} = \frac{d\int Pdx}{dy} + \frac{dY}{dy}, \qquad \text{But } \frac{du}{dy} = Q.$$

$$\therefore \ \frac{dY}{dy} = Q - \frac{d\int Pdx}{dy}, \quad \frac{dY}{dy}dy = \left(Q - \frac{d\int Pdx}{dy}\right)dy,$$

and by integration

$$Y = \int \left[Q - \frac{d\int Pdx}{dy} \right] dy.$$

This value reduces (1) to the form

$$u = \int Pdx + \int \left[Q - \frac{d\int Pdx}{dy} \right] dy, \quad . \ . \ . \ . \ (2),$$

which is the required formula.

125. It is necessary to prove, however, that the coefficient $Q - \dfrac{d\int Pdx}{dy}$ of dy, does not contain x, since otherwise, the second integration would be attended with the same difficulty as the first.

Differentiating that coefficient with respect to x, we obtain

$$\frac{dQ}{dx} - \frac{d^2\int Pdx}{dydx} = \frac{dQ}{dx} - \frac{d^2\int Pdx}{dxdy} = \frac{dQ}{dx} - \frac{dP}{dy},$$

and this is equal to zero by the condition of integrability, which is supposed to be satisfied. Hence the coefficient of dy in (2) cannot contain x.

126. This proof also establishes the truth of the converse of the first proposition, viz.: that when the condition $\frac{dP}{dy} = \frac{dQ}{dx}$ is satisfied, the integration is possible.

127. By a similar process we obtain a second formula

$$u = \int Qdy + \int\left[P - \frac{d\int Qdy}{dx}\right]dx. \ldots\ldots (3).$$

in which the coefficient of dx does not contain y.

Cor. If there were given $du = Pdx + Qdy + Rdz$, we would write

$$u = \int Pdx + V,$$

in which V is a function of y and z.

Then differentiating with respect to y, we obtain

$$\frac{dV}{dy} = \frac{du}{dy} - \frac{d\int Pdx}{dy} = Q - \frac{d\int Pdx}{dy},$$

$$\therefore \quad \frac{dV}{dy}dy = \int\left[Q - \frac{d\int Pdx}{dy}\right]dy,$$

and by integrating with respect to y and adding a function Z of z, we get

$$V = \int\left[Q - \frac{d\int Pdx}{dy}\right]dy + Z,$$

$$\therefore \quad u = \int Pdx + \int\left[Q - \frac{d\int Pdx}{dy}\right]dy + Z.$$

Now differentiating with reference to z, we obtain

$$\frac{dZ}{dz} = \frac{du}{dz} - \frac{d\int Pdx}{dz} - \frac{d\int Qdy}{dz} + \frac{d}{dz}\left[\int \frac{d\int Pdx}{dy}dy\right]$$

$$\therefore Z = \int\left[R - \frac{d\int Pdx}{dz} - \frac{d\int Qdy}{dz} + \frac{d}{dz}\left(\int \frac{d\int Pdx}{dy}dy\right)\right]dz,$$

in which the coefficient of dz is independent of x and y.

$$\therefore \quad u = \int Pdx + \int \left[Q - \frac{d \int Pdx}{dy} \right] dy$$

$$+ \int \left[R - \frac{d \int Pdx}{dz} - \frac{d \int Qdy}{dz} + \frac{d}{dz} \left(\int \frac{d \int Pdx}{dy} dy \right) \right] dz.$$

128. In practice it will be found usually more convenient, and always more instructive, to apply the method, rather than the formula explained above; especially where there are three variables.

EXAMPLES.

129. 1. Integrate $du = (3x^2 + 2axy)dx + (ax^2 + 3y^2)dy$.

Here $\qquad P = 3x^2 + 2axy, \qquad Q = ax^2 + 3y^2.$

$$\frac{dP}{dy} = 2ax = \frac{dQ}{dx}, \text{ and the expression is integrable.}$$

But $\qquad \int Pdx = \int (3x^2 + 2axy)dx = x^3 + ax^2y,$

$$\therefore \int \frac{d \int Pdx}{dy} = ax^2, \quad \text{and} \quad Q - \frac{d \int Pdx}{dy} = ax^2 + 3y^2 - ax^2 = 3y^2.$$

These values reduce (2) to the form,

$$u = x^3 + ax^2y + \int 3y^2dy = x^3 + ax^2y + y^3 + C.$$

2. Integrate $du = (3xy^2 - x^2)dx - (1 + 6y^2 - 3x^2y)dy$.

$$P = 3xy^2 - x^2, \quad Q = -(1 + 6y^2 - 3x^2y), \quad \frac{dP}{dy} = 6xy = \frac{dQ}{dx}.$$

$$u = \int Pdx = \int (3xy^2 - x^2)dx = \frac{3}{2}x^2y^2 - \frac{1}{3}x^3 + Y.$$

$$\therefore \frac{dY}{dy} = \frac{du}{dy} - 3x^2y = -(1 + 6y^2 - 3x^2y) - 3x^2y = -1 - 6y^2.$$

$$\therefore \frac{dY}{dy}dy = -dy - 6y^2dy, \quad \text{and} \quad Y = -y - 2y^3 + C.$$

$$\therefore \quad u = \frac{3}{2}x^2y^2 - \frac{1}{3}x^3 - y - 2y^3 + C.$$

3. $$du = (\sin y + y\cos x)dx + (\sin x + x\cos y)dy.$$

$$\frac{dP}{dy} = \cos y + \cos x = \frac{dQ}{dx}.$$

$$u = \int(\sin y + y\cos x)dx = x\sin y + y\sin x + Y,$$

$$\therefore \frac{dY}{dy} = \frac{du}{dy} - x\cos y - \sin x = 0, \quad \therefore Y = C,$$

and $$u = x\sin y + y\sin x + C.$$

4. $$du = \frac{ydx}{a-z} + \frac{xdy}{a-z} + \frac{xydz}{(a-z)^2}.$$

$$\frac{dP}{dy} = \frac{1}{a-z} = \frac{dQ}{dx}, \quad \frac{dP}{dz} = \frac{y}{(a-z)^2} = \frac{dR}{dx}, \quad \frac{dQ}{dz} = \frac{x}{(a-z)^2} = \frac{dR}{dy}$$

$$u = \int\frac{ydx}{a-z} = \frac{xy}{a-z} + V, \quad \therefore \frac{dV}{dy} = \frac{du}{dy} - \frac{x}{a-z} = 0, \therefore V = Z.$$

Then $$\frac{dZ}{dz} = \frac{du}{dz} - \frac{xy}{(a-z)^2} = 0, \quad \therefore Z = C.$$

$$\therefore u = \frac{xy}{a-z} + C.$$

130. In practice the preceding process may be abridged by first integrating Pdx, then integrating the terms in Qdy, which do not contain x, and finally integrating those terms in Rdz which do not contain either x or y, and adding the results. That the complete integral will be given by this process, appears immediately, from the consideration that the integration of Pdx necessarily gives *all* the terms in the integral sought except such as contain y and z without x. Hence in integrating Qdy we must not consider any term which contains x, as otherwise we would introduce into the integral new terms containing x. Similarly the integration of the selected terms in Qdy gives all the remaining terms except such as contain z only, and therefore in integrating Rdz we must neglect all terms involving both x and y.

Ex. $du = \frac{xdx + ydy + zdz}{(x^2 + y^2 + z^2)^{\frac{1}{2}}} + \frac{zdx - xdz}{x^2 + z^2} + zdz + y^2dy.$

This satisfies the conditions of integrability, and by taking the terms in Pdx we get

$$\int Pdx = \int \frac{xdx}{(x^2+y^2+z^2)^{\frac{1}{2}}} + \int \frac{zdx}{x^2+z^2} = (x^2 + y^2 + z^2)^{\frac{1}{2}} + \tan^{-1}\frac{x}{z}.$$

Now taking the terms in Qdy which do not contain x, we get

$$\int Qdy = \int y^2dy = \frac{1}{3}y^3,$$

and finally taking the terms in R which do not contain x nor y,

$$\int Rdz = \int zdz = \frac{1}{2}z^2.$$

$$\therefore u = (x^2 + y^2 + z^2)^{\frac{1}{2}} + \tan^{-1}\frac{x}{z} + \frac{1}{3}y^3 + \frac{1}{2}z^2 + C.$$

Homogeneous Exact Differentials.

131. Although the methods of integration just explained apply to all exact differentials, yet another and simpler process can be used when the expression belongs to the class called homogeneous. A differential expression is said to be homogeneous when the sum of the exponents of the variables is the same in the coefficient of every term. Thus

$$xdy + ydx, \quad x^2zdx + xz^2dx - xyzdy, \quad \text{and} \quad \frac{ax^2dx - by^2dy}{\left(x^3 - \frac{b}{a}y^3\right)^{\frac{41}{13}}},$$

are homogeneous differentials. The degree of the terms is estimated by this sum of the exponents; thus in the first expression it is 1, in the second it is 3, and in the third it is $2 - \frac{123}{13} = -\frac{97}{13}$.

132. *Prop.* If an exact differential be homogeneous, and the terms of any degree except -1, its integral may be obtained by simply replacing dx, dy, and dz, &c., by x, y, z, &c., respectively, and dividing the result by $n+1$, when n denotes the degree of the terms.

Proof. Let $du = Pdx + Qdy + Rdz +$ &c., be homogeneous and exact, in which P, Q, R, &c., are algebraic functions of x, y, z, &c., of the degree n.

This must have resulted from the differentiation of a homogeneous algebraic function

$$u = P_1x + Q_1y + R_1z + \text{\&c.} \ldots . (1),$$

of the degree $n+1$, since differentiation diminishes by unity one of the exponents in the term differentiated at every step.

Put $y = y_1x$, $z = z_1x$, &c., and substitute in P_1, Q_1, R_1, &c., which quantities contain x, y, z, &c., involved to the nth degree. Replace also y by y_1x, z by z_1x, &c., in (1); then each term in the value of u will contain the factor

$$x^{n+1}, \text{ and } \therefore u = P_2x^{n+1} \ldots . . (2),$$

in which P_2 is a function of y_1, z_1, &c., but does not contain x.

Differentiating (2) with respect to x we get

$$\frac{du}{dx} = (n+1)P_2x^n \ldots . (3).$$

A similar substitution in the value of du gives

$$du = Pdx + Qd(y_1x) + Rd(z_1x) + \text{\&c.} \ldots . (4);$$

and therefore the partial differential coefficient $\frac{du}{dx}$ derived from (4) by differentiating the products y_1x, z_1x, &c., with respect to x only, is

$$\frac{du}{dx} = P + Qy_1 + Rz_1, \text{\&c.} \ldots . (5).$$

Multiplying (3) and (5) by x and equating the results, we get

$$(n+1)P_2x^{n+1} = Px + Qy_1x + Rz_1x = Px + Qy + Rz + \text{\&c.}$$

$$\therefore u = P_2 x^{n+1} = \frac{Px + Qy + Rz + \&c.}{n+1},$$

as stated in the enunciation.

133. When $n = -1$, this formula would make $u = \infty$. In this case it is easily seen that the formula ought not to be applicable, because it is not then true that the desired integral is an algebraic function of the degree $n + 1$; but on the contrary it is transcendental.

1. To integrate $du = (2y^2x + 3y^3)dx + (2x^2y + 9xy^2 + 8y^3)dy.$

$$\frac{dP}{dy} = 4yx + 9y^2 = \frac{dQ}{dx}.$$

$\therefore$ the differential is exact; it is also homogeneous, and since

$$n = 3 \text{ or } n + 1 = 4, \quad u = \frac{Px + Qy}{n+1} + C = y^2x^2 + 3y^3x + 2y^4 + C.$$

2. $$du = \frac{ydx}{z} + \frac{(x - 2y)dy}{z} + \frac{(y^2 - xy)dz}{z^2}$$

$$\frac{dP}{dy} = \frac{1}{z} = \frac{dQ}{dx}, \quad \frac{dP}{dz} = -\frac{y}{z^2} = \frac{dR}{dx}, \quad \frac{dQ}{dz} = \frac{2y - x}{z^2} = \frac{dR}{dy}.$$

$\therefore$ The differential is exact; and being also homogeneous and of the order 0, we have $n + 1 = 1$.

$$\therefore u = \frac{Px + Qy + Rz}{n+1} + C = \frac{yx}{z} + \frac{xy - 2y^2}{z} + \frac{y^2z - xyz}{z^2} + C$$

$$= \frac{xy - y^2}{z} + C.$$

134. When there are three or more variables, the application of the test of integrability will often be very troublesome. In such cases, if the differential be homogeneous, it will be found more convenient to apply the preceding process as though the differential were known to be exact, and then to ascertain, by actual trial, whether the given expression can be reproduced by differentiation.

$$Ex.\ du = \frac{(2\sqrt{xy}-z)dx}{2s^2\sqrt{x}} + \frac{xdy}{2s^2\sqrt{y}} - \frac{2(x\sqrt{y}-z\sqrt{x})ds}{s^3} - \frac{\sqrt{x}dz}{s^2} \ldots (1)$$

This being homogeneous and of the degree $-\frac{3}{2}$, its integral, if possible, must be

$$u = -\frac{2(2\sqrt{xy}-z)x}{2s^2\sqrt{x}} - \frac{2xy}{2s^2\sqrt{y}} + \frac{4(x\sqrt{y}-z\sqrt{x})s}{s^3} + \frac{2z\sqrt{x}}{s^2} + C,$$

$$\text{or } u = \frac{-2xy^{\frac{1}{2}} + zx^{\frac{1}{2}} - xy^{\frac{1}{2}} + 4xy^{\frac{1}{2}} - 4zx^{\frac{1}{2}} + 2zx^{\frac{1}{2}}}{s^2} + C = \frac{xy^{\frac{1}{2}} - zx^{\frac{1}{2}}}{s^2} + C.$$

This, differentiated, gives

$$du = \frac{\left(y^{\frac{1}{2}} - \frac{1}{2}zx^{-\frac{1}{2}}\right)dx}{s^2} + \frac{xy^{-\frac{1}{2}}dy}{2s^2} - \frac{2\left(xy^{\frac{1}{2}} - zx^{\frac{1}{2}}\right)ds}{s^3} - \frac{x^{\frac{1}{2}}dz}{s^2},$$

which is identical with the proposed expression (1).

It must be distinctly understood that in the differential *expressions* here considered, the variables x, y, z, &c., are wholly independent of each other. If, then, the conditions of integrability be not fulfilled, the integration must be impossible, since there is no relation between the variables, by the aid of which we might hope to transform the given differential into another of an integrable form.

It would be otherwise if a relation between the variables were given in the form of a differential equation, such $Pdx + Qdy = 0$.

Here the form of the first member may be greatly modified by the introduction of a variable factor (or by other methods), and thus the integration may be facilitated.

CHAPTER II.

DIFFERENTIAL EQUATIONS.

135. A differential equation between two variables x and y is a relation involving one or more of the differential coefficients such as $\frac{dy}{dx}, \frac{d^2y}{dx^2}, \frac{d^3y}{dx^3}$, &c., or their powers $\left(\frac{dy}{dx}\right)^2, \left(\frac{dy}{dx}\right)^3, \left(\frac{d^2y}{dx^2}\right)^2$, &c.

Such equations are arranged in classes dependent upon the *order and degree* of the differential coefficient. Thus, when the equation involves only the first powers of $\frac{dy}{dx}, \frac{d^2y}{dx^2} \ldots \frac{d^ny}{dx^n}$, it is said to be of the nth *order and* 1*st degree.*

When it contains only the powers of the 1st differential coefficient, viz.: $\frac{dy}{dx}, \left(\frac{dy}{dx}\right)^2 \ldots \left(\frac{dy}{dx}\right)^n$, it is of the 1*st order and* n^{th} *degree.* And when it contains the nth powers of one or more differential coefficients, and a coefficient of the mth order, the equation is of the nth *degree and* m^{th} *order.*

136. The resolution of a differential equation consists in finding a relation between x and y and constants. This relation, called the primitive, must be such, that the given differential equation can be deduced from it, either by the direct process of differentiation, or by the elimination of a constant between the primitive and the direct differential equation. Hence the same primitive may have several differential equations of the same order. Thus the equation

$$ay + bx + c = 0 \ldots . (1)$$

gives by differentiation $a\frac{dy}{dx} + b = 0 \ldots . (2),$

and by elimination between (1) and (2) we get the indirect differential equations

$$bx\frac{dy}{dx} + c\frac{dy}{dx} - by = 0 \ldots . (3), \text{ when } a \text{ is eliminated;}$$

and $$ay + c - ax\frac{dy}{dx} = 0 \ldots . (4), \text{ when } b \text{ is eliminated.}$$

In each of the equations (2), (3), and (4), the variables are connected by the same relation as in (1), which latter is their common primitive.

137. As the integration of differential equations can be effected in comparatively few cases, it is found convenient to arrange them in the order of the difficulties presented, commencing with the simplest case.

Differential Equations of the First Order and Degree.

138. These are of the general form $P + Q\frac{dy}{dx} = 0$ or $Pdx + Qdy = 0$, in which P and Q may be functions of both x and y. The integration will obviously be possible, by the method applied to differential *expressions*, whenever $Pdx + Qdy$ is an exact differential, and the required solution will be of the form

$$F(x, y) = C,$$

where C is an arbitrary constant.

139. Again, the integration can be effected whenever the separation of the variables is possible, that is, when the equation can be reduced to the form

$$Xdx + Ydy = 0,$$

where X is a function of x only, and Y a function of y only.

The form of the solution will then be

$$Fx + \varphi y = C,$$

which requires only the integration of functions of a single variable.

The separation of the variables is possible in several cases.

140. *Case* 1*st.* Let the form be

$$Ydx + Xdy = 0,$$

in which the coefficient of dx contains only y, and that of dy contains only x.

Divide by XY, the product of the two coefficients, and there will result

$\frac{dx}{X} + \frac{dy}{Y} = 0$, in which the variables are separated.

141. *Ex.* Given $(1 + y^2)dx - x^{\frac{1}{2}}dy = 0$, to find the primitive relation between x and y.

Divide by $(1 + y^2)x^{\frac{1}{2}}$, then

$$\frac{dx}{x^{\frac{1}{2}}} - \frac{dy}{1 + y^2} = 0, \qquad \therefore\ 2x^{\frac{1}{2}} - \tan^{-1}y = C,$$

which is the required relation.

142. *Case* 2*d.* Let the form be

$$XYdx + X_1Y_1dy = 0,$$

in which each coefficient is the product of a function of x by another function of y.

Divide by X_1Y,

$\therefore\ \frac{Xdx}{X_1} + \frac{Y_1dy}{Y} = 0$, and the variables are separated.

143. *Ex.* Determine the primitive of

$$(1 - x)^2ydx - (1 + y)x^2dy = 0.$$

Divide by x^2y,

$$\therefore\ \frac{(1 - x)^2}{x^2}dx - \frac{1 + y}{y}dy = 0, \text{ or, } \frac{dx}{x^2} - \frac{2dx}{x} + dx - \frac{dy}{y} - dy = 0,$$

$$\therefore\ -\frac{1}{x} - 2\log + x - \log y - y = C.$$

144. *Case 3d.* Let the proposed equation be homogeneous, or of the form

$$(x^n y^m + ax^{n+1}y^{m-1} + bx^{n+2}y^{m-2} \ldots\ldots + px^{n+c}y^{m-c})dx$$
$$+ (ex^n y^m + fx^{n+1}y^{m-1} + gx^{n+2}y^{m-2} \ldots\ldots\ldots qx^{n+s}y^{m-s})dy = 0.$$

Put $\qquad y = xz, \qquad$ then $\qquad dy = xdz + zdx,$

and by substitution

$$x^{n+m}(z^m + az^{m-1} + bz^{m-2} \ldots\ldots + pz^{m-c})dx$$
$$+ x^{n+m}(ez^m + fz^{m-1} + gz^{m-2} \ldots\ldots + qz^{m-s})(xdz + zdx) = 0.$$

$$\therefore \frac{dx}{x} + \frac{(ez^m + fz^{m-1} + gz^{m-2} \ldots\ldots + qz^{m-s})dz}{z^m + az^{m-1} + bz^{m-2} \ldots + pz^{m-c} + ez^{m+1} + fz^m + gz^{m-1} \ldots + qz^{m-s+1}}$$

$= 0$ and the variables are separated.

145. *Ex.* To find the primitive of

$$x^2dy - y^2dx - xydx = 0.$$

Put $\qquad y = xz, \qquad$ then $\qquad dy = xdz + zdx.$

$$\therefore x^2(xdz + zdx) - x^2z^2dx - x^2zdx = 0,$$

$$\therefore \frac{dx}{x} - \frac{dz}{z^2} = 0, \qquad \text{and} \qquad \log x + \frac{1}{z} = C;$$

or, by restoring the value of $z = \frac{y}{x}$,

$$\log x + \frac{x}{y} = C.$$

2. $$xdy - ydx = dx\sqrt{x^2 - y^2}.$$

$y = xz, \qquad$ then $\qquad x(xdz + zdx) - xzdx = x(1 - z^2)^{\frac{1}{2}}dx,$

$$\therefore \frac{dx}{x} = \frac{dz}{(1 - z^2)^{\frac{1}{2}}}, \qquad \therefore \log x = \sin^{-1}z + C.$$

146. The same method of transformation may be extended to such differential equations as involve any function of $\frac{y}{x}$ unmixed with the variables, provided the equation would be otherwise homogeneous.

Ex. $$xydy - y^2dx = (x+y)^2 e^{-\frac{y}{x}}dx.$$

Put $$y = xz, \quad \text{or,} \quad \frac{y}{x} = z,$$

$$\therefore\ x^2z(xdz + zdx) - x^2z^2dx = x^2(1+z)^2e^{-z}dx,$$

$$\therefore\ \frac{dx}{x} = \frac{e^z zdz}{(1+z)^2}, \quad \text{and} \quad \log x = \frac{e^z}{1+z} + C.$$

$$\therefore\ \log x = \frac{e^{\frac{y}{x}}}{1+\frac{y}{x}} + C = \frac{xe^{\frac{y}{x}}}{x+y} + C.$$

147. *Case 4th.* Let the form be

$$(a + bx + cy)dx + (a_1 + b_1x + c_1y)dy = 0.$$

Put $a + bx + cy = v$, and $a_1 + b_1x + c_1y = u$.

Then $dv = bdx + cdy$, and $du = b_1dx + c_1dy$,

and by elimination

$$dx = \frac{c_1dv - cdu}{bc_1 - b_1c}, \quad \text{and similarly} \quad dy = \frac{bdu - b_1dv}{bc_1 - b_1c}.$$

$\therefore$ By substitution $v(c_1dv - cdu) + u(bdu - b_1dv) = 0$,

which is a homogeneous equation.

148. This method fails when $bc_1 - b_1c = 0$, because the attempt to eliminate either dx or dy causes the other to disappear also. But since we then have $c_1 = \frac{b_1c}{b}$, the proposed equation reduces to

$$(a + bx + cy)dx + (a_1 + b_1x + \frac{b_1c}{b}y)dy = 0 \ldots\ldots (1).$$

Put $bx + cy = z$, then $x = \frac{z - cy}{b}$, and $dx = \frac{dz - cdy}{b}$;

and, by substitution in (1),

$$(a+z)\frac{dz-cdy}{b}+\left(a_1+\frac{b_1z}{b}\right)dy=0.$$

$$\therefore\ dy=\frac{(a+z)dz}{ca+cz-a_1b-b_1z},$$

an equation in which the variables are separated.

149. *Ex.* Find the primitive of

$$(1+x+y)dx+(1+2x+3y)dy=0.$$

Put $\quad 1+x+y=v,\ 1+2x+3y=u$; then

$$dx+dy=dv,\ \text{and}\ 2dx+3dy=du,$$

$$\therefore\ dx=3dv-du,\ \text{and}\ dy=du-2dv.$$

$$\therefore\ v(3dv-du)+u(du-2dv)=0.$$

Now put $u=rv$, then $du=rdv+vdr$, and, consequently,

$$v(3dv-rdv-vdr)+rv(rdv+vdr-2dv)=0,$$

$$\frac{dv}{v}+\frac{(r-1)dr}{r^2-3r+3}=0,\ \text{or}\ \frac{dv}{v}+\frac{\frac{1}{2}(2r-3)dr}{\left(r-\frac{3}{2}\right)^2+\frac{3}{4}}+\frac{\frac{1}{2}dr}{\left(r-\frac{3}{2}\right)^2+\frac{3}{4}}=0,$$

$$\therefore\ \log v+\frac{1}{2}\log\left[\left(r-\frac{3}{2}\right)^2+\frac{3}{4}\right]+\frac{1}{\sqrt{3}}\tan^{-1}\left[\frac{2}{\sqrt{3}}\left(r-\frac{3}{2}\right)\right]=C.$$

or $$\frac{1}{2}\log\left[u^2-3uv+3v^2\right]+\frac{1}{\sqrt{3}}\tan^{-1}\left[\frac{2}{\sqrt{3}}\left(\frac{2u-3v}{2v}\right)\right]=C.$$

or $$\log\left(1+x+x^2+3xy+3y^2\right)^{\frac{1}{2}}+\frac{1}{\sqrt{3}}\tan^{-1}\left[\frac{1}{\sqrt{3}}\left(\frac{3y+x-1}{y+x+1}\right)\right]=C.$$

150. *Case 5th.* Let the form be $dy+Xydx=X_1dx$ . . . . . (1), in which X and X_1 are functions of x.

The peculiarity of this form is that no power of y except the first enters into it, and for this reason it is usually called a *linear* equation. Its solution is always possible.

Put $y = X_2 z$ where X_2 is an arbitrary function of x, which may be so assumed as to facilitate the integration; and z a new and undetermined variable. Then

$dy = X_2 dz + z dX_2$, and (1) can be reduced to the form

$$X_2 dz + z dX_2 + XX_2 z dx = X_1 dx \;.\;.\;.\;.\;.\;.\;. (2).$$

Now let X_2 be determined by the condition

$$z dX_2 = X_1 dx \;.\;.\;.\;.\;.\;.\;. (3),$$

and (2) will become $\quad X_2 dz + XX_2 z dx = 0.$

$$\therefore \frac{dz}{z} = -X dx \quad \text{and} \quad \log z = -\int X dx, \; \therefore z = e^{-\int X dx}.$$

This value, substituted in (3), gives

$$e^{-\int X dx} dX_2 = X_1 dx \quad \text{or} \quad dX_2 = e^{\int X dx} X_1 dx.$$

$$\therefore X_2 = \int e^{\int X dx} X_1 dx, \text{ and } y = zX_2 = e^{-\int X dx} \int e^{\int X dx} X_1 dx \;.\;.\;. (4),$$

which is the required relation between x and y.

151. Let there be now taken the more general form

$$dy + Xy dx = X_1 y^m dx \;.\;.\;.\;.\;.\;. (5).$$

This is easily reduced to the linear form (1). For put

$$m = n + 1 \quad \text{and} \quad z = y^{-n}.$$

Then $\qquad dz = -ny^{-n-1} dy$ or $dy = -\dfrac{y^m dz}{n}.$

Substituting this value of dy in the equation (5), and reducing, we get

$$-\frac{dz}{n} + \frac{X dx}{y^n} = X_1 dx, \text{ or } dz - nXz dx = -nX_1 dx \;.\;.\;.\;. (6),$$

which becomes identical with (1) when we replace

$$z \text{ by } y, \; -nX \text{ by } X \quad \text{and} \quad -nX_1 \text{ by } X_1.$$

$$\therefore z = \frac{1}{y^n} = -ne^{n\int Xdx}\int X_1 e^{-n\int Xdx}dx \ldots\ldots (7).$$

Cor. In forming the integral $\int Xdx$, it will be unnecessary to add a constant C. For if we replace $\int Xdx$ by $X_3 + C$, the formula (4) will become

$$y = e^{-X_3 - C}\int e^{X_3 + C}X_1 dx = e^{-X_3}e^{-C}\int e^{X_3}e^C. X_1 dx = e^{-X_3}\int e^{X_3}X_1 dx,$$

in which the constant C has disappeared.

152. 1. To determine the primitive of $(1 + x^2)dy - yxdx = adx$.

Here $dy - \dfrac{yxdx}{1+x^2} = \dfrac{a}{1+x^2}dx$, which, compared with the linear form, gives

$$X = -\frac{x}{1+x^2} \text{ and } X_1 = \frac{a}{1+x^2}.$$

$$\therefore \int Xdx = -\int \frac{xdx}{1+x^2} = \log \frac{1}{\sqrt{1+x^2}}.$$

$$e^{\int Xdx} = e^{\log[(1+x^2)^{-\frac{1}{2}}]} = (1+x^2)^{-\frac{1}{2}} \text{ and } e^{-\int Xdx} = (1+x^2)^{\frac{1}{2}}.$$

$$\therefore \int X_1 e^{\int Xdx}. dx = \int \frac{a(1+x^2)^{-\frac{1}{2}}}{1+x^2}dx = \int \frac{adx}{(1+x^2)^{\frac{3}{2}}} = \frac{ax}{(1+x^2)^{\frac{1}{2}}} + C.$$

$$\therefore y = (1+x^2)^{\frac{1}{2}}\left[\frac{ax}{(1+x^2)^{\frac{1}{2}}} + C\right] = ax + C(1+x^2)^{\frac{1}{2}}.$$

2. $$dy + ydx = xy^3dx.$$

Here $X = 1$, $X_1 = x$, $y^m = y^3$, or $m = 3$, and $n = 2$, and, by substitution in (7),

$$\therefore \frac{1}{y^2} = -2e^{2\int dx}. \int xe^{-2\int dx}. dx = e^{2x}[xe^{-2x} + \frac{1}{2}e^{-2x} + C].$$

or $$1 = [Ce^{2x} + x + \tfrac{1}{2}]y^2.$$

3. $$dy - \frac{a}{1-x}ydx = bdx.$$

Here $X = \frac{-a}{1-x}$, $X_1 = b$. $\therefore \int X dx = \int \frac{-a dx}{1-x} = \log(1-x)^a$,

$$e^{\int X dx} = e^{\log(1-x)^a} = (1-x)^a,$$

$$\int X_1 e^{\int X dx} . dx = \int b(1-x)^a dx = -\frac{b}{a+1}(1-x)^{a+1} + C.$$

$$\therefore y = \frac{C}{(1-x)^a} - \frac{b}{a+1}(1-x).$$

4. Find an expression for the sum of the series

$$y = \frac{x}{1} + \frac{x^3}{1.3} + \frac{x^5}{1.3.5} + \frac{x^7}{1.3.5.7} + \&c.$$

Differentiating, we obtain

$$\frac{dy}{dx} = 1 + \frac{x^2}{1} + \frac{x^4}{1.3} + \frac{x^6}{1.3.5} + \&c. = 1 + x\left(x + \frac{x^3}{1.3} + \frac{x^5}{1.3.5} + \&c.\right)$$

$$= 1 + xy. \quad \therefore dy - xydx = dx, \text{ a linear equation.}$$

Also $$\int X dx = -\int x dx = -\frac{1}{2}x^2.$$

$$\therefore y = e^{\frac{1}{2}x^2}\left(\int e^{-\frac{1}{2}x^2} dx\right), \text{ the desired expression.}$$

153. *Case 6th.* Let the form be

$$dy + by^2 dx = ax^m dx \; . \; . \; . \; . \; . \; (1).$$

This form, which is called *Riccati's equation*, involves only the second power of y. Its integration has been effected for certain values of the exponent m, but a solution applicable to all values of m has not been discovered.

154. It is obvious that when $m = 0$ the equation (1) will be integrable, for then

$$dy + by^2 dx = adx. \quad \therefore \frac{dy}{a - by^2} = dx,$$

and the variables are separated.

Thus we have $2a^{\frac{1}{2}}b^{\frac{1}{2}}dx = \dfrac{b^{\frac{1}{2}}dy}{b^{\frac{1}{2}}y + a^{\frac{1}{2}}} - \dfrac{b^{\frac{1}{2}}dy}{b^{\frac{1}{2}}y - a^{\frac{1}{2}}}.$

$$\therefore 2a^{\frac{1}{2}}b^{\frac{1}{2}}x = \log \frac{b^{\frac{1}{2}}y + a^{\frac{1}{2}}}{b^{\frac{1}{2}}y - a^{\frac{1}{2}}} + C.$$

Next suppose m to have some value other than 0.

Assume $y = \dfrac{1}{bx} + \dfrac{z}{x^2}$, where z is an undetermined variable, but obviously a function of x. Then

$$dy = -\frac{dx}{bx^2} - \frac{2zdx}{x^3} + \frac{dz}{x^2}, \quad \text{and} \quad by^2 = b\left(\frac{1}{b^2x^2} + \frac{2z}{bx^3} + \frac{z^2}{x^4}\right).$$

Substituting these values in (1), it becomes

$$\frac{dz}{x^2} + \frac{bz^2dx}{x^4} = ax^m dx \quad \text{or} \quad dz + \frac{bz^2}{x^2}dx = ax^{m+2}dx \; \ldots . \; (2).$$

Now this equation (2) will be homogeneous, and therefore integrable when $m = -2$, and thus a new integrable case is found.

Again, if $m = -4$, the variables x and z can be separated, for then

$$dz + b\frac{z^2}{x^2}dx = ax^{-2}dx. \quad \therefore \frac{dz}{a - bz^2} = \frac{dx}{x^2},$$

which is a third integrable case.

155. To obtain others, put

$$\frac{1}{z} = y_1 \quad \text{and} \quad x^{m+3} = x_1.$$

Then $$dz = d\frac{1}{y_1} = -\frac{dy_1}{y_1^2}, \quad x^{m+2}dx = \frac{dx_1}{m+3},$$

and $$\frac{dx}{x^2} = \frac{1}{m+3}x_1^{-\frac{m+4}{m+3}}dx_1.$$

Hence by substitution in (2),

$$-\frac{dy_1}{y_1^2} + \frac{b}{(m+3)y_1^2}x_1^{-\frac{m+4}{m+3}}dx_1 = \frac{adx_1}{m+3} \; \ldots . \; (3).$$

Now put $\frac{b}{m+3} = a_1$, $\frac{a}{m+3} = b_1$, and $-\frac{m+4}{m+3} = m_1$

and (3) will become, after reduction,

$$dy_1 + b_1 y_1^2 dx_1 = a_1 x_1^{m_1} dx_1 \dots (4),$$

which is identical *in form* with (1). Hence (4) must be integrable when the exponent m_1 has either of the three values 0, -2, or -4. Moreover when a relation has been obtained by integration between x_1 and y_1, a simple substitution will give the desired relation between x and y.

We have therefore to examine whether by assigning either of the values 0, -2, or -4 to m_1, any new values of m will arise.

But $m_1 = -\frac{m+4}{m+3}$. $\therefore mm_1 + 3m_1 = -m - 4$, and $m = -\frac{3m_1+4}{m_1+1}$.

$\therefore$ when $m_1 = 0$, $m = -4$, a case before considered;

and when $m_1 = -2$, $m = -2$, " " "

but when $m_1 = -4$, $m = -\frac{8}{3}$, a new case.

Hence Riccati's equation is integrable when $m = -\frac{8}{3}$ also.

156. In a manner entirely similar to that by which (1) was transformed into (4), may we transform (4) into a new equation

$$dy_2 + b_2 y_2^2 dx_2 = a_2 x_2^{m_2} dx_2 \dots (5),$$

in which $m_2 = -\frac{m_1+4}{m_1+3}$, and therefore $m_1 = -\frac{3m_2+4}{m_2+1}$.

And by repeating the process, a series of such equations may be formed; so that it will be possible to find a relation between x and y when any one of the following quantities or exponents shall be $= -4$; viz.:

$$m, \text{ or } m_1 = -\frac{m+4}{m+3}, \text{ or } m_2 = -\frac{m_1+4}{m_1+3}, \text{ or } m_3 = -\frac{m_2+4}{m_2+3}, \text{ \&c.}$$

But
$$\left.\begin{array}{llll} m = -\frac{8}{3}, & \text{when} & m_1 = -4 \\ m_1 = -\frac{8}{3}, & \text{"} & m_2 = -4 \\ m_2 = -\frac{8}{3}, & \text{"} & m_3 = -4 \end{array}\right\} \text{\&c.}$$

Hence by successive substitutions,

$$\left.\begin{array}{llll} m = -\frac{12}{5}, & \text{when} & m_2 = -4 \\ m = -\frac{16}{7}, & \text{"} & m_3 = -4 \end{array}\right\} \text{\&c.}$$

Thus Riccati's equation is integrable for all values of m included in the series.

$$-4, \quad -\frac{8}{3}, \quad -\frac{12}{5}, \quad -\frac{16}{7}, \quad -\frac{20}{9}, \quad -\frac{24}{11}, \quad \text{\&c.}$$

157. The general formula for these numbers is $-\frac{4n}{2n-1}$, in which n is any positive integer.

To prove this, suppose one of the numbers, as m_8, to be of the form required $-\frac{4n}{2n-1}$; then will the adjacent terms m_7 and m_9 be of the same form, with the number n increased or diminished by unity. For we shall have,

$$m_7 = -\frac{3m_8 + 4}{m_8 + 1} = -\frac{4 - 3\frac{4n}{2n-1}}{1 - \frac{4n}{2n-1}} = -\frac{4n+4}{2n+1} = -\frac{4(n+1)}{2(n+1)-1}$$

and

$$m_9 = -\frac{m_8 + 4}{m_8 + 3} = -\frac{4 - \frac{4n}{2n-1}}{3 - \frac{4n}{2n-1}} = -\frac{4n-4}{2n-3} = -\frac{4(n-1)}{2(n-1)-1}$$

both of which forms are similar to that given above as the general form.

But we have seen that one integrable case is that in which $m = -\frac{8}{3} = -\frac{4.2}{2.2-1}$ which being of the required form, the adjacent numbers -4 and $-\frac{12}{5}$ are also of that form; and thence the same reasoning can be extended to the other numbers in the series.

158. *Second Transformation of Riccati's Equation.*—In the given equation

$$dy + by^2dx = ax^mdx. \dots\dots (1).$$

put

$$y = \frac{1}{y_1}, \quad \text{and} \quad x^{m+1} = x_1$$

then $dy = -\frac{dy_1}{y_1^2}$, $x^mdx = \frac{dx_1}{m+1}$, and $dx = \frac{x_1^{-\frac{m}{m+1}}dx_1}{(m+1)}$.

Also, $y^2 = \frac{1}{y_1^2}$, and therefore by substitution in (1),

$$-\frac{dy_1}{y_1^2} + \frac{bx_1^{-\frac{m}{m+1}}dx_1}{(m+1)y_1^2} = \frac{a}{m+1}dx_1,$$

or,

$$dy_1 + \frac{a}{m+1}y_1^2dx_1 = \frac{b}{m+1}x_1^{-\frac{m}{m+1}}dx_1. \dots (8).$$

Now make $\frac{a}{m+1} = b_1$, $\frac{b}{m+1} = a_1$, $-\frac{m}{m+1} = m_1$,

and the equation (8) will reduce to

$$dy_1 + b_1y_1^2dx_1 = a_1x_1^{m_1}dx_1. \dots (9).$$

which is of the same form with (1).

The equation (8) will evidently be integrable whenever m_1 has any one of the values included in the series before found, that is, when m_1 or its equal $-\frac{m}{m+1}$ has the form $-\frac{4n}{2n-1}$. But if

$$-\frac{m}{m+1} = -\frac{4n}{2n-1}, \text{ then } 2mn - m = 4mn + 4n,$$

and

$$\therefore \quad m = -\frac{4n}{2n+1}.$$

Hence we have a new series of integrable cases corresponding to all values of m, included in the formula $-\frac{4n}{2n+1}$. Thus Riccati's equation is integrable whenever the exponent m can be expressed in the form $-\frac{4n}{2n \pm 1}$, the quantity n being a positive integer.

It appears also, that whenever the given value of m is found in the second series, the terms of which have the form $-\frac{4n}{2n+1}$, an application of the second transformation will transfer m to the first series, and then the repeated application of the first transformation will eventually reduce m to the value -4.

159. 1. To integrate the equation

$$dy + y^2 dx = a^2 x^{-\frac{4}{3}} dx. \dots . (1).$$

Here $$m = -\frac{4}{3} = -\frac{4 \,.\, 1}{2 \,.\, 1 + 1} = -\frac{4n}{2n+1},$$

and m is found in the second series.

$\therefore$ Put $$y = \frac{1}{y_1}, \quad \text{and} \quad x^{-\frac{4}{3}+1} = x^{-\frac{1}{3}} = x_1,$$

$$\therefore \ x = x_1^{-3}, \quad dx = -3x_1^{-4} dx_1,$$

$$x^{-\frac{4}{3}} dx = -3dx_1, \quad \text{and} \quad dy = -\frac{dy_1}{y_1^2}.$$

Hence by substitution in (1),

$$-\frac{dy_1}{y_1^2} - \frac{3}{y_1^2} x_1^{-4} dx_1 = -3a^2 dx_1,$$

or, $$dy_1 - 3a^2 y_1^2 dx_1 = -3x_1^{-4} dx_1. \dots (2).$$

Now put $$-3a^2 = b_1, \quad \text{and} \quad -3 = a_1,$$

then $$dy_1 + b_1 y_1^2 dx_1 = a_1 x_1^{-4} dx_1. \dots . (3).$$

Here $$m_1 = -4 = -\frac{4 \,.\, 1}{2 \,.\, 1 - 1} = -\frac{4n}{2n-1},$$

$$\therefore \text{ put } \ y_1 = \frac{1}{b_1 x_1} + \frac{z_1}{x_1^2},$$

$$\therefore\ dy_1 = -\frac{dx_1}{b_1x_1^2} - \frac{2z_1dx_1}{x_1^3} + \frac{dz_1}{x_1^2},\quad b_1y_1^2 = \frac{1}{b_1x^2} + \frac{2z_1}{x_1^3} + \frac{b_1z_1^2}{x_1^4}.$$

Hence by substitution

$$\frac{dz_1}{x_1^2} + \frac{b_1z_1^2dx_1}{x_1^4} = \frac{a_1dx_1}{x_1^4},\quad \text{or,}\quad \frac{dz_1}{a_1 - b_1z_1^2} = \frac{dx_1}{x_1^2},$$

or, by replacing a_1 and b_1 by their values

$$\frac{dz_1}{3(a^2z_1^2 - 1)} = \frac{dx_1}{x_1^2},\quad \text{or,}\quad \frac{3adx_1}{x_1^2} = \frac{\frac{1}{2}adz_1}{az_1 - 1} - \frac{\frac{1}{2}adz_1}{az_1 + 1},$$

$$\therefore\ \frac{3a}{x_1} = \frac{1}{2}\log(az_1 + 1) - \frac{1}{2}\log(az_1 - 1) + C = \log\left[c\sqrt{\frac{az_1 + 1}{az_1 - 1}}\right]$$

$$= \log\left[c\sqrt{\frac{3a^2x_1^2y_1 + x_1 + 3a}{3a^2x_1^2y_1 + x_1 - 3a}}\right].$$

$$\therefore\ e^{6ax^{\frac{1}{3}}} = c^2\left[\frac{3a^2x^{-\frac{2}{3}} + y(3a + x^{-\frac{1}{3}})}{3a^2x^{-\frac{2}{3}} + y(x^{-\frac{1}{3}} - 3a)}\right] = c^2\left[\frac{3a^2x^{-\frac{1}{3}} + y(1 + 3ax^{\frac{1}{3}})}{3a^2x^{-\frac{1}{3}} + y(1 - 3ax^{\frac{1}{3}})}\right],$$

or,
$$\frac{1}{c^2} = C_1 = e^{-6ax^{\frac{1}{3}}}\left[\frac{3a^2x^{-\frac{1}{3}} + y(1 + 3ax^{\frac{1}{3}})}{3a^2x^{-\frac{1}{3}} + y(1 - 3ax^{\frac{1}{3}})}\right].$$

2.
$$dy + y^2dx = 4x^{-\frac{8}{3}}dx\ \ldots\ (1).$$

Here $m = -\frac{8}{3} = -\frac{4n}{2n - 1}$. $\therefore$ Put $y = \frac{1}{bx} + \frac{z_1}{x^2} = \frac{1}{x} + \frac{z_1}{x^2}$.

Then
$$dz_1 + z_1^2x^{-2}dx = 4x^{-\frac{2}{3}}dx.$$

Now make $z_1 = \frac{1}{y_1}$, and $x^{-\frac{2}{3}+1} = x^{\frac{1}{3}} = x_1$. Then

$$-\frac{dy_1}{y_1^2} + \frac{1}{y_1^2}\cdot\frac{3dx_1}{x_1^4} = 12dx_1,\quad \text{or}\quad dy_1 + 12y_1^2dx = 3x_1^{-4}dx_1.\ \ (2).$$

Repeating the process of substitution as in the last example, we get

$$dz_2 + 12z_2^2x_1^{-2}dx_1 = 3x_1^{-2}dx_1\quad \text{or,}\quad \frac{3dx_1}{x_1^2} = -\frac{dz_2}{4z_2^2 - 1}.$$

the integral of which is

$$\frac{12}{x_1} = \log c\left[\frac{2z_2 - 1}{2z_2 + 1}\right] \quad \text{or,} \quad e^{12x^{-\frac{1}{3}}}\left[\frac{2z_2 + 1}{2z_2 - 1}\right] = c,$$

or,
$$c = e^{12x^{-\frac{1}{3}}}\left[\frac{2(y_1 x_1^2 - \frac{1}{12}x_1) + 1}{2(y_1 x_1^2 - \frac{1}{12}x_1) - 1}\right]$$

$$= e^{12x^{-\frac{1}{3}}}\left[\frac{2x^{\frac{2}{3}}(yx^2 - x)^{-1} - \frac{1}{6}x^{\frac{1}{3}} + 1}{2x^{\frac{2}{3}}(yx^2 - x)^{-1} - \frac{1}{6}x^{\frac{1}{3}} - 1}\right],$$

or,
$$c = e^{12x^{-\frac{1}{3}}}\left[\frac{y(x - 6x^{\frac{2}{3}}) - 1 + 6x^{-\frac{1}{3}} - 12x^{-\frac{2}{3}}}{y(x + 6x^{\frac{2}{3}}) - 1 - 6x^{-\frac{1}{3}} - 12x^{-\frac{2}{3}}}\right].$$

160. If the proposed form be $dy + by^2x^rdx = cx^sdx$, which differs from the form just considered, in having a power of x in the second term of the first member, it may be readily reduced to the simpler form by making $x^rdx = dz$. For then

$$x^{r+1} = (r+1)z, \text{ and } x^{s+1} = (r+1)^{\frac{s+1}{r+1}} . z^{\frac{s+1}{r+1}}.$$

$$\therefore\ x^sdx = [(r+1)z]^{\frac{s-r}{r+1}}dz, \text{ and } dy + by^2dz$$

$$= c[(r+1)z]^{\frac{s-r}{r+1}}dz = az^mdz,$$

the form in which Riccati equation has been integrated.

Of the Factors necessary to render Differential Equations exact.

161. The cases examined above embrace the principal forms in which the integration is possible by a separation of the variables. We now proceed to consider those in which the first member of an equation $Pdx + Qdy = 0$ can be rendered an exact differential by the introduction of a suitable factor.

162. If the equation $Pdx + Qdy = 0$ has been obtained by direct differentiation, it will satisfy the test of integrability, viz. :

$$\frac{dP}{dy} = \frac{dQ}{dx};$$

but if it has resulted from the elimination of a constant between the direct differential and its primitive, that condition will not be satisfied, although the same relation between x and y will be implied by the two differential equations.

163. *Prop.* To show that an indirect differential equation can always be rendered exact by the introduction of a suitable factor.

Let $$Pdx + Qdy = 0 \ \ldots\ldots. \ (1)$$

be the given equation which has resulted from the elimination of a constant c between the primitive and its direct differential; and let the primitive be solved with respect to c, giving a result of the form

$$c = F(x,y) \ \ldots\ldots. \ (2).$$

Differentiating (2), c will disappear, and we shall obtain an equation of the form,

$$P_1dx + Q_1dy = 0 \ \ldots\ldots \ (3).$$

Now, since (1) and (3) contain the same constants, combined with x and y, and since the same relation connects x and y in the two equations, the differential coefficient $\frac{dy}{dx}$ must be the same, whether derived from (1) or (3).

$$\therefore \frac{dy}{dx} = -\frac{P}{Q} = -\frac{P_1}{Q_1}, \text{ or } \frac{P}{Q} = \frac{P_1}{Q_1}, \text{ and } \therefore \frac{P_1}{P} = \frac{Q_1}{Q}$$

Hence, if we multiply (1), the first member of which is not an exact differential by $\frac{P_1}{P} = \frac{Q_1}{Q}$, we shall convert it into (3), which is exact.

Cor. If it were possible to determine this factor, the integration of every differential equation of the first order and degree, could be effected without serious difficulty, but, unfortunately, the difficulty of discovering this factor is usually insuperable.

164. *Prop.* To exhibit the condition or equation, the solution of which would give the factor necessary to render any proposed differential equation exact.

Let $Pdx + Qdy = 0$ be the given equation, and z the required factor.

Then $Pzdx + Qzdy = 0$, and the first member will be exact.

$$\therefore \frac{dPz}{dy} = \frac{dQz}{dx}, \text{ the required condition.}$$

No general method of resolving this equation is known. There are, however, several particular cases in which the factor can be found.

165. *Prop.* To show that when the factor which renders an equation integrable has been found, an indefinite number of such factors can be discovered.

Let z be the factor first found. Put

$$Pzdx + Qzdy = du.$$

Multiply by Fu, an arbitrary function of u, and there will result

$$Fu.Pzdx + Fu.Qzdy = Fu.du;$$

and, since the second member is exact, (containing u only) the first member must be exact also.

$$\therefore z.Fu = z.F\textstyle\int(Pzdx + Qzdy) \text{ is a suitable factor.}$$

166. *Prop.* To explain the process for finding the required factor, when the equation can be separated into two parts, for each of which a separate factor can be found. Let

$$Pdx + Qdy = 0 \ldots\ldots (1) \text{ be divisible into the two parts.}$$

$$(P_1dx + Q_1dy) + (P_2dx + Q_2dy) = 0 \ldots\ldots (2),$$

and let z_1 and z_2 be the factors, which will render

$P_1dx + Q_1dy$ and $P_2dx + Q_2dy$ separately integrable.

Put $z_1(P_1dx + Q_1dy) = du_1$ and $z_2(P_2dx + Q_2dy) = du_2$

Then $z_1F_1u_1$ and $z_2F_2u_2$ will also be suitable factors to render the two parts separately integrable. If therefore we can so select F_1u_1 and F_2u_2 as to fulfil the condition

$$z_1F_1u_1 = z_2F_2u_2,$$

either of those factors will render the entire equation integrable.

167. 1. To find the primitive of

$$\frac{adx}{x} + \frac{bdy}{y} - \frac{cx^n dx}{y^b} = 0 \ldots\ldots (1).$$

This can be resolved into the two parts

$$\frac{adx}{x} + \frac{bdy}{y} \text{ and } -\frac{cx^n dx}{y^b};$$

the first of which is an exact differential, and therefore $z_1 = 1$; and the second can be rendered exact by the factor $y^b = z_2$.

$$\therefore u_1 = \int\left[z_1\left(\frac{adx}{x} + \frac{bdy}{y}\right)\right] = \int\left(\frac{adx}{x} + \frac{bdy}{y}\right)$$

$$= a \log x + b \log y = \log(x^a . y^b).$$

$$u_2 = \int\left[z_2\left(-\frac{cx^n dx}{y^b}\right)\right] = \int(-cx^n dx) = \frac{-cx^{n+1}}{n+1}.$$

Hence we must endeavor to satisfy the condition $z_1F_1u_1 = z_2F_2u_2$, or

$$1 \times F_1[\log(x^a y^b)] = y^b . F_2\left(-\frac{cx^{n+1}}{n+1}\right).$$

Assume $F_1[\log(x^a y^b)] = (x^a y^b)^k$ and $F_2\left(-\frac{cx^{n+1}}{n+1}\right) = x^{k_1 n}$

in which k and k_1 are undetermined constants. Then

$$x^{ka} y^{kb} = y^b x^{k_1 n},$$

a condition which will be satisfied by making

$$kb = b \quad \text{and} \quad ka = k_1 n, \quad \text{or} \quad k = 1 \quad \text{and} \quad k_1 = \frac{a}{n}.$$

Hence $x^a y^b$ is the required factor.

Now multiply (1) by $x^a y^b$, and there will result

$$ax^{a-1} y^b dx + bx^a y^{b-1} dy - cx^{a+n} dx = 0,$$

which is exact, since

$$\frac{d(ax^{a-1} y^b - cx^{a+n})}{dy} = \frac{d(bx^a y^{b-1})}{dx};$$

and the required solution is

$$x^a y^b = \frac{cx^{a+n+1}}{a+n+1} + C.$$

2. $$\frac{1}{2} xdy - ydx - \frac{1}{2} adx = 0.$$

This can be resolved into $\frac{1}{2} xdy - ydx$, and $-\frac{1}{2} adx$, of which th first will be rendered exact by the factor $z_1 = \frac{1}{xy}$, and the secon is already exact, giving $z_2 = 1$.

$$\therefore u_1 = \int \frac{1}{xy}\left(\frac{1}{2} xdy - ydx\right) = \int \frac{1}{2} \frac{dy}{y} - \int \frac{dx}{x} = \log \frac{y^{\frac{1}{2}}}{x}.$$

$$u_2 = \int \left(-\frac{1}{2} adx\right) = -\frac{1}{2} ax,$$

and we must satisfy the conditions

$$F_1\left(\log \frac{y^{\frac{1}{2}}}{x}\right) \times \frac{1}{xy} = 1 \times F_2\left(-\frac{1}{2} ax\right).$$

Assume $F_1\left(\log \frac{y^{\frac{1}{2}}}{x}\right) = \frac{y}{x^2}$. $\therefore \frac{1}{xy} F_1\left(\log \frac{y^{\frac{1}{2}}}{x}\right) = \frac{1}{x^3}$,

and $$\therefore F_2\left(-\frac{1}{2} ax\right) = \frac{1}{x^3} \text{ also.}$$

Hence $\frac{1}{x^3}$ is the proper factor.

$$\therefore \frac{1}{2}\cdot\frac{xdy}{x^3} - \frac{ydx}{x^3} - \frac{1}{2}\cdot\frac{adx}{x^3} = 0,$$

which is exact, and the solution is

$$\frac{y}{x^2} + \frac{1}{2}\cdot\frac{a}{x^2} + C = 0 \quad \text{or} \quad y + Cx^2 + \frac{a}{2} = 0.$$

168. *Prop.* To determine the factor necessary to render a differential equation exact, when that factor is a function of one variable only.

Let $Pdx + Qdy = 0 \ldots . (1)$ be the given equation, and $z = Fx$ the required factor.

$$\therefore zPdx + zQdy = 0 \text{ will be exact, and therefore}$$

$\frac{d(zP)}{dy} = \frac{d(zQ)}{dx}$; or since z does not contain y, and therefore $\frac{dz}{dy} = 0$,

$$z\frac{dP}{dy} = z\frac{dQ}{dx} + Q\cdot\frac{dz}{dx}. \quad \therefore \frac{dz}{z} = \frac{1}{Q}\left[\frac{dP}{dy} - \frac{dQ}{dx}\right]dx.$$

Here, by hypothesis, the first member does not contain y, and therefore the second member must be independent of y also. Consequently

$$\log z = \int\left[\frac{dP}{dy} - \frac{dQ}{dx}\right]\frac{dx}{Q} = \varphi x \text{ and } z = e^{\phi x}, \text{ the required value.}$$

169. 1. Given $\quad ydx - xdy = 0 \ldots . (1).$

Suppose z to contain x only.

$$\frac{dP}{dy} = \frac{dy}{dy} = 1, \quad \frac{dQ}{dx} = \frac{d(-x)}{dx} = -1.$$

$$\therefore \left[\frac{dP}{dy} - \frac{dQ}{dx}\right]\frac{dx}{Q} = -\frac{2dx}{x}.$$

$$\therefore \log z = -\int\frac{2dx}{x} = \log\frac{1}{x^2} \quad \text{and} \quad z = \frac{1}{x^2}.$$

Multiplying (1) by $\frac{1}{x^2}$ we get

$$\frac{ydx - xdy}{x^2} = 0 \quad \text{and} \quad \frac{y}{x} = C \quad \text{or} \quad y = Cx.$$

2. The linear equation $dy + Xydx - X_1dx = 0$.

Suppose $z = Fx$. $\therefore \frac{dP}{dy} = \frac{d(Xy - X_1)}{dy} = X$, and $\frac{dQ}{dx} = 0$.

$$\therefore \log z = \int Xdx \quad \text{and} \quad z = e^{\int Xdx}, \text{ the factor sought.}$$

Multiplying (1) by this factor, we have

$$e^{\int Xdx}.dy + e^{\int Xdx}Xydx - e^{\int Xdx}X_1dx = 0, \text{ which is exact.}$$

Remark. The value of z found by the method just explained, was obtained by assuming that a factor containing x only can be discovered; but since such factor may not exist, it will be proper to apply the test of integrability to the transformed equation.

170. *Prop.* To determine the factor necessary to render a homogeneous differential equation exact.

Let $$Pdx + Qdy = 0 \ldots\ldots (1)$$

be a homogeneous differential equation, the coefficients P and Q being each of the nth degree; and let the factor z be of the mth degree. Then

$$zPdx + zQdy = 0$$

will be exact, homogeneous, and of the $(m + n)^{th}$ degree.

Hence, by the rule for integrating homogeneous exact differential expressions, we have

$$\int (zPdx + zQdy) = \frac{zPx + zQy}{m + n + 1}.$$

$$\therefore \frac{zPx + zQy}{m + n + 1} = C, \quad \text{and} \quad z = \frac{(m + n + 1)C}{Px + Qy};$$

or, since C is arbitrary, we may put $(m + n + 1)C = 1$.

$$\therefore z = \frac{1}{Px + Qy} \text{ is a suitable factor.}$$

Ex. $(xy + y^2)dx - (x^2 - xy)dy = 0 \ldots\ldots (1).$

Here $xy + y^2 = P$ and $-x^2 + xy = Q.$

$$\therefore z = \frac{1}{Px + Qy} = \frac{1}{2xy^2}.$$

Multiply (1) by z. $\therefore \frac{dx}{2y} + \frac{dx}{2x} - \frac{xdy}{2y^2} + \frac{dy}{2y} = 0.$

$$\therefore \frac{1}{2}\frac{x}{y} + \frac{1}{2}\log x + \frac{1}{2}\log y = C, \quad \text{or} \quad \frac{x}{y} + \log(xy) = C_1.$$

Geometrical Applications of Differential Equations of the first order and degree.

171. 1. Determine the curve whose tangent PT is a mean proportional between the parts AT and BT of its axis, intercepted between the tangent and two fixed points A and B.

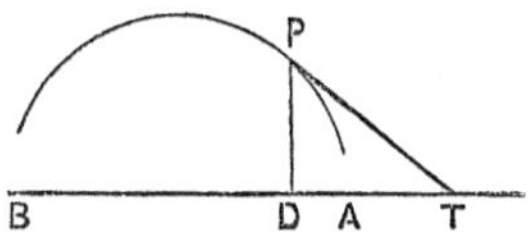

Place the origin at B, and put

$$BD = x, \quad DP = y, \quad BA = a.$$

The equation of the tangent is $y - y_1 = \frac{dy_1}{dx_1} \cdot (x - x_1),$

in which when $y = 0, \quad x - x_1 = -y_1 \frac{dx_1}{dy_1} = DT,$

and $x_1 - y_1 \frac{dx_1}{dy_1} = BD + DT = BT.$

$$\therefore AT = x_1 - y_1 \frac{dx_1}{dy_1} - a.$$

And $PT^2 = PD^2 + DT^2 = y_1^2 + y_1^2 \frac{dx_1^2}{dy_1^2}.$

Hence, by the conditions of the problem,

$$PT^2 = BT \times AT, \quad \text{or} \quad y_1^2 + y_1^2 \frac{dx_1^2}{dy_1^2}$$

$$= \left(x_1 - y_1 \frac{dx_1}{dy_1}\right)\left(x_1 - y_1 \frac{dx_1}{dy_1} - a.\right).$$

Reducing, and omitting accents, we get

$$y^2 dy = x^2 dy - 2xydx - axdy + aydx,$$

which is the differential equation of the required curve.

This may be written

$$dy + \frac{2xydx - x^2dy}{y^2} = a\frac{ydx - xdy}{y^2};$$

and, since both members are exact, we get by integration

$$y + \frac{x^2}{y} = a\frac{x}{y} + C, \quad \text{or} \quad x^2 + y^2 - ax - Cy = 0; \text{ or, finally,}$$

$$(x - \frac{1}{2}a)^2 + (y - \frac{1}{2}C)^2 = \frac{1}{4}(a^2 + C^2),$$

which is the equation of a circle whose radius is $\frac{1}{2}\sqrt{a^2 + C^2}$; and the co-ordinates of whose centre are $\frac{1}{2}a$ and $\frac{1}{2}C$, the latter co-ordinate being arbitrary.

2. Find the curve in which the subtangent is constant.

Let $x_1 y_1$ be the co-ordinates of the point of contact.

$$\text{Then, subtangent} = -y_1\frac{dx_1}{dy_1} = -a, \quad \text{or} \quad \frac{dy}{y} = \frac{dx}{a}.$$

$$\therefore \log cy = \frac{x}{a}, \qquad cy = e^{\frac{x}{a}}.$$

This is the equation of the logarithmic curve.

3. To find the curve in which the subtangent is equal to the sum of the abscissa and co-ordinate.

The differential equation of the curve will be

$$-y\frac{dx}{dy} = x + y, \quad \text{or} \quad xdy + ydx + ydy = 0.$$

$$\therefore\ y^2 + 2xy = C, \text{ a hyperbola.}$$

4. The curve in which the subnormal is constant.

Subnormal $= y\dfrac{dy}{dx} = a. \quad \therefore\ ydy = adx.$

$$\therefore\ y^2 = 2ax + C, \text{ a parabola.}$$

172. *Prop.* To find the equation of a *trajectory* or curve which shall intersect all the curves of a given family in the same angle.

Let
$$\varphi(x,y,a) = 0 \ldots\ldots (1)$$
be the equation of a class of curves, in which the parameter a may take any value; and let $t = \text{tang}\,\beta$, where β represents the constant angle of intersection.

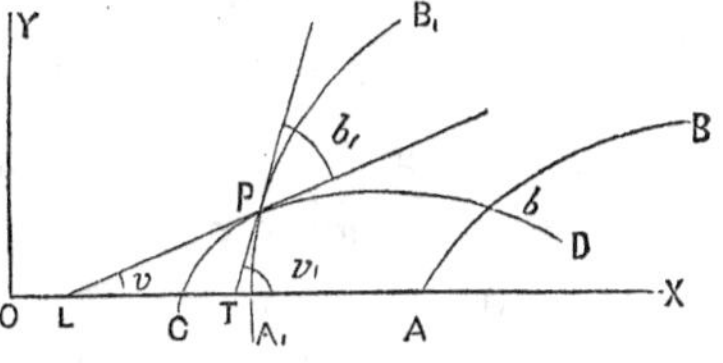

Suppose a to take a particular value, a_1 and let A_1B_1 be the particular curve in the general class resulting from this supposition. Then, if x_1y_1 denote its general co-ordinates, its equation will be

$$\varphi(x_1, y_1, a_1) = 0 \ldots. (2),$$

and the differential coefficient $\dfrac{dy_1}{dx_1}$, derived therefrom, will, when applied to the point P, express the trigonometrical tangent of the angle PTX or v_1, included between the tangent PT and the axis of x.

Also, if x and y denote the general co-ordinates of the required trajectory CPD, the differential coefficient $\dfrac{dy}{dx}$, given by its equation, will, when applied to P, give the tangent of PLX or v.

But $\beta = v_1 - v. \quad \therefore \tan\beta = \dfrac{\tan v_1 - \tan v}{1 + \tan v_1 \tan v}$,

or by substitution,

$$t = \frac{\dfrac{dy_1}{dx_1} - \dfrac{dy}{dx}}{1 + \dfrac{dy_1}{dx_1}\cdot\dfrac{dy}{dx}} \ldots. (3).$$

Now at the point P, where the curves A_1B_1 and CPD intersect, $x_1 = x$ and $y_1 = y$. Hence $\frac{dy_1}{dx_1}$ can be expressed in functions of x, y, and a_1, and therefore (3) may be written thus

$$F\left(x,\ y,\ \frac{dy}{dx},\ a_1\right) = 0 \ldots . (4).$$

But (2), when applied to P, gives

$$\varphi(x, y, a_1) = 0 \ldots . (5).$$

If then a_1 be eliminated between (4) and (5), the resulting equation, being independent of the position of the particular curve A_1B_1, will apply to all points in the required curve CPD.

173. 1. Determine the curve which cuts at right angles all straight lines drawn through a given point.

Let x_2y_2 be the co-ordinates of the given point.

The equation of one of the straight lines passing through that point will be of the form

$$y_1 - y_2 = a_1(x_1 - x_2).$$

$$\therefore \varphi(x_1y_1a_1) = y_1 - y_2 - a_1(x_1 - x_2) = 0 \quad \text{and} \quad \frac{dy_1}{dx_1} = a_1.$$

Also $t = \tan \frac{1}{2}\pi = \infty$. $\therefore \dfrac{\frac{dy_1}{dx_1} - \frac{dy}{dx}}{1 + \frac{dy_1}{dx_1} \cdot \frac{dy}{dx}} = \infty$. and consequently

$$1 + \frac{dy_1}{dx_1} \cdot \frac{dy}{dx} = 0 \quad \text{or} \quad \frac{dy}{dx} \cdot a_1 = -1 \ldots . (1).$$

Also at the point of intersection

$$y - y_2 - a_1(x - x_2) = 0 \ldots .. (2).$$

Now eliminating a_1 between (1) and (2), we get

$$x - x_2 + \frac{dy}{dx}(y - y_2) = 0, \quad \text{or} \quad xdx - x_2dx + ydy - y_2dy = 0,$$

which is the differential equation of the curve.

And by integration $\frac{1}{2}x^2 - x_2x + \frac{1}{2}y^2 - y_2y = C,$

or $$(x - x_2)^2 + (y - y_2)^2 = 2C + x_2^2 + y_2^2.$$

Hence the curve sought is a circle whose centre is at the point $x_2, y_2,$ and the radius arbitrary.

2. The curve which cuts at an angle of 45° all straight lines drawn through the origin.

Here $$\varphi(x_1, y_1, a_1) = y_1 - a_1x_1 = 0.$$

$$\therefore \frac{dy_1}{dx_1} = a_1 \quad \text{also} \quad t = \tan 45° = 1.$$

$$\therefore 1 + a_1\frac{dy}{dx} = a_1 - \frac{dy}{dx} \cdots\cdots (1), \quad \text{and} \quad y - a_1x = 0 \ldots\ldots (2).$$

Eliminating a_1, $1 + \frac{y}{x}\cdot\frac{dy}{dx} = \frac{y}{x} - \frac{dy}{dx}$ or $xdx + ydy = ydx - xdy.$

This being a homogeneous equation, it will be rendered exact by multiplying by $\frac{1}{Px + Qy} = \frac{1}{x^2 + y^2}.$

$$\therefore \frac{xdx + ydy}{x^2 + y^2} = \frac{ydx - xdy}{x^2 + y^2}. \quad \therefore \log\left[\frac{x^2 + y^2}{c^2}\right]^{\frac{1}{2}} = -\tan^{-1}\frac{y}{x}.$$

Put $y = r\sin\theta$, $x = r\cos(\pi - \theta) = -r\cos\theta$. $\therefore r = (x^2 + y^2)^{\frac{1}{2}}$

and $$\frac{y}{x} = -\tan\theta. \quad \therefore \log\frac{r}{c} = \theta \quad \text{and} \quad r = ce^{\theta},$$

the equation of the logarithmic spiral.

3. The curve which cuts at right angles all parabolas having a common vertex and coincident axes.

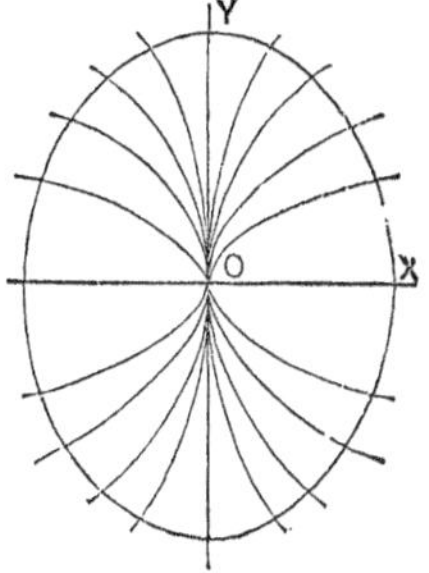

Here $\varphi(x_1, y_1, a_1) = y_1^2 - 2a_1x_1 = 0.$

$$\therefore \frac{dy_1}{dx_1} = \frac{a_1}{y_1} = \frac{a_1}{y}.$$

Also $t = \infty$. $\therefore 1 + \frac{a_1}{y}\cdot\frac{dy}{dx} = 0 \ldots . (1),$

and $$y^2 - 2a_1x = 0 \ldots . (2).$$

Eliminating a_1, $1 + \frac{y}{2x} \cdot \frac{dy}{dx} = 0$ or $2xdx + ydy = 0$.

$$\therefore x^2 + \frac{1}{2}y^2 = c^2, \quad \text{or} \quad \frac{x^2}{c^2} + \frac{y^2}{2c^2} = 1.$$

This is the equation of an ellipse whose axes have the ratio $1 : \sqrt{2}$.

CHAPTER III.

DIFFERENTIAL EQUATIONS OF THE FIRST ORDER AND OF THE HIGHER DEGREES.

174. These equations contain the several powers of the coefficient $\frac{dy}{dx}$ to the nth power inclusive where n denotes the degree of the equation. The most general form of such an equation is

$$\frac{dy^n}{dx^n} + P\frac{dy^{n-1}}{dx^{n-1}} + Q\frac{dy^{n-2}}{dx^{n-2}} + \&c. \ldots \ldots + S\frac{dy}{dx} + T = 0 \ldots . (1),$$

which equation can be derived from its primitive only in attempting to eliminate the nth power of a constant c between the primitive and its direct differential. For the direct differential contains only the first power of $\frac{dy}{dx}$, and therefore cannot be identical with (1) ; but if we suppose the primitive to contain several powers of the same constant c, as $c^1, c^2, c^3 \ldots . c^n$, and resolve with respect to c, there will result n values of c, from each of which c will disappear by differentiation ; and each of the resulting differential equations will contain only the first power of $\frac{dy}{dx}$, each being a factor of (1). Hence by multiplying together these n equations, we shall produce (1). If therefore we resolve (1) with respect to $\frac{dy}{dx}$, thereby ascertaining its n constituent factors of the first degree, then integrate each, annexing the same constant c to every result, and finally multiply the results together, the complete primitive, which includes all these separate results, will be obtained. It will be obvious, that in order to render this method applicable to all equations of the first order, it would be necessary to have a process for the solution of equations of all degrees.

Unfortunately no such process is known.

175. 1. To find the complete primitive of the equation

$$\frac{dy^2}{dx^2} = a^2 \ldots\ldots\ (1).$$

Resolving with respect to $\frac{dy}{dx}$, we get

$$\frac{dy}{dx} = +a \ldots.\ (2), \quad \text{and} \quad \frac{dy}{dx} = -a \ldots.\ (3).$$

Integrating (2) and (3), and annexing the same constant to each, we have

$$y = ax + c \ldots.\ (4), \quad \text{and} \quad y = -ax + c \ldots.\ (5),$$

either of which satisfies the given equation (1). It is also satisfied by their product.

$$(y - ax - c)(y + ax - c) = 0,$$

or

$$y^2 - 2cy - a^2x^2 + c^2 = 0 \ldots\ldots\ (6).$$

For, by differentiating (6),

$$2ydy - 2cdy - 2a^2xdx = 0, \text{ and } c = y - \frac{a^2x}{\frac{dy}{dx}}.$$

This value, substituted in (6), gives

$$y^2 - 2y^2 + \frac{2a^2xy}{\frac{dy}{dx}} - a^2x^2 + y^2 - \frac{2a^2xy}{\frac{dy}{dx}} + \frac{a^4x^2}{\frac{dy^2}{dx^2}} = 0;$$

or, by reduction, $$\frac{dy^2}{dx^2} = a^2,$$

which is identical with (1).

2. $$\frac{dy^2}{dx^2} = ax \ldots . (1).$$

$$\frac{dy}{dx} = + a^{\frac{1}{2}}x^{\frac{1}{2}}, \text{ and } \frac{dy}{dx} = - a^{\frac{1}{2}}x^{\frac{1}{2}}.$$

$\therefore$ By integration $y = \frac{2}{3}a^{\frac{1}{2}}x^{\frac{3}{2}} + c$, and $y = -\frac{2}{3}a^{\frac{1}{2}}x^{\frac{3}{2}} + c$.

$$\therefore (y - \frac{2}{3}a^{\frac{1}{2}}x^{\frac{3}{2}} - c)(y + \frac{2}{3}a^{\frac{1}{2}}x^{\frac{3}{2}} - c) = 0,$$

or $$(y - c)^2 = \frac{4}{9}ax^3 \text{ the complete primitive of (1).}$$

3. $$y\frac{dy^2}{dx^2} + 2x\frac{dy}{dx} - y = 0 \ldots . (1).$$

$$\frac{dy}{dx} = -\frac{x}{y} \pm \frac{(y^2 + x^2)^{\frac{1}{2}}}{y}, \quad \therefore dx = \pm \frac{xdx + ydy}{\sqrt{x^2 + y^2}}.$$

$$\therefore x = + (x^2 + y^2)^{\frac{1}{2}} + c, \text{ and } x = -(x^2 + y^2)^{\frac{1}{2}} + c.$$

$$\therefore (x - c - \sqrt{x^2 + y^2})(x - c + \sqrt{x^2 + y^2}) = 0, \text{ or } y^2 = c^2 - 2cx.$$

4. Determine the equation of the curve which has the property

$$s = ax + by.$$

Here $$\frac{ds}{dx} = \left[1 + \frac{dy^2}{dx^2}\right]^{\frac{1}{2}} = a + b\frac{dy}{dx}.$$

$$\therefore 1 + \frac{dy^2}{dx^2} = a^2 + 2ab\frac{dy}{dx} + \frac{b^2dy^2}{dx^2}.$$

$$\therefore \frac{dy^2}{dx^2} + \frac{2ab}{b^2 - 1}\cdot\frac{dy}{dx} = \frac{1 - a^2}{b^2 - 1}, \text{ and}$$

$$\frac{dy}{dx} = \frac{ab}{1 - b^2} \pm \frac{\sqrt{a^2 + b^2 - 1}}{b^2 - 1} = m \pm n.$$

$$\therefore y = mx + nx + c, \quad \text{and} \quad y = mx - nx + c.$$

$$\therefore\ y^2 = m^2x^2 - n^2x^2 + 2cmx + c^2.$$

This is the equation of two straight lines, which intersect on the axis of y, and which become imaginary when $a^2 + b^2 < 1$. Suppose

$$a = \sqrt{\frac{1}{2}} \text{ and } b = \sqrt{\frac{1}{2}}. \quad \therefore\ a^2 + b^2 = 1,\ ab = \frac{1}{2} \text{ and } 1 - b^2 = \frac{1}{2}.$$

$$\therefore\ m = 1 \quad \text{and} \quad n = 0, \quad \therefore\ y = x + c,$$

and the two lines become a single line, inclined to the axis of x in an angle of 45°.

176. When the proposed differential equation cannot be resolved with respect to $\frac{dy}{dx}$, its primitive may still be found in certain cases, the principal of which will be examined.

Case 1st. When the equation contains only one of the variables, and the solution with respect to that variable is possible.

Let x be the variable which enters into the equation. Put $\frac{dy}{dx} = p_1$, and resolve with respect to x. The result will be of the form

$$x = \varphi p_1 \ \ldots . \ (1).$$

But since $dy = p_1 dx$, an integration by parts will give

$$y = p_1 x - \int x dp_1 \ \ldots . \ (2).$$

Eliminating x between (1) and (2), we get

$$y = p_1 . \varphi p_1 - \int \varphi p_1 . dp_1 \ \ldots .. \ (3),$$

in which the last term is integrable as a function of a single variable.

Effecting the integration, we may unite the result thus

$$y = F p_1 \ \ldots .. \ (4).$$

Then, eliminating p_1 between (1) and (4), we obtain the desired relation between x and y.

177. This method may sometimes be applied advantageously even when the more general method is applicable, provided the differential equation can be solved more easily for x than for $\frac{dy}{dx}$.

Ex. To find the primitive of the differential equation

$$x\frac{dy^2}{dx^2} + x - 1 = 0.$$

Here $$x = \frac{1}{1+p_1^2} = \phi p_1 \ldots . (1).$$

$$\therefore y = p_1\phi p_1 - \int \phi p_1 . dp_1 = \frac{p_1}{1+p_1^2} - \int \frac{dp_1}{1+p_1^2}$$

$$= \frac{p_1}{1+p_1^2} - \tan^{-1}p_1 + C \ldots\ldots (2).$$

But from (1), $p_1 = \left(\frac{1-x}{x}\right)^{\frac{1}{2}}$, and $1 + p_1^2 = \frac{1}{x}$,

and these values reduce (2) to

$$y = (x - x^2)^{\frac{1}{2}} - \tan^{-1}\left(\frac{1-x}{x}\right)^{\frac{1}{2}} + C.$$

178. When the equation (still supposed to contain x only), cannot be resolved either for x or p_1, we may substitute xz for p_1, and we can then divide every term by a power of x, thereby depressing the degree of the equation, except in the case where there is an absolute term. If then the depressed equation can be solved for x or z, we shall have either

$$x = \phi z, \quad \text{and} \quad p_1 = \frac{dy}{dx} = z\phi z,$$

$$\therefore dy = z.\phi z.d(\phi z), \quad \text{and} \quad y = \int z.\phi z.d(\phi z), \ldots . (5),$$

or, $$z = \phi x, \quad \text{and} \quad p_1 = \frac{dy}{dx} = x\phi x,$$

$$\therefore dy = x.\phi x.dx, \quad \text{and} \quad y = \int x.\phi x.dx. \ldots . (6).$$

In the first case we eliminate z between (5) and $x = \phi z$. In the second, the desired relation is found in (6).

Ex. $$x^3 + \frac{dy^3}{dx^3} - ax\frac{dy}{dx} = 0.$$

Put $p_1 = xz$, then $x^3 + x^3z^3 - ax^2z = 0$, and $x = \frac{az}{1+z^3}$,

$$\therefore p_1 = \frac{dy}{dx} = \frac{az^2}{1+z^3}, \quad dy = \frac{az^2}{1+z^3}d\left[\frac{az}{1+z^3}\right] = \frac{a^2(z^2 - 2z^5)}{(1+z^3)^3}dz.$$

$$\therefore\ y = \frac{1}{6}a^2\frac{2z^3 - 1}{(1 + z^3)^2} + \frac{1}{3}a^2\frac{1}{1 + z^3} + C.$$

This relation, together with $x = \dfrac{az}{1 + z^3}$ expresses the relation between x and y.

179. *Case 2d.* When the equation is homogeneous with respect to x and y.

Let n denote the degree of the equation in x and y, and put $y = xz$, then the equation will be divisible by x^n, and if the transformed equation can be solved for z, we shall have a result of the form

$$z = \varphi p_1, \quad \therefore\ dz = d(\varphi p_1). \quad \text{But} \quad y = xz, \quad \therefore\ dy = xdz + zdx,$$

or, $\quad dy = xd(\varphi p_1) + \varphi p_1 dx, \quad$ or, $\quad p_1 dx = xd(\varphi p_1) + \varphi p_1 dx,$

$$\therefore\ \frac{dx}{x} = \frac{d(\varphi p_1)}{p_1 - \varphi p_1}, \quad \text{and} \quad \log x = \int\frac{d(\varphi p_1)}{p_1 - \varphi p_1} = Fp_1,$$

This combined with $y = x\varphi p_1$, gives the desired relation between x and y.

Ex. $\qquad y - xp_1 = \sqrt{1 + p_1^2}\,.\,x.$

Put $\qquad y = xz,\quad$ substitute and divide by x, then

$$z - p_1 = \sqrt{1 + p_1^2}, \quad z = p_1 + \sqrt{1 + p_1^2}, \quad dz = dp_1 + \frac{p_1 dp_1}{\sqrt{1 + p_1^2}},$$

$$p_1 dx = dy = xdz + zdx = x\left(dp_1 + \frac{p_1 dp_1}{\sqrt{1 + p_1^2}}\right) + (p_1 + \sqrt{1 + p_1^2})dx,$$

$$\therefore\ \frac{dx}{x} = -\frac{dp_1}{\sqrt{1 + p_1^2}} - \frac{p_1 dp_1}{1 + p_1^2}.$$

$$\therefore\ \log x = -\log(p_1 + \sqrt{1 + p_1^2}) - \log(1 + p_1^2)^{\frac{1}{2}} + \log c.$$

$$\therefore\ x = \frac{c}{\sqrt{1 + p_1^2}(p_1 + \sqrt{1 + p_1^2})}. \quad \text{But} \quad y = xz = \frac{c}{\sqrt{1 + p_1^2}},$$

$$\therefore\ p_1 = \frac{\sqrt{c^2 - y^2}}{y}, \quad \text{and} \quad x(c + \sqrt{c^2 - y^2}) = y^2,$$

the desired relation.

180. *Case 3d.* Let the form be

$$y = x\frac{dy}{dx} + \varphi\left(\frac{dy}{dx}\right). \quad . \; . \; . \; (1).$$

in which $\varphi\left(\frac{dy}{dx}\right)$ does not contain x or y.

By differentiation, $\frac{dy}{dx} = p_1 = p_1 + x\frac{dp_1}{dx} + \frac{d\varphi p_1}{dp_1} \cdot \frac{dp_1}{dx}$,

$$\therefore \left(x + \frac{d\varphi p_1}{dp_1}\right)\frac{dp_1}{dx} = 0.$$

This is satisfied either by making

$$x + \frac{d\varphi p_1}{dp_1} = 0, \; . \; . \; . \; . \; (2), \quad \text{or,} \quad \frac{dp_1}{dx} = 0. \; . \; . \; . \; . \; . \; (3).$$

Now the differential coefficient $\frac{d\varphi p_1}{dp_1}$ in (2), contains only p_1, since φp_1 does not contain x or y, and therefore (2) contains only x and p_1. If then, we eliminate p_1 between (1) and (2), the result will be a relation between x and y. But this relation cannot be the complete primitive, because it contains no arbitrary constant. We must then refer to the condition (3), which gives by integration

$$p_1 = C, \text{ a constant.}$$

It appears then, that in the proposed equation, which is known as *Clairault's* form, the complete primitive is obtained by simply replacing $\frac{dy}{dx}$ by an arbitrary constant.

Ex. 1. To find the primitive of

$$y - x\frac{dy}{dx} = a\left(1 + \frac{dy^2}{dx^2}\right). \; . \; . \; . \; . \; . \; (1).$$

Replacing the differential coefficient $\frac{dy}{dx}$ by C, we have.

$$y - Cx = a(1 + C^2). \; . \; . \; . \quad (2).$$

The correctness of this solution is easily verified; for by differentiating (2) we get

$$\frac{dy}{dx} - C = 0. \ldots . (3).$$

and by eliminating C between (2) and (3), we obtain (1).

2. $ydx - xdy = a(dx^3 + dy^3)^{\frac{1}{3}}$, or, $y = x\frac{dy}{dx} + \left(1 + \frac{dy^3}{dx^3}\right)^{\frac{1}{3}}a.$

Substituting C for $\frac{dy}{dx}$ we get $y = Cx + (1 + C^3)^{\frac{1}{3}}a.$

181. *Case* 4*th.* Let $y = Px + Q, \ldots . . (1).$
when P and Q are functions of p_1.

By differentiation, $dy = p_1 dx = Pdx + xdP + dQ.$

$\therefore \ (p_1 - P)dx - xdP = dQ,$ and $dx + \frac{x}{P - p_1}dP = -\frac{dQ}{P - p_1}.$

This being a linear equation, its solution is of the form

$$x = e^{-\int\frac{dP}{P-p_1}}\left[-\int e^{\int\frac{dP}{P-p_1}}.\frac{dQ}{P-p_1}\right], \quad \text{or,} \quad x = Fp_1.$$

Hence if p_1 be eliminated between this and (1), the result will be a relation between x and y.

182. 1. $$y = p_1^2 x + p_1^2 \ldots . (1).$$

$$dy = p_1 dx = p_1^2 dx + 2xp_1 dp_1 + 2p_1 dp_1.$$

$$\therefore (1 - p_1)dx - 2xdp_1 = 2dp_1.$$

$$\therefore dx + 2x\frac{dp_1}{p_1 - 1} = -\frac{2dp_1}{p_1 - 1}.$$

$$\therefore x = e^{-\int\frac{2dp_1}{p_1-1}}\left[-\int e^{\int\frac{2dp_1}{p_1-1}}.\frac{2dp_1}{p_1 - 1}\right].$$

But $$\int\frac{2dp_1}{p_1 - 1} = 2\log(p_1 - 1) = \log(p_1 - 1)^2.$$

$$\therefore e^{\int\frac{2dp_1}{p_1-1}} = e^{\log(p_1-1)^2} = (p_1 - 1)^2 \text{ and } e^{-\int\frac{2dp_1}{p_1-1}} = \frac{1}{(p_1 - 1)^2}.$$

$$\therefore x = -\frac{1}{(p_1-1)^2}\int 2(p_1-1)dp_1 = \frac{C^2}{(p_1-1)^2} - 1.$$

$$\therefore p_1 = 1 + \frac{C}{\sqrt{1+x}}; \quad \text{and from (1),} \quad p_1 = \frac{\sqrt{y}}{\sqrt{1+x}}.$$

$$\therefore \sqrt{y} = \sqrt{1+x} + C.$$

2. $$y = (1+p_1)x + p_1^2 \ldots . (1).$$

$$p_1 dx = (1+p_1)dx + xdp_1 + 2p_1 dp_1.$$

$$\therefore dx + xdp_1 = -2p_1 dp_1, \quad \text{and} \quad x = e^{-\int dp_1}\left[-\int 2e^{\int dp_1} . p_1 dp_1\right].$$

But $e^{\int dp_1} = e^{p_1}$, and $\int e^{p_1} p_1 dp_1 = e^{p_1}(p_1 - 1) + C_1$.

$$\therefore x = 2(1-p_1) + Ce^{-p_1} \quad \text{where} \quad C = -C_1;$$

and from (1), $p_1 = -\frac{1}{2}x \pm \frac{1}{2}\sqrt{4y - 4x + x^2}$.

$\therefore$ By eliminating p_1 we get

$$0 = 2 \mp \sqrt{4y - 4x + x^2} + Ce^{\frac{1}{2}x \mp \frac{1}{2}\sqrt{4y-4x+x^2}}.$$

3. $$y = x(p_1 - \sqrt{1+p_1^2}) \ldots . (1).$$

In this example $Q = 0$, and by differentiation

$$p_1 dx = p_1 dx + xdp_1 - \sqrt{1+p_1^2}\, dx - p_1 x(1+p_1^2)^{-\frac{1}{2}} dp_1.$$

$$\therefore \frac{dx}{x} = \frac{\sqrt{1+p_1^2} - p_1}{1+p_1^2} dp_1, \text{ the integral of which is}$$

$$\log x = \log \frac{(p_1 + \sqrt{1+p_1^2})c}{\sqrt{1+p_1^2}}. \quad \therefore x = \frac{(p_1+\sqrt{1+p_1^2})c}{\sqrt{1+p_1^2}},$$

and $$cp_1 + (c-x)\sqrt{1-p_1^2} = 0 \ldots . (2).$$

But from (1), $p_1 x - x\sqrt{1+p_1^2} = y$. $\quad \therefore p_1 = \dfrac{y(c-x)}{x(2c-x)}$,

and $\sqrt{1+p_1^2} = -\dfrac{cy}{x(2c-x)}$. $\quad \therefore p_1^2 = \dfrac{y^2(c-x)^2}{x^2(2c-x)^2} = \dfrac{c^2y^2}{x^2(2c-x)^2} - 1.$

$$\therefore y^2(x^2 - 2cx) = -x^2(2c-x)^2, \quad \text{or finally} \quad x^2 + y^2 = 2cx.$$

CHAPTER IV.

SINGULAR SOLUTIONS OF DIFFERENTIAL EQUATIONS.

183. Differential equations may be regarded as resulting in all cases either from the immediate differentiation of their primitives or from the elimination of constants between the primitives and their direct differentials.

184. Taking the latter and more common case, let

$$F(x, y, c) = 0 \; .\; .\; .\; . \; (1)$$

be the complete primitive of the differential equation

$$\varphi\left(x, y, \frac{dy}{dx}\right) = 0 \; .\; .\; .\; . \; (2),$$

where (2) has arisen from an elimination of the constant c between (1) and its immediate differential

$$\left[\frac{dF(x, y, c)}{dx}\right] = 0 \; .\; .\; .\; . \; (3).$$

Now if the constant c were replaced in (1) and (3) by any function of x and y, the elimination of this function would necessarily lead to the same equation (2).

If then it be possible to replace c by such a function of x and y, in equation (1) as shall give by differentiation a result entirely simi- to (3), after it has been modified by a like substitution of this function of x and y for c; then the elimination of that function would necessarily lead to (2) the proposed differential. Hence equation (1) with the value of c so replaced may be properly considered an integral of (2); although it is essentially different in form from the ordinary integral (1), in which c is an arbitrary constant.

Such a solution of a differential equation is called a *singular solution* or a *singular integral,* while the term *particular integral* is applied to each of the results obtained by substituting various constant values for c, in the general integral.

185. *Prop.* To determine the conditions necessary to render possible a singular solution of a differential equation.

Let the ordinary primitive

$$F(x, y, c) = 0 \;.\;.\;.\;.\;. \; (1)$$

be differentiated regarding c as variable, and there will result

$$\left[\frac{dF(x, y, c)}{dx}\right] + \frac{dF(x, y, c)}{dc} \cdot \left(\frac{dc}{dx}\right) = 0,$$

and to render this equation identical with

$$\left[\frac{dF(x, y, c)}{dx}\right] = 0 \;.\;.\;.\;.\;. \; (3),$$

which is obtained by supposing c constant, the necessary condition will be

$$\frac{dF(x, y, c)}{dc} \cdot \left(\frac{dc}{dx}\right) = 0 \;.\;.\;.\;.\;. \; (a).$$

Now (a) is satisfied either by making

$$\frac{dF(x,y,c)}{dc} = 0, \;.\;.\;.\;. \; (4), \qquad \text{or} \qquad \left(\frac{dc}{dx}\right) = 0, \;.\;.\;.\;. \; (5).$$

The condition (5) gives $c =$ constant, and therefore (4) can alone supply the suitable variable value of c.

The equation (4) may give several values of c, and then there will be as many singular solutions.

186. It must be observed that the value of c derived from (4), is not necessarily a function of x and y, or of either: for if c be connected with x and y only by the signs $+$ and $-$, those variables will not appear in (4), and consequently the values of c derived from (4), will be constants corresponding to particular integrals, and not singular solutions.

187. And again, the derived values of c may be functions of x and y, and yet not variable. For if the primitive (1) be solved with respect to any constant, as a, appearing in it, the result will assume the form

$$a = f(x, y, c), \dots \text{(6)};$$

and if by assigning any particular value to c, this value of a should become either identical with that of c given by (4); or if the latter be a function of the former, then c will be invariable, and therefore will not correspond to a singular solution.

188. If we solve the complete primitive (1) for x and y successively, the results may be written in the forms

$$x = f(y, c). \dots (7). \qquad y = f_1(x, c). \dots (8).$$

Which differentiated with respect to c, give (since the first members do not contain c)

$$\frac{df(y, c)}{dc} = 0, \quad \frac{df_1(x, c)}{dc} = 0, \quad \therefore \frac{dx}{dc} = 0, \dots (9), \text{ and } \frac{dy}{dc} = 0, . (10).$$

That is, if the primitive can be solved with respect to x or y, we may differentiate either of those values with respect to c, placing the result equal to zero. Thus (9) or (10) may be employed instead of (4), when more convenient, in obtaining those values of c which give singular solutions.

189. It may be observed that no differential equation of the first order and first degree can have a singular solution; for such equations have complete primitives containing only the first powers of c, and these primitives, when differentiated with respect to c, give a result (4) independent of c, which result cannot furnish a value of c.

190. The relation connecting the complete primitive with the singular solution, can be illustrated geometrically. For the former always represents a series of curves of the same class, in which c is the variable parameter, and as the process for obtaining the equation of the envelope of these curves is identical with that by which we find the singular solution, it follows that this solution must represent the envelope.

191. 1. Required the singular solution of the differential equation

$$ydx - xdy = a(dx^2 + dy^2)^{\frac{1}{2}}, \quad \text{or,} \quad y = xp_1 + a(1 + p_1^2)^{\frac{1}{2}} \dots (1).$$

This example belongs to Clairault's form, and therefore the complete integral is

$$y = cx + a(1 + c^2)^{\frac{1}{2}}, \dots (2).$$

$$\therefore \frac{dy}{dc} = x + ac(1 + c^2)^{-\frac{1}{2}} = 0, \quad \text{and} \quad c = -\frac{x}{\sqrt{a^2 - x^2}}.$$

This value substituted in (2) gives

$$y = -\frac{x^2}{\sqrt{a^2 - x^2}} + a\sqrt{1 + \frac{x^2}{a^2 - x^2}}, \quad \therefore y^2 + x^2 = a^2, \dots (3).$$

Thus the general solution (2) represents a series of straight lines all tangent to the circle represented by the singular solution (3).

2. $$yp_1^2 + 2xp_1 - y = 0.$$

The general solution of this example has been found to be (p. 396)

$$y^2 = c^2 - 2cx, \quad \therefore \frac{dy}{dc} = \frac{c - x}{\sqrt{c^2 - 2cx}} = 0,$$

$$\therefore c - x = 0, \quad \text{and} \quad c = x.$$

This value substituted in the general integral, gives

$$y^2 = x^2 - 2x^2, \quad \text{or} \quad y^2 + x^2 = 0, \text{ the singular solution.}$$

The general integral in this example represents a series of parabolas which do not intersect, and therefore the singular solution cannot, in this case, represent an envelope.

3. $$xp_1^2 - yp_1 + \frac{1}{2}q = 0, \quad \text{or,} \quad y = p_1x + \frac{q}{2p_1} \cdots\cdots (1).$$

This is Clairault's form, and therefore the general solution is

$$y = cx + \frac{q}{2c}, \ldots (2). \quad \therefore \frac{dy}{dc} = x - \frac{q}{2c^2} = 0, \text{ and } c = \sqrt{\frac{q}{2x}}.$$

This value in (2) gives

$$y = \sqrt{\frac{1}{2}qx} + \sqrt{\frac{1}{2}qx} = 2\sqrt{\frac{1}{2}qx}, \quad \text{or} \quad y^2 = 2qx.$$

Here the singular solution represents a parabola tangent to a series of straight lines represented by (2).

192. In the method of finding the singular solution of a differential equation, just explained and illustrated, it has been supposed that the general solution of the equation was known; but when it is not given we require the following proposition.

193. *Prop.* To determine the conditions by which singular solutions of differential equations may be found, without first determining their complete primitives.

Let $$u = F(x, y, c) = 0. \ldots\ldots (1),$$

be the complete primitive of the differential equation,

$$u_2 = F_2(x, y, p_1) = 0, \ldots . (2); \text{ and suppose}$$

$$u_1 = \left[\frac{dF(x,y,c)}{dx}\right] = F_1(x,y,c,p_1) = 0 \ldots\ldots (3),$$

to be the direct differential of (1).

Also let $U = f(x,y) = 0 \ldots . (4)$ be the singular solution of (2),

and $$U_1 = \left[\frac{df(x,y)}{dx}\right] = f_1(x,y) = 0 \ldots . (5),$$

the direct differential of (4).

Now, whether we eliminate c between (1) and (3), or eliminate a certain function of x and y; viz., the value of c (expressed in terms of x and y), derived from the condition

$$\frac{dF(x,y,c)}{dc} = 0,$$

between (4) and (5), the result will be (2) or its equivalent.

Let (3) be solved with reference to c, giving a result of the form

$$c = \varphi(x, y, p_1) \ldots\ldots (6),$$

and let this value be substituted in (1); we shall thus have (2) or its equivalent under the form

$$u = F(x, y, \varphi) = 0 \ldots\ldots (7),$$

where φ is put for $\varphi(x, y, p_1)$; for, by hypothesis, (2) is the result of the elimination of the constant c between (1) and (3).

Now, since (2) and (7) are equivalent, the elimination of p_1 between them must lead to an identical equation in x and y; that is, an equation, which, being true for all values of x and y, does not imply a relation between them.

Let $$p_1 = f(x, y) \ldots\ldots (8)$$

be the result obtained by solving (2) with respect to p_1.

This value substituted in (7) gives the identical equation before referred to, which can be differentiated with respect to x and y, successively, as though they were independent variables, since the equation does not imply any relation or mutual dependence between them.

Then, differentiating (7), and observing that φ contains x, y, and p_1, while $p_1 = f(x,y)$, we get

$$\frac{du}{dx} + \frac{du}{d\varphi}\cdot\frac{d\varphi}{dx} + \frac{du}{d\varphi}\cdot\frac{d\varphi}{dp_1}\cdot\frac{dp_1}{dx} = 0,$$

and $$\frac{du}{dy} + \frac{du}{d\varphi}\cdot\frac{d\varphi}{dy} + \frac{du}{d\varphi}\cdot\frac{d\varphi}{dp_1}\cdot\frac{dp_1}{dy} = 0.$$

$$\therefore \frac{dp_1}{dx} = -\left(\frac{du}{dx} + \frac{du}{d\varphi}\cdot\frac{d\varphi}{dx}\right) \div \left(\frac{du}{d\varphi}\cdot\frac{d\varphi}{dp_1}\right),$$

and
$$\frac{dp_1}{dy} = -\left(\frac{du}{dy} + \frac{du}{d\varphi}\cdot\frac{d\varphi}{dy}\right) \div \left(\frac{du}{d\varphi}\cdot\frac{d\varphi}{dp_1}\right).$$

But when the solution is singular, we have the condition

$$\frac{du}{d\varphi} = \frac{du}{dc} = 0, \therefore \frac{dp_1}{dx} = \infty, \text{ and } \frac{dp_1}{dy} = \infty,$$

or
$$\frac{dx}{dp_1} = 0 \text{ and } \frac{dy}{dp_1} = 0.$$

If p_1 be eliminated between either of the last two equations and (2), the result will be a singular solution, provided it satisfies (2). Thus we can find the singular solution without previously finding the general solution.

Or, again, from (2), we have

$$\left[\frac{du_2}{dx}\right] = \frac{du_2}{dx} + \frac{du_2}{dy}\cdot\frac{dy}{dx} + \frac{du_2}{dp_1}\left(\frac{dp_1}{dx} + \frac{dp_1}{dy}\cdot\frac{dy}{dx}\right) = 0.$$

$$\therefore \frac{du_2}{dp_1} = -\left(\frac{du_2}{dx} + \frac{du_2}{dy}\cdot\frac{dy}{dx}\right) \div \left(\frac{dp_1}{dx} + \frac{dp_1}{dy}\cdot\frac{dy}{dx}\right);$$

and, since the divisor is infinite, when the solution is singular, we shall have the condition

$$\frac{du_2}{dp_1} = 0,$$

which will give suitable values of p_1 to be substituted in (2), in order to obtain singular solutions.

194. 1. Find the singular solution of the differential equation

$$u_2 = xp_1^2 - yp_1 + b = 0 \ldots . (1),$$

without previously finding the general integral.

Differentiating u_2 with respect to p_1 and placing the result equal to zero, we get

$$\frac{du_2}{dp_1} = 2xp_1 - y = 0, \quad \therefore \; p_1 = \frac{y}{2x},$$

this substituted in (1) gives

$$\frac{y^2}{4x} - \frac{y^2}{2x} + b = 0, \quad \text{or} \quad y^2 - 4bx = 0 \; . \; . \; . \; . \; . \; (2).$$

This equation satisfies (1), as will be seen by substituting for x and p_1, their values derived from (2).

$$\frac{y^2}{4b} \cdot \left(\frac{2b}{y}\right)^2 - y\left(\frac{2b}{y}\right) + b = 0, \quad \text{or} \quad b - 2b + b = 0,$$

an identical equation. Hence (2) is a singular solution.

2. $$u_2 = y + (y - x)\frac{dy}{dx} + (a - x)\left(\frac{dy}{dx}\right)^2 = 0,$$

or $$u_2 = y + (y - x)p_1 + (a - x)p_1^2 = 0 \; . \; . \; . \; . \; (1),$$

$$\frac{du_2}{dp_1} = y - x + 2(a - x)p_1 = 0, \; \therefore \; p_1 = \frac{x - y}{2(a - x)}.$$

This value, substituted in (1), gives

$$y - \frac{(y - x)^2}{2(a - x)} + (a - x)\frac{(y - x)^2}{4(a - x)^2} = 0, \text{ or } (x + y)^2 - 4ay = 0.$$

This satisfies (1), and is therefore a singular solution.

CHAPTER V.

INTEGRATION OF DIFFERENTIAL EQUATIONS OF THE SECOND ORDER.

195. Differential equations of the second order, when presented in their most general form, include

$x, y, \frac{dy}{dx}$ and $\frac{d^2y}{dx^2}$, and may therefore be written

$$F\left(x, y, \frac{dy}{dx}, \frac{d^2y}{dx^2}\right) = 0.$$

Of these comparatively few admit of being integrated, and therefore only such particular varieties of the general form as admit of integration or reduction to a lower order will be examined.

196. *Case* 1*st.* Let the equation involve only x and $\frac{d^2y}{dx^2}$, the form being

$$F\left(x, \frac{d^2y}{dx^2}\right) = 0.$$

Then resolving the equation, if possible, with respect to $\frac{d^2y}{dx^2}$, we get

$$\frac{d^2y}{dx^2} = F_1 x = X. \quad \therefore \frac{d^2y}{dx^2} dx = X dx, \text{ and by integration,}$$

$$\frac{dy}{dx} = \int X dx = X_1 + C_1. \quad \therefore \frac{dy}{dx} dx = X_1 dx + C_1 dx,$$

and,
$$y = \int X_1 dx + \int C_1 dx = X_2 + C_1 x + C_2.$$

The constants C_1 and C_2 being arbitrary.

197. *Ex.* $\frac{d^2y}{dx^2} - ax^3 = 0$, or, $\frac{d^2y}{dx^2} = ax^3$.

$$\frac{d^2y}{dx^2}dx = ax^3dx, \quad \therefore \quad \frac{dy}{dx} = \frac{ax^4}{4} + C_1.$$

$$\frac{dy}{dx}dx = \left(\frac{ax^4}{4} + C_1\right)dx, \quad \text{and} \quad y = \frac{ax^5}{5\,.\,4} + C_1x + C_2.$$

198. *Case 2d.* Let the equation involve only y and $\frac{d^2y}{dx^2}$, the form being

$$F\left(y, \frac{d^2y}{dx^2}\right) = 0.$$

Resolving the equation, if possible, with respect to $\frac{d^2y}{dx^2}$,

$$\frac{d^2y}{dx^2} = F_1y = Y: \qquad \text{But } \frac{d^2y}{dx^2} = \frac{dp_1}{dx} = \frac{dp_1}{dy}\cdot\frac{dy}{dx} = p_1\frac{dp_1}{dy},$$

$$\therefore \quad p_1\frac{dp_1}{dy} = Y, \quad \text{and} \quad p_1\frac{dp_1}{dy}dy = Ydy,$$

$$\therefore \quad \frac{1}{2}p_1^2 = \int Ydy = Y_1 + C_1, \quad \therefore \quad \frac{dy^2}{dx^2} = 2Y_1 + C,$$

and $dx = \frac{dy}{\sqrt{2Y_1 + C}}$, and the variables are separated.

199. *Ex.* $\frac{d^2y}{dx^2} - \frac{1}{\sqrt{ay}} = 0$, or $\frac{d^2y}{dx^2} = \frac{dp_1}{dx} = \frac{1}{\sqrt{ay}}$.

Then $p_1\frac{dp_1}{dy} = \frac{1}{\sqrt{ay}}$, $p_1\frac{dp_1}{dy}\cdot dy = \frac{dy}{\sqrt{ay}}$, and $\therefore p_1^2 = 4\sqrt{\frac{y}{a}} + C_1$,

or, $$\frac{dy^2}{dx^2} = 4\frac{\sqrt{y} + \sqrt{b}}{\sqrt{a}}, \text{ by making } C_1 = 4\frac{\sqrt{b}}{\sqrt{a}}.$$

$$\therefore \quad dx = \frac{\sqrt[4]{a}\,.\,dy}{2\sqrt{\sqrt{y} + \sqrt{b}}}.$$

To integrate this, put $\sqrt{y} + \sqrt{b} = z$, $\therefore \quad y = (z - \sqrt{b})^2$.

$$dy = 2(z - \sqrt{b})dz, \quad \text{and} \quad \therefore \quad dx = \frac{\sqrt[4]{a}(z - \sqrt{b})\,dz}{\sqrt{z}}.$$

$$\therefore\ x = \sqrt[4]{a}\left(\frac{2}{3}z^{\frac{3}{2}} - \frac{1}{2}\sqrt{b}\,z^{\frac{1}{2}}\right) + C_2$$

$$= \sqrt[4]{a}\left[\frac{2}{3}(y^{\frac{1}{2}} + b^{\frac{1}{2}})^{\frac{3}{2}} - \frac{1}{2}b^{\frac{1}{2}}(y^{\frac{1}{2}} + b^{\frac{1}{2}})^{\frac{1}{2}}\right] + C_2.$$

200. *Case* 3*d.* Let the equation involve $\frac{dy}{dx}$ and $\frac{d^2y}{dx^2}$ only, the form being

$$F\left(\frac{dy}{dx}, \frac{d^2y}{dx^2}\right) = 0.$$

Resolving with respect to $\frac{d^2y}{dx^2}$ if possible, we have

$$\frac{d^2y}{dx^2} = F_1\left(\frac{dy}{dx}\right), \quad \text{or,} \quad \frac{dp_1}{dx} = F_1 p_1, \quad \text{and} \quad dx = \frac{dp_1}{F_1 p_1}.$$

This is an equation of the first order, which being integrated gives

$$x = F_2 p_1, \quad \text{and} \quad y = \int p_1 dx = \int p_1 \frac{dp_1}{F_1 p} = F_3 p_1.$$

Hence, by eliminating p_1, we obtain a relation between x and y,

201. *Ex.* $$a\frac{d^2y}{dx^2} + \left(1 + \frac{dy^2}{dx^2}\right)^{\frac{3}{2}} = 0.$$

$$\therefore\ a\frac{dp_1}{dx} = -\left(1 + \frac{dy^2}{dx^2}\right)^{\frac{3}{2}} \quad \text{and} \quad dx = \frac{-a\,dp_1}{(1 + p_1^2)^{\frac{3}{2}}}.$$

$$\therefore\ x = -\frac{ap_1}{(1+p_1^2)^{\frac{1}{2}}} + C. \quad \therefore\ dy = -\frac{ap_1 dp_1}{(1 + p_1^2)^{\frac{3}{2}}}.$$

and $$y = \frac{a}{(1 + p_1^2)^{\frac{1}{2}}} + C_1.$$

Hence, by eliminating p_1, we get

$$(C - x)^2 + (C_1 - y)^2 = a^2.$$

202. *Case* 4*th.* Let the equation involve

x, $\frac{dy}{dx}$ and $\frac{d^2y}{dx^2}$ only, being of the form

$$F\left(x, \frac{dy}{dx}, \frac{d^2y}{dx^2}\right) = 0.$$

Replacing $\frac{dy}{dx}$ and $\frac{d^2y}{dx^2}$ by p_1 and $\frac{dp_1}{dx}$, the proposed equation reduces to

$$F\left(x, p_1, \frac{dp_1}{dx}\right) = 0 \dots . (1),$$

which is of the first order between x and p_1, and must therefore be resolved, if possible, by some one of the methods applicable to such equations.

Thus, if the equation (1) can be solved with respect to x, giving

$$x = F_1 p_1 \dots . . (2),$$

we shall have, since $y = \int p_1 dx = p_1 x - \int x dp_1$,

$$y = p_1 x - \int F_1 p_1 dp_1 \dots . . (3);$$

and, by eliminating p_1 between (2) and (3), the desired relation between x and y will be obtained.

Or, again, if (1) can be solved for p_1 giving

$$p_1 = F_1 x \dots . . (4),$$

then $\qquad y = \int p_1 dx = \int F_1 x . dx$, the integral sought.

If neither of these suppositions be true, we can only resort to some one of the expedients exhibited in the foregoing chapters.

203. *Ex.* $\qquad \frac{a^2}{2x} \cdot \frac{d^2y}{dx^2} + \left(1 + \frac{dy^2}{dx^2}\right)^{\frac{3}{2}} = 0.$

By substitution

$$\frac{\frac{dp_1}{dx}}{(1 + p_1^2)^{\frac{3}{2}}} = -\frac{2x}{a^2},$$

and by integration

$$\frac{p_1}{(1 + p_1^2)^{\frac{1}{2}}} = C - \frac{x^2}{a^2} = \frac{b^2 - x^2}{a^2} \quad \text{when} \quad C = \frac{b^2}{a^2}.$$

$$\therefore \frac{1}{p_1^2} = \frac{a^4}{(b^2 - x^2)^2} - 1 = \frac{a^4 - (b^2 - x^2)^2}{(b^2 - x^2)^2};$$

$$\therefore p_1 = \frac{dy}{dx} = \frac{b^2 - x^2}{\sqrt{a^4 - (b^2 - x^2)^2}}.$$

$$\therefore y = \int \frac{(b^2 - x^2)\,dx}{\sqrt{a^4 - (b^2 - x^2)^2}}, \text{ the desired relation.}$$

204. *Case* 5*th*. Let the equation involve y, $\frac{dy}{dx}$, and $\frac{d^2y}{dx^2}$ only, the form being

$$F\left(y, \frac{dy}{dx}, \frac{d^2y}{dx^2}\right) = 0.$$

By a substitution similar to that adopted in the last case, we have

$$F\left(y, p_1, \frac{dp_1}{dx}\right) = 0.$$

But $$\frac{dp_1}{dx} = \frac{dp_1}{dy} \cdot \frac{dy}{dx} = p_1 \frac{dp_1}{dy}, \text{ and by substitution}$$

$$\therefore F\left(y, p_1, p_1 \frac{dp}{dy}\right) = F_1\left(y, p_1, \frac{dp_1}{dy}\right) = 0,$$

which is an equation of the first order between y and p_1.

205. *Ex.* $$\frac{d^2y}{dx^2} - y - m\frac{dy^2}{dx^2} = 0.$$

By substitution $$\frac{dp_1}{dx} - y - mp_1^2 = 0.$$

$$\therefore \frac{dp_1}{dx} = \frac{dp_1}{dy} \cdot \frac{dy}{dx}; \quad p_1 \frac{dp_1}{dy} - y = mp_1^2,$$

or by making $p_1^2 = 2z$, and consequently $p_1 dp_1 = dz$,

$$\frac{dz}{dy} - 2mz = y, \quad dz - 2mzdy = ydy.$$

This is a linear equation of the first order and first degree, and therefore integrable.

206. *Case* 6*th*. If we reckon (as usual) x or y as of the dimension 1, and agree to reckon $\frac{dy}{dx}$ of the dimension 0, and $\frac{d^2y}{dx^2}$ of the dimension -1, then every equation of the second order which, upon this supposition, is homogeneous, may be reduced to an equation of the first order, by making $y = vx$ and $\frac{d^2y}{dx^2} = \frac{z}{x}$.

For, if n denote the degree of the coefficients, the terms containing $\frac{d^2y}{dx^2}$ must have a factor of the degree $n+1$, and those containing $\frac{dy}{dx}$ must have factors of the degree n. Hence after substituting the assumed values of y and $\frac{d^2y}{dx^2}$, every term of the equation will necessarily be divisible by x^n, and thus x will disappear, leaving an equation between v, z, and p_1, of the general form

$$F(v, z, p_1) = 0 \; . \; . \; . \; . \; (1).$$

But $$dy = p_1 dx = vdx + xdv. \quad \therefore \; \frac{dx}{x} = \frac{dv}{p_1 - v}.$$

Also $$\frac{dp_1}{dx} = \frac{z}{x}. \quad \therefore \; \frac{dx}{x} = \frac{dp_1}{z}. \quad \therefore \; zdv = (p_1 - v)dp_1,$$

or by substituting the value of z, obtained by resolving (1), an equation of the first order will arise between v and p_1, from which p_1 may be found in terms of v. Then by eliminating p_1 from the equation

$$\frac{dx}{x} = \frac{dv}{p_1 - v},$$

and integrating, we shall get $\log x = \varphi v$.

Lastly, eliminating v between this result and $y = vx$, the desired relation between x and y will be obtained.

207. *Ex.* $$x^2 \frac{d^2y}{dx^2} - x\frac{dy}{dx} - 3y = 0.$$

Making $$\frac{d^2y}{dx^2} = \frac{z}{x} \quad \text{and} \quad y = vx \quad \text{we get}$$

$$xz - xp_1 - 3vx = 0 \quad \text{or} \quad z - p_1 - 3v = 0.$$

$$\therefore z = p_1 + 3v \quad \text{and} \quad (p_1 + 3v)dv = (p_1 - v)dp_1,$$

$$p_1 dv + v dp_1 = p_1 dp_1 - 3v dv.$$

$$C + p_1 v = \frac{1}{2}p_1^2 - \frac{3}{2}v^2, \quad \text{or} \quad p_1^2 - 2p_1 v + v^2 = 4v^2 + 2C.$$

$$p_1 - v = \sqrt{4v^2 + 2C}. \quad \text{Hence}$$

$$\frac{dx}{x} = \frac{dv}{\sqrt{4v^2 + 2C}}, \quad \text{and} \quad \log x = \frac{1}{2}\log\left[C_1(2v + \sqrt{4v^2 + 2C}\right].$$

$$\therefore x^2 = C_1(2v + \sqrt{4v^2 + 2C}). \quad \text{But } v = \frac{y}{x}.$$

$$\therefore x^2 = C_1\left(\frac{2y}{x} + \sqrt{\frac{4y^2}{x^2} + 2C}\right) \text{ and } \frac{x^2}{C_1} - \frac{2y}{x} = \sqrt{\frac{4y^2}{x^2} + 2C}.$$

$$\therefore \frac{x^4}{C_1^2} - \frac{4xy}{C_1} + \frac{4y^2}{x^2} = \frac{4y^2}{x^2} + 2C.$$

$$\therefore y = \frac{x^3}{4C_1} - \frac{2CC_1}{4x} = ax^3 - \frac{b}{x} \quad \text{when} \quad \frac{1}{4C_1} = a \quad \text{and} \quad \frac{CC_1}{2} = b.$$

CHAPTER VI.

INTEGRATION OF DIFFERENTIAL EQUATIONS OF THE HIGHER ORDERS.

208. The integration of differential equations of an order higher than the second is attended with difficulties still greater than those which have been overcome hitherto, and in consequence the number of integrable forms is very restricted. The following exhibit a few of the simplest cases.

1st. Let the form be $F\left(\frac{d^n y}{dx^n}, \frac{d^{n-1}y}{dx^{n-1}}\right) = 0.$

Put $\frac{d^{n-1}y}{dx^{n-1}} = u$, then $\frac{d^n y}{dx^n} = \frac{du}{dx}$, and by substitution

$$F\left(u, \frac{du}{dx}\right) = 0,$$

which is an equation of the first order between u and x. This being resolved, gives

$$u = F_1 x. \quad \therefore \frac{d^{n-1}y}{dx^{n-1}} = F_1 x \quad \text{and} \quad y = \int^{n-1} F_1 x \,.\, dx^{n-1}.$$

209. Next let the form be

$$\left(\frac{d^n y}{dx^n}, \frac{d^{n-2}y}{dx^{n-2}}\right) = 0.$$

Put $\frac{d^{n-2}y}{dx^{n-2}} = u$, then $\frac{d^n y}{dx^n} = \frac{d^2 u}{dx^2}$, and by substitution

$$\therefore F\left(\frac{d^2 u}{dx^2}, u\right) = 0,$$

an integrable form of the second order, which has been already examined.

210. 1. Let $$\frac{d^4y}{dx^4}\cdot\frac{d^3y}{dx^3} = 1.$$

Put $$\frac{d^3y}{dx^3} = u, \quad \frac{d^4y}{dx^4} = \frac{du}{dx}.$$

$$\therefore u\frac{du}{dx} = 1, \text{ or } dx = udu, \text{ and } x = \frac{1}{2}u^2 + C_1.$$

$$\therefore u = \sqrt{2x - 2C_1} \quad \text{or} \quad \frac{d^3y}{dx^3} = \sqrt{2x - 2C_1}.$$

$$\therefore \frac{d^3y}{dx^3}dx = (2x - 2C_1)^{\frac{1}{2}}dx, \text{ and } \frac{d^2y}{dx^2} = \frac{(2x - 2C_1)^{\frac{3}{2}}}{3} + C_2.$$

$$\frac{d^2y}{dx^2}dx = \left[\frac{(2x - 2C_1)^{\frac{3}{2}}}{3} + C_2\right]dx, \quad \frac{dy}{dx} = \frac{(2x - 2C_1)^{\frac{5}{2}}}{3\cdot 5} + \frac{C_2x}{1} + C_3,$$

and $$y = \frac{(2x - 2C_1)^{\frac{7}{2}}}{3\cdot 5\cdot 7} + \frac{C_2x^2}{1\cdot 2} + \frac{C_3x}{1} + C_4.$$

2. $$\frac{d^4y}{dx^4} = \frac{d^2y}{dx^2}.$$

Put $$\frac{d^2y}{dx^2} = u \quad \text{then} \quad \frac{d^4y}{dx^4} = \frac{d^2u}{dx^2}.$$

$$\therefore \frac{d^2u}{dx^2} = u, \quad \frac{d^2u}{dx^2}\cdot\frac{du}{dx}dx = u\frac{du}{dx}dx.$$

$$\therefore \frac{du^2}{dx^2} = u^2 + C_1 \quad \text{and} \quad dx = \frac{du}{\sqrt{u^2 + C_1}}$$

$$\therefore x = \log\frac{u + \sqrt{u^2 + C_1}}{C_1} \ldots\ldots (1).$$

Now $$p_1 = \frac{dy}{dx} = \int udx = \int\frac{udu}{\sqrt{u^2 + C_1}} = \sqrt{u^2 + C_1} + C_3,$$

and $$y = \int p_1dx = \int[(u^2 + C_1)^{\frac{1}{2}} + C_3]\frac{du}{\sqrt{u^2 + C_1}}$$

$$= u + C_3x + C_4 \ldots\ldots (2).$$

Then, to eliminate u between (1) and (2), we get from (1)

$$C_2 e^x = u + \sqrt{u^2 + C_1}, \quad C_2^2 e^{2x} - 2u C_2 e^x + u^2 = u^2 + C_1.$$

$$\therefore u = \frac{C_2^2 e^{2x} - C_1}{2 C_2 e^x},$$

which, substituted in (2), gives a result which may be written in the form

$$y = c_1 e^x + c_2 e^{-x} + c_3 x + c_4.$$

CHAPTER VII.

INTEGRATION OF SIMULTANEOUS DIFFERENTIAL EQUATIONS.

211. In the applications of the Calculus to Physical Astronomy, it occurs, not unfrequently, that several variables, as x, y, t, &c. are connected by co-existent relations, the number of such relations being one less than the number of variables; and the object proposed is, to deduce equations which shall express the values of x, y, &c. in terms of the remaining variable t. The following solution of some of the simplest cases of such equations was first given by D'Alembert.

212. *Prop.* To resolve the system of equations,

$$A \frac{dx}{dt} + B \frac{dy}{dt} + Cx + Dy = T,$$

$$A_1 \frac{dx}{dt} + B_1 \frac{dy}{dt} + C_1 x + D_1 y = T_1,$$

in which A, B, C, D, A_1, B_1, C_1, and D_1, are constants, and T and T_1 functions of t; so as to express x and y in terms of t.

Eliminating first $\frac{dy}{dt}$, and then $\frac{dx}{dt}$, we can reduce the proposed equations to the forms

$$\frac{dx}{dt} + ax + by = T_2 \ldots . (1), \quad \text{and} \quad \frac{dy}{dt} + a_1x + b_1y = T_3 \ldots (2),$$

in which T_2 and T_3 are also functions of t, and $a, b, a_1 b_1$ are constants.

Multiply (2) by an undetermined constant m, and add the resulting product to (1).

$$\therefore \frac{d}{dt}(x + my) + (a + ma_1)\left(x + \frac{b + mb_1}{a + ma_1}y\right) = T_2 + mT_3.$$

Now determine m by the condition $m = \frac{b + mb_1}{a + ma_1}$;

or $$ma + m^2a_1 - b - mb_1 = 0,$$

and suppose m_1 and m_2 to be the two values of m given by this quadratic.

Also put $a + m_1a_1 = r_1$ and $a + m_2a_1 = r_2$. Then

$$\frac{d}{dt}(x + m_1y) + r_1(x + m_1y) = T_2 + m_1T_3,$$

$$\frac{d}{dt}(x + m_2y) + r_2(x + m_2y) = T_2 + m_2T_3.$$

These being linear equations of the first order, their solutions will be

$$\left.\begin{aligned} x+m_1y &= e^{-r_1t}\left[\int e^{r_1t}(T_2+m_1T_3)dt\right] \\ x+m_2y &= e^{-r_2t}\left[\int e^{r_2t}(T_2+m_2T_3)dt\right] \end{aligned}\right\}$$ from which x and y may be found in terms of t.

213. *Ex.* Let $\frac{dx}{dt} + 4y + 5x = e^t$, and $\frac{dy}{dt} + x + 2y = e^{2t}$ be the proposed equations.

As these have the forms (1) and (2) of the last article, we multiply the second by m, and add.

$$\therefore \frac{d}{dt}(x + my) + (5 + m)\left(x + \frac{4 + 2m}{5 + m}y\right) = e^t + me^{2t}.$$

Put $$m = \frac{4 + 2m}{5 + m}, \quad \text{or} \quad m^2 + 3m = 4.$$

$$\therefore m_1 = 1, \quad \text{and} \quad m_2 = -4, \; r_1 = 5 + 1 = 6, \; r_2 = 5 - 4 = 1.$$

$$\therefore x + y = e^{-6t}\left[\int e^{6t}(e^t + e^{2t})dt\right] = e^{-6t}\left[\frac{1}{7}e^{7t} + \frac{1}{8}e^{8t} + C\right]$$
$$= \frac{1}{7}e^t + \frac{1}{8}e^{2t} + Ce^{-6t}$$
$$x - 4y = e^{-t}\left[\int e^t(e^t - 4e^{2t})dt\right] = e^{-t}\left[\frac{1}{2}e^{2t} - \frac{4}{3}e^{3t} + C_1\right]$$
$$= \frac{1}{2}e^t - \frac{4}{3}e^{2t} + C_1e^{-t},$$

from which x and y are readily found.

214. *Prop.* To integrate the system of equations,

$$\frac{dx}{dt} + (Ax + By + Cz) = T \; .\;.\;.\;. \; (1).$$

$$\frac{dy}{dt} + (A_1x + B_1y + C_1z) = T_1 \; .\;.\;.\;. \; (2).$$

$$\frac{dz}{dt} + (A_2x + B_2y + C_2z) = T_2 \; .\;.\;.\;. \; (3),$$

in which $A, B, C,$ &c. are constants, and $T, T_1 T_2$ functions of t.

Multiply (2) by m, and (3) by n, and add.

$$\therefore \frac{d}{dt}(x + my + nz) + (A + A_1m + A_2n)$$
$$\left(x + \frac{B + B_1m + B_2n}{A + A_1m + A_2n}y + \frac{C + C_1m + C_2n}{A + A_1m + A_2n}z\right) = T + T_1m + T_2n.$$

Hence, if we put

$$x + my + nz = v, \quad \text{and} \quad A + A_1m + A_2n = M,$$

and determine m and n by the conditions

$$m = \frac{B + B_1m + B_2n}{A + A_1m + A_2n}, \quad n = \frac{C + C_1m + C_2n}{A + A_1m + A_2n} \; .\;.\;.\;. \; (4),$$

the equation will assume the form

$$\frac{dv}{dt} + Mv = T + T_1m + T_2n,$$ which is a linear equation.

This, being integrated, will give a relation between v and t. Also, in finding the values of m and n from equations (4), two cubic equations will arise, and therefore each of these quantities will have three values. Denoting them by m_1, m_2, and m_3, n_1, n_2, and n_3, and representing the three values of the second member, after integration by U_1, U_2, and U_3, there will result three equations of the form

$$x + m_1 y + n_1 z = U_1,$$

$$x + m_2 y + n_2 z = U_2,$$

$$x + m_3 y + n_3 z = U_3,$$

from which x, y, and z, can be found in terms of t.

215. *Prop.* To integrate the system of equations.

$$\frac{d^2x}{dt^2} + ax + by + c = 0 \ldots . (1).$$

$$\frac{d^2y}{dt^2} + a_1x + b_1y + c_1 = 0 \ldots . (2).$$

Multiply (2) by m, and add. Then

$$\frac{d^2}{dt^2}(x + my + C) + (a + ma_1)\left(x + \frac{b + mb_1}{a + ma_1}y + \frac{c + mc_1}{a + ma_1}\right) = 0.$$

$$\text{Put } m = \frac{b + mb_1}{a + ma_1}, \quad u = x + my + C,$$

and
$$C = \frac{c + mc_1}{a + ma_1}, \quad \text{and} \quad a + ma_1 = -n^2,$$

and the equation will reduce to

$$\frac{d^2u}{dt^2} - n^2u = 0.$$

The integral of this equation is

$$u = C_1e^{nt} + C_2e^{-nt}.$$

Hence if m_1 and m_2 be the values of m, deduced from the assumed relation of m and the constants, then

$$x + m_1 y + \frac{c + m_1 c_1}{a + m_1 a_1} = C_1 e^{n_1 t} + C_2 e^{-n_1 t}.$$

$$x + m_2 y + \frac{c + m_2 c_1}{a + m_2 a_1} = C_3 e^{n_2 t} + C_4 e^{-n_2 t}.$$

216. *Ex.* $\frac{d^2 x}{dt^2} = 3x + 4y - 3, \ \frac{d^2 y}{dt^2} = 8y - x - 5.$

Here $\frac{b + mb_1}{a + ma_1} = \frac{-4 - 8m}{-3 + m} = m. \quad \therefore m^2 + 5m = -4.$

$\therefore m_1 = -1,$ and $m_2 = -4. \quad \therefore n_1 = 2$ and $n_2 = \sqrt{7}.$

$$\therefore x - y + \frac{1}{2} = C_1 e^{2t} + C_2 e^{-2t}.$$

$$x - 4y + \frac{17}{7} = C_3 e^{t\sqrt{7}} + C_4 e^{-t\sqrt{7}}.$$

$$\therefore x = \frac{1}{7} + 4C_5 e^{2t} + 4C_6 e^{-2t} - C_7 e^{t\sqrt{7}} - C_8 e^{-t\sqrt{7}},$$

$$y = \frac{9}{14} + C_5 e^{2t} + C_6 e^{-2t} - C_7 e^{t\sqrt{7}} - C_8 e^{-t\sqrt{7}}.$$

CALCULUS OF VARIATIONS.

CHAPTER I.

FIRST PRINCIPLES.

1. In the general expression $u = \varphi(x_1, x_2, x_3 \ldots\ldots x_n)$, which signifies that u is a function of several independent variables $x_1, x_2, x_3 \ldots x_n$, the value of u obviously depends upon two essentially different considerations, viz.: 1st. The values of the variables $x_1, x_2, x_3 \ldots\ldots x_n$, and 2d., the form of the function φ.

2. The consideration of the changes imparted to u by changes in the values of the independent variables, while the function φ is supposed to retain the same form, is the chief object of the Differential Calculus, and then the form of the function is supposed to be known. But there are many cases, especially in questions relating to maxima and minima, in which the form of the function necessary to fulfil some specified condition, is the principle object of inquiry. For the resolution of such questions, the ordinary methods of the Differential Calculus do not suffice, and their consideration is reserved for the *Calculus of Variations*.

3. There are, it is true, some cases in which it becomes necessary to consider the change in u due to both these causes, namely, a change in the values of the independent variables, and a change in the form of the function, but it is with the latter that the Calculus of Variations is more immediately concerned.

4. The form of a function may be so connected with the form or forms of one or more other functions, that when the latter are given, the former will become known. For example, a differential coefficient has a certain form always deducible from that of the function itself. This connection between functions is expressed by calling the original function, whose form is arbitrary, the *primitive*, and that whose form is dependent upon it, the *derived* function.

Now if the form of one or more of the primitive functions be supposed to change, the form of the derived function will undergo a corresponding change, and if the relation connecting the forms of the primitive and derived functions be invariable, the change in the form of the latter will not be arbitrary, but will be connected with the change in the form of the former by a fixed relation.

5. To trace this dependence, or *to investigate the change in a derived function resulting from an arbitrary change in the form of its primitive*, is the design of the Calculus of Variations.

6. In this, as in the Differential Calculus, it is usually necessary that the increments of the function shall admit of being indefinitely diminished, and also that such increments shall continue indefinitely small, when any values, consistent with the conditions of the question, are assigned to the variables x_1, x_2, &c.

Hence the necessity of the following proposition.

7. *Prop.* To investigate a general method of giving to a function such a change of form as shall impart to it an increment of any proposed order of magnitude, without reference to the values of the independent variables $x_1, x_2, x_3 \dots x_n$ which enter into it.

Let $u = \varphi(x_1, x_2, x_3 \dots x_n)$ be the original function, and

$u_1 = \varphi_1(x_1, x_2, x_3 \dots x_n)$, after it has undergone the required change of form; and suppose i to represent a small quantity of the same order of magnitude as that which we desire to impart to the difference $u_1 - u$, so that if $u_1 - u = ni$, the quantity n shall be neither excessively great or extremely small. Then

$$\frac{u_1 - u}{i} = \frac{\varphi_1(x_1, x_2, x_3 \ldots x_n) - \varphi(x_1, x_2, x_3 \ldots x_n)}{i}$$

must be finite for all values of $x_1, x_2, x_3 \ldots x_n$, consistent with the conditions of the question. Assume

$$\frac{\varphi_1(x_1, x_2, x_3 \ldots x_n) - \varphi(x_1, x_2, x_3 \ldots x_n)}{i} = \psi(x_1, x_2, x_3 \ldots x_n). \quad \text{Then}$$

$$u_1 - u = i.\psi(x_1, x_2, x_3 \ldots x_n) \quad \text{or} \quad u_1 = u + i.\psi(x_1, x_2, x_3 \ldots x_n);$$

in which the function ψ is subjected to no condition but that of not becoming infinite for any values of $x_1, x_2, x_3 \ldots x_n$ within the restriction of the problem.

Hence, in order to impart to a given primitive function such a change of form as shall cause it to receive an increment susceptible of indefinite diminution, we must add to it another arbitrary function of the variables (subject to the above restriction), multiplied by a constant i, which constant is to be assumed of the same order of magnitude as that proposed to be given to the increment of the function.

8. *Ex.* Suppose $u = \sin x$, where x can take any value between 0 and π, and let the increment $u_1 - u$, proposed to be given to u by a change of form, be required of the same order of magnitude with dx.

Then making $i = adx$, when a is nearly equal to unity, we may write

$$u_1 = u + i\cos x, \quad \text{or} \quad u_1 = u + i\sin 2x, \quad \text{or} \quad u_1 = u + i\sin 4x, \ \&c.;$$

but it would not be admissible to assume

$$u_1 = u + i\tan x,$$

because $\tan x$ would become infinite for one of the admissible values of x, viz., $x = \frac{1}{2}\pi$, and therefore $i \tan x$ would not be necessarily small, as required.

If the increment required to be given to u, were of the same order with dx^2 or dx^3, then we would make

$$i = a\,.\,dx^2 \qquad \text{or} \qquad i = a\,.\,dx^3.$$

9. The indefinitely small change in the value of a function produced by a change in its form, is called a *variation*, and it appears that the variation of a primitive function is entirely arbitrary, but the variation of a derived function is dependent upon that of its primitive, and therefore not arbitrary.

10. *Prop.* Let $u = \varphi(x_1, x_2, x_3 \ldots . x_n)$ be an indeterminate function of $x_1, x_2, x_3 \ldots . x_n$, and let $v = Fu$ denote a relation by which v is derived from u, that is, a relation of *form*, but *not of magnitude:* it is proposed to find the change in the value of the derived function (or the variation of v) resulting from an indefinitely small change in the form of u.

Let $\varphi(x_1, x_2, x_3, \ldots . x_n)$ be replaced by

$$\varphi(x_1, x_2, x_3 \ldots . x_n) + i\,.\,\psi\,(x_1, x_2, x_3 \ldots . x_n),$$

and let the operation denoted by the symbol F be performed on the substituted function so far as to obtain the coefficient of the first power of i in the development of

$$F[\varphi(x_1, x_2, x_3 \ldots . x_n) + i\,.\,\psi(x_1, x_2, x_3 \ldots . x_n)]$$

If the co-efficient of this term be denoted by ω, then will $i\,.\,\omega$ be the variation of v. This will appear by reasoning entirely similar to that employed in the Differential Calculus, in finding the differential of a function (p. 18).

11. The proposition enunciated above is far more general than that commonly presented for consideration. Usually the only derived functions necessary to be considered, are such as are obtained by the processes of differentiation and integration, which are represented by the symbols d and $\int$ respectively; and for these two cases, the symbol F is *distributive*, that is,

$$F(\varphi + \varphi') = F\varphi + F\varphi'.$$

Then to find the variation of $v = F\varphi$, substitute $\varphi + i.\psi$ for φ,

and since $$F(\varphi + i.\psi) = F\varphi + F(i.\psi),$$

the variation or increment given to $F\varphi$ will be $F(i.\psi)$ or $i.F\psi$, since i is a constant, and therefore not a function of x_1, x_2, &c.

12. Thus far we have supposed the function to receive the kind of increment peculiar to the calculus of variations, viz., that due to a change of form; but if the independent variable be supposed to change also, the function will receive an additional increment, and the total change imparted to the function will be the algebraic sum of the two increments resulting from the two causes.

13. The following notation is used to distinguish the increments due to one or both of these causes.

1*st*. The character δ refers to the change in the value of the function resulting from a change in its form.

2*d*. The character d refers to the change in the value of the function produced by changes in the values of the independent variables x_1, x_2, &c.

3*d*. And the character D refers to the total change resulting from both causes.

∴ If u be a determinate function of several variables, then $Du = du$.

∴ If u be an indeterminate function of invariable quantities, then $Du = \delta u$.

And if u be an indeterminate function of variable quantities, then $Du = du + \delta u$.

14. Since an independent variable admits of both species of change, we might denote that change by either character. Unless the contrary is specified this change will be indicated by d.

15. The distinction between differentiation and variation admits of a simple geometrical illustration.

Thus let $y = \varphi x$ (1) be the equation of a curve ACB and $y_1 = \varphi_1 x$ (2), that of a second curve $A_1C_1B_1$, the form and position of the second curve being supposed to differ very slightly from those of the first.

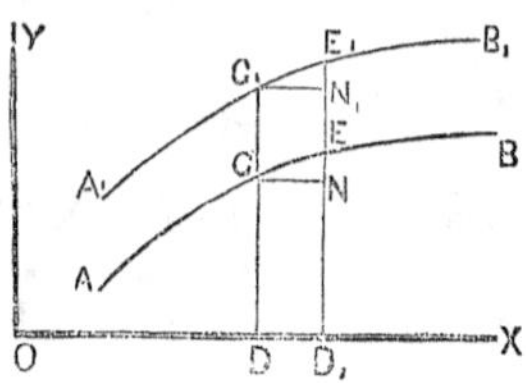

Put $OD = x$, $DD_1 = dx$, $DC = y$, and $DC_1 = y_1$. Then the change NE imparted to y by an addition $DD_1 = dx$ to x, while the point referred remains on the same curve ACB, will represent dy; the change $CC_1 = y_1 - y$, imparted to y by passing from C to a point C_1 on the second curve, (while x remains unchanged,) will represent δy; and the change NE_1 due to both causes will represent Dy.

16. *Prop.* Given $u = f(x_1, x_2, x_3 \ldots . x_n)$ a *determinate* function of several variables, to determine its total increment.

Since the form of the function is supposed invariable, we have

$$Du = du = \frac{du}{dx_1} dx_1 + \frac{du}{dx_2} . dx_2 + \ldots . + \frac{du}{dx_n} . dx_n \ldots . . [A].$$

17. *Prop.* Given $u = \varphi(x_1, x_2, x_3 \ldots . x_n)$ an *indeterminate* function of several variables, to determine its total increment.

Here the form of the function and the magnitudes of the independent variables must be supposed susceptible of change, and therefore

$$Du = du + \delta u.$$

But $$du = \frac{du}{dx_1} \cdot dx_1 + \frac{du}{dx_2} \cdot dx_2 + \ldots . + \frac{du}{dx_n} \cdot dx_n;$$

and $$\delta u = i \cdot \psi (x_1, x_2, x_3 \cdots x_n).$$ Hence,

$$Du = \frac{du}{dx_1} dx_1 + \frac{du}{dx_2} \cdot dx_2 + \ldots .$$

$$+ \frac{du}{dx_n} \cdot dx_n + i \cdot \psi\,(x_1, x_2, x_3 \ldots x_n) \ldots (B).$$

18. *Prop.* Given $u = F \cdot \varphi\,(x_1, x_2, x_3, \ldots . x_n)$, where F is the symbol of a *derived* function which fulfils the condition $F\,(\varphi + \varphi') = F\varphi + F\varphi'$, and φ is the symbol of an indeterminate function, to determine the total increment of u.

Here $\qquad \delta u = F[i \cdot \psi\,(x_1, x_2, x_3, \ldots . x_n)],$

$$= F \cdot \delta \cdot \varphi\,(x_1, x_2, x_3, \ldots . x_n).$$

$$\therefore\ Du = \frac{du}{dx_1} \cdot dx_1 + \frac{du}{dx_2} \cdot dx_2 + \ldots . + \frac{du}{dx_n} \cdot dx_n$$

$$+ F \cdot \delta \cdot \varphi\,(x_1, x_2, x_3, \ldots . x_n) \ldots . (C).$$

19. *Prop.* Given $V = f\,(x_1,\ x_2,\ x_3, \ldots . x_n, u_1, u_2, u_3, \ldots . u_n)$, where f is a *determinate* function of the quantities within the ( ); $x_1, x_2, x_3, \ldots . x_n$ being independent variables, and $u_1, u_2, u_3, \ldots u_n$ indeterminate functions of one or more of these variables, to find the total increment of V.

Here V varies in consequence of changes in the values of $x_1, x_2, x_3, \ldots x_n$, and also from the changes in the forms of $u_1, u_2, u_3, \ldots u_n$.

Now V is directly a function of x_1, and indirectly a function of x_1 through $u_1, u_2, u_3, \ldots . u_n$. Hence, if x_1 be supposed alone variable, the change in V will be

$$\frac{dV}{dx_1} dx_1 + \frac{dV}{du_1} \cdot \frac{du_1}{dx_1} \cdot dx_1 + \frac{dV}{du_2} \cdot \frac{du_2}{dx_1} \cdot dx_1 + \ldots \frac{dV}{du_n} \cdot \frac{du_n}{dx_1} \cdot dx_1;$$

and similarly, where x_2 alone varies, the change in V will be

$$\left[\frac{dV}{dx_2} + \frac{dV}{du_1} \cdot \frac{du_1}{dx_2} \cdot + \frac{dV}{du_2} \cdot \frac{du_2}{dx_2} + \ldots . + \frac{dV}{du_n} \cdot \frac{du_n}{dx_2}\right] dx_2,$$

and the other variables will furnish like expressions.

Now let the form of the function u_1 change, other things being the same, and the corresponding change in V will be

$$\frac{dV}{du_1} \cdot \delta u_1,$$

since V is a function of u_1, and the change produced in V by a change in u_1 depends only upon the *amount* of change in u_1, not on the manner in which it is received.

Introducing similar terms for the variations of $u_2, u_3, \ldots . u_n$, and adding, the total change in V will be thus expressed

$$DV = \left[\frac{dV}{dx_1} + \frac{dV}{du_1} \cdot \frac{du_1}{dx_1} + \frac{dV}{du_2} \cdot \frac{du_2}{dx_1} + \ldots . + \frac{dV}{du_n} \cdot \frac{du_n}{dx_2}\right] dx_1$$

$$+ \left[\frac{dV}{dx_2} + \frac{dV}{du_1} \cdot \frac{du_1}{dx_2} + \frac{dV}{du_2} \cdot \frac{du_2}{dx_2} + \ldots . + \frac{dV}{du_n} \cdot \frac{du_n}{dx_2}\right] dx_2$$

" " " "

$$+ \left[\frac{dV}{dx_n} + \frac{dV}{du_1} \cdot \frac{du_1}{dx_n} + \frac{dV}{du_2} \cdot \frac{du_2}{dx_n} + \ldots . + \frac{dV}{du_n} \cdot \frac{du_n}{dx_n}\right] dx_n$$

$$+ \frac{dV}{du_1} \delta u_1 + \frac{dV}{du_2} \delta u_2 + \ldots . + \frac{dV}{du_n} \cdot \delta u_n \ldots . (D),$$

the quantity in the last line being the variation proper or δV.

20. Given $U = FV$, when $V = f\ (x_1, x_2, x_3, \ldots . x_n, u_1, u_2, u_3, \ldots . u_n)$, where f is a determinate function of the quantities within the ( ), and F a derived function which satisfies the condition $F(\varphi + \varphi') = F\varphi + F\varphi'$, to find the increment of U.

First, let x_1 alone vary, and since V is a determinate function of $x_1, x_2, x_3, \ldots . x_n, u_1, u_2, u_3, \ldots . u_n$, it follows that so long as the forms of $u_1, u_2, u_3, \ldots . u_n$ remain unchanged, the quantity V will be a determinate function of the independent variables x_1, x_2, $x_3, \ldots . x_n$, and therefore the corresponding change in U will be

$$\left[\frac{dU}{dx_1}\right] dx_1, \quad \text{where} \quad \left[\frac{dU}{dx_1}\right]$$

denotes the total differential coefficient of U with respect to x_1.

And similarly when x_2 alone varies, the corresponding change in U is $\left[\frac{dU}{dx_2}\right] dx_2$; and the other variables will furnish like expressions.

Now to find the change in U due to a change in the *form* of u_1, we observe that the change in U, resulting from a change of any kind in u_1, might, at first, appear to be properly expressed, (as in the last proposition,) by $\frac{dU}{du_1} \cdot \delta u_1$. Now this would be true if U were properly a function of u_1, that is, a quantity whose magnitude is fixed by that of u_1; but such is not the case, their relation being one of form, not of magnitude; and therefore the desired increment is not $\frac{dU}{du_1} \cdot \delta u_1$. But although U is not a function of u_1, it is derived from u_1, the *form* of U being dependent upon that of V, which latter depends upon the form of u_1. And since $U = FV$, $\therefore \delta U = F \delta V$.

But, by the last proposition,

$$\delta V = \frac{dV}{du_1} \delta u_1 + \frac{dV}{du_2} \delta u_2 + \&c.$$

$\therefore F \frac{dV}{du_1} \cdot \delta u_1$, is the part of δU which results from a variation in the form of u_1.

Hence, the entire increment

$$DU = \left[\frac{dU}{dx_1}\right] dx_1 + \left[\frac{dU}{dx_2}\right] dx_2 + \dots + \left[\frac{dU}{dx_n}\right] dx_n$$

$$+ F \left[\frac{dV}{du_1} \cdot \delta u_1 + \frac{dV}{du_2} \delta u_2 + \dots + \frac{dV}{du_n} \cdot \delta u_n\right] \dots (E).$$

CHAPTER II.

APPLICATIONS OF GENERAL FORMULÆ TO FUNCTIONS OF ONE VARIABLE.

21. *Prop.* To find the total increment of the differential coefficient $\frac{d^n y}{dx^n}$, y being an indeterminate function of the single variable x.

Here the quantity proposed can vary only in two ways, viz: by a change in the magnitude of the independent variable x, and by a change in the form of the function y, the case corresponding to that of formula (C), with the number of variables reduced to one. We therefore estimate the two changes separately and add the results.

Now when x takes the increment dx,

$$u = \frac{d^n y}{dx^n} \text{ becomes } u + du = \frac{d^n y}{dx^n} + \frac{d^{n+1} y}{dx^{n+1}} dx$$

the corresponding change in u being $\frac{d^{n+1} y}{dx^{n+1}} dx$: and hence the total increment of u will be

$$Du = du + \delta u = \frac{d^{n+1} y}{dx^{n+1}} dx + \delta \frac{d^n y}{dx^n}.$$

But the symbol $\frac{d^n y}{dx^n}$ satisfies the condition $F(\varphi + \varphi') = F\varphi + F\varphi'$, and therefore

$$\delta \frac{d^n y}{dx^n} = \frac{d^n(y + \delta y)}{dx^n} - \frac{d^n y}{dx^n} = \frac{d^n y}{dx^n} + \frac{d^n \delta y}{dx^n} - \frac{d^n y}{dx^n} = \frac{d^n \delta y}{dx^n}$$

$$\therefore D \frac{d^n y}{dx^n} = \frac{d^{n+1} y}{dx^{n+1}} dx + \frac{d^n \delta y}{dx^n}.$$

22. It is to be observed that δy requires a certain restriction; for it was shown that when

$$\delta u = i \,.\, \psi(x_1, x_2 \,\&c.)$$

it is necessary to assume the function ψ of such form as not to become infinite for any values of x_1, x_2 &c., within the limits of the question. This condition is sufficient when we consider only the primitive function; but when it is necessary to take account of a function derived from the primitive, it becomes also necessary that the function similarly derived from ψ should not become infinite for any admissible values of the variables.

Thus when we say that $\delta F\phi = iF\psi$, it is to be understood that $F\psi$ remains finite for all suitable values of x_1, x_2, &c. In the present example, there being but one variable x, we have

$$\delta y = i \,.\, \psi x \,. \qquad \delta \frac{d^n y}{dx^n} = i \,.\, \frac{d^n \psi x}{dx^n}.$$

and we must so select ψ that $\frac{d^n \psi x}{dx^n}$ shall be finite for all admissible values of x.

23. *Prop.* To find the total increment of

$$V = f\left[x, y, \frac{dy}{dx}, \frac{d^2y}{dx^2} \cdots\cdots \frac{d^ny}{dx^n}\right]$$

where y is an indeterminate function of x.

This is a particular case of the general investigation which resulted in the formula [D]. To make that formula applicable to the present case, we reduce the number of variables to one, and put

$$u_1 = y, \quad u_2 = \frac{dy}{dx}, \, u_3 = \frac{d^2y}{dx^2} \cdots\cdots \&c.$$

Making the substitutions, and putting, for brevity,

$$\frac{dV}{dx} = M, \quad \frac{dV}{dy} = N, \quad \frac{dV}{d\frac{dy}{dx}} = P_1 \quad \frac{dV}{d\frac{d^2y}{dx^2}} = P_2 \cdots\cdots \frac{dV}{d\frac{d^ny}{dx^n}} = P_n$$

we get

$$DV = \left[M + N\frac{dy}{dx} + P_1\frac{d^2y}{dx^2} + P_2\frac{d^3y}{dx^3} \dots\dots + P_n\frac{d^{n+1}y}{dx^{n+1}}\right]dx$$

$$+ N\delta y + P_1\delta\frac{dy}{dx} + P_2\delta\frac{d^2y}{dx^2} \dots + P_n\delta\frac{d^ny}{dx^n};$$

or by substituting for $\delta\frac{dy}{dx}, \delta\frac{d^2y}{dx^2}$ &c.

their values given by the last proposition,

$$DV = \left[M + N\frac{dy}{dx} + P_1\frac{d^2y}{dx^2} + P_2\frac{d^3y}{dx^3}, \dots\dots + P_n\frac{d^{n+1}y}{dx^{n+1}}\right]dx$$

$$+ N\delta y + P_1\frac{d\delta y}{dx} + P_2\frac{d^2\delta y}{dx^2} \dots\dots + P_n\frac{d^n\delta y}{dx^n}.$$

24. Here δy is to be expressed as hitherto by $i \, . \, \psi x$, and therefore ψ is to be assumed of such form that neither it, nor any of its first n differential coefficients shall become infinite for any value of x consistent with the conditions of the problem.

25. *Prop.* To find the total increment of $U = \int_{x_0}^{x_1} Vdx$ when

$$V = f\left[x, y, \frac{dy}{dx}, \frac{d^2y}{dx^2} \dots\dots \frac{d^ny}{dx^n}\right].$$

It is obvious that a definite integral can change its value only in three ways, viz.:

1*st.* By a change of the superior limit x_1, while the inferior limit x_0 and the form of the differential coefficient V remain the same; 2*d.* By a change in the lower limit x_0, while the superior limit and the form of V are unchanged; and 3*d.* By a change in the form of V while the limits are invariable.

The complete variation or total increment is the algebraic sum of the three separate changes thus produced. Denote by V_1 the value of V when $x = x_1$, and suppose x_1 to take an increment dx_1. Then V_1dx_1 will be the corresponding increment received by U; for when

x_1 takes an increment, U, which consists of an indefinite number of terms, each of the form Vdx, simply receives an additional term, expressed by V_1dx_1.

And similarly, when x_0 takes an increment dx_0, the corresponding increment of U will be $-V_0dx_0$, since U will thereby be *deprived* of one term expressed by V_0dx_0.

$$\therefore DU = V_1dx_1 - V_0dx_0 + \delta\int_{x_0}^{x_1} Vdx,$$

and we must now find an expression for $\delta\int_{x_0}^{x_1} Vdx$, the change in U due to a change in the form of V. But the operation denoted by the symbol $\int_{x_0}^{x_1}$ satisfies the condition $F(\varphi + \varphi') = F\varphi + F\varphi'$.

$$\therefore \delta\int_{x_0}^{x_1} Vdx = \int_{x_0}^{x_1} (V + \delta V)\,dx - \int_{x_0}^{x_1} Vdx$$

$$= \int_{x_0}^{x_1} Vdx + \int_{x_0}^{x_1} \delta V.dx - \int_{x_0}^{x_1} Vdx = \int_{x_0}^{x_1} \delta V.dx.$$

Now as V is a determined function of $x, y, \frac{dy}{dx}, \frac{d^2y}{dx^2}$, &c., its form (considered as a function of x), can vary only by a change in the form of the function y.

Hence the variation of V, found as in the last proposition, is

$$\delta V = N\delta y + P_1\frac{d\delta y}{dx} + P_2\frac{d^2\delta y}{dx^2} + \cdots\cdot + P_n\frac{d^n\delta y}{dx^n}.$$

$$\therefore \delta\int_{x_0}^{x_1} Vdx = \int_{x_0}^{x_1}\left[N\delta y + P_1\frac{d\delta y}{dx} + P_2\frac{d^2\delta y}{dx^2} + \cdots + P_n\frac{d^n\delta y}{dx^n}\right]dx.$$

Now, by applying the formula for the integration by parts to the second member, we get

$$\int_{x_0}^{x_1} P_1\frac{d\delta y}{dx}\cdot dx = [P_1\delta y]_1 - [P_1\delta y]_0 - \int_{x_0}^{x_1}\frac{dP_1}{dx}\cdot\delta y\,.\,dx,$$

in which $[P_1\delta y]_1$ and $[P_1\delta y]_0$ represent the values of $P_1\delta y$ at the superior and inferior limits respectively. Similarly

$$\int_{x_0}^{x_1} P_2 \frac{d^2\delta y}{dx^2} \cdot dx = \left[P_2 \frac{d\delta y}{dx}\right]_1 - \left[P_2 \frac{d\delta y}{dx}\right]_0 - \int_{x_0}^{x_1} \frac{dP_2}{dx} \cdot \frac{d\delta y}{dx} \cdot dx;$$

or, by applying a similar process to the last term,

$$\int_{x_0}^{x_1} P_2 \frac{d^2\delta y}{dx^2} dx = \left[P_2 \frac{d\delta y}{dx}\right]_1 - \left[P_2 \frac{d\delta y}{dx}\right]_0 - \left[\frac{dP_2}{dx} \cdot \delta y\right]_1$$

$$+ \left[\frac{dP_2}{dx} \cdot \delta y\right]_0 + \int_{x_0}^{x_1} \frac{d^2 P_2}{dx^2} \delta y \, . \, dx.$$

$$= \left[P_2 \frac{d\delta y}{dx} - \frac{dP_2}{dx} \delta y\right]_1 - \left[P_2 \frac{d\delta y}{dx} - \frac{dP_2}{dx} \cdot \delta y\right]_0$$

$$+ \int_{x_0}^{x_1} \frac{d^2 P_2}{dx^2} \cdot \delta y \, . \, dx.$$

And if we integrate n times successively the term

$\int_{x_0}^{x_1} P_n \cdot \frac{d^n \delta y}{dx^n} dx$, there will result

$$\int_{x_0}^{x_1} P_n \frac{d^n \delta y}{dx^n} \cdot dx = \left[P_n \frac{d^{n-1}\delta y}{dx^{n-1}} - \frac{dP_n}{dx} \cdot \frac{d^{n-2}\delta y}{dx^{n-2}} + \&c.\right.$$

$$\left. \ldots\ldots + (-1)^{n-1} \frac{d^{n-1} P_n}{dx^{n-1}} \delta y\right]_1$$

$$- \left[P_n \frac{d^{n-1}\delta y}{dx^{n-1}} - \&c.\right]_0 + (-1)^n \int_{x_0}^{x_1} \frac{d^n P_n}{dx^n} \cdot \delta y \, . \, dx.$$

Now collecting the coefficients of δy, $\frac{d\delta y}{dx}$, &c, we get

$$\delta \int_{x_0}^{x_1} V dx = \left[P_1 - \frac{dP_2}{dx} + \frac{d^2 P_3}{dx^2} - \&c. \ldots + (-1)^{n-1} \frac{d^{n-1} P_n}{dx^{n-1}}\right]_1 . \delta y_1$$

$$- \left[P_1 - \frac{dP_2}{dx} + \&c.\right]_0 . \delta y_0$$

$$+ \left[P_2 - \frac{dP_3}{dx} + \&c.\right]_1 . \left[\frac{d\delta y}{dx}\right]_1 - \left[P_2 - \&c.\right]_0 . \left[\frac{d\delta y}{dx}\right]_0 + \&c.$$

$$+ \left[P_n \frac{d^{n-1}\delta y}{dx^{n-1}}\right]_1 - \left[P_n \frac{d^{n-1}\delta y}{dx^{n-1}}\right]_0$$

$$+ \int_{x_0}^{x_1} \left[N - \frac{dP_1}{dx} + \frac{d^2 P_2}{dx_2} - \&c. \ldots + (-1)^n \cdot \frac{d^n P_n}{dx^n}\right] \delta y \, . \, dx.$$

$$\therefore D\int_{x_0}^{x_1} Vdx = V_1 dx_1 - V_0 dx_0 + \left[P_1 - \frac{dP_2}{dx} + \&c.\right]_1 \delta y_1$$

$$- \left[P_1 - \frac{dP_2}{dx} + \&c.\right]_0 \delta y_0$$

$$+ \left[P_2 - \&c.\right]_1 . \left[\frac{d\delta y}{dx}\right]_1 - \left[P_2 - \&c.\right]_0 . \left[\frac{d\delta y}{dx}\right]_0 + \&c.$$

$$+ \int_{x_0}^{x_1} \left[N - \frac{dP_1}{dx} + \frac{d^2P_2}{dx^2} - \&c. \ldots + (-1)^n \frac{d^n P_n}{dx^n}\right] \delta y \,.\, dx,$$

which is the expression required.

26. The value of $D\int_{x_0}^{x_1} Vdx$ found in the above propositions, contains three parts essentially different from each other, viz.:

1st. The terms $V_1 dx_1 - V_0 dx_0$, which are independent of the change in the form of V, but depend exclusively on the variations of the limits.

2d. The terms $[P_1 - \&c.]_1 \delta y$, which depend upon the form of the function, not for every value of x; but for limiting values alone.

3d. The terms within the sign of integration

$$\int_{x_0}^{x_1} \left[N - \frac{dP_1}{dx}, \&c.\right] \delta y \,.\, dx,$$

which depend upon the general change in the form of the function.

27. The nature of this difference becomes more apparent by observing that $\delta y = i \,.\, \psi x$. For it is plain that the terms in the first class are wholly independent of the form of the function ψ: that those in the second class do not require for their determination a knowledge of the *form* of the function ψ, but only the values of that function and its first $n-1$ differential coefficients, at the limits; and that the terms of the third class depend upon the form of the function ψ, and cannot be determined so long as that form remains arbitrary.

28. *Prop.* To find the total increment of $U = \int_{x_0}^{x_1} Vdx$, when

$$V = f\left[x, y, \frac{dy}{dx}, \frac{d^2y}{dx^2} \cdots \frac{d^ny}{dx^n}, x_1, y_1, \left(\frac{dy}{dx}\right)_1, \left(\frac{d^2y}{dx^2}\right)_1 \cdots x_0, y_0, \left(\frac{dy}{dx}\right)_0, \left(\frac{d^2y}{dx^2}\right)_0 \&c.\right],$$

the quantity V being supposed to contain *explicitly* the limiting values of one or more of the quantities, $x, y, \frac{dy}{dx}$, &c.

Since x_1, y_1 and x_0, y_0 are connected by the same general relation as x and y, the integral $\int_{x_0}^{x_1} Vdx$ can be varied only in the three methods explained in the last proposition.

Now when x_1 receives the increment dx_1, the form of the function y remaining unchanged, the increment received by U will be

$$\left[V_1 + \int_{x_0}^{x_1} \left\{\frac{dV}{dx_1} + \frac{dV}{dy_1} \cdot \left(\frac{dy}{dx}\right)_1 + \frac{dV}{d\left(\frac{dy}{dx}\right)_1} \cdot \left(\frac{d^2y}{dx^2}\right)_1 + \&c.\right\} dx\right] dx_1.$$

Similarly, when x_0 receives an increment dx_0, the change in U will be

$$\left[-V_0 + \int_{x_0}^{x_1} \left\{\frac{dV}{dx_0} + \frac{dV}{dy_0} \cdot \left(\frac{dy}{dx}\right)_0 + \frac{dV}{d\left(\frac{dy}{dx}\right)_0} \cdot \left(\frac{d^2y}{dx^2}\right)_0 + \&c.\right\} dx\right] dx_0.$$

Now let the form of the function y change, while other things remain the same, and the corresponding change in U will be

$$\delta U = \int_{x_0}^{x_1} \left[N\delta y + P_1 \frac{d\delta y}{dx} + P_2 \frac{d^2\delta y}{dx^2} + \cdots + P_n \frac{d^n\delta y}{dx^n}\right] dx$$
$$+ \delta y_1 \int_{x_0}^{x_1} \frac{dV}{dy_1} dx + \delta y_0 \int_{x_0}^{x_1} \frac{dV}{dy_0} \cdot dx$$
$$+ \left(\frac{d\delta y}{dx}\right)_1 \cdot \int_{x_0}^{x_1} \frac{dV}{d\left(\frac{dy}{dx}\right)_1} dx + \left(\frac{d\delta y}{dx}\right)_0 \cdot \int_{x_0}^{x_1} \frac{dV}{d\left(\frac{dy}{dx}\right)_0} \cdot dx$$
$$+ \left(\frac{d^2\delta y}{dx^2}\right)_1 \cdot \int_{x_0}^{x_1} \frac{dV}{d\left(\frac{d^2y}{dx^2}\right)_1} dx + \left(\frac{d^2\delta y}{dx^2}\right)_0 \int_{x_0}^{x_1} \frac{dV}{d\left(\frac{d^2y}{dx^2}\right)_0} \cdot dx + \&c.$$

Put $\frac{dV}{dx_1} = m_1$, $\frac{dV}{dy_1} = n_1$, $\frac{dV}{d\left(\frac{dy}{dx}\right)_1} = p_1$, $\frac{dV}{d\left(\frac{d^2y}{dx^2}\right)_1} = q_1$, &c.

$\frac{dV}{dx_0} = m_0$, $\frac{dV}{dy_0} = n_0$, $\frac{dV}{d\left(\frac{dy}{dx}\right)_0} = p_0$, $\frac{dV}{d\left(\frac{d^2y}{dx^2}\right)_0} = q_0$, &c.

Now integrating by parts, as in the last proposition, and collecting the terms, we obtain

$$DU = \left[V_1 + \int_{x_0}^{x_1} \left\{ m_1 + n_1 \left(\frac{dy}{dx}\right)_1 + p_1 \left(\frac{d^2y}{dx^2}\right)_1 + q_1 \left(\frac{d^3y}{dx^3}\right)_1 + \text{\&c.} \right\} dx \right] dx_1$$

$$+ \left[- V_0 + \int_{x^0}^{x_1} \left\{ m_0 + n_0 \left(\frac{dy}{dx}\right)_0 + p_0 \left(\frac{d^2y}{dx^2}\right)_0 + q_0 \left(\frac{d^3y}{dx^3}\right)_0 + \text{\&c.} \right\} dx \right] dx_0$$

$$+ \left[\left\{ P_1 - \frac{dP_2}{dx} + \text{\&c.} \right\}_1 + \int_{x_0}^{x_1} n_1 dx \right] \delta y_1$$

$$- \left[\left\{ P_1 - \frac{dP_2}{dx} + \text{\&c.} \right\}_0 + \int_{x_0}^{x_1} n_0 dx \right] \delta y_0$$

$$+ \left[\left\{ P_2 - \text{\&c.} \right\}_1 + \int_{x_0}^{x_1} p_1 dx \right]_1 \left(\frac{d\delta y}{dx}\right)_1 - \left[\left\{ P_2 - \text{\&c.} \right\}_0 + \int_{x_0}^{x_1} p_0 dx \right]_0 \cdot \left(\frac{d\delta y}{dx}\right)_0$$

&c., &c., &c., &c.

$$+ \int_{x_0}^{x_1} \left[N - \frac{dP_1}{dx} + \frac{d^2 P_2}{dx_2} - \text{\&c.} \right] \delta y \cdot dx.$$

29. *Prop.* To find the total increment of $U = \int_{x_0}^{x_1} V dx$, in which

$$V = f\left[x, y, \frac{dy}{dx}, \frac{d^2y}{dx^2} \cdots\cdots \frac{d^ny}{dx^n}; z, \frac{dz}{dx}, \frac{d^2z}{dx^2}, \cdots\cdots \frac{d^mz}{dx^m}\right]$$

y and z being indeterminate functions of x.

Put $\frac{dV}{dx} = M$, $\frac{dV}{dy} = N$, $\frac{dV}{d\frac{dy}{dx}} = P_1$, $\frac{dV}{d\frac{d^2y}{dx^2}} = P_2, \cdots \frac{dV}{d\frac{d^ny}{dx^n}} = P_n$;

$$\frac{dV}{dz} = N', \quad \frac{dV}{d\frac{dz}{dx}} = P_1', \quad \frac{dV}{d\frac{d^2y}{dz^2}} = P_2', \cdots\cdots \frac{dV}{d\frac{d^mz}{dx^m}} = P_m'$$

Then, since the value of U can change only in four ways, viz.: 1st. By a change in the value of x_1; 2d, by a change in the value of x_0; 3d, by a change in the form of the function y; and 4th, by a change in the form of the function z; we shall obtain by reasoning, as in a preceding proposition, where y was the only function,

$$DU = V_1 dx_1 - V_0 dx_0 + \left[P_1 - \frac{dP_2}{dx} + \&c.\right]_1 \delta y_1$$

$$- \left[P_1 - \frac{dP_2}{dx} + \&c.\right]_0 \cdot \delta y_0$$

$$+ \left[P_2 - \&c.\right]_1 \cdot \left(\frac{d\delta y}{dx}\right)_1 - \left[P_2 - \&c.\right]_0 \cdot \left(\frac{d\delta y}{dx}\right)_0 + \&c. \cdot$$

$$+ \left(P_n \frac{d^{n-1}\delta y}{dx^{n-1}}\right)_1 - \left(P_n \frac{d^{n-1}\delta y}{dx^{n-1}}\right)_0$$

$$+ \int_{x_0}^{x_1} \left[N - \frac{dP_1}{dx} + \frac{d^2P_2}{dx^2} - \&c. \cdots\cdots\right.$$

$$\left. + (-1)^n \cdot \frac{d^nP_n}{dx^n}\right] \delta y \cdot dx$$

$$+ \left[P_1' - \frac{dP_2'}{dx} + \&c.\right]_1 \cdot \delta z_1 - \left[P_1' - \frac{dP_2'}{dx}\right.$$

$$\left. + \&c.\right]_0 \cdot \delta z_0 + \left[P_2' - \&c.\right]_1 \left(\frac{d\delta z}{dx}\right)_1$$

$$- \left[P_2' - \&c.\right]_0 \cdot \left(\frac{d\delta z}{dx}\right)_0$$

$$+ \&c. \cdots\cdots + \left(P_m' \frac{d^{m-1}\delta z}{dx^{m-1}}\right)_0 - \left(P_m' \frac{d^{m-1}\delta z}{dx^{m-1}}\right)_0$$

$$+ \int_{x_0}^{x_1} \left[N' - \frac{dP_1'}{dx} + \frac{d^2 P_2'}{dx^2} - \&c. \cdots\cdots \right.$$

$$\left. + (-1)^m \cdot \frac{d^m P_m'}{dx^m} \right] \delta z \cdot dx \cdots\cdots (a).$$

And if there be several indeterminate functions of x in the value of U, each will introduce a set of similar terms in DU or δU.

30. *Remark.* The results just obtained are equally true, whether the functions x, y, z, &c., are entirely independent of each other, or are connected by one or more equations of condition.

31. *Prop.* To find the total increment of $U = \int_{x_0}^{x_1} V dx$, in which

$$V = f\left[x, y, \frac{dy}{dx}, \frac{d^2y}{dx^2} \cdots\cdots \frac{d^ny}{dx^n}, z, \frac{dz}{dx}, \frac{d^2z}{dx^2}, \cdots\cdots \frac{d^mz}{dx^m}\right],$$

the functions y and z being connected by the relation $L = 0$, which relation may, or may not, be a differential equation.

The equation (a) of the last proposition is immediately applicable to this case, but since z and y are connected by a given relation, δz and δy are not both arbitrary, one being dependent upon the other.

32. If the equation $L = 0$ can be resolved with respect to one of the variables (as z), giving a result of the form $z = Fy$, the several differential coefficients $\frac{dz}{dx}$, $\frac{d^2z}{dx^2}$, &c., can be formed by simple differentiation, and these values, substituted in that of V, will render it a function of x, y, and their differential coefficients. Thus, the case will become the same as that considered in a previous proposition.

But since the equation $L = 0$ is often a differential equation which cannot be integrated, this method is frequently inapplicable. It will now be shown that by another method (due to Lagrange) one of the variations δy or δz can be removed from under the sign of integration.

Put $\frac{dL}{dy} = \alpha$, $\frac{dL}{d\frac{dy}{dx}} = \beta$, $\frac{dL}{d\frac{d^2y}{dx^2}} = \gamma$, &c., $\frac{dL}{dz} = \alpha'$,

$$\frac{dL}{d\frac{dz}{dx}} = \beta', \quad \frac{dL}{d\frac{d^2z}{dx^2}} = \gamma', \text{ \&c.}$$

Now, since the equation $L = 0$ is true for all forms of y and z consistent with the conditions of the question, we must have $\delta L = 0$.

$$\therefore \alpha\delta y + \beta\frac{d\delta y}{dx} + \gamma\frac{d^2\delta y}{dx^2} + \text{\&c.}$$

$$+ \alpha'\delta z + \beta'\frac{d\delta z}{dx} + \gamma'\frac{d^2\delta z}{dz^2} + \text{\&c.} = 0 \ldots\ldots (b).$$

When this equation can be integrated so as to give a value of either δy or δz in terms of the other, (as for example that of δz in terms of δy), we can form the values of $\frac{d\delta z}{dx}$, $\frac{d^2\delta z}{dx^2}$ &c., by differentiation, and then substitute them in the value of δU, as determined in the last proposition, thus effecting the desired transformation. But as this integration is rarely possible, it is usually necessary to adopt the method referred to above, which will be now explained.

33. The value of δV being

$$\delta V = N\delta y + P_1\frac{d\delta y}{dx} + P_2\frac{d^2\delta y}{dx^2} + \text{\&c.} + N'\delta z$$

$$+ P_1'\frac{d\delta z}{dx} + P_2'\frac{d^2\delta z}{dx^2} + \text{\&c.}$$

we can (without disturbing the equality here expressed) add to the second member of this equation, the value of δL multiplied by an arbitrary quantity λ, since $\lambda . \delta L = 0$. Hence we may write

$$\delta V = (N + \lambda\alpha)\delta y + (P_1 + \lambda\beta)\frac{d\delta y}{dx} + (P_2 + \lambda\gamma)\frac{d^2\delta y}{dx^2} + \text{\&c.}$$

$$+ (N' + \lambda\alpha')\delta z + (P_1' + \lambda\beta')\frac{d\delta z}{dx} + (P_2' + \lambda\gamma')\frac{d^2\delta z}{dx^2} + \text{\&c.}$$

$$\therefore \delta U = (P_1 + \lambda\beta - \frac{d(P_2 + \lambda\gamma)}{dx} + \&c.)_1 \delta y_1$$

$$- (P_1 + \lambda\beta - \frac{d(P_2 + \lambda\gamma)}{dx} + \&c.)_0 . \delta y_0$$

$$+ (P_2 + \lambda\gamma - \&c.)_1 . \left(\frac{d\delta y}{dx}\right)_1$$

$$- (P_2 + \lambda\gamma - \&c.)_0 . \left(\frac{d\delta y}{dx}\right)_0 + \&c.$$

$$+ \int_{x_0}^{x_1} (N + \lambda\alpha - \frac{d(P_1 + \lambda\beta)}{dx} + \&c.) \delta y . dx$$

$$+ (P_1' + \lambda\beta' - \frac{d(P_2' + \lambda\gamma')}{dx} + \&c.)_1 . \delta z_1$$

$$- (P_1' + \lambda\beta' - \frac{d(P_2' + \lambda\gamma')}{dx} + \&c.)_0 . \delta z_0$$

$$+ (P_2' + \lambda\gamma' - \&c.)_1 . \left(\frac{d\delta z}{dx}\right)_1$$

$$- (P_2' + \lambda\gamma' - \&c.)_0 . \left(\frac{d\delta z}{dx}\right)_0 + \&c.$$

$$+ \int_{x_0}^{x_1} [N' + \lambda\alpha' - \frac{d(P_1' + \lambda\beta')}{dx} + \&c.] \delta z . dx.$$

Now let it be required to determine an expression for δU containing but one of the variations δy, δz, under the sign of integration. If the value of λ be determined by the condition

$$N' + \lambda\alpha' - \frac{d(P_1' + \lambda\beta')}{dx} + \&c. = 0$$

the variation δz will disappear from under the sign of integration, and similarly, if λ be determined by the condition

$$N + \lambda\alpha - \frac{d(P_1 + \lambda\beta)}{dx} + \&c. = 0.$$

δy will disappear from under the sign of integration.

The following example exhibits an application of this method.

34. *Prop.* To find the total increment of $U=\int_{x_0}^{x_1} Vdx$ in which

$$V=f\left[x, y, \frac{dy}{dx}, \frac{d^2y}{dx^2} \cdots \cdots \frac{d^ny}{dx^n}, \int vdx\right]$$

and
$$v=f_1\left[x, y, \frac{dy}{dx}, \frac{d^2y}{dx^2} \cdots \cdots \frac{d^my}{dx^m}\right].$$

Put
$$\frac{dV}{dx}=M, \ \frac{dV}{dy}=N, \ \frac{dV}{d\frac{dy}{dx}}=P_1, \ \frac{dV}{d\frac{d^2y}{dx^2}}=P_2 \text{ \&c.}$$

$$\frac{dv}{dx}=m, \ \frac{dv}{dy}=n, \ \frac{dv}{d\frac{dy}{dx}}=p_1, \ \frac{dv}{d\frac{d^2y}{dx^2}}=p_2 \text{ \&c.} \int vdx=z, \ \frac{dV}{dz}=N'$$

The equation $L=0$ becomes in this case

$$v-\frac{dz}{dx}=0 \qquad \text{since} \qquad \int vdx=z. \quad \text{Hence}$$

$$\frac{dL}{dy}=\frac{dv}{dy} \quad \text{or} \quad \alpha=n, \quad \text{and similarly} \quad \beta=p_1, \gamma=p_2, \text{ \&c.}$$

Also
$$\frac{dL}{dz}=\frac{dv}{dz} \text{ or } \alpha'=0 \text{ and similarly } \beta'=-1, \gamma'=0 \text{ \&c.}$$

And by substituting these values in the formula of the last proposition, we obtain

$$\begin{aligned}
\delta U = & \left[P_1+\lambda p_1-\frac{d(P_2+\lambda p_2)}{dx}+\text{\&c.}\right]_1 . \delta y_1 \\
& -\left[P_1+\lambda p_1-\frac{d(P_2+\lambda p_2)}{dx}+\text{\&c.}\right]_0 . \delta y_0 \\
& +\left[P_2+\lambda p_2-\text{\&c.}\right]_1 . \left(\frac{d\delta y}{dx}\right)_1-\left[P_2+\lambda p_2-\text{\&c.}\right]_0\left(\frac{d\delta y}{dx}\right)_0+\text{\&c.} \\
& +\int_{x_0}^{x_1}\left[N+\lambda n-\frac{d(P_1+\lambda p_1)}{dx}+\frac{d^2(P_2+\lambda p_2)}{dx^2}-\text{\&c.}\right] \delta y . dx \\
& -(\lambda_1\delta z_1-\lambda_0\delta z_0)+\int_{x_0}^{x_1}\left[N'+\frac{d\lambda}{dx}\right]\delta z . dx.
\end{aligned}$$

Since $P_1' = 0$, $P_2' = 0$ &c., there will be no terms containing

$$\left(\frac{d\delta z}{dx}\right)_1 \left(\frac{d\delta z}{dx}\right)_0 \text{ \&c.}$$

By adding $V_1 dx_1 - V_0 dx_0$ to the expression for δU just found, we shall obtain the total increment DU, and in order to reduce DU to form in which δy shall be the only variation remaining under the sign of integration, we determine λ by the condition

$$N' + \frac{d\lambda}{dx} = 0,$$

which gives $$\lambda = -\int N' dx.$$

Denoting this value by i we obtain

$$DU = V_1 dx_1 - V_0 dx_0 + [P_1 + i \,.\, p_1 - \frac{d(P_2 + ip_2)}{dx} + \text{\&c.}]_1 \delta y_1$$

$$- [P_1 + ip_1 - \frac{d(P_2 + ip_2)}{dx} + \text{\&c.}]_0 \delta y_0$$

$$+ [P_2 + ip_2 - \text{\&c.}]_1 \,.\, \left(\frac{d\delta y}{dx}\right)_1 - [P_2 + ip_2 - \text{\&c.}]_0 \,.\, \left(\frac{d\delta y}{dy}\right)_0 + \text{\&c.}$$

$$- (i_1 \delta z_1 - i_0 \delta z_0)$$

$$+ \int_{x_0}^{x_1} [N + in - \frac{d(P_1 + ip_1)}{dx} + \frac{d^2(P_2 + ip_2)}{dx^2} - \text{\&c.}]\, \delta y \,.\, dx.$$

CHAPTER III.

SUCCESSIVE VARIATION.

35. Thus far no condition has been imposed as to the invariability of form of the function ψ or δy. The conclusions arrived at are equally true, whether that form be variable or invariable.

Thus if the symbol F satisfy the condition

$$F(\varphi + \varphi') = F\varphi + F\varphi',$$

it is equally true that

$$\delta F\varphi = F\delta\varphi = Fi\,.\,\psi,$$

whether the form of ψ be constant or variable. But this condition ceases to be immaterial when it is necessary to take account of the *second variation*, that is, the variation of the variation. Thus in the case just referred to, we should always have

$$\delta^2 F\varphi = F\delta^2\varphi = Fi\delta\psi.$$

But this, when the form of ψ is supposed invariable, reduces to

$$\delta^2 F\varphi = F0.$$

Now $F0 = 0$, since by the nature of the function F, we have

$$F(\varphi + 0) = F\varphi + F0$$

$$\therefore\ F.0 = F(\varphi + 0) - F\varphi = F\varphi - F\varphi = 0. \qquad \therefore\ \delta^2 F\varphi = 0.$$

Hence for convenience we agree that the variation δu of any function u, although of arbitrary form, shall yet preserve that form invariable, so as in all cases to satisfy the condition

$$\delta^2 u = 0.$$

36. We may notice here a striking analogy between a primitive function and an independent variable, the first increment of each being arbitrary, and the second equal to zero.

37. *Prop.* To find the second variation of the differential coefficient $\frac{d^ny}{dx^n}$. It has been already shown that

$$\delta\frac{d^ny}{dx^n}=\frac{d^n\delta y}{dx^n},\quad\text{and}\quad\therefore\ \delta^2\frac{d^ny}{dx^n}=\delta\left[\delta\frac{d^ny}{dx^n}\right]=\delta\frac{d^n\delta y}{dx^n}=\frac{d^n\delta^2y}{dx^2}.$$

But since y is a primitive function $\quad\delta^2y=0.$

$$\therefore\ \frac{d^n\delta^2y}{dx^n}=0,\quad\text{and consequently}\quad\delta^2\frac{d^ny}{dx^n}=0.$$

38. *Prop.* To find the second variation of

$$V=f\left[x,y,\frac{dy}{dx},\ldots\ldots\frac{d^ny}{dx^n}\right].$$

We have already found

$$\delta V=\frac{dV}{dy}\delta y+\frac{dV}{d\frac{dy}{dx}}\cdot\frac{d\delta y}{dx}+\ldots\ldots+\frac{dV}{d\frac{d^ny}{dx^n}}\cdot\frac{d^n\delta y}{dx^n}.$$

$$\therefore\ \delta^2V=\delta\left[\frac{dV}{dy}\delta y\right]+\delta\left[\frac{dV}{d\frac{dy}{dx}}\cdot\frac{d\delta y}{dx}\right]+\&c.$$

But $\qquad\delta^2y=0,\quad\frac{d\delta^2y}{dx}=0,\ \&c.$

$$\therefore\ \delta\left[\frac{dV}{dy}\delta y\right]=\delta y\delta\frac{dV}{dy};$$

and, by determining the value of $\delta\frac{dV}{dy}$ in a manner similar to that in which δV was found, we get

$$\delta\frac{dV}{dy}=\frac{d^2V}{dy^2}\delta y+\frac{d^2V}{dyd\frac{dy}{dx}}\cdot\frac{d\delta y}{dx}\ldots\ldots+\frac{d^2V}{dyd\frac{d^ny}{dx^n}}\cdot\frac{d^n\delta y}{dx^n}.$$

Similarly $$\delta\left[\frac{dV}{d\frac{dy}{dx}}\cdot\frac{d\delta y}{dx}\right]=\frac{d\delta y}{dx}\cdot\delta\frac{dV}{d\frac{dy}{dx}},$$ and

$$\delta\frac{dV}{d\frac{dy}{dx}}=\frac{d^2V}{dyd\frac{dy}{dx}}\delta y+\frac{d^2V}{\left[d\frac{dy}{dx}\right]^2}\frac{d\delta y}{dx}+\&c.$$

&c. &c. &c.

Hence, by substitution, we at length find

$$\delta^2V=\frac{d^2V}{dy^2}\delta y^2+2\frac{d^2V}{dyd\frac{dy}{dx}}\delta y.\frac{d\delta y}{dx}+\frac{d^2V}{\left[d\frac{dy}{dx}\right]^2}\cdot\left[\frac{d\delta y}{dx}\right]^2+\&c.$$

39. *Prop.* To find the second variation of $\int Vdx$, when

$$V=f\left[x, y, \frac{dy}{dx}, \frac{d^2y}{dx^2}\dots\frac{d^ny}{dx^n}\right].$$

It has been shown that $\delta\int Vdx=\int\delta Vdx$, and similarly we get

$$\delta^2\int Vdx=\delta\left[\delta\int Vdx\right]=\delta\int\delta Vdx=\int\delta^2Vdx.$$

Substituting for δ^2V, its value found in the last proposition, we obtain

$$\delta^2\int Vdx=\int\left\{\frac{d^2V}{dy^2}\delta y^2+2\frac{d^2V}{dyd\frac{dy}{dx}}\delta y.\frac{d\delta y}{dx}+\frac{d^2V}{\left[d\frac{dy}{dx}\right]^2}\cdot\left[\frac{d\delta y}{dx}\right]^2+\&c.\right\}dx.$$

By similar methods, the third and higher variations could be deduced, but the results are of little practical value.

CHAPTER IV.

MAXIMA AND MINIMA.

40. The Calculus of Variations is applied with great advantage in resolving questions of maxima and minima, to which the ordinary methods of the Differential Calculus are not applicable.

41. A maximum value of a function is one which exceeds other values of that function, produced by infinitely small changes in any or all of its varying elements.

In the Differential Calculus, these changes in the values of the function are produced by changes in the values of the independent variables, while the form of the function remains the same; but in the Calculus of Variations the change in the value of the function is due to a change in its form.

42. The problem of maxima and minima, as resolved in the Differential Calculus, is the following:

Given $u = fx$, where x is an independent variable, and f a function of determinate form, to find what values of x will render u a maximum or minimum.

In the Calculus of Variations, the corresponding problem is this:

Let φ denote a function of indeterminate form, and $u = F\varphi$ a function derived therefrom, to find what form of φ will render u a maximum or minimum.

43. The mode of resolving this latter problem is as follows:

Let $\varphi + i.\psi$ be substituted for φ in the derived function, and let $F(\varphi + i.\psi)$ be developed in terms of the ascending powers of i.

Then, by a course of reasoning, entirely similar to that employed in the Differential Calculus, it will appear that when φ has the form proper to render $F\varphi$ a maximum or minimum, the coefficient of the first power of i must reduce to zero, and that of the second power of i must be negative for a maximum, but positive for a minimum. In other words, if the form of φ alone be supposed to change, we must have $\delta u = 0$. But when, from the nature of the question, both the form of φ and the value of x are liable to variation, we must have

$$Du = 0.$$

44. The application of this theory will now be explained, observing that in the present state of this Calculus, the functions to which it is applied are, almost exclusively, those having the form of a definite integral, such as

$$\int_{x_0}^{x_1} V dx.$$

45. *Prop.* Let $y = \varphi x$ be an indeterminate function of a single variable x, and let it be proposed to find the form of φ, which shall render

$$u = f\left[x, y, \frac{dy}{dx}, \frac{d^2y}{dx^2}, \&c.\right]$$

a maximum or minimum, the symbol f denoting a determinate function.

Let $$du = M\,dx + N\frac{dy}{dx}dx + P_1\frac{d^2y}{dx^2}dx + P_2\frac{d^3y}{dx^3}dx + \&c.$$

Then $$\delta u = N\delta y + P_1\frac{d\delta y}{dx} + P_2\frac{d^2\delta y}{dx^2} + \&c.$$

and if the form of φ be such as will render u a maximum or minimum for any given value of x, we must have

$$\delta u = 0, \quad \text{or} \quad N\delta y + P_1\frac{d\delta y}{dx} + P_2\frac{d^2\delta y}{dx^2} + \&c. = 0.$$

This equation cannot in general be satisfied without destroying the independent character assigned to the form of the function ψ or δy. For, unless the coefficients N, P_1, P_2, &c., be separately equal to zero, the equation

$$N\delta y + P_1 \frac{d\delta y}{dx} + P_2 \frac{d^2\delta y}{dx^2} + \&c. = 0.$$

will establish a relation between the form of the function ψ or δy, and that of φ or y, which is inadmissible. Nor is it possible in general to satisfy the separate conditions $N=0$, $P_1=0$, $P_2=0$, &c., since each of these equations establishes a relation between x and y, or in other words, determines the form of y.

Hence unless all these equations should concur in giving the same form to y, they would contradict each other: and since this concurrence does not usually take place, the problem does not ordinarily admit of a solution.

46. If in the last proposition the value of u should contain but one of the quantities $y, \frac{dy}{dx}, \frac{d^2y}{dx^2}$, &c., or if by the nature of the proposed question, the value of all but one of these be fixed for each value of x, the equation

$$N\delta y + P_1 \frac{d\delta y}{dx} + P_2 \frac{d^2\delta y}{dx^2} + \&c. = 0,$$

will be reduced to a single term, and can therefore be satisfied.

47. *Example.* Let $u = f\left(x, y, \frac{dy}{dx}\right)$, and let it be required to determine what form attributed to the function y will render u a maximum or minimum, it being understood that the value of y is to be given for each value of x.

In this case, since y is constant for the same value of x, $\delta y = 0$, and the equation

$$N\delta y + P_1 \frac{d\delta y}{dx} + P_2 \frac{d^2\delta y}{dx^2} + \&c. = 0 \quad \text{reduces to}$$

$$P_1 = 0.$$

The following geometrical application will render this example more intelligible.

Prop. To determine a curve such that, if at each point P a tangent be drawn and produced to cut two given lines, DC and D_1C_1, parallel to the axis of y, the rectangle $DC \times D_1C_1$ of the parts intercepted between the tangent and the axis of x shall be a maximum or minimum; it being understood that the curve is to be compared only with such other curves as pass through that point.

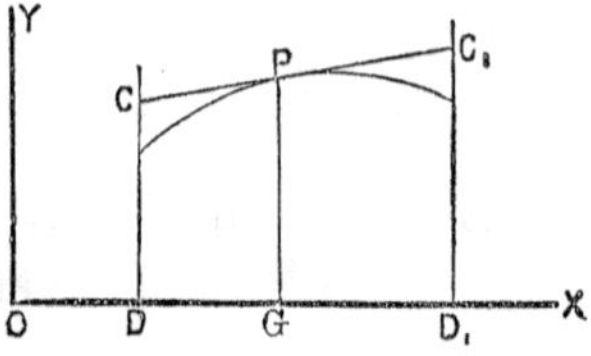

Let O be the origin, OX and OY the axes.

Put $$OD = a, \;\; OD_1 = a_1 \;\; OG = x, \;\; GP = y.$$

Then we shall have

$$DC = y - (x - a)\frac{dy}{dx}, \quad \text{and} \quad D_1C_1 = y + (a_1 - x)\frac{dy}{dx} = y - (x - a_1)\frac{dy}{dx}$$

$$\therefore V = DC \times D_1C_1 = \left[y - (x-a)\frac{dy}{dx}\right] \times \left[y - (x - a_1)\frac{dy}{dx}\right] = f\left(x, y, \frac{dy}{dx}\right)$$

$$\therefore \delta V = N\delta y + P_1\frac{d\delta y}{dx}, \quad \text{where} \quad N = \frac{dV}{dy} \quad \text{and} \quad P_1 = \frac{dV}{d\frac{dy}{dx}}.$$

or $$\delta V = \left[2y + (a + a_1 - 2x)\frac{dy}{dx}\right]\delta y + [2(x - a_1)(x - a)\frac{dy}{dx}$$
$$+ y(a + a_1 - 2x)]\frac{d\delta y}{dx}.$$

But since it is proposed that the curve shall at each point be compared with such curves only as pass through the same point, we must have

$$\delta y = 0$$

and therefore the condition $\delta V = 0$, which is necessary for a maximum or minimum, becomes

$$2(x-a)(x-a_1)\frac{dy}{dx}+y(a+a_1-2x)=0$$

$$\therefore\ 2\frac{dy}{y}-\frac{dx}{x-a}-\frac{dx}{x-a_1}=0,$$

whence by integration,

$$2\log y-\log(x-a)-\log(x-a_1)=\log c.$$

or
$$\log(y^2)=\log[c(x-a)(x-a_1)]$$

$$\therefore\ y^2=c(x-a)(x-a_1),$$

the quantity c being an arbitrary constant.

This equation obviously represents an ellipse or hyperbola according as c is negative or positive.

Passing now to the second variation, we have

$$\delta^2 V=\frac{d^2V}{dy^2}\delta y^2+2\frac{d^2V}{dy\,.\,d\frac{dy}{dx}}\delta y\,.\,\frac{d\delta y}{dx}+\frac{d^2V}{\left[d\frac{dy}{dx}\right]^2}\cdot\left[\frac{d\delta y}{dx}\right]^2\ \&c.$$

and since in the present case $V=f\left(x,y,\frac{dy}{dx}\right)$ and $\delta y=0$

we shall have
$$\delta^2 V=\frac{d^2V}{\left[d\frac{dy}{dx}\right]^2}\cdot\left[\frac{d\delta y}{dx}\right]^2$$

or
$$\delta^2 V=2(x-a)(x-a_1)\left[\frac{d\delta y}{dx}\right]^2$$

or by putting for $(x-a)(x-a_1)$ its value $\frac{y^2}{c}$

$$\delta^2 V=\frac{y^2}{c}\left[\frac{d\delta y}{dx}\right]^2$$

The sign of this quantity is the same as that of c. Hence the curve is an ellipse when V is a maximum, and a hyperbola when V is a minimum. In the first case the curve lies entirely within the lines CD and C_1D_1; and in the second entirely exterior to those lines.

48. *Prop.* To find the form of the function y, and the values of the limits x_0 and x_1, which shall render the definite integral $U = \int_{x_0}^{x_1} V dx$ a maximum or minimum, when

$$V = f\left[x, y, \frac{dy}{dx}, \frac{d^2y}{dx^2} \cdots \frac{d^ny}{dx^n}\right];$$

the character f denoting, as usual, a determinate function.

Here, we have

$$DU = V_1 dx_1 - V_0 dx_0 + \left[P_1 - \frac{dP_2}{dx} + \&c.\right]_1 \delta y_1$$

$$- \left[P_1 - \frac{dP_2}{dx} + \&c.\right]_0 \cdot \delta y_0$$

$$+ \left[P_2 - \&c.\right]_1 \cdot \left[\frac{d\delta y}{dx}\right]_1 - \left[P_2 - \&c.\right]_0 \cdot \left[\frac{d\delta y}{dx}\right]_0 + \&c.$$

$$+ \int_{x_0}^{x_1} \left[N - \frac{dP_1}{dx} + \frac{d^2P_2}{dx^2} - \&c. \cdots\right.$$

$$\left. + (-1)^n \frac{d^nP_n}{dx^n}\right] \delta y \, . \, dx = 0 \cdots (A).$$

Two cases may occur in the attempt to satisfy this equation, viz.:

1st. The variation δy, or the form of the function ψ, may be wholly unrestricted (except by the general condition always applicable to this function); or,

2d. It may be necessary to assume the function ψ of such form as will satisfy some given condition or conditions.

In the first case, the object proposed is to determine among *all possible functions*, that one which shall render u a maximum or minimum. In the second case, the derived function is required to belong to a *particular class*, each individual of which fulfils certain given conditions.

Maxima and minima belonging to the first of these divisions are called *absolute*, and those belonging to the second division are termed *relative*. Taking the first of these divisions, put for brevity

$$a_1 = V_1 dx_1 + \left[P_1 - \frac{dP_2}{dx} + \&c.\right]_1 \cdot \delta y_1$$

$$+ \left[P_2 - \frac{dP_3}{dx} + \&c.\right]_1 \cdot \left[\frac{d\delta y}{dx}\right]_1 + \&c.$$

$$a_0 = V_0 dx_0 + \left[P_1 - \frac{dP_2}{dx} + \&c.\right]_0 \cdot \delta y_0$$

$$+ \left[P_2 - \frac{dP_3}{dx} + \&c.\right]_0 \cdot \left[\frac{d\delta y}{dx}\right]_0 + \&c.$$

$$b = N - \frac{dP_1}{dx} + \frac{d^2P_2}{dx^2} - \&c. + (-1)^n \frac{d^nP_n}{dx^n},$$

and equation (A) will reduce to the form

$$a_1 - a_0 + \int_{x_0}^{x_1} b \cdot \delta y \cdot dx = 0 \cdot \cdot \cdot \cdot (B).$$

This equation cannot be satisfied so long as the form of δy or ψ remains unrestricted, unless we have the two independent conditions:

$$a_1 - a_0 = 0, \quad \text{and } b = 0.$$

For, if $a_1 - a$ be not equal to zero, we must have

$$a_1 - a_0 = - \int_{x_0}^{x_1} b\delta y \cdot dx,$$

a condition manifestly impossible, since the value of the definite integral in the second member cannot possibly remain invariable; while we are at liberty to change arbitrarily the form of the quantity to be integrated; but the value of $a_1 - a_0$, which depends only upon the values which certain quantities have at the limits, will not necessarily vary with a change in the form of δy. Hence, we must have

$$a_1 - a_0 = 0, \quad \text{and} \quad \int_{x_0}^{x_1} b\delta y \, . \, dx = 0.$$

Now this last equation cannot be true for every form of δy, unless $b = 0$, or

$$N - \frac{dP_1}{dx} + \frac{d^2P}{dx^2} - \&c. \ldots . + (-1)^n \frac{d^nP_n}{dx^n} = 0,$$

a differential equation which serves to determine the form of the function y.

49. The two equations, $a_1 - a_0 = 0$, and $b = 0$, differ essentially in their signification, the latter establishing a general relation between the variables x and y, while the former connects the particular values which these quantities have at the limits of integration.

50. Without this distinction, the solution of the problem would be impossible, since there could not be *two* general relations between x and y.

51. The coefficients of the increments in the equation $a_1 - a_0 = 0$ being constant, and the increments themselves either entirely arbitrary, or restricted by a limited number of conditions, that equation will be equivalent to as many distinct equations as can be formed by placing equal to zero each of the coefficients of those increments which remain arbitrary, after we have eliminated all such increments as are restricted by the given conditions. We now proceed to show that the equations thus formed, together with that obtained by integrating the differential equation $b = 0$, will just suffice for the complete solution of the problem when a solution is possible.

52. The differential equation $b = 0$, or

$$N - \frac{dP_1}{dx} + \frac{d^2P_2}{dx_2} - \&c. \cdots\cdots + (-1)^n \frac{d^nP_n}{dx^n} = 0 \cdots\cdot (C),$$

is in general of the $2n^{th}$ order. For since V contains $\frac{d^ny}{dx^n}$, the quantity $P_n = \frac{dV}{d\frac{d^ny}{dx^n}}$ will usually contain $\frac{d^ny}{dx^n}$ also; and therefore

$\frac{d^nP_n}{dx^n}$ will usually contain $\frac{d^{2n}y}{dx^{2n}}$.

Hence the integral of (C) will usually contain $2n$ arbitrary constants.

But if the limiting values of $x, y, \frac{dy}{dx}, \frac{d^2y}{dx^2} \cdots\cdot \frac{d^{n-1}y}{dx^{n-1}}$ be entirely unrestricted, the equation $a_1 - a_0 = 0$ will contain $2n + 2$ arbitrary increments, viz.:

$$dx_1, \delta y_1, \delta\left[\frac{dy}{dx}\right]_1, \ldots \delta\left[\frac{d^{n-1}y}{dx^{n-1}}\right]_1, dx_0, \delta y_0, \delta\left[\frac{dy}{dx}\right]_0 \ldots \delta\left[\frac{d^{n-1}y}{dx^{n-1}}\right]_0,$$

in which case that equation cannot be satisfied, since there would be formed, by placing the coefficient of each arbitrary increment equal to zero $2n + 2$ equations, while there are but $2n$ constants whose values are to be determined.

This result might have been anticipated, for it is evident that if the form of the function y, and the limits of integration be entirely unrestricted, the integral may have any value from 0 to ∞, and, therefore, cannot admit of a maximum or minimum.

53. The nature of the restriction imposed upon the limits must depend in each case upon the conditions of the proposed problem.

1st. Let the limiting values of x, viz., x_0 and x_1 be given; that is, let it be proposed to find such a form of the function y as will render $\int V dx$, when taken between fixed limits, a maximum or minimum.

Here we have $dx_1 = 0$, and $dx_0 = 0$, and the equation $a_1 - a = 0$ is now equivalent to the following separate equations:

$$\left[P_1 - \frac{dP_2}{dx} + \&c.\right]_1 = 0, \left[P_1 - \frac{dP_2}{dx} + \&c.\right]_0 = 0,$$

$$[P_2 - \&c.]_1 = 0, \ [P_2 - \&c.]_0 = 0, \ \&c. \ \&c. \ \&c. \ldots [P_n]_1 = 0, \ [P_n]_0 = 0.$$

The number of these equations is $2n$, the same as that of the constants remaining to be determined; and hence the solution is in this case complete.

2d. Let the limiting values of both x and y be given.

Then $dx_1 = 0$, $\delta y_1 = 0$, $dx_0 = 0$, $\delta y_0 = 0$, and the equation $a_1 - a_0 = 0$ is equivalent to $2n - 2$ separate equations, viz.: those formed by placing equal to zero the coefficients of the following increments:

$$\delta\left[\frac{dy}{dx}\right]_1, \delta\left[\frac{dy}{dx}\right]_0, \delta\left[\frac{d^2y}{dx^2}\right]_1, \delta\left[\frac{d^2y}{dx^2}\right]_0 \ldots \delta\left[\frac{d^{n-1}y}{dx^{n-1}}\right]_1, \delta\left[\frac{d^{n-1}y}{dx^{n-1}}\right]_0.$$

But there are now two additional equations resulting from the substitution of the given limiting values of x and y in the general solution of the differential equation $b = 0$. For let the integral of that equation be

$$f[x, y, c_1, c_2 \ldots c_{2n}] = 0,$$

where $c_1, c_2 \ldots c_{2n}$ are the $2n$ arbitrary constants. Then we shall have the $2n$ equations

$$f[x_1, y_1, c_1, c_2 \ldots c_{2n}] = 0, \quad f[x_0, y_0, c_1, c_2 \ldots c_{2n}] = 0,$$

$$\left[P_2 - \frac{dP_3}{dx} + \&c.\right]_1 = 0, \quad \left[P_2 - \frac{dP_3}{dx} + \&c.\right]_0 = 0,$$

$$[P_3 - \&c.]_1 = 0, \ [P_3 - \&c.]_0 = 0, \ \&c. \ \&c. \ldots [P_n]_1 = 0, \ [P_n]_0 = 0,$$

with which to determine the $2n$ constants.

3d. Similarly, if the limiting values of x, y, and $\frac{dy}{dx}$ were given the new condition, would remove two of the preceding equations, viz.:

$$\left[P_2 - \frac{dP_3}{dx} + \&c.\right]_1 = 0 \quad \text{and} \quad \left[P_2 - \frac{dP_3}{dx} + \&c.\right]_0 = 0,$$

but two new conditions would be derived from the substitution of the limiting values of $\frac{dy}{dx}$ in the equation obtained by differentiating the general solution.

$$f[x, y, c_1, c_2, \ldots c_{2n}] = 0.$$

For let $$f_1\left[x, y, \frac{dy}{dx}, c_1, c_2, \ldots c_{2n}\right] = 0$$

be the result of a differentiation with respect to x. Then we shall have

$$f_1\left[x_1, y_1, \left(\frac{dy}{dx}\right)_1, c_1, c_2 \ldots c_{2n}\right] = 0,$$

and $$f_1\left[x_0, y_0, \left(\frac{dy}{dx}\right)_0, c_1, c_2 \ldots c_{2n}\right] = 0.$$

54. Similarly, if the limiting values of $\frac{d^2y}{dx^2}$ were given, two more equations would disappear from the group obtained by making $a_1 - a_0 = 0$; and, on the other hand, two new equations would result from the substitution of the limiting values of $\frac{d^2y}{dx^2}$ in the equation obtained by differentiating the general solution twice; thus preserving the total number of equations equal to $2n$, the same as that of the constants to be determined. And, in general, whatever may be the number of the quantities having given limits, the total number of equations will be $2n$, and therefore just sufficient for the complete solution of the problem.

55. When the limiting values of $x, y, \frac{dy}{dx}$, &c., are not absolutely fixed, but simply connected by one or more equations of condition, the variations of the quantities so connected are not independent, and therefore two or more of the equations, resulting from the condition $a_1 - a_0 = 0$, will be replaced by a single equation. Thus the total number of equations deducible from $a_1 - a_0 = 0$ will be diminished; but, on the other hand, a number of new equations, just sufficient to supply the deficiency, will arise from the equations of condition. To illustrate this, take the following

Example. Let the limiting values of x and y be connected by the equations

$$y_1 = f_1 x_1 \quad \text{and} \quad y_0 = f_0 x_0.$$

The quantities $dx_1, \delta y_1, dx_0, \delta y_0$ will be connected by the following relations:

$$\left[\frac{dy}{dx}\right]_1 \cdot dx_1 + \delta y_1 = f_1' x_1 \,.\, dx_1, \quad \left[\frac{dy}{dx}\right]_0 \cdot dx_0 + \delta y_0 = f_0' x_0 \,.\, dx_0.$$

Now, substituting the values of δy_1 and δy_0, derived from these equations in $a_1 - a_0 = 0$, and placing equal to zero the coefficient of each remaining variation, the following equations will result:

$$V_1 + \left[P_1 - \frac{dP_2}{dx} + \&c.\right]_1 \times \left(f_1'x_1 - \left[\frac{dy}{dx}\right]_1\right) = 0$$

$$\left[P_2 - \frac{dP_3}{dx} + \&c.\right]_1 = 0, \ \&c.$$

$$V_0 + \left[P_1 - \frac{dP_2}{dx} + \&c.\right]_0 \times \left(f_0'x_0 - \left[\frac{dy}{dx}\right]_0\right) = 0,$$

$$\left[P_2 - \frac{dP_3}{dx} + \&c.\right]_0 = 0, \ \&c.$$

The other equations being the same as heretofore.

These equations, ($2n$ in number,) in connection with the four following, viz.:

$$y_1 = f_1 x_0, \quad y_1 = f_0 x_0, \quad f(x_1, y_1, c_1, c_2, \cdots\cdot c_{2n}) = 0$$
$$f(x_0, y_0, c_1, c_2, \cdots\cdot c_{2n}) = 0$$

will just suffice for determining the $2n + 4$ quantities

$$x_1, y_1, x_0, y_0, c_1, c_2, \cdots\cdot c_{2n}.$$

56. And if the limiting values of x and $\frac{dy}{dx}$ were also connected by the relations

$$\left[\frac{dy}{dx}\right]_1 = f_1'x_1, \qquad \left[\frac{dy}{dx}\right]_0 = f_0'x_0,$$

we should have

$$\left[\frac{d^2y}{dx^2}\right]_1 \cdot dx_1 + \delta\left[\frac{dy}{dx}\right]_1 = f_1''x_1 dx_1, \text{ and } \left[\frac{d^2y}{dx^2}\right]_0 + \delta\left[\frac{dy}{dx}\right]_0 = f_0''x_0 dx_0.$$

Hence, the first three terms in each of the quantities, a_1 and a_0, will reduce to one, and the number of equations deducible from $a_1 - a_0 = 0$ will be reduced to $2n - 2$. But we shall have in addition six other equations, viz.: the four used in the preceding case, and the two following:

$$f'[x_1, y_1, f_1'x_1, c_1, c_2, \cdots c_{2n}] = 0, \quad f'[x_0, y_0, f_0'x_0, c_1, c_2, \cdots c_{2n}] = 0,$$

which are obtained by differentiating the general solution

$$f(x, y, c_1, c_2, \cdots c_{2n}) = 0,$$

and substituting in the result the limiting values of x, y, and $\frac{dy}{dx}$.

Thus the total number of equations will be $2n + 4$, which is just sufficient.

And the same result will be found true when the restrictions imposed upon the limiting values of the several variations are more numerous.

57. The exceptions to the preceding theory will now be considered.

58. *Case* 1*st*. Let V be a linear function of the highest differential coefficient $\frac{d^n y}{dx^n}$.

Then P_n will not contain this coefficient, and therefore $\frac{d^n P_n}{dx^n}$ cannot be of an order higher than $2n - 1$. Hence, the equation

$$N - \frac{dP_1}{dx} + \frac{d^2 P_2}{dx^2} - \&c. \cdots + (-1)^n \frac{d^n P_n}{dx^n} = 0$$

cannot be of an order higher than $2n - 1$, and its solution will contain $2n - 1$ disposable constants. Thus the equation $a_1 - a_0 = 0$, which is equivalent to $2n$ equations, cannot, in this case, be satisfied.

59. It may even be proved that the equation $b = 0$ cannot, in this case, be of an order higher than $2n - 2$.

For, put $\frac{d^n y}{dx^n} = v$. Then $V = \theta v + \theta'$,

where θ and θ' are functions of $x, y, \frac{dy}{dx}, \frac{d^2y}{dx^2}, \cdots \frac{d^{n-1}y}{dx^{n-1}}$.

It has been shown already that the equation $b = 0$ does not, in this case, contain $\frac{d^{2n}y}{dx^{2n}}$, and therefore it is only necessary to prove that it does not contain the coefficient $\frac{d^{2n-1}y}{dx^{2n-1}}$.

Now, this coefficient cannot occur, unless it be in one of two terms,

viz:
$$\frac{d^{n-1}P_{n-1}}{dx^{n-1}} \quad \text{or} \quad \frac{d^nP_n}{dx^n}.$$

But $V = \theta v + \theta' \quad \therefore P_n = \dfrac{dV}{d\dfrac{d^ny}{dx^n}} = \dfrac{dV}{dv} = \theta$, and $\dfrac{d^nP_n}{dx^n} = \dfrac{d^n\theta}{dx^n}$.

Now to find the coefficient of $\dfrac{d^{2n-1}y}{dx^{2n-1}}$ in $\dfrac{d^nP_n}{dx^n}$, we must form the values of $\left(\dfrac{d\theta}{dx}\right), \left(\dfrac{d^2\theta}{dx^2}\right), \dots \left(\dfrac{d^n\theta}{dx^n}\right)$, and reject, in each, every term except that of the highest order.

Making
$$\frac{d^{n-1}y}{dx^{n-1}} = u, \quad \text{we have}$$

$$\left(\frac{d\theta}{dx}\right) = \frac{d\theta}{dx} + \frac{d\theta}{dy}\cdot\frac{dy}{dx} + \frac{d\theta}{d\dfrac{dy}{dx}}\cdot\frac{d^2y}{dx^2} + \text{\&c.} \dots + \frac{d\theta}{du}\cdot\frac{du}{dx}.$$

Here, the last term $\dfrac{d\theta}{du}\cdot\dfrac{du}{dx} = \dfrac{d\theta}{du}\cdot\dfrac{d^ny}{dx^n}$ is the only term to be retained, because all the others are of an order less than n. And similarly the only term in $\left(\dfrac{d^2\theta}{dx^2}\right)$ of the order $n+1$ is

$$\frac{d\theta}{du}\cdot\frac{d^2u}{dx^2} = \frac{d\theta}{du}\cdot\frac{d^{n+1}y}{dx^{n+1}}.$$

In the same manner, it appears that the only term in $\left(\dfrac{d^n\theta}{dx^n}\right)$ of the order $2n-1$ is

$$\frac{d\theta}{du}\cdot\frac{d^nu}{dx^n} = \frac{d\theta}{du}\cdot\frac{d^{2n-1}y}{dx^{2n-1}}.$$

Again, since $V = \theta v + \theta'$, $\therefore P_{n-1} = \dfrac{dV}{d\dfrac{d^{n-1}y}{dx^{n-1}}} = \dfrac{dV}{du} = v\dfrac{d\theta}{du} + \dfrac{d\theta'}{du}$.

Hence, by forming the values of $\dfrac{dP_{n-1}}{dx}, \dfrac{d^2P_{n-1}}{dx^2} \dots \dfrac{d^{n-1}P_{n-1}}{dx^{n-1}}$, retaining only the terms of the highest order in each successive differentiation, it will be seen that the only term of the order $2n-1$, in

$$\frac{d^{n-1}P_{n-1}}{dx^{n-1}} \quad \text{is} \quad \frac{d\theta}{du}\cdot\frac{d^{n-1}v}{dx^{n-1}} = \frac{d\theta}{du}\cdot\frac{d^{2n-1}y}{dx^{2n-1}},$$

and, since this term is precisely the same as the term of the same order in $\frac{d^nP_n}{dx^n}$, the two will disappear in

$$\frac{d^{n-1}P_{n-1}}{dx^{n-1}} - \frac{d^nP_n}{dx^n}.$$

$\therefore$ the equation $b = 0$ is not of an order higher than $2n - 2$.

60. *Case 2d.* Let $V = y \cdot fx + F(x, p_1)$, where $p_1 = \frac{dy}{dx}$.

Here $\quad N = \frac{dV}{dy} = fx, \quad$ and $P_1 = \frac{dV}{dp_1} = \frac{dF(x, p_1)}{dp_1}$

and since V is in this case a function of x, y, and $\frac{dy}{dx}$ only, the equation $b = 0$ will become simply

$$N - \frac{dP_1}{dx} = 0, \quad \text{or,}\ fx = \frac{dP_1}{dx}$$

and is immediately integrable, giving

$$P_1 = \int fx \cdot dx = f_1x + c.$$

Substituting the value of P_1, derived from the proposed equation, we shall have an equation involving x_1, p_1, &c., which, solved with respect to p_1, will give a result of the form

$$p_1 = \varphi(x, c) \qquad \text{or} \qquad \frac{dy}{dx} = \varphi(x, c)$$

$$\therefore\ y = \varphi_1(x, c) + c_1 \ldots\ldots (1).$$

Now suppose the limiting values of x given, those of y being indeterminate:

The equation $\quad a_1 - a = 0 \quad$ is then equivalent to the two equations

$[P_1]_1 = 0$, and $[P_1]_0 = 0$ or $f_1x_1 + c = 0$, (2) and $f_1x_0 + c = 0$ (3)

The two equations, (2) and (3) contain but one arbitrary constant c, and therefore cannot usually be satisfied, although the general solution (1) contains the proper number of constants. Hence the proposed problem does not admit of a solution.

61. If in the case just considered $fx = 0$, so that $V = F(x,p_1)$ the two equations (2) and (3) become identical, and the solution is then possible: but it belongs to the indeterminate class, since one of the constants remains entirely arbitrary.

62. The results just obtained are not peculiar to functions of the first order, such as that just considered for if V be supposed of such form as will give

$$N = \frac{dV}{dy} \doteq fx,$$

and if the limiting values of x only be given, similar reasoning will apply. The equation $b = 0$ will, in this instance, as in the preceding, be immediately integrable, giving

$$P_1 - \frac{dP_2}{dx} + \&c. = f_1x + c$$

and the first two equations resulting from the equation $a_1 - a_0 = 0$, are

$$f_1x_1 + c = 0, \quad \text{and} \quad f_1x_0 + c = 0.$$

These two equations cannot usually be satisfied except when $f_1x = 0$, in which case y does not appear in the value of V.

And in general if $\frac{d^sy}{dx^s}$ be the lowest differential coefficient appearing in V, the form of V being such that $\frac{dV}{d\frac{d^sy}{dx^s}} = fx_1$, and if the limiting values of x and of those coefficients which are higher than the sth be alone given, we may prove, in like manner, that the problem will not admit of a solution.

Case 3*d*. Let $N = 0$, and let the limiting values of x only be given.

In this case the equation $b = 0$ becomes

$$\frac{dP_1}{dx} - \frac{d^2P_2}{dx^2} + \frac{d^3P_3}{dx^3} + \&c. = 0$$

and is integrable, giving

$$P_1 - \frac{dP_2}{dx} + \frac{d^2P_3}{dx^2} + \&c. = c$$

and the two conditions furnished by placing equal to zero the coefficients δy_1 and δy_0, viz.:

$$[P_1 - \frac{dP_2}{dx} + \&c.]_1 = 0 \quad \text{and} \quad [P_1 - \frac{dP_2}{dx} + \&c.]_0 = 0$$

are equivalent to the single condition $c = 0$.

Hence the equation $a_1 - a_0 = 0$ is equivalent to but $2n - 1$ equations, instead of $2n$, and the problem is indeterminate. This result might have been expected, for since y does not appear in V, nor in the conditions fulfilled at the limits, the coefficient $\frac{dy}{dx}$ might have been taken as the principal function, instead of y, and then the equations given by $DU = 0$ would have been just sufficient to establish a relation between x and $\frac{dy}{dx}$, without arbitrary constants, which relation, when integrated, must give an equation between x and y, containing one arbitrary constant.

63. If, in the last case, one of the limiting values of y were given, the problem would again become determinate. Similarly, when $N = 0$ and $P_1 = 0$, and both limiting values of y and $\frac{dy}{dx}$ are indeterminate, the solution will contain *two* arbitrary constants, and will be rendered determinate by assigning at least one limiting value to y and $\frac{dy}{dx}$.

And generally, if the first m terms of the equation

$$N - \frac{dP_1}{dx} + \frac{d^2P_2}{dx^2} - \&c. = 0$$

be wanting, and if there be no conditions fixing the limiting values of $y, \frac{dy}{dx}, \ldots . \frac{d^{n-1}y}{dx^{n-1}}$, the solution will contain m arbitrary constants.

The preceding cases afford the principal examples of exception to the general theory. We now return to the consideration of that theory.

64. As it will sometimes be possible to integrate the equation

$$N - \frac{dP_1}{dx^2} + \frac{d^2P}{dx^2} - \text{\&c.} = 0$$

one or more times without determining the form of the function V, and as the consideration of these cases will greatly facilitate the application of the theory to particular examples, we proceed to examine some of these cases, arranging them in two classes.

65. 1*st Case.* Let the first m of the quantities $y, \frac{dy}{dx}, \frac{d^2y}{dx^2}$, &c. be wanting in V, or let

$$V = f\left[x, \frac{d^my}{dx^m} \ldots . \frac{d^ny}{dx^n}\right]$$

Then the first m terms of the equation

$$N \quad \frac{dP_1}{dx} + \text{\&c.} = 0$$

will be wanting, and that equation will reduce to

$$\frac{d^mP_m}{dx^m} - \frac{d^{m+1}P_{m+1}}{dx^{m+1}} + \text{\&c.} = 0$$

which gives, when integrated, m times,

$$P_m - \frac{dP_{m+1}}{dx} + \text{\&c.} = c_0 + c_1x + c_2x^2 + \ldots . c_{m-1}x^{m-1},$$

a differential equation of the order $2n - m$.

66. *Case 2d.* Let the independent variable x be wanting in V, or let

$$V = f\left[y, \frac{dy}{dx}, \frac{d^2y}{dx^2} \cdots\cdots \frac{d^ny}{dx^n}\right].$$

In this case, we have

$$dV = Ndy + P_1 d\frac{dy}{dx} + P_2 d\frac{d^2y}{dx^2} + \&c.$$

$$= \left[N\frac{dy}{dx} + P_1\frac{d^2y}{dx^2} + P_2\frac{d^3y}{dx^3} - \cdots\cdots + P_n\frac{d^{n+1}y}{dx^{n+1}}\right]dx;$$

or, by substituting for N, its value derived from the equation,

$$N - \frac{dP_1}{dx} + \frac{d^2P_2}{dx^2} - \&c. = 0, \text{ we get}$$

$$dV = \left[P_1\frac{d^2y}{dx^2} + \frac{dy}{dx}\cdot\frac{dP_1}{dx}\right]dx + \left[P_2\frac{d^3y}{dx^3} - \frac{dy}{dx}\cdot\frac{d^2P_2}{dx^2}\right]dx + \&c.$$

$$+ \left[P_n\frac{d^{n+1}y}{dx^{n+1}} - (-1)^n\cdot\frac{dy}{dx}\cdot\frac{d^nP_n}{dx^n}\right]dx.$$

$$\therefore V = c + \int\left[P_1\frac{d^2y}{dx^2} + \frac{dy}{dx}\cdot\frac{dP_1}{dx}\right]dx + \int\left[P_2\frac{d^3y}{dx^3} - \frac{dy}{dx}\cdot\frac{d^2P_2}{dx^2}\right]dx + \&c.$$

$$+ \int\left[P_n\frac{d^{n+1}y}{dx^{n+1}} - (-1)^n\cdot\frac{dy}{dx}\cdot\frac{d^nP_n}{dx^n}\right]dx.$$

But the quantity $\int P_n\frac{d^{n+1}y}{dx^{n+1}}dx$ gives, by an integration by parts,

$$\int P_n\frac{d^{n+1}y}{dx^{n+1}}dx = P_n\frac{d^ny}{dx^n} - \frac{dP_n}{dx}\cdot\frac{d^{n-1}y}{dx^{n-1}} + \&c.$$

$$\cdots\cdots + (-1)^n\int\frac{dy}{dx}\cdot\frac{d^nP_n}{dx^n}dx.$$

$$\therefore \int\left[P_n\frac{d^{n+1}y}{dx^{n+1}} - (-1)^n\frac{dy}{dx}\cdot\frac{d^nP_n}{dx^n}\right]dx = P_n\frac{d^ny}{dx^n} - \frac{dP_n}{dx}\cdot\frac{d^{n-1}y}{dx^{n-1}} + \&c.$$

$$\cdots\cdots + (-1)^{n-1}\cdot\frac{dy}{dx}\cdot\frac{d^{n-1}P_n}{dx^{n-1}}.$$

$$\therefore V = c + P_1\frac{dy}{dx} + \left[P_2\frac{d^2y}{dx^2} - \frac{dy}{dx}\cdot\frac{dP_2}{dx}\right]$$

$$+ \left[P_3\frac{d^3y}{dx^3} - \frac{dP_3}{dx}\cdot\frac{d^2y}{dx^2} + \frac{d^2P_3}{dx^2}\cdot\frac{dy}{dx}\right] + \text{\&c.}$$

$$+ P_n\frac{d^ny}{dx^n} - \frac{dP_n}{dx}\cdot\frac{d^{n-1}y}{dx^{n-1}} + \text{\&c.} \ldots\ldots + (-1)^{n-1}\cdot\frac{d^{n-1}P_n}{dx^{n-1}}\cdot\frac{dy}{dx}\cdots(D),$$

which is a differential equation of an order not higher than $2n - 1$.

Thus it appears that when V does not contain the independent variable x, the equation $b = 0$ can be reduced at least one order.

67. The following are the most important applications of formula (D):

1st. Let $$V = f\left(\frac{dy}{dx}\right) \cdots\cdots (a).$$

Here $V = c + P_1\frac{dy}{dx}$ by formula (D), since $P_2 = 0$, $P_3 = 0$, &c.

But V is a function of $\frac{dy}{dx}$. $\therefore P_1 = \frac{dV}{d\frac{dy}{dx}}$ is also a function of $\frac{dy}{dx}$.

Hence by substituting for V and P_1 their values, and then solving with respect to $\frac{dy}{dx}$, the result would take the form

$$\frac{dy}{dx} = c_1. \quad \therefore y = c_1x + c_2$$

Here y is a linear function of x, and this result shows that linear functions have the property of giving a maximum or minimum value to every function of $\frac{dy}{dx}$ which admits of such a value.

2d. Let $$V = f\left(y, \frac{dy}{dx}\right) \cdots\cdots (b).$$

Then $$V = c + P_1\frac{dy}{dx}.$$

3d. Let $$V = f\left(y, \frac{d^2y}{dx^2}\right) \cdots\cdots (c).$$

Then $$V = c + P_2 \frac{d^2y}{dx^2} - \frac{dy}{dx} \cdot \frac{dP_2}{dx}.$$

68. *Case* 3*d.* Let the function V belong at the same time to both of the preceding classes, that is, let the independent variable x, and the first first m of the quantities $y, \frac{dy}{dx}, \frac{d^2y}{dx^2}$, &c., be wanting in V.

The equation $b = 0$ gives, as in the first case by integration,

$$P_m - \frac{dP_{m+1}}{dx} + \&c. = c_0 + c_1 x + c_2 x^2 + \&c. \ldots\ldots c_{m-1} x^{m-1}.$$

$$\therefore\ P_m = \frac{dP_{m+1}}{dx} - \&c. + c_0 + c_1 x + c_2 x^2 + \&c. \ldots\ldots c_{m-1} x^{m-1}.$$

This value substituted in

$$dV = \left[P_m \frac{d^{m+1}y}{dx^{m+1}} + P_{m+1} \cdot \frac{d^{m+2}y}{dx^{m+2}} + \&c. \cdots\cdots + P_n \frac{d^{n+1}y}{dx^{n+1}}\right] dx,$$

the differential of the given relation

$$V = f\left[\frac{d^my}{dx^m}, \frac{d^{m+1}y}{dx^{m+1}} \cdots\cdots \frac{d^ny}{dx^n}\right], \text{ gives}$$

$$dV = \left[P_{m+1} \frac{d^{m+2}y}{dx^{m+2}} + \frac{dP_{m+1}}{dx} \cdot \frac{d^{m+1}y}{dx^{m+1}}\right] dx$$

$$+ \left[P_{m+2} \frac{d^{m+3}y}{dx^{m+3}} - \frac{d^2P_{m+2}}{dx^2} \cdot \frac{d^{m+1}y}{dx^{m+1}}\right] dx$$

$$+ \quad \&c. \qquad \&c.$$

$$+ \left[P_n \frac{d^{n+1}y}{dx^{n+1}} - (-1)^{n-1} \cdot \frac{d^{n-m}P_n}{dx^{n-m}} \cdot \frac{d^{m+1}y}{dx^{m+1}}\right] dx$$

$$+ \left[c_0 + c_1 x + c_2 x^2 + \&c. \ldots\ldots c_{m-1} \cdot x^{m-1}\right] \frac{d^{m+1}y}{dx^{m+1}} dx.$$

Integrating by parts, we get

$$V=c+P_{m+1}\cdot\frac{d^{m+1}y}{dx^{m+1}}+\left[P_{m+2}\cdot\frac{d^{m+2}y}{dx^{m+2}}-\frac{dP_{m+2}}{dx}\cdot\frac{d^{m+1}y}{dx^{m+1}}\right]+\text{\&c.}$$

$$+P_n\frac{d^ny}{dx^n}-\frac{dP_n}{dx}\cdot\frac{d^{n-1}y}{dx^{n-1}}+\text{\&c.}+(-1)^{n-m-1}\cdot\frac{d^{n-m-1}P_n}{dx^{n-m-1}}\cdot\frac{d^{m+1}y}{dx^{m+1}}$$

$$+\int\left[c_0+c_1x+c_2x^2+\text{\&c.}\ldots\ldots+c_{m-1}x^{m-1}\right]\frac{d^{m+1}y}{dx^{m+1}}\,dx\ldots\ldots(E).$$

But since in general

$$\int x^r\cdot\frac{d^{m+1}y}{dx^{m+1}}dx=x^r\cdot\frac{d^my}{dx^m}-r\,.\,x^{r-1}\cdot\frac{d^{m-1}y}{dx^{m-1}}+r(r-1)x^{r-2}\frac{d^{m-2}y}{dx^{m-2}}\ \text{\&c.}$$

$$+(-1)^r\,.\,r(r-1)(r-2)\ldots\ldots2\,.\,1\,\frac{d^{m-r}y}{dx^{m-r}},$$

if we put successively r equal to $(1, 2, 3, \ldots . m-1)$, and substitute the resulting values of the integrals,

$$\int x\frac{d^{m+1}y}{dx^{m+1}}dx,\quad \int x^2\frac{d^{m+1}y}{dx^{m+1}}dx,\ \ldots\ \int x^{m-1}\frac{d^{m+1}y}{dx^{m+1}}dx,$$

in equation (E) it will be a differential equation of the order $2n-m-1$; that is, the original differential equation will have had its order reduced by $m+1$ degrees.

69. Suppose for example that

$$V=f\left(\frac{dy}{dx},\frac{d^2y}{dx^2}\right)\ldots\ldots(1)$$

Then the equation $b=0$, becomes

$$\frac{dP_1}{dx}-\frac{d^2P_2}{dx^2}=0,$$

whence by integration $P_1=\dfrac{dP_2}{dx}+c.$

and this value substituted in the differential of (1) viz.:

$$dV=\left[P_1\frac{d^2y}{dx^2}+P_2\frac{d^3y}{dx^3}\right]dx$$

gives $$dV = \left[c + \frac{dP_2}{dx}\right]\frac{d^2y}{dx^2} \cdot dx + P_2 \frac{d^3y}{dx^3}dx$$

$$\therefore\ V = c' + c\frac{dy}{dx} + P_2 \frac{d^2y}{dx^2}$$

a differential equation of the second order as it should be, since

$$2n - m - 1 = 2.$$

Relative Maxima and Minima of One Variable.

70. *Prop.* To determine the form of the function $y = \varphi x$ which will render $\int Vdx$ (taken between certain limits) a maximum or minimum, when y is selected from those functions which satisfy the additional condition $\int V'dx = c$ (between the same limits); the quantities V and V' being functions of $x, y, \frac{dy}{dx}, \frac{d^2y}{dx^2}$, &c.

The condition $\int Vdx =$ a maximum or minimum, gives

$$D\int Vdx = 0 \cdots\cdots (1).$$

And the condition $\int V'dx = c$, gives

$$D\int V'dx = 0 \cdots\cdots (2).$$

Multiply (2) by an arbitrary quantity λ, and add the result to (1); then $D\int Vdx + \lambda . D\int V'dx = 0$ or $D\int(V + \lambda V')dx = 0 \cdots\cdot (3)$

and equation (3) will include all the conditions involved in the problem, and will imply that both (1) and (2) are necessarily true.

For since by hypothesis λ is an arbitrary quantity, we may write

$$D\int(V + \lambda_1 V')dx = 0 \quad \text{and} \quad D\int(V + \lambda_2 V')dx = 0$$

$$\therefore\ D\int(\lambda_1 - \lambda_2)V'dx = 0 \quad \text{or} \quad (\lambda_1 - \lambda_2)D\int Vdx = 0.$$

Now λ_1 and λ_2 are not equal, and therefore $\lambda_1 - \lambda_2$ is not equal to zero. Hence we must have

$$D\int V'dx = 0, \quad \text{and} \therefore \text{from (3)} \quad D\int Vdx = 0 \text{ also.}$$

Thus (3) includes all the conditions required; and therefore if we replace V by $V+\lambda V'$, the problem can be solved as one of absolute maxima or minima.

The formula (3) expanded and applied to the limits x_0 and x_1, gives

$$V_1 dx_1 - V_0 dx_0 + \delta\int_{x_0}^{x_1} V dx + \lambda(V_1' dx_1 - V_0' dx_0) + \delta\int_{x_0}^{x_1}\lambda V' dx = 0.$$

71. *Cor.* It may be shown in nearly the same manner, that when $\int V dx =$ a maximum or minimum, and also

$$\int V' dx = c \qquad \text{and} \qquad \int V'' dx = c',$$

the problem may be solved as a case of absolute maxima and minima by replacing V by $V + \lambda V' + \lambda' V''$ where λ and λ' are arbitrary constants.

Applications.

72. We will now illustrate the principles already explained by a few examples.

1. To find the nature of the line (lying entirely in one plane) which is the shortest distance between two given points.

Let $x_0 y_0$ be the co-ordinates of the point A, and $x_1 y_1$ those of B. The general value of the length of the arc of a plane curve AB is $\int\left(1+\frac{dy_2}{dx^2}\right)^{\frac{1}{2}} dx$ taken between the proper limits. Hence in the present case we shall have

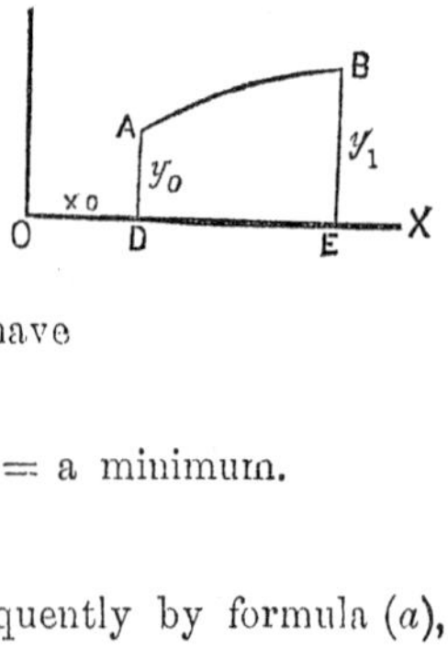

$$U = \int_{x_0}^{x_1} V dx = \int_{x_0}^{x_1}\left(1+\frac{dy^2}{dx^2}\right)^{\frac{1}{2}} dx = \text{a minimum.}$$

Here $V = \left(1+\frac{dy^2}{dx^2}\right)^{\frac{1}{2}} = f\left(\frac{dy}{dx}\right)$, and consequently by formula (a), the solution of the equation $b = 0$ becomes

$$y = cx + c'.$$

and the shortest path from A to B is a straight line.

The equation

$$a_1 - a_0 = 0 \quad \text{or} \quad V_1 dx_1 - V_0 dx_0 + P_1 \delta y_1 - P_0 \delta y_0 = 0$$

disappears in this case, since

$$dx_0 = 0,\ dx_1 = 0,\ \delta y_1 = 0, \quad \text{and} \quad \delta y_0 = 0,$$

the limiting values of both x and y being fixed.

To determine the values of the constants c and c' we have the two equations

$$y_1 = cx_1 + c', \qquad \text{and} \qquad y_0 = cx_0 + c';$$

thus the solution of the problem is complete.

2. To find the line of shortest distance between two given curves.

Let the equation of the curve AB be $y_0 = F_0 x_0 \ldots\ldots (1)$.

and that of the curve CD,

$$y_1 = F_1 x_1 \cdots\cdot (2).$$

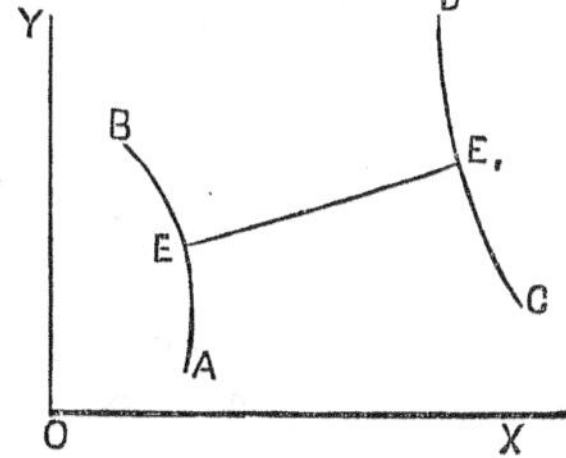

As in example 1,

$$V = \left(1 + \frac{dy^2}{dx^2}\right)^{\frac{1}{2}} = f\left(\frac{dy}{dx}\right);$$

$$\therefore\ y = cx + c',$$

and the shortest distance is still a straight line.

To determine the values of the constants c and c', and the limiting values x_0, y_0, x_1, y_1, we proceed, as follows:

From (1) and (2) we get the following conditions connecting dx_0, δy_0; dx_1 and δy_1, viz.:

$$\delta y_0 + \left[\frac{dy}{dx}\right]_0 \cdot dx_0 = t_0 dx_0, \quad \text{and} \quad \delta y_1 + \left[\frac{dy}{dx}\right]_1 dx_1 = t_1 dx_1,$$

in which $$t_0 = \frac{dF_0 x_0}{dx_0}, \quad \text{and} \quad t_1 = \frac{dF_1 x_1}{dx_1}$$

Also $$\left[\frac{dy}{dx}\right]_0 = c, \quad \text{and} \quad \left[\frac{dy}{dx}\right]_1 = c;$$

$$\therefore\ \delta y_0 = (t_0 - c)\, dx_0, \qquad \delta y_1 = (t_1 - c)\, dx_1.$$

Substituting these values in the equation $a_1 - a_0 = 0$, and replacing V_1, V_0, P_1, P_0 by their values, we get

$$(1 + c^2)^{\frac{1}{2}} dx_1 - (1 + c^2)^{\frac{1}{2}} dx_0 + c\,(1 + c^2)^{-\frac{1}{2}}\,(t_1 - c) dx_1$$
$$- c\,(1 + c^2)^{-\frac{1}{2}}\,(t_0 - c)\,dx_0 = 0.$$

Now, placing equal to zero the coefficient of dx_0 and dx_1, the only arbitrary increments remaining in the equation, we get

$$(1 + c^2)^{\frac{1}{2}} + c\,(1 + c^2)^{-\frac{1}{2}}(t_1 - c) = 0, \quad \text{and}$$

$$(1 + c^2)^{\frac{1}{2}} + c\,(1 + c^2)^{-\frac{1}{2}}\,(t_0 - c) = 0;$$

or, $\quad 1 + ct_1 = 0 \cdots (3), \quad$ and $\quad 1 + ct_0 = 0 \cdots (4).$

These two equations, with the following

$$y_0 = cx_0 + c', \quad y_1 = cx_1 + c', \quad y_0 = F_0 x_0, \quad y_1 = F'_1 x_1,$$

suffice to determine the six quantities, c, c', x_0, y_0, x_1, y_1.

The equations (3) and (4) show that the shortest line EE' cuts both curves at right angles.

73. In the preceding example, suppose the given curves to become straight lines perpendicular to the axis of x. Then $dx_0 = 0$, and $dx_1 = 0$, since the extremities of the shortest line will necessarily have invariable abscissæ.

Also $\quad t_0 = \dfrac{dy_0}{dx_0} = \infty, \quad$ and $\quad t_1 = \dfrac{dy_1}{dx_1} = \infty\,; \quad \therefore c = \dfrac{1}{t_0} = 0\,;$

and as c' is now indeterminate, the required line of shortest distance may pass through any point of AB.

This is an example of Exception 2.

3. To find the form of the function y, which shall render

$$U = \int_{x_0}^{x_1} y^n \left(1 + \frac{dy^2}{dx^2}\right)^{\frac{1}{2}} dx$$

a maximum or minimum.

Here we have

$$V = y^n \left(1 + \frac{dy^2}{dx^2}\right)^{\frac{1}{2}} = f\left(y, \frac{dy}{dx}\right),$$

and therefore, by formula (b),

$$V = P_1 \frac{dy}{dx} + c.$$

But
$$P_1 = y^n \left(1 + \frac{dy^2}{dx^2}\right)^{-\frac{1}{2}} \frac{dy}{dx};$$

$$\therefore y^n \left(1 + \frac{dy^2}{dx^2}\right)^{\frac{1}{2}} = \frac{y^n \dfrac{dy^2}{dx^2}}{\left(1 + \dfrac{dy^2}{dx^2}\right)^{\frac{1}{2}}} + c;$$

or,
$$y^n = c\left(1 + \frac{dy^2}{dx^2}\right)^{\frac{1}{2}}$$

Making $c = l^n$, and solving with respect to dx, we get

$$dx = \frac{l^n dy}{\sqrt{y^{2n} - l^{2n}}} = l^n \left(y^{2n} - l^{2n}\right)^{-\frac{1}{2}} \cdot dy.$$

This comes under the binomial form, and therefore is integrable when

$$\frac{1}{2n} = i, \quad \text{or,} \quad \frac{1}{2n} - \frac{1}{2} = i,$$

an integer or zero; that is, when n has one of the following values, viz.:

$$1, \frac{1}{2}, \frac{1}{3}, \frac{1}{4}, \text{\&c.}; \quad \text{or,} \ -1, -\frac{1}{2}, -\frac{1}{3}, -\frac{1}{4}, \text{\&c.}$$

As a particular case of this problem, suppose $n = -\frac{1}{2}$;

$$\therefore dx = \frac{1}{\sqrt{l}} \left(\frac{1}{y} - \frac{1}{l}\right)^{-\frac{1}{2}} dy = \sqrt{\frac{y}{l-y}} \cdot dy;$$

or,
$$dx = \frac{y\,dy}{\sqrt{ly - y^2}} = \frac{\frac{1}{2} l\,dy}{\sqrt{ly - y^2}} - \frac{\frac{1}{2}(l - 2y)\,dy}{\sqrt{ly - y^2}};$$

$$\therefore\ x + c = \frac{1}{2} l \cdot \overset{-1}{\text{versin}} \frac{2y}{l} - (ly - y^2)^{\frac{1}{2}}.$$

If the limiting values of x and y be given, then

$$dx_0 = 0,\quad \delta y_0 = 0,\quad dx_1 = 0,\quad \delta y_1 = 0,$$

and the equation $a_1 - a_0 = 0$ disappears.

To find the two constants c and l, we have the two equations

$$x_0 + c = \frac{1}{2} l \cdot \overset{-1}{\text{versin}} \frac{2y_0}{l} - \sqrt{ly_0 - y_0^2},$$

$$x_1 + c = \frac{1}{2} l \cdot \overset{-1}{\text{versin}} \frac{2y_1}{l} - \sqrt{ly_1 - y_1^2};$$

and if $\quad x_0 = 0,\ $ and $\ y_0 = 0,\ $ then $\ c = 0,\ $ and

$$x = \frac{1}{2} l \cdot \overset{-1}{\text{versin}} \frac{2y}{l} - \sqrt{ly - y^2} \quad \cdots\cdots (1).$$

74. The equation (1) of this last example exhibits the solution of the celebrated problem of the *Brachystochrone*, or the curve of *swiftest descent*.

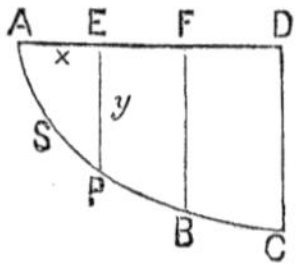

Thus, let A and B be two points in the same vertical plane, and let it be proposed to determine the nature of the curve APB, along which a heavy body will descend from A to B (under the influence of the force of gravity alone) in the shortest possible time.

Denoting by t the time occupied in passing from A to any point P in the unknown path, the co-ordinates of which point are x and y; by s the variable arc AP, and by g the velocity acquired by a heavy body falling vertically during a unit of time; then it is shown by the principles of Mechanics, that the velocity acquired by the body in descending along the curve, (when it has reached the point P,) will be expressed by

$$\sqrt{2gy},\ \text{and also by}\ \frac{ds}{dt} = \frac{dx}{dt}\left(1 + \frac{dy^2}{dx^2}\right)^{\frac{1}{2}}$$

$$\therefore \sqrt{2gy} = \frac{dx}{dt}\sqrt{1+\frac{dy^2}{dx^2}} \text{ and } t = \frac{1}{\sqrt{2g}}\int y^{-\frac{1}{2}}\left(1+\frac{dy^2}{dx^2}\right)^{\frac{1}{2}} dx$$

$$\therefore \int y^{-\frac{1}{2}}\left(1+\frac{dy^2}{dx^2}\right) dx = \text{a minimum between the limits}$$

$$x = x_0 = 0, \quad \text{and} \quad x = x_1 = AF.$$

The equation (1) represents a cycloid, the axis $DC = l$ being vertical, and the extremity of the base coincident with A, the point of departure.

75. 4. Through two given points A and B, draw a curve, of given length, so that the area included between the chord AB and the curve APB may be the greatest possible.

This is a problem of *relative* maxima and minima, since the curve is to be selected from a particular class, viz.: those which have a given length l, or which fulfil the condition

$$\int_{x_0}^{x_1}\left(1+\frac{dy^2}{dx^2}\right)^{\frac{1}{2}} dx = l = \int V'dx.$$

Also $\int Vdx = \int_{x_0}^{x_1} ydx =$ a maximum.

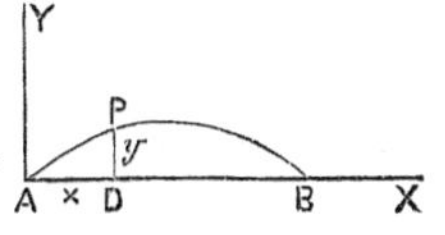

Therefore by the method of relative maxima and minima, we have

$$D = D\int(V + \lambda V')dx = D\int_{x_0}^{x_1}\left[y + \lambda\left(1+\frac{dy^2}{dx^2}\right)^{\frac{1}{2}}\right]dx$$

$$= V_1dx_1 - V_0dx_0 + \delta\int_{x_0}^{x_1} Vdx$$

$$+ \lambda(V_1'dx_1 - V_0'.dx_0 + \delta\int_{x_0}^{x_1} V'dx).$$

Here the limiting values of both x and y are invariable, giving

$$dx_0 = 0, \quad \delta y_0 = 0, \quad dx_1 = 0, \quad \delta y_1 = 0.$$

Also $$V+\lambda V' = y+\lambda\left(1+\frac{dy^2}{dx^2}\right)^{\frac{1}{2}} = f\left(y,\ \frac{dy}{dx}\right)$$

Hence the equation $a_1 - a_0$ disappears, and formula (b) gives

$$V+\lambda V' = c + \frac{\lambda\left(\frac{dy^2}{dx^2}\right)}{\left(1+\frac{dy^2}{dx^2}\right)^{\frac{1}{2}}} = y+\lambda\left(1+\frac{dy^2}{dx^2}\right)^{\frac{1}{2}}$$

$$\therefore c\left(1+\frac{dy^2}{dx^2}\right)^{\frac{1}{2}} + \lambda\frac{dy^2}{dx^2} = y\left(1+\frac{dy^2}{dx^2}\right)^{\frac{1}{2}} + \lambda\left(1+\frac{dy^2}{dx^2}\right).$$

$$\therefore (y-c)\left(1+\frac{dy^2}{dx^2}\right)^{\frac{1}{2}} = \lambda \quad \text{and} \quad \frac{dy^2}{dx^2} = \frac{\lambda^2}{(y-c)^2} - 1.$$

or $$dx = \frac{(y-c)\,dy}{\sqrt{\lambda^2-(y-c)^2}}, \qquad \text{whence}$$

$$x = -\left[\lambda^2-(y-c)^2\right]^{\frac{1}{2}} + c' \quad \text{or} \quad (x-c')^2+(y-c)^2 = \lambda^2$$

and the required curve is the arc of a circle.

To determine the constants c, c', and λ we have the three equations

$$(x_0-c')^2+(y_0-c)^2=\lambda^2, \quad (x_1-c')^2+(y_1-c)^2=\lambda^2 \quad \text{and}$$

$$\frac{\frac{1}{2}\text{chord } AB}{\lambda} = \sin\left(\frac{\frac{1}{2}l}{\lambda}\right),$$

or when the origin is at A and the chord AB coincides with the axis of x,

$$c'^2+c^2=\lambda^2, \ (x_1-c')^2+c^2=\lambda^2, \ \text{and} \ \frac{x_1}{2\lambda} = \sin\frac{l}{2\lambda}.$$

76. 5. Given the length l of the curve joining two fixed points A and B, to find the form of the curve when the surface generated by its revolution about the axis AB is the greatest possible.

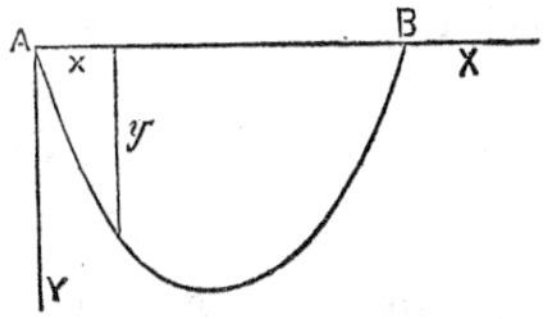

Here $\int V dx = \int_{x_0}^{x_1} 2\pi y\left(1 + \frac{dy^2}{dx^2}\right)^{\frac{1}{2}} dx$

$= \text{a maximum},$

and $\int V' dx = \int_{x_0}^{x_1} \left(1 + \frac{dy^2}{dx^2}\right)^{\frac{1}{2}} dx = l.$

$$\therefore\ DU = D\int (V + \lambda V')dx = 0$$ and

$$V + \lambda V' = 2\pi y\left(1 + \frac{dy^2}{dx^2}\right)^{\frac{1}{2}} + \lambda\left(1 + \frac{dy^2}{dx^2}\right)^{\frac{1}{2}} = f\left(y, \frac{dy}{dx}\right).$$

The equation $a_1 - a_0 = 0$ disappears, and (b) gives

$$V + \lambda V' = c + (2\pi y + \lambda)\left(1 + \frac{dy^2}{dx^2}\right)^{-\frac{1}{2}} \frac{dy^2}{dx^2} = \left(1 + \frac{dy^2}{dx^2}\right)^{\frac{1}{2}} (2\pi y + \lambda).$$

$$\therefore\ c\left(1 + \frac{dy^2}{dx^2}\right)^{\frac{1}{2}} = 2\pi y + \lambda,\quad \frac{dy^2}{dx^2} = \frac{(2\pi y + \lambda)^2}{c^2} - 1.$$

$$\therefore\ dx = \frac{c\,dy}{\sqrt{(2\pi y + \lambda)^2 - c^2}}.$$

To integrate this put $2\pi y + \lambda = z$ and $\sqrt{z^2 - c^2} = z - t$,

then $-c^2 = -2zt + t^2$, $z = \frac{c^2 + t^2}{2t}$, and $dy = \frac{dz}{2\pi} = \frac{t^2 - c^2}{4\pi t^2} dt$,

and $\sqrt{(2\pi y + \lambda)^2 - c^2} = \frac{c^2 - t^2}{2t}$. $\qquad \therefore\ dx = -\frac{c}{2\pi}\cdot\frac{dt}{t}.$

$$\therefore\ x = \frac{c}{2\pi}\log\frac{c'}{t} = \frac{c}{2\pi}\log\frac{c'}{2\pi y + \lambda - \sqrt{(2\pi y + \lambda)^2 - c^2}}$$

$$= \frac{c}{2\pi}\log\frac{2\pi y + \lambda + \sqrt{(2\pi y + \lambda)^2 - c^2}}{\frac{c^2}{c'}}$$

$$= C\log\frac{y + C' + \sqrt{(y + C')^2 - C^2}}{C''}$$

in which $C = \frac{c}{2\pi}$, $C' = \frac{\lambda}{2\pi}$, and $C'' = \frac{c^2}{2\pi c'}$.

This is the equation of the Catenary, which therefore is the required curve.

77. *Prop.* To find the form of the function and the values of the limits x_0 and x_1 which shall render

$U = V' + \int_{x_0}^{x_1} V dx$ a maximum or minimum, where

$$V = f\left(x, y, \frac{dy}{dx}, \frac{d^2y}{dx^2} \cdots\cdots \frac{d^ny}{dx^n}\right),$$ and

$$V' = f'\left[x_0, y_0, \left(\frac{dy}{dx}\right)_0 \cdots\cdots \left(\frac{d^{n'}y}{dx^{n'}}\right), x_1, y_1, \left(\frac{dy}{dx}\right)_1 \cdots\cdots \left(\frac{d^{n''}y}{dx^{n''}}\right)_1\right]$$

The general equation $b = 0$, being derived exclusively from the terms under the sign of integration, must be the same as in the last proposition, and therefore it will be necessary to consider only those terms which refer to the limits:

$$\text{Put } dV = M'dx_0 + N'dy_0 + P_1'd\left(\frac{dy}{dx}\right)_0 + P_2'd\left(\frac{d^2y}{dx^2}\right)_0 + \&c. \cdots\cdots$$

$$+ P_n'd\left(\frac{d^{n'}y}{dx^{n'}}\right) + M''dx_1 + N''dy_1 + P_1''d\left(\frac{dy}{dx}\right)_1$$

$$+ P_2''d\left(\frac{d^2y}{dx^2}\right)_1 + \&c. \cdots\cdots + P''_n{}''d\left(\frac{d^{n''}y}{dx^{n''}}\right)$$

Then the additional terms in DU, resulting from V', are

$$M'dx_0 + N'\delta y_0 + P_1'\left(\frac{d\delta y}{dx}\right)_0 + P_2'\left(\frac{d^2\delta y}{dx^2}\right)_0 \cdots\cdots + P_n''\left(\frac{d^{n'}\delta y}{dx^{n'}}\right)$$

$$+ M''dx_1 + N''\delta y_1 + P_1''\left(\frac{d\delta y}{dx}\right)_1 + P_2''\left(\frac{d^2\delta y}{dx^2}\right)_1 \cdots + P_n'''\left(\frac{d^{n''}\delta y}{dx^{n''}}\right);$$

and the first member of the equation $a_1 - a_0 = 0$ will be increased by these terms, which, being of the same form with the terms previously found in that equation, there will be no difference in the manner of discussing it in its modified form.

It must be remembered, however, that the possibility of satisfying the condition $DU = 0$, depends upon the fact that the number of independent increments in the equation $a_1 - a_0 = 0$, does not usually exceed the number of arbitrary constants in the integral of the equation $b = 0$. Hence if, in any particular case, the number of independent increments should be greater than the number of constants, the solution would be impossible.

Now in the case at present under consideration, the number of increments.

$$dx_0, \delta y_0, \left(\frac{d\delta y}{dx}\right)_0 \cdots\cdots \left(\frac{d^{n'}\delta y}{dx^{n'}}\right)_0$$

relating to the inferior limit is $n' + 2$; and the number of increments already found to exist in a_0 is $n + 1$.

If then $n' + 2 > n + 1$, or $n' > n - 1$, the solution of the problem will be impossible.

Similar remarks apply to the superior limit; and we conclude that when the new function V' contains any coefficient of an order higher than $n - 1$, the function U will not admit of a maximum or minimum.

78. *Prop.* To find the form of the function y and the values of the limits x_0 and x_1, which shall render $U = \int_{x_0}^{x_1} Vdx$ a maximum or minimum, where

$$V = f\left[x, y, \frac{dy}{dx} \cdots \frac{d^n y}{dx^n}, x_0, y_0, \left(\frac{dy}{dx}\right)_0 \cdots \left(\frac{d^{n'}y}{dx^{n'}}\right)_0, x_1, y_1,\right.$$

$$\left.\left(\frac{dy}{dx}\right)_1 \cdots\cdots \left(\frac{d^{n''}y}{dx^{n''}}\right)_1\right].$$

The general equation $DU = 0$ becomes in this case (p. 441)

$$\left[V_1+\int_{x_0}^{x_1}\left\{m_1+n_1\left(\frac{dy}{dx}\right)_1+p_1\left(\frac{d^2y}{dx^2}\right)_1+q_1\left(\frac{d^3y}{dx^3}\right)_1+\&c.\right\}dx\right]dx_1$$

$$+\left[-V_0+\int_{x_0}^{x_1}\left\{m_0+n_0\left(\frac{dy}{dx}\right)_0+p_0\left(\frac{d^2y}{dx^2}\right)_0+q_0\left(\frac{d^3y}{dx^3}\right)_0+\&c.\right\}dx\right]dx_0$$

$$+\left[\left(P_1-\frac{dP_2}{dx}+\&c.\right)_1+\int_{x_0}^{x_1}n_1dx\right]\delta y_1-\left[\left(P_1-\frac{dP_2}{dx}+\&c.\right)_0\right.$$

$$\left.-\int_{x_0}^{x_1}n_0dx\right]\delta y_0$$

$$+\left[\left(P_2-\frac{dP_3}{dx}+\&c.\right)_1+\int_{x_0}^{x_1}p_1dx\right]\left(\frac{d\delta y}{dx}\right)_1-\left[\left(P_2-\frac{dP}{dx}+\&c.\right)_0\right.$$

$$\left.+\int_{x_0}^{x_1}p_0dx\right]\left(\frac{d\delta y}{dx}\right)_0+\&c.$$

$$+\int_{x_0}^{x_1}\left[N-\frac{dP_1}{dx}+\frac{d^2P_2}{dx^2}-\&c.\right]\delta y\,.\,dx=0.$$

This being written in the form

$$a_1-a_0+\int_{x_0}^{x_1}b\delta ydx=0,$$

shows that b is the same as before, and therefore the form of the function y is not changed by supposing V to contain explicitly the limiting values of x, y, $\frac{dy}{dx}$, &c.

Also the terms in $a_1-a_0=0$ are of the same nature as if V did not contain the limits, forming a series

$$A_1dx_1+B_1\delta y_1+C_1\left(\frac{d\delta y}{dx}\right)_1+\&c.+A_0dx_0+B_0\delta y_0+C_0\left(\frac{d\delta y}{dx}\right)_0\&c.$$

A_1, B_1, C_1, &c., A_0, B_0, C_0, &c., being constants. For in the expressions

$$\int_{x_0}^{x_1}m_1dx,\ \int_{x_0}^{x_1}m_0dx,\ \&c.,$$

the same supposition is made as in the terms

$$\left(P_1-\frac{dP_2}{dx}+\&c.\right)_1,\ \left(P_1-\frac{dP_2}{dx}+\&c.\right)_0,$$

and the other coefficients of the several increments in the equation $a_1 - a_0 = 0$, where V did not contain the limits; viz.: that the value of y, derived from the equation $b = 0$, has been substituted in m_1, m_0, &c. This substitution being effected, and the definite integrals

$$\int_{x_0}^{x_1} m_1 dx, \quad \int_{x_0}^{x_1} m_0 dx, \text{ \&c.},$$

being formed, the quantities A_1, B_1, A_0, B_0, &c., will become entirely constant.

Thus the mode of treating the equation $DU = 0$ is in all respects the same as in the case previously considered.

The following examples will illustrate the cases considered in the last two propositions.

79. *Ex.* Having given the area c of the figure BAA_1B_1, bounded by the axis of x, by two ordinates passing through the given points B and B_1, and by a curve ACA_1, to find the nature of the curve and the values of the extreme ordinates BA and B_1A_1, when the perimeter of the figure is a minimum. Put $OB = x_0$, $OB_1 = x_1$, $BA = y_0$, $B_1A_1 = y_1$. Then, since

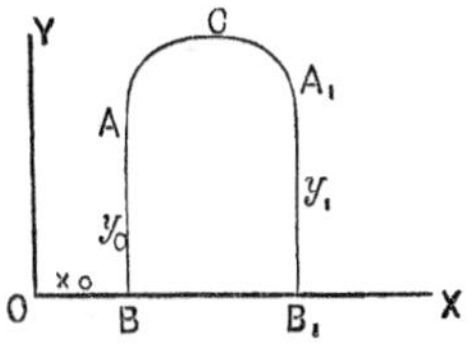

$$BB_1 = x_1 - x_0$$

is constant, we have

$$BA + B_1A_1 + ACA_1 = y_0 + y_1$$

$$+ \int_{x_0}^{x_1} \left(1 + \frac{dy^2}{dx^2}\right)^{\frac{1}{2}} dx = V'' + \int_{x_0}^{x_1} V dx = \text{a minimum.}$$

Also $$\int_{x_0}^{x_1} V' dx = \int_{x_0}^{x_1} y dx = c.$$

$$\therefore \; U = V'' + \int_{x_0}^{x_1} (V + \lambda V') dx = \text{a minimum.}$$

Here U contains a term V'', exterior to the sign of integration, involving the limiting values of y, and, therefore, by the method applicable to such cases, combined with that of relative maxima and minima, we have

$$DU = D\left[y_0 + y_1 + \int_{x_0}^{x_1}\left\{\left(1+\frac{dy^2}{dx^2}\right)^{\frac{1}{2}} + \lambda y\right\} dx\right] = 0.$$

Now $V + \lambda V' = f\left(y, \frac{dy}{dx}\right)$, and therefore by formula (b)

$$V + \lambda V' = c_1 + P_1\frac{dy}{dx}.$$

But
$$P_1 = \frac{d(V+\lambda V')}{d\frac{dy}{dx}} = \left(1+\frac{dy^2}{dx^2}\right)^{-\frac{1}{2}} \cdot \frac{dy}{dx}.$$

$$\therefore c_1 + \frac{dy^2}{dx^2}\left(1+\frac{dy^2}{dx^2}\right)^{-\frac{1}{2}} = \left(1+\frac{dy^2}{dx^2}\right)^{\frac{1}{2}} + \lambda y:$$

or,
$$\left(1+\frac{dy^2}{dx^2}\right)^{\frac{1}{2}}(c_1 - \lambda y) = 1.$$

Put $\frac{c_1}{\lambda} = \beta$, and $\frac{1}{\lambda} = \alpha$, then $\left(1+\frac{dy^2}{dx^2}\right)(\beta - y)^2 = \alpha^2$;

$$\therefore dx = \frac{(\beta - y)\,dy}{\sqrt{\alpha^2 - (\beta - y)^2}}, \quad \text{and} \quad x = c_2 + [\alpha^2 - (\beta - y)^2]^{\frac{1}{2}}$$

or, $(x - c_2)^2 + (y - \beta)^2 = \alpha^2$, the equation of a circle.

Hence, the curve ACA_1, is a circular arc.

To determine the values of the ordinates y_0 and y_1, and that of α, the radius of the circle, we recur to the equation

$a_1 - a_0 = 0$, which becomes, in the present case

$$(V+\lambda V')_1 dx_1 - (V+\lambda V')_0 dx_0 + (P_1)_1 \delta y_1 - (P_1)_0 \delta y_0$$
$$+ N''\delta y_1 + N'\delta y_0 = 0, \quad (1),$$

since $V + \lambda V'$ does not contain P_2, P_3, &c., and V'' contains only y_0 and y_1.

Also, since the points B and B_1 are given, $dx_0 = 0$, and $dx_1 = 0$. Thus, (1) is equivalent to the two conditions

$$(P_1)_1 + N'' = 0, \qquad (P_1)_0 - N' = 0.$$

But $$N' = \frac{dV''}{dy_0} = 1, \quad \text{and} \quad N'' = \frac{dV''}{dy_1} = 1.$$

Hence, by substituting the values of N', N'' and P_1, we obtain

$$\left[\frac{dy}{dx}\left(1+\frac{dy^2}{dx^2}\right)^{-\frac{1}{2}}\right]_1 + 1 = 0, \quad \text{and} \quad \left[\frac{dy}{dx}\left(1+\frac{dy^2}{dx^2}\right)^{-\frac{1}{2}}\right]_0 - 1 = 0;$$

$$\therefore \left(\frac{dy}{dx}\right)_0 = +\infty, \quad \text{and} \quad \left(\frac{dy}{dx}\right)_1 = -\infty.$$

And therefore the arc ACA_1 is a semicircle, the tangents at A and A_1 being perpendicular to OX.

Also, $$\text{radius } a = \frac{1}{2}(x_1 - x_0), \quad \text{and} \quad y_1 = y_0.$$

But area $BAA_1B_1 = 2a \cdot y_0 + \frac{1}{2}\pi a^2 = c$, and $\therefore y_0$ becomes known, thus making the solution complete.

80. *Ex.* To find the curve of swiftest descent from one given curve to another, the motion being supposed to commence at the upper curve.

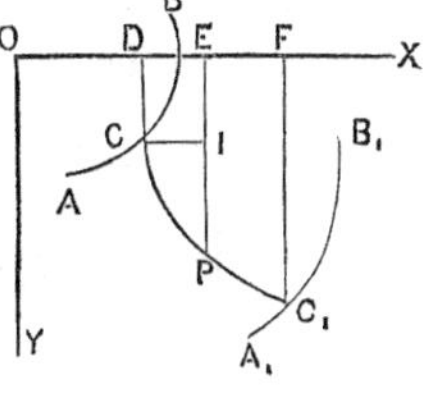

Let AB and A_1B_1 be the given curves, and CC_1 the curve required.

Put $OD = x_0$, $DC = y_0$, $OE = x$, $EP = y$, $OF = x_1$, $FC_1 = y_1$, $CP = s$. Then, by the principles of Mechanics (before cited), the velocity acquired by the body in descending from C to P along the curve CPC_1, is expressed by

$$\sqrt{2g \times IP} = \sqrt{2g(y - y_0)}; \quad \text{and also by } \frac{ds}{dt} = \frac{dx}{dt}\sqrt{1 + \frac{dy^2}{dx^2}};$$

$$\therefore dt = [2g(y - y_0)]^{-\frac{1}{2}} \cdot \left[1 + \frac{dy^2}{dx^2}\right]^{\frac{1}{2}} dx.$$

$$\therefore\ U = \int_{x_0}^{x_1} (y - y_0)^{-\frac{1}{2}} \cdot \left(1 + \frac{dy^2}{dx^2}\right)^{\frac{1}{2}} dx = \int_{x_0}^{x_1} V dx = \text{a minimum.}$$

Here V contains the limit y_0 explicitly; and therefore DU will contain the additional terms

$$\left[\int_{x_0}^{x_1} n_0 \left(\frac{dy}{dx}\right)_0 dx\right] dx_0 + \left[\int_{x_0}^{x_1} n_0\, dx\right] \delta y_0$$

which terms appear in the equation $a_1 - a_0 = 0$, but not in the equation $b = 0$.

Also, since $V = f\left(y, \frac{dy}{dx}\right)$, we have, by formula ($b$),

$$V = c + P_1 \frac{dy}{dx};$$

$$\therefore\ (y - y_0)^{-\frac{1}{2}} \cdot \left(1 + \frac{dy^2}{dx^2}\right)^{\frac{1}{2}} = c + \frac{dy^2}{dx^2}\left[(y - y_0)\left(1 + \frac{dy^2}{dx^2}\right)\right]^{-\frac{1}{2}};$$

or, $$\left[(y - y_0)\left(1 + \frac{dy^2}{dx^2}\right)\right]^{\frac{1}{2}} = \frac{1}{c} = (2C)^{\frac{1}{2}}$$

$$\therefore\ \frac{dy}{dx} = \left[\frac{2C - (y - y_0)}{y - y_0}\right]^{\frac{1}{2}}$$

This is the differential equation of a cycloid having the axis parallel to y, the cusp or extremity of the base at the upper point x_0, y_0, and the diameter of the generating circle $= 2C$.

The equation $a_1 - a_0 = 0$ gives, in this case,

$$V_1 dx_1 - V_0 dx_0 + (P_1)_1 \delta y_1 - (P_1)_0 \delta y_0 + \left(\int_{x_0}^{x_1} n_0 \left(\frac{dy}{dx}\right)_0 dx\right) dx_0$$
$$+ \left(\int_{x_0}^{x_1} n_0\, dx\right) \delta y_0 = 0 \cdots\cdots (1).$$

But $n_0 = \dfrac{dV}{dy_0} = -\dfrac{dV}{dy} = -N = -\dfrac{dP_1}{dx}$, since $N - \dfrac{dP_1}{dx} = 0$

$$\therefore\ \int n_0\, dx = -\int \frac{dP_1}{dx} dx = -P_1 + c_1.$$

$$\therefore \int_{x_0}^{x_1} n_0\,dx = (P_1)_0 - (P_1)_1, \quad \text{and} \quad \int \left(\frac{dy}{dx}\right)_0 n_0\,dx$$

$$= \left(\frac{dy}{dx}\right)_0 [(P_1)_0 - (P_1)_1].$$

Again, if the differential equations of the two given curves be

$$\frac{dy_0}{dx_0} = t_0, \quad \text{and} \quad \frac{dy_1}{dx_1} = t_1,$$

we shall have the following conditions connecting the values of dx_0, δy_0, dx_1, and δy_1, viz.:

$$\delta y_0 + \left(\frac{dy}{dx}\right)_0 dx_0 = t_0 dx_0, \quad \text{and} \quad \delta y_1 + \left(\frac{dy}{dx}\right)_1 dx_1 = t_1 dx_1.$$

Now substituting the values of δy_0, δy_1, $\int_{x_0}^{x_1} n_0 dx$, and $\int_{x_0}^{x_1} \left(\frac{dy}{dx}\right)_0 n_0 dx$

in (1), and placing the coefficients of dx_0 and dx_1, separately, equal to zero, we get

$$V_1 + (P_1)_1 \left[t_1 - \left(\frac{dy}{dx}\right)_1\right] = 0, \quad \text{and}$$

$$V_0 + (P_1)_0 \left[t_0 - \left(\frac{dy}{dx}\right)_0\right] - [(P_1)_0 - (P_1)_1] \left(\frac{dy}{dx}\right)_0$$

$$- [(P_1)_0 - (P_1)_1] \left[t_0 - \left(\frac{dy}{dx}\right)_0\right] = 0, \quad \text{or,}$$

$$\left[\left(1 + \frac{dy^2}{dx^2}\right)^{\frac{1}{2}} \cdot (y - y_0)^{-\frac{1}{2}}\right]_1 + \left(\frac{dy}{dx}\right)_1 \left[\left(1 + \frac{dy^2}{dx^2}\right)^{-\frac{1}{2}} \cdot\right.$$

$$\left.(y - y_0)^{-\frac{1}{2}}\right]_1 \cdot \left[t_1 - \left(\frac{dy}{dx}\right)_1\right] = 0 \cdots\cdots (2);$$

and $$\left[\left(1 + \frac{dy^2}{dx^2}\right)^{\frac{1}{2}} \cdot (y - y_0)^{-\frac{1}{2}}\right]_0 - \left(\frac{dy^2}{dx^2}\right)_0 \left[\left(1 + \frac{dy^2}{dx^2}\right)^{-\frac{1}{2}} \cdot\right.$$

$$\left.(y - y_0)^{-\frac{1}{2}}\right]_0 + t_0 \left(\frac{dy}{dx}\right)_1 \left[\left(1 + \frac{dy^2}{dx^2}\right)^{-\frac{1}{2}} \cdot (y - y_0)^{-\frac{1}{2}}\right]_1 = 0 \cdot\cdot (3).$$

From (2) we obtain $1+t_1\left(\frac{dy}{dx}\right)_1=0$; and therefore the cycloid intersects the second curve at right angles.

Also, from (3) we get $1+t_0\left(\frac{dy}{dx}\right)_1=0$; $\therefore t_1=t_0$, and the tangents to the two curves, at the points of intersection with the cycloid, are parallel. The co-ordinates of those points are readily found.

81. *Prop.* To determine the forms of the functions y and z, and the values of the limits x_1 and x_0, which shall render

$$U=\int_{x_0}^{x_1}Vdx \text{ a maximum or minimum, where}$$

$$V=f\left[x, y, \frac{dy}{dx}, \frac{d^2y}{dx^2}, \dots\dots \frac{d^ny}{dx^n}, \quad z, \frac{dz}{dx}, \frac{d^2z}{dx^2} \dots\dots \frac{d^mz}{dx^m}\right].$$

The equation $DU=0$ becomes in this case

$$V_1dx_1-V_0dx_0+\left[P_1-\frac{dP_2}{dx}+\&c.\right]_1\delta y_1-\left[P_1-\frac{dP_2}{dx}+\&c.\right]_0\cdot\delta y_0$$

$$+[P_2-\&c.]_1\left(\frac{d\delta y}{dx}\right)_1-[P_2-\&c.]_0\left(\frac{d\delta y}{dx}\right)_0\dots\dots+\left[P_n\frac{d^{n-1}\delta y}{dx^{n-1}}\right]_1$$

$$-\left[P_n\frac{d^{n-1}\delta y}{dx^{n-1}}\right]_0+\int_{x_0}^{x_1}\left[N-\frac{dP_1}{dx}+\frac{d^2P_2}{dx^2}-\&c.\right.$$

$$\left.\dots\dots+(-1)^n\cdot\frac{d^nP_n}{dx^n}\right]\delta y dx.$$

$$+\left[P_1'-\frac{dP_2'}{dx}+\&c.\right]_1\delta z_1-\left[P_1'-\frac{dP_2'}{dx}+\&c.\right]_0\cdot\delta z_0$$

$$+[P_2'-\&c.]_1\left(\frac{d\delta z}{dx}\right)_1-[P_2'-\&c.]_0\left(\frac{d\delta z}{dx}\right)_0\&c.$$

$$+\left[P_m'\frac{d^{m-1}\delta z}{dx^{m-1}}\right]_1-\left[P_m'\frac{d^{m-1}\delta z}{dx^{m-1}}\right]_0+\int_{x_0}^{x_1}\left[N'-\frac{dP_1'}{dx}+\frac{d^2P_2'}{dx^2}-\&c.\right.$$

$$\left.\dots\dots+(-1)^m\cdot\frac{d^mP_m'}{dx^m}\right]\delta z\,.\,dx=0.$$

If the functions y and z be independent of each other, their variations δy and δz will also be independent; and, by reasoning as in previous propositions, it will appear that we shall have the conditions

$$N - \frac{dP_1}{dx} + \frac{d^2P_2}{dx^2} - \&c. \ldots . + (-1)^n \cdot \frac{d^nP_n}{dx^n} = 0,$$

$$N' - \frac{dP_1'}{dx} + \frac{d^2P'_2}{dx^2} - \&c. \ldots + (-1)^m \cdot \frac{d^mP_m'}{dx^m} = 0 \ldots (1).$$

And for the equation of the limits

$$V_1 dx_1 - V_0 dx_0 + \left[P_1 - \frac{dP_2}{dx} + \&c.\right]_1 \delta y_1 - \left[P_1 - \frac{dP_2}{dx} + \&c.\right]_0 \delta y_0$$

$$+ [P_2 - \&c.]_1 \left(\frac{d\delta y}{dx}\right)_1 - [P_2 - \&c.]_0 \left(\frac{d\delta y}{dx}\right)_0 \&c. \&c.$$

$$+ \left[P_1' - \frac{dP_2'}{dx} + \&c.\right]_1 \cdot \delta z_1 - \left[P_1' - \frac{dP_2'}{dx} + \&c.\right]_0 \cdot \delta z_0$$

$$+ [P_2' - \&c.] \left(\frac{d\delta z}{dx}\right)_1 - [P_2' - \&c.]_0 \left(\frac{d\delta z}{dx}\right)_0 \cdot \&c. \&c. = 0, \ldots . . (2).$$

The mode of treating these equations is exactly the same as that employed when V contained but one function, and by reasoning, as in that case, it may be readily shown that the number of equations applicable to the solution of the problem will not, in general, be affected by any equations of condition restricting the limits. For every such equation of condition will diminish by unity the number of terms in (2), either by reducing to zero the variation which appears in such term; or, by uniting two terms in one, and thereby diminishing by unity the number of equations deducible from (2).

But the given equation of condition will just supply the place of that which has disappeared.

Thus it will suffice to prove that (1) and (2) furnish the requisite number of equations in a single case, as when the limits of x are alone fixed.

Now the first of equations (1) is of the order $2n$ in y, and $m+n$ in z, and the second of equations (1) is of the order $m+n$ in y, and $2m$ in z. They are therefore of the forms

$$F_1\left[x, y, \frac{dy}{dx} \cdots \frac{d^{2n}y}{dx^{2n}}, z, \frac{dz}{dx} \cdots \frac{d^{m+n}z}{dx^{m+n}}\right] = 0 \ldots (3).$$

$$F_2\left[x, y, \frac{dy}{dx} \cdots \frac{d^{m+n}y}{dx^{m+n}}, z, \frac{dz}{dx} \cdots \frac{d^{2m}z}{dx^{2m}}\right] = 0 \ldots (4).$$

If, then, we differentiate (3) $2m$ times, and (4) $m+n$ times, we shall have $3m+n+2$ equations with which to eliminate the $3m+n$ quantities $z, \frac{dz}{dx} \cdots \frac{d^{3m+n}z}{dx^{3m+n}}$, and the resulting equation will be of the order $2m+2n$ in y. The integral of this equation will contain $2m+2n$ constants. But the number of equations given by (2) is exactly $2n+2m$, viz.: the $2n$ equations,

$$\left[P_1 - \frac{dP_2}{dx} + \&c.\right]_1 = 0, \left[P_1 - \frac{dP_2}{dx} + \&c.\right]_0 = 0, [P_2 - \&c.]_1 = 0, \&c.;$$

and the $2m$ equations,

$$\left[P_1' - \frac{dP_2'}{dx} + \&c.\right]_1 = 0, \left[P_1' - \frac{dP_2'}{dx} + \&c.\right]_0 = 0,$$

$$[P_2' - \&c.]_1 = 0, \&c.$$

Hence the problem is in general determinate, but there are exceptions entirely similar to those considered in the case of a single dependent function y.

82. If the functions y and z be connected by an equation $L=0$, and if it be possible to resolve that equation with respect to y or z, so as to obtain a result of the form $z = f\left(x, y, \frac{dy}{dx}, \&c.\right)$, the values of $\frac{dz}{dx}, \frac{d^2z}{dx^2}$, &c., can be formed by differentiation, and substituted in that of V, which will then contain x, y, and the differential coefficients of y with respect to x, thus presenting a case already considered.

83. But since the proposed equation $L = 0$ is often a differential equation difficult to be integrated, we are often compelled to adopt the method already noticed, (Page 444) in which by the introduction of a new indeterminate quantity λ, and a suitable determination of its value, we are enabled to obtain an expression for δU which shall contain but one of the variations δy and δz under the sign of integration.

Thus, if we denote by θ, the sum of the terms exterior to the sign of integration in the value of δU, (Page 445) there will result

$$\delta U = \theta + \int_{x_0}^{x_1} [N + \lambda\alpha - \frac{d(P_1 + \lambda\beta)}{dx} + \&c.]\, \delta y dx$$
$$+ \int_{x_0}^{x_1} [N' + \lambda\alpha' - \frac{d(P_1' + \lambda\beta')}{dx} + \&c.]\, \delta z dx;$$

and if we so assume the quantity λ as to fulfil the condition

$$N' + \lambda\alpha' + \frac{d(P_1' + \lambda\beta')}{dx} + \&c. = 0,$$

it will appear by reasoning, similar to that employed when y was the only function, that the condition $\delta U = 0$ cannot be satisfied (so long as the form of δy is arbitrary) unless we have the two conditions

$$\theta = 0 \quad \text{and} \quad N + \lambda\alpha - \frac{d(P_1 + \lambda\beta)}{dx} + \&c. = 0.$$

Hence, we have for the solution of the problem, the three general equations

$$L = 0, \; N + \lambda\alpha - \frac{d(P_1 + \lambda\beta)}{dx} + \&c. = 0,$$

and
$$N' + \lambda\alpha' - \frac{d(P_1' + \lambda\beta')}{dx} + \&c. = 0.$$

which are just sufficient to determine the three unknown quantities, λ, y and z.

84. We will now give, in conclusion, examples to illustrate the cases and methods above explained.

Ex. To find the nature of the line which is the shortest distance between two given points in space, there being no restriction by which the line is required to be confined to one plane.

The general value of the length of the arc of a curve of double curvature is

$$\int \left(1 + \frac{dy^2}{dx^2} + \frac{dz^2}{dx^2}\right)^{\frac{1}{2}} dx$$

taken between the proper limits.

Hence in the present case we shall have

$$U = \int_{x_0}^{x_1} \left(1 + \frac{dy^2}{dx^2} + \frac{dz^2}{dx^2}\right)^{\frac{1}{2}} dx = \text{a minimum.}$$

Here $V = \left(1 + \frac{dy^2}{dx^2} + \frac{dz^2}{dx^2}\right)^{\frac{1}{2}}$, $N = \frac{dV}{dy} = 0$, $N' = \frac{dV}{dz} = 0$

$$P_1 = \frac{dV}{d\frac{dy}{dx}} = \frac{\frac{dy}{dx}}{\sqrt{1 + \frac{dy^2}{dx^2} + \frac{dz^2}{dx^2}}}, \quad P_1' = \frac{dV}{d\frac{dz}{dx}} = \frac{\frac{dz}{dx}}{\sqrt{1 + \frac{dy^2}{dx^2} + \frac{dz^2}{dx^2}}}$$

$$P_2 = 0, \quad P_2' = 0, \text{ \&c.}$$

Hence the equations

$$N - \frac{dP_1}{dx} + \text{\&c.} = 0 \quad \text{and} \quad N' - \frac{dP_1'}{dx} + \text{\&c.} = 0$$

become $$\frac{dP_1}{dx} = 0 \quad \text{and} \quad \frac{dP_1'}{dx} = 0$$

or $$P_1 = \frac{\frac{dy}{dx}}{\sqrt{1 + \frac{dy^2}{dx^2} + \frac{dz^2}{dx^2}}} = c \quad \text{and} \quad P_1' = \frac{\frac{dz}{dx}}{\sqrt{1 + \frac{dy^2}{dx^2} + \frac{dz^2}{dx^2}}} = c$$

Eliminating first $\frac{dz}{dx}$ and then $\frac{dy}{dx}$ between these two equations, we readily obtain results of the forms

$$\frac{dy}{dx} = m \quad \text{and} \quad \frac{dz}{dx} = n \text{ in which } m \text{ and } n \text{ are constants,}$$

$$\therefore\ y = mx + p, \qquad \text{and} \qquad z = nx + q.$$

These are the equations of a straight line, which therefore is the shortest distance required.

To find the values of the constants m, n, p, and q, we introduce the given limits $x_0, y_0, z_0, x_1, y_1, z_1$, and thus get

$$y_0 = mx_0 + p, \quad z_0 = nx_0 + q, \quad y_1 = mx_1 + p, \quad z_1 = nx_1 + q,$$

which suffice to determine m, n, p and q.

85. If the limiting values of x only were given, those of y and z remaining indeterminate, the terms exterior to the sign of integration would give

$$(P_1)_1 = 0, \ (P_1)_0 = 0, \ (P_1')_1 = 0, \ (P_1')_0 = 0,$$

which are equivalent to the two equations

$$m = 0 \quad \text{and} \quad n = 0,$$

thus leaving the other two constants p and q indeterminate, and presenting one of the cases of exception already noticed.

86. *Ex.* To find the shortest distance between two given surfaces.

Let the equation of the first surface be $f_0(x_0, y_0, z_0) = 0 \cdots\cdots (1)$
and that of the second surface $f_1(x_1, y_1, z_1) = 0 \cdots\cdots (2)$

As in the last example $V = \left(1 + \dfrac{dy^2}{dx^2} + \dfrac{dz^2}{dx^2}\right)^{\frac{1}{2}}$

and we immediately deduce as before

$$y = mx + p \cdots\cdots (3), \quad z = nx + q \cdots\cdots (4)$$

which show that the shortest path is still a straight line.

To fix the co-ordinates of the extremities of this line we form the complete increment of (1) and (2) thus:

$$\left[\frac{df_0}{dx_0}+\frac{df_0}{dy_0}\left(\frac{dy}{dx}\right)_0+\frac{df_0}{dz_0}\left(\frac{dz}{dx}\right)\right]dx_0+\frac{df_0}{dy_0}.\delta y_0+\frac{df_0}{dz_0}.\delta z_0=0\cdots\cdots(5)$$

$$\left[\frac{df_1}{dx_1}+\frac{df_1}{dy_1}\cdot\left(\frac{dy}{dx}\right)_1+\frac{df_1}{dz_1}\cdot\left(\frac{dz}{dx}\right)_1\right]dx_1+\frac{df_1}{dy_1}.\delta y_1+\frac{df_1}{dz_1}.\delta z_1=0\cdots(6)$$

Put for brevity

$$m_0=\frac{\frac{df_0}{dy_0}}{\frac{df_0}{dx_0}},\ m_1=\frac{\frac{df_1}{dy_1}}{\frac{df_1}{dx_1}},\ n_0=\frac{\frac{df_0}{dz_0}}{\frac{df_0}{dx_0}},\ n_1=\frac{\frac{df_1}{dz_1}}{\frac{df_1}{dx_1}}.$$

and substitute for $\left(\frac{dy}{dx}\right)_0$, $\left(\frac{dy}{dx}\right)_1$, $\left(\frac{dz}{dx}\right)_0$, $\left(\frac{dz}{dx}\right)_1$

their values derived from equations (3) and (4). We shall thus obtain

$$(1+mm_0+nn_0)\,dx_0+m_0\delta y_0+n_0\delta z_0=0$$

$$(1+mm_1+nn_1)\,dx_1+m_1\delta y_1+n_1\delta z_1=0.$$

Now eliminating, by the aid of these equations, dx_0 and dx_1, from the equations

$$V_0dx_0+(P_1)_0\delta y_0+(P_1')_0\delta z_0=0$$

$$V_1dx_1+(P_1)_1\delta y_1+(P_1')_1\delta z_1=0,$$

and placing equal to zero the coefficients of δy_0, δz_0, δy_1, δz_1, we obtain

$$m_0V_0-(P_1)_0(1+mm_0+nn_0)=0\cdots\cdots(7)$$

$$m_1V_1-(P_1)_1(1+mm_1+nn_1)=0\cdots\cdots(8)$$

$$n_0V_0-(P_1')_0(1+mm_0+nn_0)=0\cdots\cdots(9)$$

$$n_1V_1-(P_1')_1(1+mm_1+nn_1)=0\cdots\cdots(10).$$

If now we replace V_0 and $(P_1)_0$ &c. in (7), (8), (9) and (10), by their values

$$(1+m^2+n^2)^{\frac{1}{2}},\quad \frac{m}{\sqrt{1+m^2+n^2}}\ \text{\&c.}$$

we readily find from (7) and (9) $m = m_0,\ n = n_0 \cdots\cdots$ (11)

and from (8) and (10), $m = m_1$ and $n = n_0 \cdots\cdots$ (12).

Now eliminating $x_0, y_0, z_0, x_1, y_1, z_1$, which quantities occur in the values of m_0, n_0, m_1, and n_1, by means of the six equations,

$$y_0 = mx_0 + p, \quad y_1 = mx_1 + p,$$

$$z_0 = nx_0 + q, \quad z_1 = nx_1 + q,$$

$$f_0(x_0, y_0, z_0) = 0, \quad f_1(x_1, y_1, z_1) = 0,$$

there will remain the four equations (11) and (12) with which to compute the values of m, n, p, and q; thus the line of shortest distance will be fixed in position; and, by combining its equations with those of the given surfaces, we can find the values of

$$x_1\, y_1\, z_1 \quad x_0\, y_0\, z_0.$$

87. The equations (11) and (12) show that the line of shortest distance is normal to both surfaces. For the assumed values of m_0 and n_0 indicate that they represent the tangents of the angles formed by the projections of the normal to the first surface on the planes of xy and xz with the axis of x; while m and n denote the tangents of the corresponding angles formed by the projections of the line of shortest distance.

A similar remark applies to the quantities m_1 and n_1, and the normal to the second surface.

88. *Ex.* To find the shortest distance traced on the surface of a given sphere between two given points in the surface.

Here the quantity to be rendered a minimum is the same as in the last two examples, viz.:

$$U = \int_{x_0}^{x_1} \left(1 + \frac{dy^2}{dx^2} + \frac{dz^2}{dx^2}\right)^{\frac{1}{2}} dx \ldots . (1);$$

but since the path is restricted to the surface of a given sphere, the co-ordinates x, y, and z, of any point in the required path, will be connected by the relation

$$x^2+y^2+z^2=r^2, \quad \text{or} \quad L=x+y\frac{dy}{dx}+z\frac{dz}{dx}=0 \ldots . (2).$$

Hence the variations of y and z will not be independent of each other.

Now we might form from (2) the value of $\frac{dz}{dx}$, which, substituted in (1), would reduce V to a form in which it would no longer contain the function z, or its differential coefficient, or we may adopt the method of Lagrange, which is usually the easier. Taking the second method, we have

$$V=\left(1+\frac{dy^2}{dx^2}+\frac{dz^2}{dx^2}\right)^{\frac{1}{2}}.$$

$$\therefore\ P_1=\frac{\frac{dy}{dx}}{\sqrt{1+\frac{dy^2}{dx^2}+\frac{dz^2}{dx^2}}}, \quad P_1'=\frac{\frac{dz}{dx}}{\sqrt{1+\frac{dy^2}{dx^2}+\frac{dz^2}{dx^2}}}$$

$$N=\frac{dV}{dy}=0, \quad N'=\frac{dV}{dz}=0$$

$$\alpha=\frac{dL}{dy}=\frac{dy}{dx}, \quad \beta=\frac{dL}{d\frac{dy}{dx}}=y, \quad \alpha'=\frac{dz}{dx}, \quad \beta'=z.$$

Hence the equations $\quad N+\lambda\alpha-\frac{d(P_1+\lambda\beta)}{dx}+\&c.=0,$

and $\quad N'+\lambda\alpha'-\frac{d(P_1'+\lambda\beta')}{dx}+\&c.=0,$

become, in this case,

$$\lambda\frac{dy}{dx}-\frac{dP_1}{dx}-\lambda\frac{dy}{dx}-y\frac{d\lambda}{dx}=0,$$

$$\lambda\frac{dz}{dx}-\frac{dP_1'}{dx}-\lambda\frac{dz}{dx}-z\frac{d\lambda}{dx}=0,$$

or
$$y\frac{d\lambda}{dx}+\frac{d}{dx}\left(\frac{\frac{dy}{dx}}{\sqrt{1+\frac{dy^2}{dx^2}+\frac{dz^2}{dx^2}}}\right)=0\ \ldots\ldots\ (3),$$

$$z\frac{d\lambda}{dx}+\frac{d}{dx}\left(\frac{\frac{dz}{dx}}{\sqrt{1+\frac{dy^2}{dx^2}+\frac{dz^2}{dx^2}}}\right)=0\ \ldots\ldots\ (4).$$

Eliminating $\frac{d\lambda}{dx}$ between (3) and (4), we get

$$z\frac{d}{dx}\left(\frac{\frac{dy}{dx}}{\sqrt{1+\frac{dy^2}{dx^2}+\frac{dz^2}{dx^2}}}\right)-y\frac{d}{dx}\left(\frac{\frac{dz}{dx}}{\sqrt{1+\frac{dy^2}{dx^2}+\frac{dz^2}{dx^2}}}\right)=0;$$

and by integration

$$\frac{z\frac{dy}{dx}-y\frac{dz}{dx}}{\sqrt{1+\frac{dy^2}{dx^2}+\frac{dz^2}{dx^2}}}=c\ \ldots\ldots\ (5);$$

or, by changing the independent variable from x to s, (5) becomes

$$z\frac{dy}{ds}-y\frac{dz}{ds}=c\ \ldots\ldots\ (6).$$

By similar reasoning we may obtain

$$y\frac{dx}{ds}-x\frac{dy}{ds}=c_1\ \ldots\ldots\ (7),\quad \text{and}\quad x\frac{dz}{ds}-z\frac{dx}{ds}=c_2\ \ldots\ldots\ (8).$$

Multiplying (6) by x, (7) by z, and (8) by y, and adding, we get

$$cx+c_1z+c_2y=0,\quad \text{or}\quad z+\frac{c}{c_1}x+\frac{c_2}{c_1}y=0\ \ldots\ldots\ (9),$$

the equation of a plane passing through the origin.

Thus the required line of shortest distance on the surface of the sphere, is confined to a plane passing through the centre, and is, consequently, a great circle.

The equation $a_1 - a_0 = 0$ in this case disappears, since

$$dx_0 = 0, \quad dx_1 = 0, \quad \delta y_0 = 0, \quad \delta y_1 = 0, \quad \delta z_0 = 0, \quad \text{and} \quad \delta z_1 = 0.$$

The constants $\frac{c}{c_1}$ and $\frac{c_2}{c_1}$ are found by substituting

$$x_0, y_0, z_0, \quad \text{and} \quad x_1, y_1, z_1, \quad \text{for} \quad x, y, \text{ and } z \text{ in } (9).$$

89. If the limiting values of x only were given, or the problem that in which it is required to find on the surface of the sphere, the shortest path between two parallel sections, the variations δy_0, δy_1, δz_0, δz_1, would not reduce to zero, and the equation $a_1 - a_0 = 0$ would give the four conditions

$$(P_1 + \lambda\beta)_0 = 0, \ (P_1 + \lambda\beta)_1 = 0, \ (P_1' + \lambda\beta')_0 = 0, \ (P_1' + \lambda\beta')_1 = 0,$$

or,

$$\left(\frac{\frac{dy}{dx}}{\sqrt{1 + \frac{dy^2}{dx^2} + \frac{dz^2}{dx^2}}}\right)_0 + \lambda_0 y_0 = 0 \cdots\cdots (10);$$

and

$$\left(\frac{\frac{dz}{dx}}{\sqrt{1 + \frac{dy^2}{dx^2} + \frac{dz^2}{dx^2}}}\right)_0 + \lambda_0 z_0 = 0 \cdots\cdots (11);$$

which apply to the inferior limit, with two similar equations for the superior limit.

Eliminating λ_0 between (10) and (11), there results

$$\left(\frac{z\frac{dy}{dx} - y\frac{dz}{dx}}{\sqrt{1 + \frac{dy^2}{dx^2} + \frac{dz^2}{dx^2}}}\right)_0 = 0.$$

Hence, the constant $c = 0$ in (5); and that equation becomes

$$\therefore z\frac{dy}{dx} - y\frac{dz}{dx} = 0; \quad \text{or,} \quad \frac{dy}{y} = \frac{dz}{z}.$$

$$\therefore \log y = \log z + \log m = \log mz; \quad \text{and} \quad y = mz.$$

This is the equation of a plane passing through the axis of x, and forming an arbitrary angle ($\tan^{-1} m$) with the plane of xz. Hence, the required path is the arc of any great circle perpendicular to the planes of the parallel sections.

THE END.

www.ingramcontent.com/pod-product-compliance
Lightning Source LLC
LaVergne TN
LVHW021318110826
845150LV00003B/607

* 9 7 8 1 4 2 5 5 5 7 4 1 6 *